NONLINEAR PARTIAL
DIFFERENTIAL EQUATIONS
IN ENGINEERING

MATHEMATICS IN SCIENCE AND ENGINEERING

A SERIES OF MONOGRAPHS AND TEXTBOOKS

Edited by Richard Bellman

THE RAND COPORATION, SANTA MONICA, CALIFORNIA

1. TRACY Y. THOMAS. Concepts from Tensor Analysis and Differential Geometry. Second Edition. 1965
2. TRACY Y. THOMAS. Plastic Flow and Fracture in Solids. 1961
3. RUTHERFORD ARIS. The Optimal Design of Chemical Reactors: A Study in Dynamic Programming. 1961
4. JOSEPH LA SALLE and SOLOMON LEFSCHETZ. Stability by Liapunov's Direct Method with Applications. 1961
5. GEORGE LEITMANN (ed.). Optimization Techniques: With Applications to Aerospace Systems. 1962
6. RICHARD BELLMAN and KENNETH L. COOKE. Differential-Difference Equations. 1963
7. FRANK A. HAIGHT. Mathematical Theories of Traffic Flow. 1963
8. F. V. ATKINSON. Discrete and Continuous Boundary Problems. 1964
9. A. JEFFREY and T. TANIUTI. Non-Linear Wave Propagation: With Applications to Physics and Magnetohydrodynamics. 1964
10. JULIUS T. TOU. Optimum Design of Digital Control Systems. 1963
11. HARLEY FLANDERS. Differential Forms: With Applications to the Physical Sciences. 1963
12. SANFORD M. ROBERTS. Dynamic Programming in Chemical Engineering and Process Control. 1964
13. SOLOMON LEFSCHETZ. Stability of Nonlinear Control Systems. 1965
14. DIMITRIS N. CHORAFAS. Systems and Simulation. 1965
15. A. A. PERVOZVANSKII. Random Processes in Nonlinear Control Systems. 1965
16. MARSHALL C. PEASE, III. Methods of Matrix Algebra. 1965
17. V. E. BENEŠ. Mathematical Theory of Connecting Networks and Telephone Traffic. 1965
18. WILLIAM F. AMES. Nonlinear Partial Differential Equations in Engineering. 1965

MATHEMATICS IN SCIENCE AND ENGINEERING

In preparation

A. Halanay. Differential Equations: Stability, Oscillations, Time Lags

R. E. Murphy. Adaptive Processes in Economic Systems

Dimitris N. Chorafas. Control Systems Functions and Programming Approaches

J. Aczél. Functional Equations

A. A. Fel'dbaum. Optimal Control Systems

David Sworder. Optimal Adaptive Control Systems

S. E. Dreyfus. Dynamic Programming and the Calculus of Variations.

M. Namik Oguztoreli. Time Lag Control Processes

NONLINEAR PARTIAL DIFFERENTIAL EQUATIONS IN ENGINEERING

by W. F. AMES

DEPARTMENT OF STATISTICS AND COMPUTER SCIENCE
AND
DEPARTMENT OF MECHANICAL ENGINEERING
UNIVERSITY OF DELAWARE
NEWARK, DELAWARE

1965

ACADEMIC PRESS • New York • London

COPYRIGHT © 1965, BY ACADEMIC PRESS INC.
ALL RIGHTS RESERVED.
NO PART OF THIS BOOK MAY BE REPRODUCED IN ANY FORM,
BY PHOTOSTAT, MICROFILM, OR ANY OTHER MEANS, WITHOUT
WRITTEN PERMISSION FROM THE PUBLISHERS.

ACADEMIC PRESS INC.
111 Fifth Avenue, New York, New York 10003

United Kingdom Edition published by
ACADEMIC PRESS INC. (LONDON) LTD.
Berkeley Square House, London W.1

LIBRARY OF CONGRESS CATALOG CARD NUMBER: 65-22767

PRINTED IN THE UNITED STATES OF AMERICA

Preface

This book is almost entirely concerned with methods of solution for nonlinear partial differential equations. In particular, it provides the first unified treatment of a great variety of procedures ranging in character from the completely analytic to the completely numerical. The literature contains a variety of excellent textbooks and reference works concentrating on specific technical subjects and mathematical topics which, in part, concern us here. The present volume consolidates this scattered information, draws from a wide range of technical journals, and reflects some of my own research and that of my students.

The exposition is sometimes by means of general theory but more often is largely through examples. These are drawn from a variety of scientific disciplines including fluid mechanics, vibrations, elasticity, diffusion, heat transfer, and plasma physics. Since the occurrence of these mathematical models is not limited to these fields, the techniques discussed will certainly find further applications in a multitude of other disciplines, as well as continued use in these studies.

This work is partially the outgrowth of two courses that I have taught, since 1959, in the Engineering School of the University of Delaware. The research and writing were continued and completed while I held a National Science Foundation Science Faculty Fellowship. During this period I was in residence at the Stanford University Department of Chemical Engineering. The Stanford Libraries were of inestimable value in this effort.

The book is approximately divided in thirds—in the first I consider the origin of some equations and develop analytic methods; in the second a variety of approximate methods such as asymptotic processes, perturbation procedures, and weighted residual methods are studied; the final third is devoted to an extensive study and collection of numerical procedures specifically aimed at these equations.

I am greatly indebted to all of the tireless researchers in this very difficult subject. Over 600 references attest to this fact. My thanks go to R. L. Pigford, E. A. Erdelyi, J. R. Ferron, L. G. Clark, M. H. Cobble, and J. P. Hartnett, colleagues and friends, for encouragement in time of adversity. I am further indebted to a number of other colleagues for helpful comments, criticisms, and corrections including in particular, A. Acrivos and C. E. McQueary.

This book could not have been written without the understanding, help, and encouragement of my wife Theresa, and I therefore dedicate it to her.

WILLIAM F. AMES

April, 1965
Newark, Delaware

Contents

PREFACE   vii

CHAPTER 1. The Origin of Nonlinear Differential Equations

1.0	Introduction	1
1.1	What is Nonlinearity?	2
1.2	Equations from Diffusion Theory	4
1.3	Equations from Fluid Mechanics	8
1.4	Equations from Solid Mechanics	10
1.5	Miscellaneous Examples	13
1.6	Selected References	17
	References	17

CHAPTER 2. Transformation and General Solutions

2.0	Introduction	20
2.1	Transformations on Dependent Variables	21
2.2	Transformations on Independent Variables	31
2.3	Mixed Transformations	35
2.4	The Unknown Function Approach	47
2.5	General Solutions	49
2.6	General Solutions of First-Order Equations	50
2.7	General Solutions of Second-Order Equations	58
2.8	Table of General Solutions	65
	References	69

CHAPTER 3. Exact Methods of Solution

3.0	Introduction	71
3.1	The Quasi-Linear System	72
3.2	An Example of the Quasi-Linear Theory	78
3.3	The Poisson-Euler-Darboux Equation	84
3.4	Remarks on the PED Equation	88
3.5	One-Dimensional Anisentropic Flows	90
3.6	An Alternate Approach to Anisentropic Flow	94
3.7	General Solution for Anisentropic Flow	100
3.8	Vibration of a Nonlinear String	103
3.9	Other Examples of the Quasi-Linear Theory	109
3.10	Direct Separation of Variables	109
3.11	Other Solutions Obtained by *Ad Hoc* Assumptions	117
	References	121

ix

CHAPTER 4. Further Analytic Methods

4.0	Introduction	123
4.1	An *Ad Hoc* Solution from Magneto-Gas Dynamics	123
4.2	The Utility of Lagrangian Coordinates	126
4.3	Similarity Variables	133
4.4	Similarity via One-Parameter Groups	135
4.5	Extensions of the Similarity Procedure	141
4.6	Similarity via Separation of Variables	144
4.7	Similarity and Conservation Laws	150
4.8	General Comments on Transformation Groups	156
4.9	Similarity Applied to Moving Boundary Problems	158
4.10	Similarity Considerations in Three Dimensions	162
4.11	General Discussion of Similarity	166
4.12	Integral Equation Methods	167
4.13	The Hodograph	171
4.14	Simple Examples of Hodograph Application	173
4.15	The Hodograph in More Complicated Problems	177
4.16	Utilization of the General Solutions of Chapter 2	180
4.17	Similar Solutions in Heat and Mass Transfer	183
4.18	Similarity Integrals in Compressible Gases	186
4.19	Some Disjoint Remarks	190
	References	192

CHAPTER 5. Approximate Methods

5.0	Introduction	195
5.1	Perturbation Concepts	196
5.2	Regular Perturbations in Vibration Theory	197
5.3	Perturbation and Plasma Oscillations	198
5.4	Perturbation in Elasticity	204
5.5	Other Applications	207
5.6	Perturbation about Exact Solutions	208
5.7	The Singular Perturbation Problem	211
5.8	Singular Perturbations in Viscous Flow	215
5.9	The "Inner-Outer" Expansion (a Motivation)	219
5.10	The Inner and Outer Expansions	222
5.11	Examples	226
5.12	Higher Approximations for Flow past a Sphere	231
5.13	Asymptotic Approximations	237
5.14	Asymptotic Solutions in Diffusion with Reaction	240
5.15	Weighted Residual Methods: General Discussion	243
5.16	Examples of the Use of Weighted Residual Methods	249
5.17	Comments on the Methods of Weighted Residuals	261
5.18	Mathematical Problems of Approximate Methods	262
	References	267

CHAPTER 6. Further Approximate Methods

6.0	Introduction	271
6.1	Integral Methods in Fluid Mechanics	271
6.2	Nonlinear Boundary Conditions	278
6.3	Integral Equations and Boundary Layer Theory	280
6.4	Iterative Solutions for $\nabla^2 u = bu^2$	284
6.5	The Maximum Operation	287
6.6	Equations of Elliptic Type and the Maximum Operation	289
6.7	Other Applications of the Maximum Operation	292
6.8	Series Expansions	295
6.9	Goertler's Series	299
6.10	Series Solutions in Elasticity	301
6.11	"Traveling Wave" Solutions by Series	305
	References	312

CHAPTER 7. Numerical Methods

7.0	Introduction	315
7.1	Terminology and Computational Molecules	316

A. Parabolic Equations

7.2	Explicit Methods for Parabolic Systems	320
7.3	Some Nonlinear Examples	324
7.4	Alternate Explicit Methods	326
7.5	The Quasi-Linear Parabolic Equation	330
7.6	Singularities	330
7.7	A Treatment of Singularities (Example)	334
7.8	Implicit Procedures	338
7.9	A Second-Order Method for $Lu = f(x, t, u)$	343
7.10	Predictor Corrector Methods	345
7.11	Traveling Wave Solutions	348
7.12	Finite Differences Applied to the Boundary Layer Equations	349
7.13	Other Nonlinear Parabolic Examples	355

B. Elliptic Equations

7.14	Finite Difference Formula for Elliptic Equations in Two Dimensions	365
7.15	Linear Elliptic Equations	370
7.16	Methods of Solution of $Au = v$	373
7.17	Point Iterative Methods	375
7.18	Block Iterative Methods	384
7.19	Examples of Nonlinear Elliptic Equations	389
7.20	Singularities	411

C. Hyperbolic Equations

7.21	Method of Characteristics	416
7.22	The Supersonic Nozzle	423
7.23	Properties of Hyperbolic Systems	426

7.24 One-Dimensional Isentropic Flow 432
7.25 Method of Characteristics: Numerical Computation 435
7.26 Finite Difference Methods: General Discussion 437
7.27 Explicit Methods 438
7.28 Explicit Methods in Nonlinear Second-Order Systems 440
7.29 Implicit Methods for Second-Order Equations 443
7.30 "Hybrid" Methods for a Nonlinear First-Order System 445
7.31 Finite Difference Schemes in One-Dimensional Flow 448
7.32 Conservation Equations 453
7.33 Interfaces 454
7.34 Shocks 456
7.35 Additional Methods 461

D. Mixed Systems

7.36 The Role of Mixed Systems 462
7.37 Hydrodynamic Flow and Radiation Diffusion 462
7.38 Nonlinear Vibrations of a Moving Threadline 464
References 467

CHAPTER 8. Some Theoretical Considerations

8.0 Introduction 474
8.1 Well-Posed Problems 475
8.2 Existence and Uniqueness in Viscous Incompressible Flow 478
8.3 Existence and Uniqueness in Boundary Layer Theory 482
8.4 Existence and Uniqueness in Quasi-Linear Parabolic Equations 486
8.5 Uniqueness Questions for Quasi-Linear Elliptic Equations 487
References 489

APPENDIX. Elements of Group Theory

A.1 Basic Definitions 491
A.2 Groups of Transformations 492

AUTHOR INDEX 495
SUBJECT INDEX 501

NONLINEAR PARTIAL
DIFFERENTIAL EQUATIONS
IN ENGINEERING

CHAPTER 1

The Origin of Nonlinear Partial Differential Equations

1.0 INTRODUCTION

The assumption of linearity underlies, as a fundamental postulate, a considerable domain of mathematics. Therefore, the mathematical tools available to the natural scientist are essentially linear. However, nature, with scant regard for the desires of the mathematician, is essentially nonlinear—that is to say, the mathematical models believed to best approximate her (they are of course all approximations) are nonlinear. There is no general extant theory for nonlinear partial differential equations of any "order" of nonlinearity. We do not mean to imply by this statement that there will be two theories, one for the linear system and one for the nonlinear systems. Indeed there will be many theories for the multitude of nonlinearities that can exist.

Since suitable formulations of natural laws are inherently nonlinear it would seem that some forays into the "nonlinear morass" would have been made. This is true but the going is rough and slow since guiding principles, at this early stage, are almost entirely missing. It is the purpose of this work to discuss the origin of nonlinear partial differential equations, show how some are related, and illustrate various methods of attack, analytic, approximate and numerical, together with examples. The chapters are in the above order.

1.1 WHAT IS NONLINEARITY?

Consider the attitudes we have taken in our engineering calculations. Perhaps the most common procedure is to mention nonlinearities merely to dismiss them. This results in many phrases, found in books and journals alike, such as "assume the fluid is inviscid" or "given a perfect insulator" or "for constant thermal conductivity" or "in a homogeneous and isotropic medium the equations become..." and so on. Pedagogically we often give the impression that everything is ideal, isotropic, frictionless, rigid, inviscid, incompressible, and the like.

What do all these phrases mean and why do we mumble them? Recall that the derivative, which we denote by D, had the fundamental property that "the derivative of a sum of two functions was equal to the sum of the derivatives of the functions"; i.e.,

$$D(f+g) = Df + Dg. \tag{1.1}$$

This property is also possessed by the integral, all difference operators (e.g., the forward difference $\Delta f(x) = f(x+h) - f(x)$) in short, by the basic operations that underlie our fundamental mathematical tools, the differential equations, integral equations, difference equations, and combinations of them.

In general we say an operator L is linear if

$$L(f+g) = Lf + Lg. \tag{1.2}$$

Clearly the logarithm operator does not possess this property since

$$\log f + \log g = \log fg \neq \log(f+g). \tag{1.3}$$

Indeed the truth of such a statement as Eq. (1.3) would completely undermine the utility of the logarithm.

Consider the implications of linearity upon solutions of the linear heat conduction equation

$$\frac{\partial^2 T}{\partial x^2} = \alpha \frac{\partial T}{\partial t} \tag{1.4}$$

where α is constant. If T_1 and T_2 are both solutions of Eq. (1.4), i.e.,

$$\frac{\partial^2 T_1}{\partial x^2} = \alpha \frac{\partial T_1}{\partial t}, \quad \frac{\partial^2 T_2}{\partial x^2} = \alpha \frac{\partial T_2}{\partial t},$$

1.1 WHAT IS NONLINEARITY?

then it follows, by the *linearity* of the first and second partial derivatives, that $T_1 + T_2$ (in fact $AT_1 + BT_2$, where A and B are arbitrary constants) are solutions. This statement is the foundation of the *principle of superposition* which was essentially responsible for the great successes of the past in constructing effective theories for (linearized) physical phenomena. In accordance with this principle, elementary solutions of the pertinent mathematical equations could be combined to yield more flexible ones, namely ones which could satisfy the auxiliary conditions that describe the particular physical phenomena.

If we turn to the more physically reasonable model of heat conduction

$$\frac{\partial}{\partial x}\left[k(T)\frac{\partial T}{\partial x}\right] = c_v \frac{\partial T}{\partial t}, \tag{1.5}$$

where the conductivity k depends upon the temperature, the situation is radically changed. For purposes of discussion suppose that

$$k(T) = k_0 T \tag{1.6}$$

and that k_0 and c_v are constant. With $\alpha = c_v/k_0$, Eq. (1.5) becomes

$$T\frac{\partial^2 T}{\partial x^2} + \left(\frac{\partial T}{\partial x}\right)^2 = \alpha \frac{\partial T}{\partial t}. \tag{1.7}$$

If two solutions $T_1(x, t)$ and $T_2(x, t)$ of Eq. (1.7) have been found, is $T_1 + T_2$ also a solution? Substituting $T_1 + T_2$ for T we have

$$(T_1 + T_2)\frac{\partial^2(T_1 + T_2)}{\partial x^2} + \left[\frac{\partial(T_1 + T_2)}{\partial x}\right]^2 \stackrel{?}{=} \alpha \frac{\partial(T_1 + T_2)}{\partial t}, \tag{1.8}$$

which reduces, after algebra, to

$$\left[T_1\frac{\partial^2 T_1}{\partial x^2} + \left(\frac{\partial T_1}{\partial x}\right)^2 - \alpha\frac{\partial T_1}{\partial t}\right] + \left[T_2\frac{\partial^2 T_2}{\partial x^2} + \left(\frac{\partial T_2}{\partial x}\right)^2 - \alpha\frac{\partial T_2}{\partial t}\right]$$

$$+ T_1\frac{\partial^2 T_2}{\partial x^2} + T_2\frac{\partial^2 T_1}{\partial x^2} + 2\frac{\partial T_1}{\partial x}\frac{\partial T_2}{\partial x} \stackrel{?}{=} 0. \tag{1.9}$$

The first two bracketed expressions of equation (1.9) vanish by assumption but the last three terms are not zero. Hence $T_1 + T_2$ is not a solution. Thus the principle of superposition no longer holds and we have lost a powerful tool. It is the loss of this principle and the lack of an effective replacement that constitutes a formidable hurdle.

On occasion we can develop a transformation which reduces the nonlinear problem to a linear one and therefore utilize the principle of superposition. In fact, as will be discussed at length later, transformations are powerful tools in this subject.

1.2 EQUATIONS FROM DIFFUSION THEORY

Diffusion in an isotropic medium, where the diffusion coefficient depends upon the concentration C, is governed by the equation

$$\frac{\partial C}{\partial t} = \text{div}[D(C) \text{ grad } C] = \nabla \cdot [D(C) \nabla C] \tag{1.10}$$

expressed in vector notation. Of considerable interest, because of the variety of problems reducible to that form, is the case where

$$D(C) = D_0[C/C_0]^n. \tag{1.11}$$

In addition to the obvious case of diffusion the following physical problems may be cast into the form of Eq. (1.10) utilizing Eq. (1.11) for D.

a. Heat Conduction in an Isotropic Medium

If $c_v(T)$ is the specific heat per unit volume and $k(T)$ is the thermal conductivity the pertinent equation is

$$c_v(T) \frac{\partial T}{\partial t} = \text{div}(k(T) \text{ grad } T). \tag{1.12}$$

If $c_v = c_{v_0} T^r$, $k = k_0 T^s$, c_{v_0}, k_0, r, s are constants, and $r \neq -1$ the equation has the form

$$T^r \frac{\partial T}{\partial t} = \text{div}\left[\frac{k_0}{c_{v_0}} T^s \text{ grad } T\right] \tag{1.13}$$

and since

$$T^r \frac{\partial T}{\partial t} = \frac{1}{r+1} \frac{\partial T^{r+1}}{\partial t} \tag{1.14}$$

a simplification occurs if we set $U = T^{r+1}$. With this transformation, Eq. (1.13) becomes

$$\frac{\partial U}{\partial t} = \text{div}\left[\frac{k_0}{c_{v_0}} U^{(s-r)/(r+1)} \text{ grad } U\right] \tag{1.15}$$

1.2 EQUATIONS FROM DIFFUSION THEORY

which is the form given by Eq. (1.10) with $n = (s - r)/(r + 1)$. Normally the case $n \geq 0$ is of physical interest. If $s = r$, that is the conductivity and specific heat have the same form, Eq. (1.15) becomes

$$\frac{\partial U}{\partial t} = \text{div}\left[\frac{k_0}{c_{v_0}} \text{grad } U\right] \quad (1.16)$$

which is linear.

b. Fluid Flow through Porous Media

According to Muskat [1][†] the mathematical model for flow through a porous media is based on the empirical relation (D'Arcy's law):

$$\mu \mathbf{V} = -\alpha \text{ grad } P, \quad (1.17)$$

conservation of mass (continuity):

$$\text{div}(\rho \mathbf{V}) = -\epsilon \, \partial \rho / \partial t, \quad (1.18)$$

and the equation of state:

$$\rho/\rho_0 = (P/P_0)^{1/\gamma}, \quad (1.19)$$

where $P, \rho, \mathbf{V}$ are pressure, density and velocity respectively, and $\mu, \alpha, \epsilon, \gamma$ are physical constants representing fluid viscosity, permeability, porosity and ratio of specific heats (c_p/c_v), respectively. The equation for density ρ can be obtained by starting with Eq. (1.18) and successively eliminating the velocity thereby obtaining

$$\text{div}[(\alpha/\epsilon\mu)\rho \text{ grad } P] = \frac{\partial \rho}{\partial t}.$$

Utilizing Eq. (1.19) for P the result is

$$\text{div}[(\gamma\alpha P_0/\epsilon\mu\rho_0\beta)\rho^\gamma \text{ grad } \rho] = \frac{\partial \rho}{\partial t} \quad (1.20)$$

which is of the form given by Eq. (1.10) with $n = \gamma$.

[†] Numbers in brackets refer to the References at the end of each chapter.

c. Boundary Layer Flow over a Semi-Infinite Flat Plate

In this example we shall utilize a transformation which will have great utility later in our work. Von Mises [2] described a transformation which bears his name and has been widely utilized as described for example in Goldstein [3] and Pai [4]. If $\nu = \mu/\rho_0$ is the constant kinematic viscosity, then the boundary layer equations are [5] (note: $-1/\rho\, dp/dx$ is omitted for this discussion)

$$u\left(\frac{\partial u}{\partial x}\right)_y + v\left(\frac{\partial u}{\partial y}\right)_x = \nu\left(\frac{\partial^2 u}{\partial y^2}\right)_x$$
$$\left(\frac{\partial u}{\partial x}\right)_y + \left(\frac{\partial v}{\partial y}\right)_x = 0 \qquad (1.21)$$

where u, v are the velocity components in the x, y directions, respectively, and the subscripts are introduced to emphasize that x, y are the independent variables, that is, $(\partial u/\partial y)_x$ is the derivative of u with respect to y with x held constant. The physical problem is illustrated in Figure 1-1.

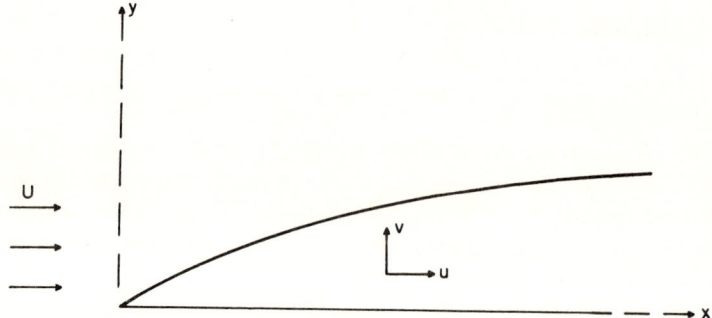

FIG. 1-1. Schematic of boundary layer development.

A standard procedure involves the introduction of a stream function ψ so that the second equation (continuity) of Eq. (1.21) is automatically satisfied. Thus define ψ by setting

$$u = \frac{\partial \psi}{\partial y}, \qquad v = -\frac{\partial \psi}{\partial x}. \qquad (1.22)$$

We now perform *von Mises' transformation*, which is basically the introduction of ψ as a new independent variable (replacing y). Since u can then be expressed in two ways

$$u = u(x, y) \tag{1.23}$$

or

$$u = u(x, \psi), \quad \psi = \psi(x, y) \tag{1.24}$$

we can calculate $(\partial u/\partial x)_y$ in two ways. From the first and second representations we have, respectively,

$$\left(\frac{\partial u}{\partial x}\right)_y = \left(\frac{\partial u}{\partial x}\right)_\psi + \left(\frac{\partial u}{\partial \psi}\right)_x \left(\frac{\partial \psi}{\partial x}\right)_y = \left(\frac{\partial u}{\partial x}\right)_\psi - v\left(\frac{\partial u}{\partial \psi}\right)_x \tag{1.25}$$

upon insertion of $-v$ for $\partial \psi/\partial x$. Similarly

$$\left(\frac{\partial u}{\partial y}\right)_x = \left(\frac{\partial u}{\partial \psi}\right)_x \left(\frac{\partial \psi}{\partial y}\right)_x = u\left(\frac{\partial u}{\partial \psi}\right)_x \tag{1.26}$$

and

$$\left(\frac{\partial^2 u}{\partial y^2}\right)_x = \frac{\partial}{\partial y}\left[u\left(\frac{\partial u}{\partial \psi}\right)_x\right]_x = u\frac{\partial}{\partial \psi}\left[u\left(\frac{\partial u}{\partial \psi}\right)_x\right]_x. \tag{1.27}$$

Upon substituting these relations in the first boundary layer equation (1.21) we have

$$u\left(\frac{\partial u}{\partial x}\right)_\psi = vu\frac{\partial}{\partial \psi}\left[u\left(\frac{\partial u}{\partial \psi}\right)_x\right]_x. \tag{1.28}$$

If we recall that the *two* independent variables are now x and ψ, the subscripts can be dropped leaving

$$\frac{\partial u}{\partial x} = \frac{\partial}{\partial \psi}\left[vu\frac{\partial u}{\partial \psi}\right], \tag{1.29}$$

which is formally equivalent to Eq. (1.10) with $n = 1$.

d. Persistence of Solar Prominences

Various authors, notably Severnyi [6] and Rosseland *et al.* [7], consider the variation of temperature in the prominence by means of the heat conduction equation (1.5). The coefficients c_v and $k(T)$ are taken from the kinetic theory relations (see Hirschfelder *et al.* [8] or Spitzer [9]) as $c_v = 3KN$, $k = \bar{v}KNL/2$ where K is Boltzmann's constant and $\bar{v}$, N, and L represent the mean thermal velocity, mean number per unit volume, and mean free path of the negative ions in the gas. If, further, temperature is defined as "kinetic temperature" via the relation

$$\tfrac{1}{2}m(\bar{v})^2 = \tfrac{3}{2}KT$$

and

$$L = 9K^2T^2/4\pi Ne^2,$$

where e and m are electron charge and mass, respectively, then the equation becomes

$$\frac{\partial T}{\partial t} = \text{div}(\lambda T^{5/2} \text{ grad } T) \tag{1.30}$$

with $\lambda = 9K^{5/2}/8\pi N(3m)^{1/2}$. Formally, Eq. (1.30) is equivalent to Eq. (1.10) with $n = 5/2$.

Finally we remark that nothing has yet been said about auxiliary conditions and what complications may arise when transformations are applied. Different boundary conditions may require diverse methods of attack.

1.3 EQUATIONS FROM FLUID MECHANICS

Fluid mechanics is especially rich in nonlinearities as has been already observed in the previous discussion of the boundary layer and flow through porous media. The inherent nonlinearities of the mathematical model of this physical system has stimulated much of our present day knowledge of nonlinear problems. The general governing equations for fluid mechanics including electric and magnetic body forces may be found in numerous references, for example Pai [10] and Goldstein [3]. Since no general methods exist for obtaining solutions for the full system they are not reproduced here. Instead we remark some of the systems for which methods of solution are extant.

a. The Equations of Hydrodynamics (Ideal Fluid)

In this system we assume an inviscid and incompressible fluid. For such a system the velocity distribution can be determined by the so called "Eulerian" equations of motion. Denoting by u_i, $i = 1, 2, 3$, the components of velocity in the directions denoted by x_i, the pressure by p,

1.3 EQUATIONS FROM FLUID MECHANICS

the constant density by ρ and the components of the body force by F_i, we have the three equations of momentum[†]:

$$\frac{\partial u_i}{\partial t} + u_j \frac{\partial u_i}{\partial x_j} = -\frac{1}{\rho}\frac{\partial p}{\partial x_i} + F_i, \qquad (1.31)$$

the equation of continuity:

$$\frac{\partial u_i}{\partial x_i} = 0.$$

These equations are nonlinear in the u_i's, with the four unknowns u_1, u_2, u_3, and p.

b. The Equations of Gas Dynamics

This system differs from that of (a) in that the incompressible assumption is relaxed and in addition we wish to calculate the entropy S or internal energy e. The full system is then the three equations (1.31) augmented by a new equation of continuity (conservation of mass):

$$\frac{\partial \rho}{\partial t} + \frac{\partial}{\partial x_i}(\rho u_i) = 0; \qquad (1.32)$$

the changes of state are adiabatic[‡]

$$\frac{\partial S}{\partial t} + u_i \frac{\partial S}{\partial x_i} = 0; \qquad (1.33)$$

and the equation of state

$$p = f(\rho, S). \qquad (1.34)$$

This nonlinear system constitutes six equations for the six unknowns u_1, u_2, u_3, p, ρ, and S. Generally it is more convenient to regard the system as one consisting of only five equations in five unknowns since p is easily eliminated by means of Eq. (1.34).

[†] Note that the repeated index notation of tensor analysis is used herein. Thus

$$u_j \frac{\partial u_1}{\partial x_j} = u_1 \frac{\partial u_1}{\partial x_1} + u_2 \frac{\partial u_1}{\partial x_2} + u_3 \frac{\partial u_1}{\partial x_3}.$$

[‡] Note that we are using the Courant-Friedrichs [11] terminology here and restrict the use of "homentropic" for the concept of constant entropy throughout the medium and "isentropic" for constant entropy along a streamline.

c. Flow of Viscous Incompressible Fluid

The dynamical equations for the flow of a viscous incompressible fluid are formed by augmenting Eqs. (1.31) by a viscous term. Thus the velocity components u_1, u_2, u_3, and pressure p are determined from the Navier-Stokes equations

$$\frac{\partial u_i}{\partial t} + u_j \frac{\partial u_i}{\partial x_j} = -\frac{1}{\rho}\frac{\partial p}{\partial x_i} + \nu\, \nabla^2 u_i, \tag{1.35}$$

$i = 1, 2, 3$, and

$$\frac{\partial u_i}{\partial x_i} = 0 \tag{1.36}$$

where ν is the kinematic viscosity $= \mu/\rho$ and ∇^2 is the usual terminology for the Laplacian operator (div grad).

d. Electrostatic Plasma Oscillations

As a last example in this area we give an example from Chandrasekhar [12] wherein the so-called "dispersion relation" is required for electrostatic plasma oscillation. If $f(\mathbf{x}, \mathbf{v}, t)$ is the distribution function of the negative ions, where $\mathbf{x}$ is the space vector, $\mathbf{v}$ the velocity vector, and t time, then f satisfies the Vlasov-Boltzmann equation

$$\frac{\partial f}{\partial t} + \mathbf{v} \cdot \nabla_x f + (e\mathbf{E}/m) \cdot \nabla_v f = \left(\frac{\partial f}{\partial t}\right)_{\text{coll}} \tag{1.37}$$

and

$$\nabla \cdot \mathbf{E} = 4\pi e(n - n_0). \tag{1.38}$$

The notation ∇_x means the gradient of f with respect to the space coordinates, e and m are the electron charge and mass, respectively, $\mathbf{E}$ is the electric field, and $n - n_0$ is the density referred to an equilibrium density n_0.

1.4 EQUATIONS FROM SOLID MECHANICS

a. Oscillations of an Elastic String

The nonlinear vibration problem of the elastic string, without dissipation, gives rise to a difficult system considered by Carrier [13] and others. If u, v are the displacements in the x, y directions, respectively,

1.4 EQUATIONS FROM SOLID MECHANICS

T the tension, ρ the density, A the cross-sectional area, and θ the angle of the string with the horizontal, the equations of motion are

$$\frac{\partial}{\partial x}(T \sin \theta) = \rho A \frac{\partial^2 u}{\partial t^2} \tag{1.39}$$

$$\frac{\partial}{\partial x}(T \cos \theta) = \rho A \frac{\partial^2 v}{\partial t^2} \tag{1.40}$$

where it is assumed from the stress-strain relations that

$$T = T_0 + EA\{[(1 + v_x)^2 + (u_x)^2]^{\frac{1}{2}} - 1\}. \tag{1.41}$$

A schematic of the physical problem is given in Fig. 1-2 and from this it is clear that

$$\tan \theta = u_x/(1 + v_x).$$

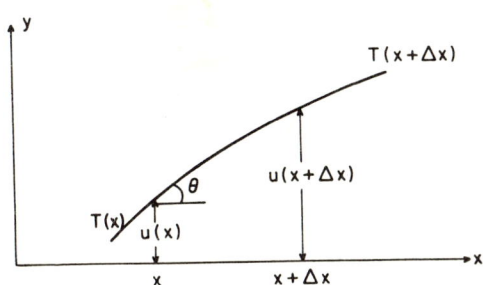

FIG. 1-2. Differential element of a string.

Addition of a reasonable viscous dissipation term of the form $|\partial u/\partial t| \partial u/\partial t$ creates an additional nonlinearity. The generalization of this system for an oscillating moving threadline has been considered by Zaiser and the present author and will be discussed later. Zabusky [14] discusses a related problem to that of Carrier but replaces Eq. (1.41) by an alternate form.

b. Large Deflection of Plates

The large normal deflection w of a rectangular plate of thickness t, modulus E, and bending stiffness B per unit length is obtained (when the squares and products of the deflections and strain components are included) from the system

$$\nabla^2 F = E\left[\left(\frac{\partial^2 w}{\partial x\, \partial y}\right)^2 - \frac{\partial^2 w}{\partial x^2}\frac{\partial^2 w}{\partial y^2}\right], \tag{1.42}$$

$$\nabla^2 w = \frac{t}{B}\left[\frac{\partial^2 F}{\partial y^2}\frac{\partial^2 w}{\partial x^2} + \frac{\partial^2 F}{\partial x^2}\frac{\partial^2 w}{\partial y^2} - 2\frac{\partial^2 F}{\partial x\, \partial y}\frac{\partial^2 w}{\partial x\, \partial y}\right]. \tag{1.43}$$

In these equations the function $F(x, y)$ is an Airy stress function defined, so that the equations of equilibrium are always satisfied, in terms of the normal stresses σ_x, σ_y and the shearing stress τ_{xy} as

$$\sigma_x = \frac{\partial^2 F}{\partial y^2}, \quad \sigma_y = \frac{\partial^2 F}{\partial x^2}, \quad \text{and} \quad \tau_{xy} = -\frac{\partial^2 F}{\partial x\, \partial y}. \tag{1.44}$$

Von Kármán [15] first gave these equations in 1910 and they bear his name. Nowinski [16, 17] has applied several generalizations of these equations to orthotropic circular plates and spinning disks.

c. Stress Distribution in Plastic Bodies

Instead of assuming the body to be elastic we now assume it to be "ideally" plastic (see Hoffman and Sachs [18]), so that the stresses satisfy a von Mises-Hencky condition of the form

$$(\sigma_x - \sigma_y)^2 + 4\tau_{xy}^2 = \tfrac{4}{3}\sigma_0^2 = \text{(constant)}. \tag{1.45}$$

Introduction of the stress function F, defined by Eq. (1.44), into Eq. (1.45) demonstrates that F satisfies the equation

$$\left[\frac{\partial^2 F}{\partial x^2} - \frac{\partial^2 F}{\partial y^2}\right]^2 + 4\left(\frac{\partial^2 F}{\partial x\, \partial y}\right)^2 = \tfrac{4}{3}\sigma_0^2. \tag{1.46}$$

The stress distribution in a mass of sand which is in the limiting state of equilibrium, i.e., just about to collapse, depends upon a similar equation. If the sand is under a large external load and the influence of its own weight is neglected, the stress function F satisfies the equation

$$\left(\frac{\partial^2 F}{\partial x^2} + \frac{\partial^2 F}{\partial y^2}\right)^2 - \frac{4}{1 - f^2}\left[\frac{\partial^2 F}{\partial x^2}\frac{\partial^2 F}{\partial y^2} - \frac{\partial^2 F}{\partial x\, \partial y}\right]^2 = 0 \tag{1.47}$$

where f is the friction coefficient between the sand particles. A theory of earth pressure, developed originally by Rankine has Eq. (1.47) as its foundation and has also been applied to other problems in soil mechanics.

d. Restricted Form of Plateau's Problem [19]

Plateau's problem is not, strictly speaking, a problem from solid mechanics but its foundation lies in that area. In its restricted form the problem can be stated thusly: Let R be a simply connected plane region with boundary C and let $f(x, y)$ represent the height of a given space curve γ above the point (x, y) on C. The problem is to construct a surface S of minimum surface area passing through γ. If the height of S above (x, y) for (x, y) in R is denoted by $h(x, y)$, then the surface area of S is given by

$$A = \iint_R (1 + h_x^2 + h_y^2)^{\frac{1}{2}} \, dx \, dy. \tag{1.48}$$

The necessary condition for this integral to be minimum [19] is that h should satisfy the Euler-Lagrange equation

$$(1 + h_y^2)h_{xx} - 2h_x h_y h_{xy} + (1 + h_x^2)h_{yy} = 0 \tag{1.49}$$

and $h(x, y) = f(x, y)$ on C. The subscript notation is the common one for the partial derivatives—i.e., $h_x = \partial h/\partial x$. The Euler-Lagrange equation for variational problems is usually nonlinear.

1.5 MISCELLANEOUS EXAMPLES

When engineering problems are considered which have a multiplicity of "fundamental laws" in operation (the usual case) it is the rule rather than the exception for the mathematical model to be a nonlinear partial differential equation. One excellent example is furnished by recent work of Ferron [20] and Pearson [21].

a. Diffusion with Second-Order Reaction

When one substance A diffuses into a medium containing another substance B with which it reacts according to the reaction

$$A + B \xrightarrow{k} C$$

then a mathematical model for the process is the set of coupled equations

$$\frac{\partial a}{\partial t} = D_a \frac{\partial^2 a}{\partial x^2} - kab$$
$$\frac{\partial b}{\partial t} = D_b \frac{\partial^2 b}{\partial x^2} - kab. \qquad (1.50)$$

In Eq. (1.50) a and b are the concentrations of components A and B, respectively, k is the reaction rate constant, and the D's represent diffusion coefficients.

Katz et al. [22] and Reese and Eyring [23] have considered the problem of diffusion accompanied by an immobilizing reaction which is bimolecular. Their model has the same type of product nonlinearity (ab) as Eq. (1.50).

A second example whose difficulty led to the development of a new numerical technique is due to Crank and Nicolson [24].

b. Flow of Heat Generated by a Chemical Reaction

The reaction rate constant depends upon temperature T and the usual assumed relation is the Arrhenius expression

$$k = e^{-A/T}. \qquad (1.51)$$

If one studies the flow of heat generated by a chemical reaction in a medium where Eq. (1.51) holds then the mathematical model is

$$u_t = u_{xx} - aw_t \qquad (1.52)$$
$$w_t = -be^{-A/u}w \qquad (1.53)$$

(a, b, A positive constants), $0 \leqslant x \leqslant 1$, subject to the (additional nonlinearity) nonlinear boundary conditions

$$u(x, 0) = f(x), \quad w(x, 0) = g(x)$$
$$u_x(0, t) = h(u), \quad u_x(1, t) = L(u).$$

From the form of the system it is seen that Eq. (1.52) is the conduction equation modified by the effect of the chemical reaction as described by Eq. (1.53).

The last two examples indicate something previously undescribed, i.e., the nonlinearities may result from the auxiliary conditions. Such nonlinearities occur in problems where the boundary conditions depend (nonlinearly) upon the dependent variable and or its derivatives. In addition, the boundaries may be unknown as occurs in diffusion where the concentration front changes.

c. Heat Conduction with Radioactive Decay

Jones [25] and Douglas and Jones [26] considered a heat conduction problem in which the medium is undergoing radioactive decay, so that its thermal conductivity varies with the degree of decay. In turn, the degree of decay may be related to the time, so that the thermal conductivity may be considered a function (unknown) of time. Thus the mathematical model would be

$$\frac{\partial u}{\partial t} = a(t)\frac{\partial^2 u}{\partial x^2}, \qquad 0 < x < \infty, \qquad 0 < t < T$$

$$u(x, 0) = 0, \qquad 0 \leqslant x < \infty \qquad (1.54)$$

$$u(0, t) = f(t), \qquad 0 \leqslant t < T; \quad f(0) = 0$$

$$-a(t)\frac{\partial u}{\partial x}(0, t) = g(t).$$

In this problem both $u(x, t)$ and $a(t)$ are unknown and are to be determined from the data $f(t)$ and $g(t)$. As is seen in Jones [25] the problem of determining $a(t)$ is equivalent to obtaining the solution of a nonlinear integral equation.

d. Flow Problems with a Free Boundary

As we have seen the velocity distribution in an ideal incompressible fluid is governed by the Eulerian equations of motion (1.31) and the equation of continuity (1.32). If we assume that the flow is without vorticity (irrotational), i.e., the velocity components u_i are derivatives of a velocity potential ϕ ($u_1 = \phi_{x_1}$, $u_2 = \phi_{x_2}$, $u_3 = \phi_{x_3}$) and the forces F_i are conservative (a force potential V exists), then the equations of motion take the form

$$\frac{\partial^2 \phi}{\partial t\, \partial x_i} + \frac{\partial \phi}{\partial x_j}\frac{\partial^2 \phi}{\partial x_i\, \partial x_j} = -\frac{1}{\rho}\frac{\partial p}{\partial x_i} + \frac{\partial V}{\partial x_i}, \qquad (1.55)$$

$i = 1, 2, 3$, and continuity becomes

$$\nabla^2 \phi = 0. \qquad (1.56)$$

A first integral of Eq. (1.55) is possible, yielding

$$\frac{\partial \phi}{\partial t} + \frac{1}{2} \frac{\partial \phi}{\partial x_j} \frac{\partial \phi}{\partial x_j} + \frac{P}{\rho} - V = \text{constant}. \tag{1.57}$$

Thus we see that in the case of irrotational flow the determination of the velocity distribution is reduced to solving Laplace's equation (1.56) and the pressure is given by Eq. (1.57).

We now suppose that the fluid is bounded by a (unknown) free surface, i.e., by another fluid of negligible specific gravity which is at rest. In such a case the boundary condition must express the fact that the pressure p is constant along the free boundary. Thus from (1.57) we have

$$\frac{\partial \phi}{\partial t} + \frac{1}{2} \frac{\partial \phi}{\partial x_j} \frac{\partial \phi}{\partial x_j} - V = \text{constant}, \tag{1.58}$$

which is a nonlinear boundary condition owing both to the second term and to the fact that the boundary is itself unknown.

e. Circulating Fuel Reactors

The neutron density u and temperature v in a circulating fuel reactor satisfy nonlinear partial differential equations of the form

$$\begin{aligned} u_t &= u_{xx} + A(v)u \\ u &= cv_x + v_t, \quad 0 \leqslant x \leqslant a, \quad t \geqslant 0. \end{aligned} \tag{1.59}$$

These equations are dimensionless; t is time, x is a space variable, and c is the positive constant speed at which the fuel flows through the reactor. $A(v)$ is a given function of v describing the temperature density interaction. A common assumption is that of $A(v) = bv$. Elimination of u results in the equation for v:

$$\left(\frac{\partial}{\partial t} + c \frac{\partial}{\partial x}\right) [v_{xx} + A(v)v - v_t] = 0, \tag{1.60}$$

which implies that there exists a function f for which

$$v_t = v_{xx} + vA(v) + f(x - ct). \tag{1.61}$$

The reactor boundary conditions are assumed to be $u(0, t) = u(a, t) = 0$ and $v(0, t) = 0$. The initial state will be described as

$$u(x, 0) = u_0(x), \quad v(x, 0) = v_0(x)$$

in terms of known functions u_0 and v_0. These conditions allow us to find the functional form of f since from Eq. (1.61), at $t = 0$,

$$f(x) = v_t(x, 0) - v_{xx}(x, 0) - v(x, 0)A(v(x, 0))$$
$$= u_0(x) - cv_0'(x) - v_0''(x) - v_0(x)A(v_0(x)),$$

and so Eq. (1.61) is completely defined.

1.6 SELECTED REFERENCES

In addition to the references cited we here append a selected list of books which, in the author's opinion, contain important formulations in various fields. The review paper of von Kármán [27] supplies an adequate bibliography up to 1940. Some overlap occurs but most of the listed references are post-1940.

References Cited

1. Muskat, M., "The Flow of Homogeneous Fluids through Porous Media." McGraw-Hill, New York, 1937; Edwards, Ann Arbor, Michigan, 1946.
2. von Mises, R., Bemerkungen zur Hydrodynamik, *Z. Angew. Math. Mech.* **7**, 425 (1927).
3. Goldstein, S., "Modern Developments in Fluid Dynamics," Vol. 1. Oxford Univ. Press, London and New York, 1938.
4. Pai, S. I., "Viscous Flow Theory," Vol. 1: Laminar Flow. Van Nostrand, Princeton, New Jersey, 1956.
5. Schlichting, H., "Boundary Layer Theory." McGraw-Hill, New York, 1955.
6. Severnyi, A. B., Investigations of light variations in solar prominences, *Dokl. Akad. Nauk SSSR* **73**, 475 (1950).
7. Rosseland, S., Jensen, E., and Tandberg-Hanssen, E., Some considerations on thermal conduction and magnetic fields in prominences, *in* "Electromagnetic Phenomena in Cosmical Physics" (B. Lehnert, ed.), p. 150. Cambridge Univ. Press, London and New York, 1958.
8. Hirschfelder, J. O., Curtiss, C. F., and Bird, R. B., "Molecular Theory of Gases and Liquids." Wiley, New York, 1954.

9. Spitzer, J. R., Jr., "Physics of Fully Ionized Gases." Wiley (Interscience), New York, 1956.
10. Pai, S. I., "Magnetogas-Dynamics and Plasma Dynamics." Springer, Vienna (Prentice-Hall, Englewood Cliffs, New Jersey), 1962.
11. Courant, R., and Friedrichs, K. O., "Supersonic Flow and Shock Waves." Wiley (Interscience), New York, 1948.
12. Chandrasekhar, S., "Plasma Physics." Univ. of Chicago Press, Chicago, Illinois, 1960.
13. Carrier, G. F., On the non-linear vibration problem of the elastic string, *Quart. Appl. Math.* **3**, 157 (1945).
14. Zabusky, N. J., Elastic solution for the vibrations of a nonlinear continuous model string, *J. Math. Phys.* **3**, 1028 (1962).
15. von Kármán, T., *Encycl. Math. Wiss.* **4**, 349 (1910).
16. Nowinski, J. L., Nonlinear vibrations of elastic circular plates exhibiting rectilinear orthotropy, *Z. Angew. Math. Phys.* **14**, 112 (1963).
17. Nowinski, J. L., "Nonlinear Transverse Vibrations of a Spinning Disk," ASME Paper No. 63-APMW-15 (1963).
18. Hoffman, O., and Sachs, G., "Introduction to the Theory of Plasticity for Engineers." McGraw-Hill, New York, 1953.
19. Courant, R., and Hilbert, D., Methods of Mathematical Physics," Vol. 1. Wiley (Interscience), New York, 1953.
20. Ferron, J. R., Approximations for mass transfer with chemical reaction, *A. I. C. E. Journal* in press (1965).
21. Pearson, J. R. A., Diffusion with second order reaction, *Appl. Sci. Res. Sect. A* **11**, 321 (1963).
22. Katz, S. M., Kuba, E. T., and Wakelin, J. H., The chemical attack on polymeric materials as modified by diffusion, *Textile Res. J.* **20**, 754 (1950).
23. Reese, C. E., and Eyring, H., Mechanical properties and the structure of hair, *Textile Res. J.* **20**, 743 (1950).
24. Crank, J., and Nicolson, P., A practical method for numerical evaluation of solutions of partial differential equations of the heat conduction type, *Proc. Cambridge Phil. Soc.* **43**, 50 (1947).
25. Jones, B. F., Jr., The determination of a coefficient in a parabolic differential equation, Part I. Existence and uniqueness, *J. Math. Mech.* **11**, 907 (1962).
26. Douglas, J., Jr., and Jones, B. F., The determination of a coefficient in a parabolic differential equation, Part II. Numerical approximation, *J. Math. Mech.* **11**, 919 (1962).
27. von Kármán, T., The engineer grapples with nonlinear problems, *Bull. Am. Math. Soc.* **46**, 615 (1940).

Selected Bibliography (Post-1940)

28. C. E. Eringen, "Nonlinear Theory of Continuous Media." McGraw-Hill, New York, 1962.
29. S. Prager, Transport processes with chemical reactions, *Chem. Eng. Progr.* **55**, 1 (1959).
30. G. Sandri, Kinetic equations in the presence of external fields, *Aeron. Res. Assoc. Princeton, N.Y. Rept.* No. 58 (1963).

1.6 SELECTED REFERENCES

31. S. Chandrasekhar, "Hydrodynamic and Hydromagnetic Stability." Oxford Univ. Press, London and New York, 1961.
32. G. Birkhoff and E. H. Zarantonello, "Jets, Wakes, and Cavities." Academic Press, New York, 1957.
33. S. Chapman and T. G. Cowling, "The Mathematical Theory of Nonuniform Gases." Cambridge Univ. Press, London and New York, 1938 (2nd ed., 1951).
34. M. Kruskal, "The Theory of Neutral and Ionized Gases." Wiley, New York, 1960.
35. O. A. Ladyzhenskaya, "Mathematical Problems in the Dynamics of Viscous Incompressible Fluids." Acad. Sci. U.S.S.R., Moscow, 1961.
36. S. Chandrasekhar, "Radiative Transfer." Dover, New York, 1960.
37. A. E. Green and J. E. Adkins, "Large Elastic Deformations." Oxford Univ. Press, London and New York, 1960.
38. A. E. Green and W. Zerna, "Theoretical Elasticity." Oxford Univ. Press, London and New York, 1954.
39. "Problems of Continuum Mechanics" (in Honor of N. I. Muskhelishvili). S.I.A.M., Philadelphia, Pennsylvania, 1961.
40. M. Van Dyke, "Perturbation Methods in Fluid Mechanics." Academic Press, New York, 1964.
41. A. G. Hansen, "Similarity Analyses of Boundary Value Problems in Engineering." Prentice-Hall, Englewood Cliffs, New Jersey, 1964.
42. H. Fujita, "The Mathematical Theory of Sedimentation Analysis." Academic Press, New York, 1962.

CHAPTER 2

Transformation and General Solutions

2.0 INTRODUCTION

While there is no existing general theory for nonlinear partial differential equations many special cases have yielded to appropriate changes of variable. In fact transformations are perhaps the most powerful general analytic tool currently available in this area. Typically these linearize the systems (for example the Kirchhoff and hodograph transformations), reduce the equation to a nonlinear ordinary differential equation (for example the similarity transformation), transform the system to one already solved, or perform some other reduction of complexity. In general these changes can be classified into three groups: Class I includes those which are transformations only of the dependent variables; Class II includes transformations only of the independent variable (s); Class III consists of transformations of both dependent and independent variables. We shall call this last class "mixed."

A substantial number of these operations, having sufficient generality, will be included in this chapter. Insofar as possible motivation for the transformation and suggestions for its generalization will be given. However, the author is the first to admit that some of the developments have a substantial "art" content. Imagination, ingenuity, and good fortune play an important role. Later chapters will draw heavily upon the information developed in these pages.

The second part of this chapter concerns the development of general solutions of equations. General solutions involve arbitrary functions of specific functions of the dependent and independent variables, e.g., $f(x + ct)$. A knowledge of these general solutions is extremely important in the process of obtaining approximate solutions as well as acting as a guide to analytic methods.

2.1 TRANSFORMATIONS ON DEPENDENT VARIABLES

a. Kirchhoff Transformation [1]

The pseudo-Laplacian

$$\text{div}[K(C) \text{ grad } C] = \nabla \cdot [K(C) \nabla C] = 0 \qquad (2.1)$$

is the governing steady state equation for a number of important physical processes in an isotropic medium where the physical parameters depend upon the dependent variable. Such processes include diffusion, heat conduction, transmission lines, etc.

Our goal is the construction of a transformation which linearizes this equation. Hence we try to construct a new dependent variable $\psi = \psi(C)$ so that Eq. (2.1) is linear in this new variable. Calculating the gradient yields

$$\nabla \psi = \frac{d\psi}{dC} \nabla C, \qquad (2.2)$$

and when the right-hand side of Eq. (2.2) is compared with Eq. (2.1) it is clear that the proper choice of ψ is such that

$$\frac{d\psi}{dC} = K(C) \qquad (2.3)$$

or in integral form

$$\psi = \int_{C_0}^{C} K(t) \, dt \qquad (2.4)$$

where C_0 is an arbitrary reference. Equation (2.1) then reduces to the linear Laplace's equation in ψ

$$\nabla^2 \psi = 0. \qquad (2.5)$$

We must be concerned also with the boundary conditions and what effect this transformation has upon them. If the physical problem is a Dirichlet problem, that is, a boundary value problem where the boundary conditions involve only the prescription of the function, so that we may write the boundary conditions as

$$B_i[C] = f \tag{2.6}$$

then the transfer to ψ of these boundary conditions is usually easily carried out. For example if $K(C) = C^s$, and the boundary conditions have the form $C(\text{boundary}) = f$, then

$$\psi = \frac{1}{s+1}[C^{s+1} - C_0^{s+1}]$$

and the transformed boundary value problem becomes

$$\nabla^2 \psi = 0 \tag{2.7}$$

subject to the boundary condition

$$\psi(\text{boundary}) = \frac{1}{s+1}[f^{s+1} - C_0^{s+1}] \tag{2.8}$$

which is again of Dirichlet form.

If the boundary conditions are of Neumann type, that is the normal derivative is prescribed on the boundary, say

$$\frac{\partial C}{\partial n}(\text{boundary}) = g \tag{2.9}$$

the Kirchhoff transformation may introduce nonlinearities in the boundary conditions. With $C_0 = 0$ the case $g = \text{constant} = \alpha$ gives rise to the transformed problem, Eq. (2.7), with the Neumann condition

$$\psi^{-s/(s+1)} \frac{\partial \psi}{\partial n}(\text{boundary}) = \text{constant.} \tag{2.10}$$

We usually also require that C remains finite and $\lim_{x \to \text{boundary}} C^s \, \partial C/\partial x$ exists.

We remark that a transformation on the dependent variable has the feature that the physical range of the independent variable is unchanged. However, the boundary conditions may become complicated. Lastly we note that the use of the Kirchhoff transformation in reducing Eq. (2.1) to Eq. (2.5) offers an opportunity to extend the relatively well developed numerical procedures for Laplace's equation to a large class of associated nonlinear problems.

b. A Functional Transformation

If the linear parabolic equation

$$\psi_t = \lambda \nabla^2 \psi \tag{2.11}$$

is subjected to the transformation

$$\psi = F(u) \tag{2.12}$$

where, for the moment F is unspecified, Eq. (2.11) becomes

$$u_t = \lambda (F''/F')(\nabla u)^2 + \lambda \nabla^2 u. \tag{2.13}$$

The prime notation refers to differentiation with respect to u. Equation (2.13) can easily be transformed to Eq. (2.11) by the inverse transformation $u = F^{-1}(\psi)$. Equation (2.13) has considerable importance in physical problems and great potential.

The burning of a gas in a rocket is described by Forsythe and Wasow [2] in terms of the equation

$$u_t = -\tfrac{1}{2} u_x^2 + \lambda u_{xx} + d(x, t) \tag{2.14}$$

for $-\infty < x < \infty$, $t > 0$, where λ is a positive constant and $d(x, t)$ is periodic of period 2π in x. The auxiliary conditions are

$$u(x, 0) = f(x), \quad u(x + 2\pi, t) = u(x, t) \tag{2.15}$$

where $f(x)$ is periodic of period 2π. While the general case of Eq. (2.14) is not completely reducible to Eq. (2.13) the case $d(x, t) = g(t)$ is reducible. Setting $u = v(x, t) + \int_0^t g(r)\, dr$ the equation for v becomes

$$v_t = -\tfrac{1}{2} v_x^2 + \lambda v_{xx} \tag{2.16}$$

with the auxiliarity conditions

$$v(x, 0) = f(x), \quad v(x + 2\pi, t) = v(x, t). \tag{2.17}$$

The one-dimensional equation

$$u_t + uu_x = \lambda u_{xx} \tag{2.18}$$

has been utilized by Burgers [3] as a mathematical model of turbulence and by Lagerstrom, Cole, and Trilling [4] in the approximate theory for weak nonstationary shock waves in a real fluid. A useful solution of Eq. (2.18) has been developed by Hopf [5] and Cole [6] via a special case of the $\psi = F(u)$ transformation used to develop Eq. (2.13) from Eq. (2.11). If Eq. (2.18) is modified by the introduction of the function v defined by setting $u = \partial v/\partial x$, then Eq. (2.18) becomes

$$v_{xt} + v_x v_{xx} = \lambda v_{xxx} \tag{2.19}$$

which upon integration with respect to x (discarding an arbitrary function of t) becomes

$$v_t + \tfrac{1}{2} v_x^2 = \lambda v_{xx}$$

i.e., Eq. (2.16). Furthermore Eq. (2.16) is equivalent to Eq. (2.13) if $\lambda(F''/F') = -\tfrac{1}{2}$ or

$$F = A \exp[-v/2\lambda] + B. \tag{2.20}$$

Conversely, v satisfies Eq. (2.16) providing $v = F^{-1}(\psi)$ where ψ satisfies the one-dimensional diffusion Eq. (2.11). In the case of the Burger's equation (with $A = 1$, $B = 0$) we have $v = -2\lambda \ln \psi$ or

$$u = -2\lambda \frac{\partial}{\partial x}(\ln \psi) \tag{2.21}$$

which transformation, although a special case, we will call the Hopf transformation.

An easy generalization to higher dimensions reads as follows: If ψ satisfies the diffusion equation (2.11), then

$$\mathbf{u} = -2\nu \nabla(\ln \psi), \tag{2.22}$$

where $\mathbf{u} = (u_1, u_2, u_3)$, satisfies the equation (three-dimensional momentum equations without the pressure term)

$$\frac{\partial u_i}{\partial t} + u_j \frac{\partial u_i}{\partial x_j} = \nu \nabla^2 u_i \tag{2.23}$$

2.1 TRANSFORMATIONS ON DEPENDENT VARIABLES

$i = 1, 2, 3$. However, the continuity equation $\partial u_j/\partial x_j = 0$ is generally *not* satisfied, thereby limiting the physical utility of the transformation (2.22). However, this limitation does not obviate its utility for the development of numerical procedures.[†]

Returning to the rocket problem, Eq. (2.14), we note that under the inverse transformation $u = F^{-1}(\psi) = -2\lambda \ln \psi$ Eqs. (2.14) and (2.15) reduce to

$$\psi_t = \lambda \psi_{xx} - (2\lambda)^{-1} d(x, t)\psi \qquad (2.24)$$

with

$$\psi(x, 0) = \exp[-f(x)/2\lambda]$$
$$\psi(x + 2\pi, t) = \psi(x, t). \qquad (2.25)$$

Although we may encounter difficulties in extracting the boundary conditions the use of the functional transformation is certainly not limited to the situation considered above. Several other examples will be briefly explored.

First, since Burger's equation can be solved, as indicated, the next step in the "hierarchy" of nonlinearity may be obtained by setting $u = F(U)$ in Eq. (2.18). The result is $(\lambda = 1)$

$$U_t + F(U)U_x = (F''/F')U_x^2 + U_{xx}. \qquad (2.26)$$

One may continue in this vein to generate further equations whose "grandfather" is the linear diffusion equation.

Alternatively if we initiate our considerations from the wave equation

$$\psi_{tt} = \lambda \nabla^2 \psi, \qquad (2.27)$$

the transformation $\psi = F(u)$ yields

$$u_{tt} = \frac{F''}{F'} [\lambda(\nabla u)^2 - u_t^2] + \lambda \nabla^2 u. \qquad (2.28)$$

We may summarize the method with the sentence: "Beginning with

---

[†] Cole [6] proceeded by a related method. Starting from Eq. (2.16) set $v = F[\theta(x, t)]$ where both F and θ may be selected to simplify the equation. Under this substitution Eq. (2.16) becomes $F'[\theta_t - \lambda \theta_{xx}] = [\lambda F'' - \frac{1}{2}(F')^2]\theta_x^2$. We now choose θ as a solution of $\theta_t = \lambda \theta_{xx}$, using the appropriate auxiliary conditions. Since $\theta_x \neq 0$, F must be a solution of $2\lambda F'' = (F')^2$. Thus $F(\theta) = -2\lambda \ln(\theta - C_1) + C_2$. With $C_1 = C_2 = 0$ we arrive at the general form (2.21) for u.

the linear partial differential equation $L\psi = 0$, set $\psi = F(u)$, thereby generating an equation

$$LF(u) = 0 \qquad (2.29)$$

equivalent to $L\psi = 0$ under $\psi = F(u)$".

c. Auxiliary Functions

The classical (linear) theories of elasticity and potential flow use the technique of introducing an auxiliary function which automatically satisfies one or more of the essential equations. In elasticity the stress function automatically satisfies the equilibrium equations. The stream function of fluid mechanics automatically satisfies the continuity equation. On occasion the auxiliary function approach can have utility for the nonlinear problem also. Several examples will be used to illustrate the procedure.

1. STEADY, TWO-DIMENSIONAL, IRROTATIONAL, AND ISENTROPIC FLOW OF A POLYTROPIC GAS

Let ρ be the density, p the pressure, and u, v the velocity components of the flow so that the equation of continuity is

$$\frac{\partial u}{\partial x} + \frac{\partial v}{\partial y} + \frac{1}{\rho}\left[u\frac{\partial \rho}{\partial x} + v\frac{\partial \rho}{\partial y}\right] = 0, \qquad (2.30)$$

and since the motion is irrotational there is a velocity potential $\phi(x, y)$ such that

$$u = \frac{\partial \phi}{\partial x}, \qquad v = \frac{\partial \phi}{\partial y},$$

and these imply that

$$\frac{\partial v}{\partial x} - \frac{\partial u}{\partial y} = 0. \qquad (2.31)$$

When the pressure and density are related by the state equation $p = k\rho^\gamma$ (γ = ratio of specific heats), then Bernouilli's integral takes the form [7]

$$\frac{\gamma k}{\gamma - 1}\rho^{\gamma-1} + \frac{1}{2}(u^2 + v^2) = \frac{\gamma k}{\gamma - 1}\rho_0^{\gamma-1} + \frac{1}{2}V_0^2 \qquad (2.32)$$

2.1 TRANSFORMATIONS ON DEPENDENT VARIABLES

where V_0 is the velocity at some reference position where the density is ρ_0. If c_0 is the velocity of sound at the reference, then

$$c_0^2 = \left(\frac{dp}{d\rho}\right)_0 = k\gamma\rho_0^{\gamma-1} \tag{2.33}$$

so

$$\left(\frac{\rho}{\rho_0}\right)^{\gamma-1} = 1 - \frac{\gamma-1}{2c_0^2}(u^2 + v^2 - V_0^2) = c^2/c_0^2, \tag{2.34}$$

where c = velocity of sound at the point (x, y) in the flow field. Equation (2.34) when differentiated with respect to x and y gives, after some algebra,

$$\frac{c^2}{\rho}\frac{\partial \rho}{\partial x} = -\left(u\frac{\partial u}{\partial x} + v\frac{\partial v}{\partial x}\right) \tag{2.35}$$

$$\frac{c^2}{\rho}\frac{\partial \rho}{\partial y} = -\left(u\frac{\partial u}{\partial y} + v\frac{\partial v}{\partial y}\right). \tag{2.36}$$

and when these results are substituted into the equation of continuity (2.30) it takes the form

$$c^2\left(\frac{\partial u}{\partial x} + \frac{\partial v}{\partial y}\right) = u\left(u\frac{\partial u}{\partial x} + v\frac{\partial v}{\partial x}\right) + v\left(u\frac{\partial u}{\partial y} + v\frac{\partial v}{\partial y}\right)$$

or

$$(c^2 - u^2)u_x - uv(u_y + v_x) + (c^2 - v^2)v_y = 0. \tag{2.37}$$

If the potential function ϕ is introduced into Eq. (2.37) and it is recalled that c^2 is a function of $u^2 + v^2 = \phi_x^2 + \phi_y^2$, then the equation for ϕ becomes

$$(c^2 - \phi_x^2)\phi_{xx} - 2\phi_x\phi_y\phi_{xy} + (c^2 - \phi_y^2)\phi_{yy} = 0. \tag{2.38}$$

Further, a stream function $\psi(x, y)$ can be introduced, such that

$$\psi_x = -\rho v, \quad \psi_y = \rho u$$

and upon substitution of these Eq. (2.31) we have the second-order partial equation

$$\frac{\partial}{\partial x}\left(\frac{1}{\rho}\psi_x\right) + \frac{\partial}{\partial x}\left(\frac{1}{\rho}\psi_y\right) = 0. \tag{2.39}$$

As a matter of record the velocity potential $\phi(x, y, z, t)$ for the three-dimensional irrotational flow of an inviscid compressible fluid satisfies the nonlinear partial differential equation

$$\phi_{xx}(\phi_x^2 - c^2) + \phi_{yy}(\phi_y^2 - c^2) + \phi_{zz}(\phi_z^2 - c^2) + 2[\phi_{xy}\phi_x\phi_y + \phi_{yz}\phi_y\phi_z \\ + \phi_{zx}\phi_z\phi_x + \phi_{xt}\phi_x + \phi_{yt}\phi_y + \phi_{zt}\phi_z] = 0. \quad (2.40)$$

2. THREE-DIMENSIONAL BOUNDARY LAYER THEORY

The boundary layer equations for an incompressible fluid in three dimensions contains a continuity equation (among others)

$$\frac{\partial u}{\partial x} + \frac{\partial v}{\partial y} + \frac{\partial w}{\partial z} = 0 \quad (2.41)$$

where u, v, w are the velocity components in the boundary layer. It is sometimes convenient in such problems to introduce *two* stream functions $\phi(x, y, z)$, $\psi(x, y, z)$ by the relations

$$u = \frac{\partial \phi}{\partial y}, \quad w = \frac{\partial \psi}{\partial y}. \quad (2.42)$$

Substituting these into Eq. (2.41) gives

$$\frac{\partial^2 \phi}{\partial x \, \partial y} + \frac{\partial^2 \psi}{\partial y \, \partial z} = -\frac{\partial v}{\partial y} \quad (2.43)$$

which "forces" the definition of v as

$$v = -\left(\frac{\partial \phi}{\partial x} + \frac{\partial \psi}{\partial z}\right). \quad (2.44)$$

d. The Riemann Invariants

All equations of second order can be factored into two equations of first order. A number of fluid mechanics problems, plasticity problems and vibration problems have "natural" mathematical models of the form

$$F_1 \frac{\partial u}{\partial x} + F_2 \frac{\partial u}{\partial y} + F_3 \frac{\partial v}{\partial x} + F_4 \frac{\partial v}{\partial y} = F(u, v, x, y) \\ G_1 \frac{\partial u}{\partial x} + G_2 \frac{\partial u}{\partial y} + G_3 \frac{\partial v}{\partial x} + G_4 \frac{\partial v}{\partial y} = G(u, v, x, y). \quad (2.45)$$

2.1 TRANSFORMATIONS ON DEPENDENT VARIABLES

If the F_i, G_i, $i = 1, 2, 3, 4$, are functions of u, v, x, and y only they are classified as *quasi linear*. For this system a rather general theory exists which will be discussed in considerable detail in a later chapter. For the moment let us demonstrate the introduction of the Riemann invariants—which are transformations of the dependent variables—for a nonlinear vibration problem.

The large amplitude oscillations of a vibrating string have been previously described by Eqs. (1.39) and (1.40). We shall consider the longitudinal vibrations only so that $\theta = 0$. Thus the equation for the horizontal displacement y becomes

$$\frac{\partial T}{\partial x} = \rho A \frac{\partial^2 y}{\partial t^2}. \tag{2.46}$$

When a linear stress-strain relationship of the form $T = T_0(1 + y_x)$ is assumed the classical wave equation results. If on the other hand we assume that[†]

$$T = T_0[1 + y_x + F^2(y_x)], \tag{2.47}$$

then substitution of this relation into Eq. (2.46) the governing equation becomes

$$y_{tt} = [F(y_x)]^2 y_{xx} \tag{2.48}$$

in dimensionless form.

We first factor Eq. (2.48) by setting

$$u = y_x, \quad v = y_t \tag{2.49}$$

so that the equation takes the form (of Eq. (2.45))

$$\begin{aligned} u_t - v_x &= 0 \\ v_t - F^2(u)u_x &= 0. \end{aligned} \tag{2.50}$$

If we multiply the first by $F(u)$ and add to the second we have

$$v_t + F(u)u_t - F(u)v_x - F^2(u)u_x = 0. \tag{2.51}$$

On the other hand if we subtract the second equation from $F(u)$ times the first equation, there results

$$-v_t + F(u)u_t - F(u)v_x + F^2(u)u_x = 0. \tag{2.52}$$

[†] The first correction would seem to be of the form ϵy_x^2 but the method applies for arbitrary $F^2(y_x)$, where F^2 is used, for convenience, in factoring later.

Now we wish to introduce two new dependent variables, following Riemann, so that the resulting equations have essentially the same form. Examination of Eq. (2.51) suggests that if we let $r_t = v_t + F(u)u_t$ and $r_x = v_x + F(u)u_x$ then Eq. (2.51) takes the form

$$r_t - F(u)r_x = 0. \qquad (2.53)$$

In a similar way if s is introduced into Eq. (2.52) by $s_t = -v_t + F(u)u_t$, $s_x = -v_x + F(u)u_x$, then that equation becomes

$$s_t + F(u)s_x = 0. \qquad (2.54)$$

By integrating the defining relations for s and r it is seen that

$$r = v + \int_{u_0}^{u} F(\eta)\, d\eta = v + B(u) \qquad (2.55)$$

$$s = -v + \int_{u_0}^{u} F(\eta)\, d\eta = -v + B(u), \qquad (2.56)$$

where

$$B(u) = \int_{u_0}^{u} F(\eta)\, d\eta. \qquad (2.57)$$

Finally, we see that $2B(u) = r + s$ or that

$$u = B^{-1}[\tfrac{1}{2}(r+s)] \qquad (2.58)$$

where B^{-1} means the inverse function. Similarly

$$v = \tfrac{1}{2}(r - s). \qquad (2.59)$$

Equation (2.58) allows us to write Eqs. (2.53) and (2.54) in the form

$$\begin{aligned} r_t - F[B^{-1}\{\tfrac{1}{2}(r+s)\}]r_x &= 0 \\ s_t + F[B^{-1}\{\tfrac{1}{2}(r+s)\}]s_x &= 0 \end{aligned} \qquad (2.60)$$

which depend only on r and s!

The quantities r and s are known as the Riemann invariants since r is invariant in the characteristic direction $dx/dt = -F$ and s is invariant in the characteristic direction $dx/dt = +F$.

2.2 TRANSFORMATIONS ON INDEPENDENT VARIABLES

a. The von Mises Transformation

The utility of the von Mises transformation has already been demonstrated in Chapter 1, Eq. (1.23)–(1.29). A generalization of this idea has been exploited by Kalman [7] in his discussion of the oscillations of a zero-temperature plasma. The plasma is considered to be one dimensional with an induced electric field as the only body force. With ρ = density, v = velocity, E = electric field, q and m the electron charge and mass, respectively, and ρ_e the equilibrium plasma density, then the governing equations are the continuity equation:

$$\frac{\partial}{\partial t}(\rho/\rho_e) + \frac{\partial}{\partial x}\left(\frac{\rho v}{\rho_e}\right) = 0, \tag{2.61}$$

momentum:

$$\frac{\partial v}{\partial t} + v\frac{\partial v}{\partial x} = \frac{q}{m}E, \tag{2.62}$$

and electric field:

$$\frac{q}{m}\frac{\partial E}{\partial x} = \frac{q^2 \rho_e}{m\epsilon_0}\left(\frac{\rho}{\rho_e} - 1\right) = \omega_p^2\left(\frac{\rho}{\rho_e} - 1\right). \tag{2.63}$$

We first introduce a stream function ψ, so that continuity is automatically satisfied, by the relations

$$\frac{\partial \psi}{\partial t} = \frac{\rho v}{\rho_e}, \qquad \frac{\partial \psi}{\partial x} = -\frac{\rho}{\rho_e}. \tag{2.64}$$

Since $\psi = \psi(x, t)$ it is clear that $d\psi/dt = 0$ since

$$\frac{d\psi}{dt} = \frac{\partial \psi}{\partial x}\frac{dx}{dt} + \frac{\partial \psi}{\partial t} = -\frac{\rho}{\rho_e}v + \frac{\rho v}{\rho_e} = 0.$$

This relation will be important as we change variables from the old (x, t) system to the new (ψ, t) system. We think now that $v = v(\psi, t)$, $x = x(\psi, t)$ and $E = E(\psi, t)$ and remark that this transformation will be satisfactory if the Jacobian $J[(x, t)/(\psi, t)] = \partial x/\partial \psi \neq 0$. Recalling that

$$v = \frac{dx}{dt} = \frac{d}{dt}x(\psi, t) = \frac{\partial x}{\partial t}\bigg)_\psi$$

we have
$$\left.\frac{\partial v}{\partial t}\right)_\psi = \left.\frac{\partial^2 x}{\partial t^2}\right)_\psi$$

on the one hand, but we also have

$$\frac{dv}{dt} = \underbrace{\frac{\partial v}{\partial \psi}\frac{d\psi}{dt} + \left.\frac{\partial v}{\partial t}\right)_\psi}_{\text{in }(\psi,\,t)\text{ system}} = \underbrace{v\left.\frac{\partial v}{\partial x}\right)_t + \left.\frac{\partial v}{\partial t}\right)_x}_{\text{in }(x,\,t)\text{ system}}. \tag{2.65}$$

However, $d\psi/dt = 0$ so Eq. (2.65) gives the Lagrangian type relation

$$\left.\frac{\partial v}{\partial t}\right)_\psi = v\left.\frac{\partial v}{\partial x}\right)_t + \left.\frac{\partial v}{\partial t}\right)_x \tag{2.66}$$

which provides the necessary mechanism to change the momentum Eq. (2.62) to

$$\left.\frac{\partial v}{\partial t}\right)_\psi = \left.\frac{\partial^2 x}{\partial t^2}\right)_\psi = \frac{q}{m}E. \tag{2.67}$$

However, the field equation (2.63) becomes upon use of Eq. (2.64)

$$\frac{q}{m}\frac{\partial E}{\partial x} = -\omega_p^2\left(\frac{\partial \psi}{\partial x}+1\right), \tag{2.68}$$

and since no differentiation occurs with respect to t it is integrable to

$$\frac{q}{m}E = -\omega_p^2[\psi + x + g(t)], \tag{2.69}$$

where $g(t)$ is an arbitrary function of integration. Thus

$$\frac{\partial^2 x}{\partial t^2} + \omega_p^2 x = -\omega_p^2[\psi + g(t)] \tag{2.70}$$

is the linear equation for x, thought of as a function of ψ and t. Since there is no differentiation with respect to ψ we may integrate twice (with ψ appearing as a parameter) with respect to t and get

$$x(\psi, t) = A(\psi)\sin\omega_p t + B(\psi)\cos\omega_p t - \psi - G(t). \tag{2.71}$$

2.2 TRANSFORMATIONS ON INDEPENDENT VARIABLES

Here $G(t)$ is a particular solution of the inhomogeneous equation

$$\frac{d^2G}{dt^2} + \omega_p^2 G(t) = \omega_p^2 g(t).$$

Finally we obtain

$$V(\psi, t) = \omega_p[A(\psi) \sin \omega_p t - B(\psi) \cos \omega_p t] - G'(t) \qquad (2.72)$$

$$E(\psi, t) = -\frac{m}{q} \omega_p^2 [A(\psi) \sin \omega_p t + B(\psi) \cos \omega_p t - G(t) + g(t)] \qquad (2.73)$$

$$\rho(\psi, t) = \rho_e \left[1 - \frac{dA}{d\psi} \sin \omega_p t - \frac{dB}{d\psi} \cos \omega_p t\right]^{-1}. \qquad (2.74)$$

The functions $A(\psi)$, $B(\psi)$, and $g(t)$ must be determined from the initial and boundary conditions.

b. A "Similarity" Transformation (Boltzmann [8])

When the integration domains are semi- or doubly infinite and the auxiliary conditions are such that two of them can be consolidated into one, the diffusion equation

$$\frac{\partial C}{\partial t} = \frac{\partial}{\partial x}\left[D(C)\frac{\partial C}{\partial x}\right] \qquad (2.75)$$

can be reduced to a nonlinear ordinary differential equation. Moreover, this initial value problem becomes a boundary value problem under the "similarity" (Boltzmann) transformation. What we attempt to do with this transformation on the independent variables is to amalgamate the two into one. An obvious procedure is to try

$$\eta = x^\alpha t^\beta \qquad (2.76)$$

where α and β are to be determined (if possible) so that the resulting ordinary differential equation in η is free of x and t. Thus we have

$$\frac{\partial C}{\partial x} = \frac{dC}{d\eta}\frac{\partial \eta}{\partial x} = \alpha x^{\alpha-1} t^\beta C'$$

$$\frac{\partial C}{\partial t} = \beta x^\alpha t^{\beta-1} C' \qquad (2.77)$$

$$\frac{\partial}{\partial x}\left[D\frac{\partial C}{\partial x}\right] = \alpha(\alpha - 1)x^{\alpha-2} t^\beta D(C)C' + \alpha^2 x^{2(\alpha-1)} t^{2\beta}(DC')'. \qquad (2.78)$$

Substituting the last two results into Eq. (2.75) we find that

$$\beta x^{\alpha} t^{\beta-1} C' = \alpha(\alpha - 1)x^{\alpha-2} t^{\beta} D(C)C' + \alpha^2 x^{2(\alpha-1)} t^{2\beta}(DC')'. \qquad (2.79)$$

Introducing η, wherever possible, we have

$$\beta \frac{x^2}{t} \eta C' = \alpha(\alpha - 1)\eta DC' + \alpha^2 \eta^2 (DC')'. \qquad (2.80)$$

Now the right-hand side of Eq. (2.80) is free of x and t, thus for the full equation to be free x^2/t must be a function only of η. This yields no unique answer, for all that is required is that

$$\frac{x^2}{t} = f(\eta), \qquad (2.81)$$

which determines the values of α and β; e.g., if $f(\eta) = \eta$ then $\alpha = 2$, $\beta = -1$; if $f(\eta) = \eta^2$ then $\alpha = 1$, $\beta = -\frac{1}{2}$. The latter choice has some advantages for the first term on the right-hand side of Eq. (2.80) vanishes when $\alpha = 1$. The transformation

$$\eta = x/t^{\frac{1}{2}} \qquad (2.82)$$

is called the Boltzmann transformation, and, under it, Eq. (2.75) becomes the second-order nonlinear ordinary differential equation

$$\frac{d}{d\eta}\left[D(C)\frac{dC}{d\eta}\right] + \frac{\eta}{2}\frac{dC}{d\eta} = 0. \qquad (2.83)$$

When can this transformation be effectively used? The original initial value problem whose describing equation is (2.75) requires two boundary conditions and one initial condition for a unique solution. But the transformed equation (2.83) requires only two boundary conditions. The way out of this dilemma is to require that a consolidation of auxiliary conditions must take place. Such consolidation can occur in a semi-infinite domain. For suppose $0 \leqslant x < \infty$, $0 \leqslant t < \infty$, with the boundary conditions

$$C(0, t) = C_0 \qquad (2.84)$$
$$C(\infty, t) = C_1 \qquad (2.85)$$

and the initial condition

$$C(x, 0) = C_2. \qquad (2.86)$$

Under the transformation $\eta = x/t^{\frac{1}{2}}$ the first condition becomes (since $t > 0$)

$$C(0) = C_0 \tag{2.87}$$

and the second and third conditions, Eqs. (2.85) and (2.86), consolidate only if $C_1 = C_2$, in which case the second auxiliary condition for for Eq. (2.83) becomes

$$C(\infty) = C_1. \tag{2.88}$$

Doubly infinite domains may be treated similarly. A considerable number of applications of this transformation are given by Crank [9]. In Section 4.3 we present a theory of similarity development.

2.3 MIXED TRANSFORMATIONS

The mixed transformation in which both sets of variables are involved appears to be the most productive of the three general classes. Here we see the interchange of dependent and independent variables, the Legendre transformation, the general similarity transformation and others.

a. Hodograph Transformation

We previously defined the quasi-linear system in two dependent variables u, v by Eqs. (2.45). If the coefficients of that system F_i, G_i, $i = 1, 2, 3, 4$, are functions only of u and v and $F = G \equiv 0$, then the equations are called *reducible*. In this case for any region where the Jacobian

$$J\left(\frac{u, v}{x, y}\right) = u_x v_y - u_y v_x \neq 0$$

the system can be transformed into an equivalent linear system by interchanging the roles of dependent and independent variables.

To carry out this interchange we think of x and y as functions of u and v. Hence

$$\begin{aligned} x &= x(u, v) \\ y &= y(u, v), \end{aligned} \tag{2.89}$$

so that differentiating both with respect to x yields

$$1 = x_u u_x + x_v v_x$$
$$0 = y_u u_x + y_v v_x. \tag{2.90}$$

We solve these two equations for u_x and v_x and so obtain

$$u_x = \frac{y_v}{x_u y_v - x_v y_u} = \frac{y_v}{j} = J y_v \tag{2.91}$$

$$v_x = -\frac{y_u}{j} = -J y_u \tag{2.92}$$

where $j = x_u y_v - x_v y_u = J^{-1}$.

In a similar fashion, upon differentiating the system equations (2.89) with respect to y and solving, we find that

$$u_y = -J x_v \tag{2.93}$$
$$v_y = J x_u \tag{2.94}$$

so that substituting Eqs. (2.91)–(2.94) into Eq. (2.45) gives the *linear* partial differential equations for $x(u, v)$ and $y(u, v)$ as

$$F_1 y_v - F_2 x_v - F_3 y_u + F_4 x_u = 0$$
$$G_1 y_v - G_2 x_v - G_3 y_u + G_4 x_u = 0. \tag{2.95}$$

Conversely every solution x, y of Eqs. (2.95) leads to a solution of our modified system equations (2.45) if the Jacobian $j = x_u y_v - x_v y_u \neq 0$. Solutions for which $j = 0$ cannot be obtained via the hodograph transformation.

Reducible equations occur in one-dimensional flow, in two dimensional steady flow, in singular perturbations, in vibration theory, and in plasticity.

The nonlinear vibration problem described in Section 2.1, Eqs. (2.50), namely, the equations

$$u_t - v_x = 0$$
$$v_t - F^2(u) u_x = 0,$$

are reducible. If we identify t with y then the linearized form obtained from the hodograph transformation is

$$x_v - y_u = 0$$
$$x_u - F^2(u) y_v = 0 \tag{2.96}$$

2.3 MIXED TRANSFORMATIONS

or upon eliminating x we have the linear equation for y:

$$y_{uu} = F^2(u)y_{vv}. \qquad (2.97)$$

Previously we considered the steady irrotational, isentropic flow of a gas and found the governing equations (2.31) and (2.37) to be

$$v_x - u_y = 0$$

and

$$(u^2 - c^2)u_x + uv(u_y + v_x) + (v^2 - c^2)v_y = 0.$$

The hodograph transformation linearizes these equations to the form

$$x_v - y_u = 0 \qquad (2.98)$$

and

$$(c^2 - u^2)y_v + uv(x_v + y_u) + (c^2 - v^2)x_u = 0.$$

The first equation suggests the introduction of a stream function $X(u, v)$ defined by

$$X_u = x, \quad X_v = y \qquad (2.99)$$

so that the second equation takes the form

$$(c^2 - u^2)X_{vv} + 2uvX_{uv} + (c^2 - v^2)X_{uu} = 0. \qquad (2.100)$$

Yet a third example from plasticity, as given by Hill [10], is the set of equations

$$p_x + (2k \cos 2\psi)\psi_x + (2k \sin 2\psi)\psi_y = 0$$
$$p_y + (2k \sin 2\psi)\psi_x - (2k \cos 2\psi)\psi_y = 0 \qquad (2.101)$$

describing the two-dimensional flow of a perfectly plastic bar being squeezed out between two parallel plates. Here k is a constant denoting the yield stress and p is the compressive stress in the direction ψ which is the orientation angle of the maximum shear.

b. The Legendre Transformation

The function $X(u, v)$ of Eq. (2.100) is the Legendre transformation

$$X(u, v) = ux + yv - \phi(x, y) \qquad (2.102)$$

of the potential function $\phi(x, y)$ defined by the relations $u = \phi_x$, $v = \phi_y$. Its utility arises since one can go directly from Eqs. (2.31) and (2.37) to (2.100) by use of this transformation.

This transformation is suggested by a geometrical interpretation of the differential equation, in two independent variables, if we represent the integral surface by its tangent plane coordinates instead of by (x, y) coordinates. A surface in (x, y, u) space may be described in two ways. One may give the surface as a point set determined by the function $u(x, y)$ or one may prescribe it as the envelope of its tangent planes. If $\bar{x}, \bar{y}, \bar{u}$ are the running coordinates of a plane whose equation is

$$\bar{u} - \xi\bar{x} - \eta\bar{y} + \omega = 0$$

then the coefficients ξ, η, ω may be called the descriptive *coordinates of this plane*. Now a plane tangent to the surface $u(x, y)$ at (x, y, u) has the equation

$$\bar{u} - u - (\bar{x} - x)u_x - (\bar{y} - y)u_y = 0,$$

and its plane coordinates are therefore

$$\xi = u_x, \qquad \eta = u_y, \qquad \omega = xu_x + yu_y - u. \qquad (2.103)$$

The surface can be equally well described if ω is given as $\omega = \omega(\xi, \eta)$. This dependence is found from $u(x, y)$ by obtaining the values of x and y as functions of ξ and η from $\xi = u_x, \eta = u_y$ and setting them into

$$\omega = xu_x + yu_y - u = x\xi + y\eta - u. \qquad (2.103a)$$

In order to determine the point coordinates from the tangent plane coordinates we differentiate partially in (2.103a) and find

$$\omega_\xi = x + \xi\frac{\partial x}{\partial \xi} + \eta\frac{\partial y}{\partial \xi} - u_x\frac{\partial x}{\partial \xi} - u_y\frac{\partial y}{\partial \xi} = x$$

and $\omega_\eta = y$. The dual character of the relation between point and tangent plane coordinates is nicely summarized by

$$\omega(\xi, \eta) + u(x, y) = x\xi + y\eta$$
$$u_x = \xi, \qquad \omega_\xi = x, \qquad (2.104)$$
$$u_y = \eta, \qquad \omega_\eta = y.$$

This Legendre transformation is quite different in character from a simple coordinate transformation. Clearly a single point is not assigned to another point but Eq. (2.104) assigns to every surface element (x, y, u, u_x, u_y) a surface element $(\xi, \eta, \omega, \omega_\xi, \omega_\eta)$.

2.3 MIXED TRANSFORMATIONS

The Legendre transformation is always feasible if the two equations $\xi = u_x$, $\eta = u_y$ can be solved for x and y. This is possible whenever the Jacobian

$$J = u_{xx}u_{yy} - u_{xy}^2 \neq 0.$$

It fails for *developable* surfaces, characterized by $J = 0$.

Application of the Legendre transformation to second-order partial differential equations requires the second derivatives of $u(x, y)$ and $\omega(\xi, \eta)$. Differentiating $\xi = u_x$ and $\eta = u_y$ with respect to ξ and η we have

$$1 = u_{xx}\frac{\partial x}{\partial \xi} + u_{xy}\frac{\partial y}{\partial \xi} = u_{xx}\omega_{\xi\xi} + u_{xy}\omega_{\xi\eta}$$

$$0 = u_{xy}\omega_{\xi\xi} + u_{yy}\omega_{\xi\eta}$$

$$0 = u_{xx}\omega_{\xi\eta} + u_{xy}\omega_{\eta\eta}$$

$$1 = u_{xy}\omega_{\xi\eta} + u_{yy}\omega_{\eta\eta}$$

where we have used the expression $x = \omega_\xi$, $y = \omega_\eta$ from Eq. (2.104). In matrix notation we find

$$\begin{bmatrix} u_{xx} & u_{xy} \\ u_{xy} & u_{yy} \end{bmatrix} \begin{bmatrix} \omega_{\xi\xi} & \omega_{\xi\eta} \\ \omega_{\xi\eta} & \omega_{\eta\eta} \end{bmatrix} = \begin{bmatrix} 1 & 0 \\ 0 & 1 \end{bmatrix} = I$$

and taking determinants we find

$$\omega_{\xi\xi}\omega_{\eta\eta} - \omega_{\xi\eta}^2 = 1/J.$$

This system may be easily solved to obtain

$$u_{xx} = J\omega_{\eta\eta}, \qquad u_{xy} = -J\omega_{\xi\eta}, \qquad u_{yy} = J\omega_{\xi\xi}. \qquad (2.105)$$

Application of Eqs. (2.104) and (2.105) facilitates the transformation.

Legendre's transformation is directly extendable to functions of n independent variables taking the form

$$u(x_1, x_2, ..., x_n) + \omega(\xi_1, \xi_2, ..., \xi_n) = \sum_{i=1}^{n} x_i\xi_i$$

with

$$\frac{\partial u}{\partial x_i} = \xi_i, \qquad \frac{\partial \omega}{\partial \xi_i} = x_i; \qquad i = 1, 2, ..., n.$$

In illustration of the application of the Legendre transformation we return to our steady flow problem. Upon inserting the potential function ϕ into Eq. (2.37) we have the equation

$$(\phi_x{}^2 - c^2)\phi_{xx} + 2\phi_x\phi_y\phi_{xy} + (\phi_y{}^2 - c^2)\phi_{yy} = 0.$$

If we take u and v as the new independent variables and X as the new dependent variable, defined by Eq. (2.102), then by virtue of $x = X_u$, $y = X_v$ the new equation is Eq. (2.100).

It is worth noting that a Legendre transform $Y(\rho u, \rho v)$ can also be introduced for the stream function ψ (defined by $\psi_x = \rho v$, $\psi_y = \rho u$) via the relation

$$Y(\rho u, \rho v) = \rho u y - \rho v x - \psi(x, y).$$

The reciprocal relations

$$Y_{\rho u} = y, \qquad Y_{\rho v} = -x$$

correspond to those used to define ψ.

c. Molenbroek-Chaplygin Transformation [11]

An attack on the problem of subsonic flow in a compressible fluid was suggested by Molenbroek and Chaplygin. Their transformation resembles that of Legendre but instead of using the potential function $\phi(x, y)$ and its Legendre transform $X(u, v)$ this procedure uses $\phi(x, y)$ and the stream function (in the same coordinate system) $\psi(x, y)$, which satisfies identically the continuity equation

$$(\rho u)_x + (\rho v)_y = 0.$$

As before we define ψ so that $\rho u = \psi_y$, $-\rho v = \psi_x$. The total differentials of ϕ and ψ are then

$$\begin{aligned} d\phi &= \phi_x\, dx + \phi_y\, dy = u\, dx + v\, dy \\ d\psi &= -\rho v\, dx + \rho u\, dy. \end{aligned} \qquad (2.106)$$

We next introduce polar coordinates (q, θ) in the (u, v) plane defined by the relations

$$\begin{aligned} q &= (u^2 + v^2)^{\frac{1}{2}} \\ \theta &= \tan^{-1}(v/u) \end{aligned} \qquad (2.107)$$

$$u = q\cos\theta, \qquad v = q\sin\theta. \qquad (2.108)$$

2.3 MIXED TRANSFORMATIONS

Since ψ and ϕ are then functions of q and θ we can then write

$$d\phi = \phi_q\, dq + \phi_\theta\, d\theta, \qquad d\psi = \psi_q\, dq + \psi_\theta\, d\theta. \tag{2.109}$$

When Eqs. (2.108) and (2.09) are substituted into Eq. (2.106) and these are solved for dx and dy we have

$$\begin{aligned}
dx &= [(\cos\theta/q)\phi_q - (\sin\theta/q\rho)\psi_q]\, dq \\
&\quad + [(\cos\theta/q)\phi_\theta - (\sin\theta/q\rho)\psi_\theta]\, d\theta \\
dy &= [(\sin\theta/q)\phi_q + (\cos\theta/\rho q)\psi_q]\, dq \\
&\quad + [(\sin\theta/q)\phi_\theta + (\cos\theta/\rho q)\psi_\theta]\, d\theta.
\end{aligned} \tag{2.110}$$

On the other hand, we can write

$$dx = x_q\, dq + x_\theta\, d\theta, \qquad dy = y_q\, dq + y_\theta\, d\theta \tag{2.111}$$

so that comparing Eqs. (2.111) with (2.110) gives

$$x_q = (\cos\theta/q)\phi_q - (\sin\theta/q\rho)\psi_q \tag{2.112}$$

$$x_\theta = (\cos\theta/q)\phi_\theta - (\sin\theta/q\rho)\psi_\theta \tag{2.113}$$

and similar relations for y_q and y_θ. Since $x_{q\theta} = x_{\theta q}$ (by assumption) we have upon differentiating and equating the cross partial derivatives, in both x and y,

$$-\sin\theta\, \phi_q - (\cos\theta/\rho)\psi_q = -(\cos\theta/q)\phi_\theta + \frac{\sin\theta}{\rho q}\left(1 + \frac{q}{\rho}\frac{d\rho}{dq}\right)\psi_\theta \tag{2.114}$$

and

$$\cos\theta\, \phi_q - (\sin\theta/\rho)\psi_q = -(\sin\theta/q)\phi_\theta - \frac{\cos\theta}{\rho q}\left(1 + \frac{q}{\rho}\frac{d\rho}{dq}\right)\psi_\theta, \tag{2.115}$$

where we recall that $\rho = f(q)$. Upon multiplying Eq. (2.114) by $\cos\theta$ and Eq. (2.115) by $\sin\theta$ and adding we have

$$\frac{1}{\rho}\frac{\partial\psi}{\partial q} = \frac{1}{q}\frac{\partial\phi}{\partial\theta}. \tag{2.116}$$

If we multiply Eq. (2.114) by $\sin\theta$ and Eq. (2.115) by $\cos\theta$ and subtract the first from the second we have

$$\frac{\partial\phi}{\partial q} = -\frac{1}{\rho q}\left(1 + \frac{q}{\rho}\frac{d\rho}{dq}\right)\frac{\partial\psi}{\partial\theta}. \tag{2.117}$$

To evaluate the term $(q/\rho)\,d\rho/dq$ we recall Eq. (2.34) which implies that

$$\frac{c^2}{\rho}\,d\rho + \tfrac{1}{2}d(u^2 + v^2) = 0 \tag{2.118}$$

or

$$(q/\rho)\,d\rho/dq = -q^2/c^2. \tag{2.119}$$

Substituting Eq. (2.119) into Eq. (2.117) we find

$$\frac{\partial \phi}{\partial q} = -\frac{1}{\rho q}\left(1 - \frac{q^2}{c^2}\right)\frac{\partial \psi}{\partial \theta} \tag{2.120}$$

and eliminating, for example ϕ, between Eq. (2.120) and Eq. (2.116) the equation for the stream function ψ has the linear form

$$\frac{\partial}{\partial q}\left(\frac{q}{\rho}\frac{\partial \psi}{\partial q}\right) + \left(\frac{1 - q^2/c^2}{q\rho}\right)\frac{\partial^2 \psi}{\partial \theta^2} = 0 \tag{2.121}$$

where we recall that c is given as a function of q by means of the defining relations $c^2 = dp/d\rho$, $dp/\rho + q\,dq = 0$, and $p = f(\rho)$. This last relation is the equation of state.

Equation (2.121) can be reduced to Laplace's equation by the proper (artificial) choice of the relation $p = f(\rho)$. Comparing Eq. (2.121) with Laplace's equation in (q, θ) polar coordinates,

$$q\frac{\partial}{\partial q}\left(q\frac{\partial \psi}{\partial q}\right) + \frac{\partial^2 \psi}{\partial \theta^2} = 0, \tag{2.122}$$

we note that the forms are identical if

$$\frac{dq}{q} = \frac{1}{a}\frac{\rho\,dq}{q} = a\frac{(1 - q^2/c^2)}{\rho q}\,dq \tag{2.123}$$

where a is an arbitrary constant. Thus we must choose $p = f(\rho)$ so that

$$\rho^2 = a^2(1 - q^2/c^2)$$

or

$$q^2 = c^2 - \frac{\rho^2 c^2}{a^2}. \tag{2.124}$$

Since, from Eq. (2.118) $d(q^2)/d\rho = -2c^2/\rho$, the differentiation of Eq. (2.124) gives the relation

$$-\frac{2c^2}{\rho} = \frac{d(q^2)}{d\rho} = \frac{dc^2}{d\rho}(1 - \rho^2/a^2) - \frac{2\rho c^2}{a^2}.$$

Solving for $dc^2/d\rho$ we find that

$$\frac{dc^2}{d\rho} = -\frac{2c^2}{\rho}, \tag{2.125}$$

or

$$c^2 = \frac{dp}{d\rho} = \bar{B}/\rho^2 \tag{2.126}$$

by integration. Finally

$$p = A - B/\rho$$

and with $V = 1/\rho$ (V the specific volume) we can write

$$p - p_0 = \text{constant}(V_0 - V). \tag{2.127}$$

Thus the linear equation (2.121) reduces to Laplace's equation for ψ (and also for ϕ) *if* the p, V curve is replaced by a *straight line*.

A second remark is pertinent here because of its relation to recent work of Tricomi [12] and others [13, 14]. The form of Eq. (2.121) suggests setting

$$d\sigma = \frac{\rho \, dq}{q} \tag{2.128}$$

whereupon Eq. (2.121) takes the form

$$\frac{\partial^2 \psi}{\partial \sigma^2} + K(\sigma) \frac{\partial^2 \psi}{\partial \theta^2} = 0 \tag{2.129}$$

where

$$K(\sigma) = \frac{c^2 - q^2}{c^2 \rho^2}. \tag{2.130}$$

which may be expressed in terms of σ via the intervention of a hypergeometric function. Equation (2.129) is known as *Chaplygin's equation*.

d. The General Similarity Transformation

In Section 2.2 the Boltzmann transformation was exhibited as a simple example of a similarity transformation. Mathematically, as we saw in that diffusion example, the similarity assumption makes it possible to transform the partial differential equations into total differential equations. A physical interpretation of similarity is also possible but must be given in the context of the problem discussed.

To demonstrate the applicability of the similarity transformation we shall consider the two dimensional equations of the boundary layer for an incompressible fluid. Further, the important question is, under what conditions do similar solutions exist?

The fundamental boundary layer equations (1.21) are augmented by a "velocity at infinity" term so that in this case they are

$$uu_x + vu_y = u_\infty u_{\infty x} + vu_{yy}.$$
$$u_x + v_y = 0, \qquad (2.131)$$

with the boundary conditions

$$u(x, 0) = v(x, 0) = 0$$
$$u(x, \infty) = u_\infty(x).$$

As before we introduce a stream function ψ defined by $\psi_y = u$, $\psi_x = -v$ so that continuity is automatically satisfied. Equation (2.131) becomes

$$\psi_y \psi_{xy} - \psi_x \psi_{yy} = u_\infty u_{\infty x} + v\psi_{yyy}. \qquad (2.132)$$

Our goal is to transform this equation into an ordinary differential equation. Thus we try to consolidate the independent variables x and y appropriately into a single independent variable η. In addition ψ is changed into some new function. Such a general transformation might be of the form

$$\eta = Ag(x)h(y)$$
$$\psi = Bk(x)l(y)f(\eta) \qquad (2.133)$$

and by trying this transformation the possible forms of g, h, k, l, and f may be determined and permissible flow fields $u_\infty(x)$ discovered. It is doubtful if the true power of this important transformation has been completely realized (see Section 4.3 for general theory). But at the same time we emphasize that it does not give the complete answer, since many solutions are not available by this procedure.

After some calculation and usually from the general theory we discover appropriate forms (which are also dimensionless) as say

$$\xi = x/L, \qquad \eta = \frac{y \sqrt{R_e}}{Lg(x)} \qquad (2.134)$$

and

$$f(\xi, \eta) = \frac{\psi(x, y) \sqrt{R_e}}{L u_\infty(x) g(x)} \qquad (2.135)$$

2.3 MIXED TRANSFORMATIONS

where $R_e = U_\infty L/\nu$, U_∞, L are the Reynolds number, characteristic velocity, and flow length, respectively. We have purposely left $g(x)$ arbitrary to demonstrate a portion of the calculation. Other details may be found in the original papers of Goldstein [15], Mangler [16], and Falkner and Skan [17]. Using these dimensionless quantities we calculate the velocity components as

$$u = \psi_y = u_\infty f_\eta \tag{2.136}$$

$$-v = \psi_x = \frac{L}{\sqrt{R_e}} \frac{\partial}{\partial x}(u_\infty g f)$$

$$= \frac{L}{\sqrt{R_e}} \left\{ f \frac{d}{dx}(u_\infty g) + u_\infty g \frac{\partial}{\partial x} f \right\}$$

$$= \frac{L}{\sqrt{R_e}} f \frac{d}{dx}(u_\infty g) + \frac{u_\infty g}{\sqrt{R_e}} \left\{ \frac{\partial f}{\partial \xi} - L \frac{g'}{g} \eta \frac{\partial f}{\partial \eta} \right\} \tag{2.137}$$

since

$$\frac{\partial \eta}{\partial x} = -\frac{y\sqrt{R_e}}{L} \frac{g'(x)}{g^2(x)} = -\eta \frac{g'}{g}.$$

In addition

$$\frac{\partial^2 \psi}{\partial y^2} = u_\infty^2 \frac{\partial^2 f}{\partial \eta^2}, \quad \frac{\partial^3 \psi}{\partial y^3} = u_\infty^3 \frac{\partial^3 f}{\partial \eta^3} \tag{2.138}$$

so that upon substitution of these results, Eq. (2.132) takes the form

$$f_{\eta\eta\eta} + \alpha f f_{\eta\eta} + \beta[1 - (f_\eta)^2] = \frac{u_\infty(x)}{U_\infty} g^2(x)[f_\eta f_{\eta\xi} - f_\xi f_{\eta\eta}] \tag{2.139}$$

where

$$\alpha = \frac{Lg(x)}{U_\infty} \frac{d}{dx}(u_\infty g), \quad \beta = \frac{L}{U_\infty} g^2(x) \frac{du_\infty}{dx}. \tag{2.140}$$

Similar solutions exist, that is, Eq. (2.139) is a total differential equation, if f is independent of ξ and both α and β are independent of x, i.e., they are constants. This means that $f_\xi = 0$, $f_{\eta\xi} = (f_\xi)_\eta = 0$ so the total differential equation for f is

$$f''' + \alpha f f'' + \beta[1 - (f')^2] = 0 \tag{2.141}$$

with boundary conditions

$$\text{at} \quad \eta = 0, \quad f = f' = 0$$
$$\text{at} \quad \eta = \infty, \quad f' = 1. \tag{2.142}$$

We next answer the question, What form must $g(x)$ have for similar solutions (here we are assuming u_∞ is a known relation)? From the relations (2.140) one can easily obtain

$$\alpha - \beta = \frac{L}{U_\infty} u_\infty gg' \tag{2.143}$$

$$2\alpha - \beta = \frac{L}{U_\infty} \frac{d}{dx}[g^2 u_\infty]. \tag{2.144}$$

From the last result we can infer, providing $2\alpha - \beta \neq 0$ and the constant of integration is discarded, that

$$(2\alpha - \beta)\frac{x}{L} = \frac{u_\infty}{U_\infty} g^2(x). \tag{2.145}$$

Now Eq. (2.143) can be rewritten as

$$(\alpha - \beta)\frac{u_\infty'}{u_\infty} = \left(\frac{L}{U_\infty} g^2 u_\infty'\right)\frac{g'}{g} = \beta \frac{g'}{g} \tag{2.146}$$

which, in view of the constancy of α and β, is integrable to the form

$$g = \left\{\frac{1}{C}\left(\frac{u_\infty}{U_\infty}\right)^{\alpha-\beta}\right\}^{1/\beta}, \tag{2.147}$$

where C is a constant of integration. Elimination of g between Eqs. (2.145) and (2.147) results in the expression

$$\frac{u_\infty(x)}{U_\infty} = C^{2/(2\alpha-\beta)}\left\{(2\alpha-\beta)\frac{x}{L}\right\}^{\beta/(2\alpha-\beta)} \tag{2.148}$$

where we recall that $2\alpha - \beta \neq 0$. Finally the form of g is

$$g(x) = \left\{(2\alpha-\beta)\frac{x}{L}\frac{U_\infty}{u_\infty}\right\}^{\frac{1}{2}}. \tag{2.149}$$

This analysis has disclosed that a reduction to an ordinary differential equation as described by the "similarity" transformation [Eqs. (2.134) and (2.135)] is possible if $u_\infty(x)$ is a "power" function as expressed by Eq. (2.148). Of additional interest is the dependency of the final transformation [via $g(x)$] on the flow character at ∞. Unfortunately, the resulting total differential equation is nonlinear, of a relatively intractable form, necessitating approximate and numerical methods for its solution. This drawback suggests that alternate methods may be fruitful.

As an example of a mathematically attainable solution (but not necessarily physically useful) we seek solutions of Eq. (2.132) of the form (traveling wave type)

$$F(x + \lambda y) \tag{2.150}$$

where λ is constant. Upon insertion into Eq. (2.132) we find that

$$\nu\lambda^3 F''' + u_\infty u_\infty' = 0 \tag{2.151}$$

whose integrated form is

$$F(r) = -\frac{1}{\nu\lambda^3}\iiint u_\infty u_\infty'\,(dr)^3 + c_1 r^2 + c_2 r + c_3 \ (r = x + \lambda y). \tag{2.152}$$

This form may not be useful in any physical example but it casts doubt on the wisdom of restricting our attention to the single transformation as suggested by Eq. (2.133).

2.4 THE UNKNOWN FUNCTION APPROACH

The development of the Hopf transformation by the Cole method [see footnote following Eq. (2.23)] utilized an unknown functional approach. A modification of this technique which assumes a certain functional grouping of the independent variables, and therefore is a generalization of the idea of similarity as previously used, has considerable utility. To discuss this procedure we shall again revert to an example.

A number of physically interesting problems (cf. Chapter 1) had the equation

$$w_t = (w^n w_x)_x \tag{2.153}$$

as their mathematical model. We shall consider two different approaches with the remark that these were chosen at random and others are possible.

First, what form must f have so that $f(x + \lambda t)$, λ a constant, is a solution of Eq. (2.153). We note, before continuing, that no mention of auxiliary conditions has been made. It is quite probable that this requirement of satisfaction of auxiliary conditions will dictate the choice of grouping of variables. To continue, we substitute $f(x + \lambda t)$ into Eq. (2.153) obtaining the total differential equation

$$(f^n f')' - \lambda f' = 0, \tag{2.154}$$

where the prime indicates differentiation with respect to the grouping $\eta = x + \lambda t$. One integration and discard of the constant of integration (which may be kept) gives the first integral as

$$f' = \frac{\lambda f}{f^n} \tag{2.155}$$

and a second integration yields

$$f = \{n[\lambda(x + \lambda t) + D]\}^{1/n} \tag{2.156}$$

where D is an arbitrary constant.

Second, occasionally a solution may be constructed by direct separation of variables. To that end let us assume

$$w = T(t)X(x) \tag{2.157}$$

so that upon substitution into Eq. (2.153) the separation ordinary differential equations are

$$T' - \lambda T^{n+1} = 0$$

and

$$[X X']' - \lambda X = 0 \tag{2.158}$$

which are easily solved by classical methods.

Third, let us assume that $n = 1$ and construct the functions $F(t)$ and $G(t)$ and determine p so that

$$w = F(t) + G(t)x^p \tag{2.159}$$

is a solution. Performing the differentiations and grouping gives the relation

$$F' + G'x^p = FGp(p-1)x^{p-2} + G^2 p(2p-1)x^{2p-2} \tag{2.160}$$

with unknowns F and G. With $p = 2$ the three exponents p, $p - 2$, and $2p - 2$ are 2, 0, 2, respectively, so the equation may be regrouped as

$$F' - 2FG = (6G^2 - G')x^2, \tag{2.161}$$

and if we require G to be such that

$$G' - 6G^2 = 0 \tag{2.162}$$

then F may be determined from

$$F' - 2FG = 0. \tag{2.163}$$

Performing these integrations we find that

$$G = (C - 6t)^{-1}, \quad F = D(C - 6t)^{-\frac{1}{6}}$$

and

$$w = (C - 6t)^{-1}x^2 + D(C - 6t)^{-\frac{1}{6}} \tag{2.164}$$

with C and D as constants to be determined from stated auxiliary conditions. It is immediately obvious from these results that such solutions have limited utility but indicate a direction of attack.

Last, if one attempts to find a solution of the form $w = F(t) + H(x)$, except for the trivial solution $x + t$ when $n = 1$, it is seen to be not possible.

2.5 GENERAL SOLUTIONS

D'Alembert and Euler in 1747 obtained solutions of the form $y = f(x + ct) + g(x - ct)$ with f, g arbitrary functions for the classical (linear) wave equation $c^2 y_{xx} = y_{tt}$. This is (perhaps) the first example of a *general* solution of a partial differential equation. A substantial number of researchers (mostly in pure mathematics) have extended this point of view—we mention Monge, Bateman, Laplace, Backlund, and others. However, in engineering applications this powerful tool has not been utilized to the degree that might be expected. Its utility for linear partial differential equations was completely submerged by the special methods of that subject. However, these procedures have not shown direction in nonlinear equations while in fact the general solution technique has opened a few doors.

Any relationship of the form

$$f(x, y, z, z_x, z_y) = 0 \tag{2.165}$$

where x, y are the independent variables and z is the dependent variable is a partial differential equation of first order. Any relation of the type

$$F(x, y, z, a, b) = 0 \qquad (2.166)$$

which contains two arbitrary constants a and b and is a solution of Eq. (2.165), is said to be a *complete solution* or a *complete integral* of that equation. On the other hand, a relation of the form

$$F(u, v) = 0. \qquad (2.167)$$

where F is an arbitrary function connecting two *known* functions u and v of x, y, and z and providing a solution of the partial differential Eq. (2.165) is called a *general* solution or general integral.

2.6 GENERAL SOLUTIONS OF FIRST-ORDER EQUATIONS

a. The Quasi-Linear Case

Equations of first order having the Lagrange form

$$Pz_x + Qz_y = R \qquad (2.168)$$

where P, Q, and R are functions of x, y, and z (but not z_x and z_y) have been previously called quasi-linear. Some authors use the term linear, but in this context it means linear in the first partial derivatives. The method we discuss for this equation generalizes immediately to n independent variables.

We shall not give the pertinent proofs unless they are constructive. Detailed proofs may be found in Sneddon, Forsyth, et al. [18, 19, 20].

Theorem 1. *The general solution of the quasi-linear partial differential equation* (2.168) *is*

$$F(u, v) = 0 \qquad (2.169)$$

where F is an arbitrary, sufficiently differentiable function and $u(x, y, z) = a$ and $v(x, y, z) = b$ form independent solutions of the Lagrange system

$$\frac{dx}{P} = \frac{dy}{Q} = \frac{dz}{R}. \qquad (2.170)$$

Suppose that we have found the two solutions $u = a$ and $v = b$ of the Lagrange system. We shall now indicate how the general solution

2.6 GENERAL SOLUTIONS OF FIRST-ORDER EQUATIONS

$F(u, v) = 0$ may be used to determine the integral surface which passes through a given curve whose parametric representation is

$$x = x(t), \quad y = y(t), \quad z = z(t)$$

where t is a parameter. In such a case the two solutions u, v must be such that

$$u\{x(t), y(t), z(t)\} = a, \quad v\{x(t), y(t), z(t)\} = b.$$

Consequently, we have two equations from which we may eliminate the parameter t to obtain the solution.

As an example of this theory we consider the one dimensional, time dependent flow of a compressible fluid with the assumption of constant pressure p. With u, ρ, and e representing the fluid velocity, density, and internal energy per unit volume the basic equations are

$$u_t + uu_x = 0$$
$$\rho_t + (\rho u)_x = 0 \qquad (2.171)$$
$$e_t + (eu)_x + pu_x = 0.$$

The change of variable $\epsilon = e + p$ provides the vehicle to replace the third of these by

$$\epsilon_t + (\epsilon u)_x = 0$$

which has the same form as the second equation of (2.171). Equations (2.171) are augmented by the initial conditions

$$u(x, 0) = f(x)$$
$$\rho(x, 0) = g(x) \qquad (2.172)$$
$$\epsilon(x, 0) = h(x)$$

having all necessary differentiable properties. Clearly the first equation may be solved independently and that result used in completing the solutions for ρ and ϵ.

The Lagrange system for $u_t + uu_x = 0$ is

$$\frac{dt}{1} = \frac{dx}{u} = \frac{du}{0}$$

with solutions $u = a$ and $x - ut = b$. According to Theorem 1 the general solution is

$$F(a, b) = F(u, x - ut) = 0 \qquad (2.173)$$

but in particular we may choose a special form of relation (2.173) as $a = G(b)$, i.e.,

$$u = G(x - ut) \qquad (2.174)$$

where G is arbitrary. This is not always possible but applies to this case. The form of G is determined by the initial condition from Eq. (2.172). Thus

$$u(x, 0) = G(x) = f(x),$$

so the desired solution, implicit as it may be, is

$$u(x, t) = f(x - ut) \qquad (2.175)$$

where $f(x)$ is the initial data.

This knowledge of u may now be utilized to complete the solution of system (2.171) subject to the initial conditions as given by Eqs. (2.172). Lagrange's system can be reapplied to the second equation or we can note that u_x appears in both equations. Now calculating u_x we have

$$u_x = f'(x - tu)[1 - tu_x]$$

or

$$u_x = \frac{f'(x - tu)}{1 + tf'(x - tu)} \qquad (2.176)$$

where the prime refers to differentiation with respect to the grouping $x - tu$. This form of u_x suggests trying solutions for ρ of the form

$$\rho(x, t) = \frac{H(x - tu)}{1 + tf'(x - tu)} \qquad (2.177)$$

and applying the initial condition

$$g(x) = \rho(x, 0) = H(x).$$

Thus

$$\rho(x, t) = \frac{g(x - tu)}{1 + tf'(x - tu)}, \qquad (2.178)$$

2.6 GENERAL SOLUTIONS OF FIRST-ORDER EQUATIONS

and in a similar fashion

$$\epsilon(x, t) = \frac{h(x - tu)}{1 + tf'(x - tu)}. \qquad (2.179)$$

For completeness we verify that $\rho(x, t)$, as given by Eq. (2.178), satisfies $\rho_t + (\rho u)_x = 0$. Differentiating with respect to t we have

$$\rho_t = -[1 + tf']^{-2}\{[1 + tf'](u + tu_t)g' + gf' - tgf''(u + tu_t)\}.$$

Now

$$u + tu_t = f - \frac{tff'}{1 + tf'} = \frac{f}{1 + tf'}$$

and therefore

$$\rho_t = -[1 + tf']^{-3}\{(fg)'(1 + tf') - tgff''\}. \qquad (2.180)$$

Forming ρu and differentiating with respect to x yields

$$(\rho u)_x = [1 + tf']^{-2}\{[1 + tf']gu_x + ug'(1 - tu_x) - ug(1 + tu_x)tf''\}.$$

However,

$$1 - tu_x = \frac{1}{1 + tf'},$$

thus

$$(\rho u)_x = [1 + tf']^{-3}\{(fg)'(1 + tf') - tgff''\},$$

and $\rho_t = -(\rho u)_x$.

In practice we can compute u from $u = f(x - tu)$, being careful about multivalued problems, via some iteration procedure such as

$$u_{i+1} = f(x - tu_i).$$

This result may then be directly applied to obtain ρ and ϵ. Iteration methods will be discussed in Chapter 6.

These results have been discussed for the simple system Equations (2.171). However, the methods are equally applicable to "conservation" laws (see Lax [21]) of the form[†]

$$u_t + F(u)u_x = 0 \qquad (2.181)$$

[†] By means of these procedures we can easily show that the general solution of $0 = g_y(x, y, u)u_x - g_x(x, y, u)u_y = J(u, g)/(x, y)$ is $u(x, y) = G[g(x, y, u)]$, i.e., an implicit solution. Equation (2.181) is a special case of the equation $\alpha(u)u_x - \beta(u)u_y = 0$, whose solution is implicitly defined by $u = G[\alpha(u)y + \beta(u)x]$ or by $\alpha(u)y + \beta(u)x = w(u)$.

or the more general situation

$$u_t + [A(u)]_x = 0.$$

Thus the solutions for Eq. (2.181) with initial condition $u(x, 0) = f(x)$ is

$$u(x, t) = f[x - tF(u)]. \tag{2.182}$$

If we augment Eq. (2.181) with the generalized continuity equation

$$\rho_t + [\rho F(u)]_x = 0, \quad \rho(x, 0) = g(x) \tag{2.183}$$

the resulting solution is

$$\rho(x, t) = \frac{g(x - tu)}{1 + tF'(u)f'[x - tF(u)]}.$$

The reader is referred to the fundamental paper of Noh and Protter [22] for more details.

b. The General Nonlinear Case

There are at least three methods for obtaining the *complete* solutions of the first order equation

$$F(x, y, z, p, q) = 0 \tag{2.184}$$

where $p = z_x$, $q = z_y$. These bear the names Cauchy's method of characteristics (which is essentially geometric in its argument), Jacobi's method, and Charpit's method (see Forsyth [19]). The first of these will be discussed in detail in Chapters 5 and 7 and the last two are related—Charpit's method describes the fundamental ideas so it is discussed herein.

Two first order partial differential equations

$$F(x, y, z, p, q) = 0 \tag{2.185}$$

$$G(x, y, z, p, q) = 0 \tag{2.186}$$

are said to be *compatible* if every solution of the first is also a solution of the second. Compatibility is a generalization of the idea of exactness in the theory of ordinary differential equations. The condition that the two equations be compatible is that $[F, G] \equiv 0$, where

$$[F, G] \equiv \frac{\partial(F, G)}{\partial(x, p)} + p \frac{\partial(F, G)}{\partial(z, p)} + \frac{\partial(F, G)}{\partial(y, q)} + q \frac{\partial(F, G)}{\partial(z, q)}. \tag{2.187}$$

2.6 GENERAL SOLUTIONS OF FIRST-ORDER EQUATIONS

The symbol

$$\frac{\partial(F, G)}{\partial(x, y)} = \begin{vmatrix} F_x & G_x \\ F_y & G_y \end{vmatrix} = F_x G_y - F_y G_x$$

is used to denote the Jacobian of F and G with respect to x and y.

Since the demonstration is purely algebraic we give it here. If G is such that

$$\frac{\partial(F, G)}{\partial(p, q)} \neq 0 \tag{2.188}$$

then we can solve the two equations (2.185) and (2.186) to obtain the explicit expressions

$$p = \phi(x, y, z), \quad q = \psi(x, y, z). \tag{2.189}$$

The condition of compatibility then is equivalent to the condition that the differential equations (2.189) be completely integrable, i.e., that

$$dz = p\,dx + q\,dy$$

or

$$\phi\,dx + \psi\,dy - dz = 0 \tag{2.190}$$

should be integrable.

It is well known (Sneddon [8]) that the necessary and sufficient condition for the integrability of the Pfaffian equation

$$\mathbf{X} \cdot d\mathbf{r} = 0$$

is that $X \cdot \text{curl } \mathbf{X} \equiv 0$. Applying this to Eq. (2.190) we find that the integrability condition is

$$\psi_x + \phi \psi_z = \phi_y + \psi \phi_z. \tag{2.191}$$

The remainder of the proofs shows that condition (2.191) is equivalent to Eq. (2.187).

Charpit's method of solving the equation

$$f(x, y, z, p, q) = 0 \tag{2.192}$$

is based on the preceding remarks. The idea is to introduce a second first-order partial differential equation

$$g(x, y, z, p, q, a) = 0 \tag{2.193}$$

involving an arbitrary constant a and such that:

(i) Eqs. (2.192) and (2.193) can be solved to give

$$p = p(x, y, z, a), \qquad q = q(x, y, z, a),$$

(ii) the resulting equation

$$dz = p\,dx + q\,dy \qquad (2.194)$$

is integrable.

When such a function g has been found, the solution

$$F(x, y, z, a, b) = 0 \qquad (2.195)$$

of Eq. (2.194) containing two parameters will be a *complete* integral of Eq. (2.192). Our basic problem is thus the determination of g but this was essentially done for us by the condition of compatibility. All we need do is find $g = 0$ compatible with $f = 0$. Expanding Eq. (2.187) for f and g we have

$$f_p g_x + f_q g_y + (pf_p + qf_q)g_z - (f_x + pf_z)g_p - (f_y + qf_z)g_q = 0 \qquad (2.196)$$

which is a first-order partial differential equation for g.

The generalized Lagrange subsidiary equations for this equation are

$$\frac{dx}{f_p} = \frac{dy}{f_q} = \frac{dz}{pf_p + qf_q} = \frac{dp}{-(f_x + pf_z)} = \frac{dq}{-(f_y + qf_z)} \qquad (2.197)$$

and these are called the equations of Charpit. They are equivalent to the characteristic equations of Cauchy (see Chapter 5). Once g is determined the problem reduces to that of solving for p and q and then integrating $dz = p(x, y, z, a)\,dx + q(x, y, z, a)\,dy$ by the methods of Pfaffian differential equations (see Sneddon [18, Chapter 1]).

If in the complete solution (2.195), one of the constants, say b, is replaced by a known function of the other, say $b = \phi(a)$, then

$$F(x, y, z, a, \phi(a)) = 0 \qquad (2.198)$$

is a one-parameter family. The totality of solutions obtained by varying $\phi(a)$ is called the *general solution* of Eq. (2.184).

A substantial number of first-order partial differential equations and their complete and general solutions are discussed and tabulated in Kamke [23].

2.6 GENERAL SOLUTIONS OF FIRST-ORDER EQUATIONS

We shall complete this area of relatively low physical utility by examining some special types of first-order equations

Type I. *Equations involving only p and q.* If the equation is of the form

$$f(p, q) = 0 \tag{2.199}$$

Charpit's equations become

$$\frac{dx}{f_p} = \frac{dy}{f_q} = \frac{dz}{pf_p + qf_q} = \frac{dp}{0} = \frac{dq}{0} \tag{2.200}$$

so that an obvious solution $p = a$ and the corresponding value of q is then obtained from Eq. (2.199) by

$$f(a, q) = 0 \tag{2.201}$$

so that

$$q = Q(a)$$

is a constant. The complete solution is then found from

$$dz = p\, dx + q\, dy$$
$$= a\, dx + Q(a)\, dy$$

as

$$z = ax + Q(a)y + b \tag{2.202}$$

where b is a constant.

We might wish to take $dq = 0$ to furnish our first solution in some problems where the computation is extreme.

Type II. *Independent variables missing.* In this case the equation is

$$f(z, p, q) = 0 \tag{2.203}$$

and the Charpit equations are

$$\frac{dx}{f_p} = \frac{dy}{f_q} = \frac{dz}{pf_p + qf_q} = \frac{dp}{-pf_z} = \frac{dq}{-qf_z}. \tag{2.204}$$

The last two yield the relation

$$p = aq$$

and therefore
$$f(z, aq, q) = 0$$
from which a complete integral follows when one solves for q, evaluates $p = aq$ and substitutes into (2.194).

Type III. *Separable equations.* In the event that Eq. (2.184) is expressible in the form
$$f(x, p) = g(y, q) \qquad (2.205)$$
we say the equation is *separable*. In this case Charpit's equations become
$$\frac{dx}{f_p} = \frac{dy}{-g_q} = \frac{dz}{pf_p - qg_q} = \frac{dp}{-f_x} = \frac{dq}{-g_y} \qquad (2.206)$$
so that we have the ordinary differential equation
$$\frac{dp}{dx} = -\frac{f_x}{f_p}$$
in x and p or in the form
$$f_p \, dp + f_x \, dx = 0.$$
We see that the solution is $f(x, p) = a$, a constant. Similarly $g(y, q) = a$. We determine p and q from these two relations and proceed via $dz = p \, dx + q \, dy$.

2.7 GENERAL SOLUTIONS OF SECOND-ORDER EQUATIONS

In this section we continue the use of $p = z_x$ and $q = z_y$, and further introduce the notation $r = z_{xx}$, $s = z_{xy}$ and $t = z_{yy}$. In this notation a second-order equation takes the form
$$F(x, y, z, p, q, r, s, t) = 0. \qquad (2.207)$$
Only special cases of this equation have been integrated by general methods and a few of these are given below. There is little mathematical interest in this area today but hopefully the pressure from science and engineering will breathe new life into the subject—since the author

2.7 GENERAL SOLUTIONS OF SECOND-ORDER EQUATIONS

believes that results from this area will lead to windows onto the nonlinear morass.

To introduce the ideas let us consider a typical example whose motivation lies in the theory of the boundary layer. From Eq. (2.132) we have the expression for the stream function as

$$\psi_y\psi_{xy} - \psi_x\psi_{yy} = \nu\psi_{yyy}.$$

We neglect the term on the right-hand side and consider the equation

$$\psi_y\psi_{xy} - \psi_x\psi_{yy} = 0 \qquad (2.208)$$

of nonlinear terms only. This is quite a different twist from the usual procedure for here the linear term is depressed! Now we shall see that a function $\psi(x, y)$ which satisfies Eq. (2.208) can be of the form[†]

$$f[y + g(x)], \qquad (2.209)$$

where the functions f and g are arbitrary. We verify this statement by calculation and get

$$f'f''g' - f'g'f'' \equiv 0.$$

Now that we know that $f[y + g(x)]$ identically satisfies Eq. (2.208) we can bring back the linear term $\nu\psi_{yyy}$ and ask, What form of f will allow satisfaction of the complete Eq. (2.132)? Clearly this requires that $f'''(\eta) = 0$, $\eta = y + g(x)$ and thus upon integration

$$f[y + g(x)] = a[y + g(x)]^2 + b[y + g(x)] + c \qquad (2.210)$$

where a, b, c are constants. Equation (2.210) satisfies the boundary layer equation for arbitrary $g(x)$. This special form may not, in fact, be useful for a given physical problem but it shows the direction that the general solution may give.

[†] One motivation for the choice $f(y + g(x))$ lies in a fundamental property of the boundary layer equations. If $u(x)$ and $v(x)$ are a solution of the boundary layer equations, then

$$u^* = u[x, y + g(x)], \qquad v^*(x, y) = v[x, y + g(x)] - g'(x)u[x, y + g(x)]$$

constitute another solution, where $g(x)$ is any arbitrary function of x. On the other hand the choice $f(y + g(x))$ is obtained in the next section as a result of general theory.

Since the development of general solutions is a vast subject we shall discuss herein only one method, the so-called Monge method, and shall note that the treatise of Forsyth [9] goes into great detail in this area. Upon completion of the Monge method we shall tabulate some of the interesting equations and their general solutions. The Monge method is as follows.

The second-order equation can be written as

$$F(x, y, z, p, q, r, s, t) = 0 \tag{2.211}$$

where p, q, r, s, t are the partial derivatives of z. Monge's method consists in establishing one or two first integrals in the form

$$\eta = f(\phi) \tag{2.212}$$

where η and ϕ are known functions of $x, y, z, p,$ and q and the function f is arbitrary. Not every equation of the form (2.211) has a first integral of the type (2.212). In fact any second-order equation which has a first integral of the type $\eta = f(\phi)$ must be expressible in the form of Monge's equation

$$R_1 r + S_1 s + T_1 t + U_1(rt - s^2) = V_1, \tag{2.213}$$

where $R_1, S_1, T_1, U_1,$ and V_1 are functions of $x, y, z, p,$ and q given by the relations involving various Jacobians

$$R_1 = \frac{\partial(\phi, \eta)}{\partial(p, y)} + q \frac{\partial(\phi, \eta)}{\partial(p, z)}, \qquad T_1 = \frac{\partial(\phi, \eta)}{\partial(x, q)} + p \frac{\partial(\phi, \eta)}{\partial(z, q)} \tag{2.214}$$

$$S_1 = \frac{\partial(\phi, \eta)}{\partial(q, y)} + q \frac{\partial(\phi, \eta)}{\partial(q, z)} - \frac{\partial(\phi, \eta)}{\partial(p, x)} - p \frac{\partial(\phi, \eta)}{\partial(p, z)} \tag{2.215}$$

$$U_1 = \frac{\partial(\phi, \eta)}{\partial(p, q)}, \qquad V_1 = q \frac{\partial(\phi, \eta)}{\partial(z, x)} + p \frac{\partial(\phi, \eta)}{\partial(y, z)} + \frac{\partial(\phi, \eta)}{\partial(y, x)}. \tag{2.216}$$

Equation (2.213) takes the form

$$R_1 r + S_1 s + T_1 t = V_1 \tag{2.217}$$

if and only if $U_1 \equiv 0$, i.e., if and only if

$$\frac{\partial(\phi, \eta)}{\partial(p, q)} = \phi_p \eta_q - \phi_q \eta_p = 0.$$

We now assume an equation of the form

$$Rr + Ss + Tt + U(rt - s^2) = V \tag{2.218}$$

2.7 GENERAL SOLUTIONS OF SECOND-ORDER EQUATIONS

has a first integral of the form $\eta = f(\phi)$. We now establish a procedure for finding the first integral. Since p, q, r, s, and t are the partial derivatives of z, which is a function of x and y, we have

$$dp = \frac{\partial p}{\partial x} dx + \frac{\partial p}{\partial y} dy = r\, dx + s\, dy, \qquad dq = s\, dx + t\, dy. \qquad (2.218)$$

Upon solving for r and t in these equations and substituting into Eq. (2.218) we find

$$R\, dp\, dy + T\, dq\, dx + U\, dp\, dq - V\, dx\, dy$$
$$= s(R\, dy^2 - S\, dx\, dy + T\, dx^2 + U\, dp\, dx + U\, dq\, dy). \qquad (2.219)$$

Let us suppose that

$$\phi(x, y, z, p, q) = a$$
$$\eta(x, y, z, p, q) = b \qquad (2.220)$$

are two integrals of the set of equations (Monge's equations)

$$R\, dp\, dy + T\, dq\, dx + U\, dp\, dq - V\, dx\, dy = 0 \qquad (2.221)$$

$$R\, dy^2 - S\, dx\, dy + T\, dx^2 + U(dp\, dx + dq\, dy) = 0 \qquad (2.222)$$

$$dz = p\, dx + q\, dy, \qquad (2.223)$$

then the equations

$$d\phi = 0, \qquad d\eta = 0 \qquad (2.224)$$

are equivalent to the set (2.221)–(2.223). Upon elimination of dz between Eqs. (2.223) and (2.224) the solutions for dp and dq are

$$dp = -\frac{T_1}{U_1} dx - \frac{1}{U_1} \left\{ \frac{\partial(\phi, \eta)}{\partial(y, q)} + q\, \frac{\partial(\phi, \eta)}{\partial(z, q)} \right\} dy \qquad (2.225)$$

$$dq = \frac{1}{U_1} \left\{ \frac{\partial(\phi, \eta)}{\partial(x, p)} + p\, \frac{\partial(\phi, \eta)}{\partial(z, p)} \right\} dx - \frac{R_1}{U_1} dy \qquad (2.226)$$

where R_1, S_1, T_1 are defined by Eqs. (2.214)–(2.216). Substituting these relations for dp and dq we see that $dp\, dx + dq\, dy$ of Eq. (2.222) becomes

$$dp\, dx + dq\, dy = -\frac{T_1}{U_1} dx^2 - \frac{R_1}{U_1} dy^2 + \frac{1}{U_1} \left\{ \frac{\partial(\phi, \eta)}{\partial(q, y)} \right.$$
$$\left. + \frac{\partial(\phi, \eta)}{\partial(q, z)} q - \frac{\partial(\phi, \eta)}{\partial(p, x)} - \frac{\partial(\phi, \eta)}{\partial(p, z)} p \right\} dx\, dy, \qquad (2.227)$$

which is equivalent to

$$R_1 \, dy^2 + T_1 \, dx^2 + U_1(dp \, dx + dq \, dy) - S_1 \, dx \, dy = 0. \qquad (2.228)$$

Similarly

$$R_1 \, dp \, dy + T_1 \, dq \, dx + U_1 \, dp \, dq - V_1 \, dx \, dy = 0 \qquad (2.229)$$

and comparing Eqs. (2.228) and (2.229) with Eqs. (2.221) and (2.222) we have

$$\frac{R_1}{R} = \frac{S_1}{S} = \frac{T_1}{T} = \frac{U_1}{U} = \frac{V_1}{V}$$

so that Eq. (2.218) is equivalent to Eq. (2.213) which we know has a first integral of the form $\eta = f(\phi)$. This first integral $\eta = f(\phi)$ can be derived by making one of the functions η obtained from a solution $\eta = a$, a function of a second solution ϕ. The procedure of obtaining a first integral of Eq. (2.218) reduces to that of solving the *Monge equations* (2.221)–(2.223).

Often the solution of the Monge equations can be done by inspection. The simplified boundary layer equation is of this type, for Eq. (2.208) in the present notation becomes $qs - pt = 0$ so that $R = 0$, $S = q$, $T = -p$, $U = V = 0$. The Monge equations are

$$p \, dq \, dx = 0 \quad \text{and} \quad q \, dx \, dy + p \, dx^2 = 0.$$

These imply $dq = 0$ or $q = a$ and the second yields $d\psi = q \, dy + p \, dx = 0$ or $\psi = b$. If we write the first integral as

$$q = f(\psi),$$

we may complete our solution by Lagrange's method. The auxiliary equations are

$$\frac{dx}{0} = \frac{dy}{1} = \frac{d\psi}{f(\psi)},$$

with integrals $x = c_1$, $F(\psi) - y = c_2$, $F(\psi) = \int d\psi/f(\psi)$. Thus

$$G(c_1, c_2) = 0$$

constitutes the general solution. If in particular we select

$$c_2 = \eta(c_1)$$

2.7 GENERAL SOLUTIONS OF SECOND-ORDER EQUATIONS

we have
$$F(\psi) = y + \eta(x),$$
or taking inverses
$$\psi = G\{y + \eta(x)\}, \quad G = F^{-1}.$$

This verifies the general solution, Eq. (2.209).

When the Monge equations cannot be solved by inspection the following procedure may be used. A parameter $\lambda = \lambda(x, y, z, p, q)$ is introduced so as to obtain a factorable combination

$$\Omega = \lambda[R(dy)^2 - S\,dx\,dy + T(dx)^2 + U(dx\,dp + dy\,dq)] \\ + R\,dy\,dp + T\,dx\,dq + U\,dp\,dq - V\,dx\,dy = 0, \quad (2.230)$$

which we write as

$$\Omega = (a\,dy + b\,dx + c\,dp)(\alpha\,dy + \beta\,dx + \gamma\,dq)$$
$$= a\alpha\,(dy)^2 + (a\beta + b\alpha)\,dx\,dy + b\beta\,(dx)^2 + c\beta\,dx\,dp$$
$$+ a\gamma\,dy\,dq + c\alpha\,dy\,dp + b\gamma\,dx\,dq + c\gamma\,dp\,dq = 0.$$

Since, by assumption, this is an identity, we have by comparing coefficients

$$a\alpha = \lambda R, \quad a\beta + b\alpha = -S\lambda - V, \quad b\beta = \lambda T,$$
$$c\beta = \lambda U = a\gamma, \quad c\alpha = R, \quad b\gamma = T, \quad \text{and} \quad c\gamma = U.$$

The first relation is satisfied by taking $a = \lambda$ and $\alpha = R$; this choice determines $b = T/U$, $\beta = \lambda U$, $c = 1$ and $\gamma = U$. The final relation takes the form

$$\lambda^2 U + \frac{TR}{U} = -S\lambda - V$$

or
$$U^2\lambda^2 + SU\lambda + (TR + UV) = 0 \quad (2.231)$$

which is a quadratic for the determination of the parameter λ. Unless $S^2 - 4(RT + UV) = 0$, this quadratic will have two distinct roots λ_1 and λ_2. Thus Eq. (2.230) is factored as

$$(\lambda_1 U\,dy + T\,dx + U\,dp)(R\,dy + \lambda_1 U\,dx + U\,dq) = 0 \quad (2.232)$$
$$(\lambda_2 U\,dy + T\,dx + U\,dp)(R\,dy + \lambda_2 U\,dx + U\,dq) = 0. \quad (2.233)$$

2. TRANSFORMATION AND GENERAL SOLUTIONS

There are four systems to be considered but two of these can be eliminated. The system $\lambda_1 U\, dy + T\, dx + U\, dp = 0$, $\lambda_2 U\, dy + T\, dx + U\, dp = 0$ implies $(\lambda_1 - \lambda_2) U\, dy = 0$ and, hence, unless $\lambda_1 = \lambda_2$, $U\, dy \equiv 0$. Similarly the system $R\, dy + \lambda_1 U\, dx + U\, dq = 0$, $R\, dy + \lambda_2 U\, dx + U\, dq = 0$ implies $U\, dx \equiv 0$. Therefore only two systems need be considered:

$$\lambda_1 U\, dy + T\, dx + U\, dp = 0$$
$$R\, dy + \lambda_2 U\, dx + U\, dq = 0 \qquad (2.234a)$$

and

$$\lambda_2 U\, dy + T\, dx + U\, dp = 0$$
$$R\, dy + \lambda_1 U\, dx + U\, dq = 0. \qquad (2.234b)$$

From each of these pairs we derive two integrals of the form $\phi(x, y, z, p, q) = a$, $\eta(x, y, z, p, q) = b$, and hence two first integrals

$$\eta_1 = f_1(\phi_1), \qquad \eta_2 = f_2(\phi_2),$$

which can often be solved to find p and q as functions of x, y and z. When these are substituted into $dz = p\, dx + q\, dy$ Forsyth has shown that this is always integrable—and this integral involving two arbitrary functions will be the desired general solution.

When it is possible to find only one integral $\eta = f(\phi)$ the final integral may be obtained by the Lagrange or Charpit method.

The Plateau problem for the minimal surface area is governed by Eq. (1.49). If we set $p = h_x$, $q = h_y$, $r = h_{xx}$, $s = h_{xy}$, $t = h_{yy}$ that equation takes the form

$$q^2 r - 2pqs + p^2 t = -(r + t). \qquad (2.235)$$

A solution for this equation is constructed, with the linear terms on the right-hand side of the equation depressed, leaving the equation

$$q^2 r - 2pqs + p^2 t = 0. \qquad (2.236)$$

Then $R = q^2$, $S = -2pq$, $T = p^2$, $U = V = 0$, and the Monge equations become

$$q^2\, dp\, dy + p^2\, dq\, dx = 0$$
$$(p\, dx + q\, dy)^2 = 0.$$

Since $dz = p\,dx + q\,dy$, the second of these equations gives $dz = 0$ with the integral $z = a$. Also the second equation can be rewritten as $p\,dx = -q\,dy$ and using this in the first gives $q\,dp = p\,dq$ with the solution $p = bq$. Our first integral is therefore

$$b = \psi(a)$$

or

$$p = q\psi(z) \tag{2.237}$$

where ψ is an arbitrary function. This is a quasi-linear equation with the Lagrange system

$$\frac{dx}{1} = \frac{dy}{-\psi(z)} = \frac{dz}{0}$$

with integrals

$$z = a, \quad y + x\psi(a) = b$$

so that the general solution is

$$y + x\psi(z) = \phi(z) \tag{2.238}$$

where ϕ is also arbitrary. The *implicit* nature of this general solution is (unfortunately) often unavoidable. (Later we shall characterize those equations whose general solutions are explicit.) We can go one step further and rewrite the solution as

$$\psi(z) = \frac{1}{x}[\phi(z) - y]$$

or

$$z = H\left\{\frac{1}{x}[\phi(z) - y]\right\}$$

where $H = \psi^{-1}$. Forsyth [19, Vol. 6, p. 277] obtain a general solution of the complete equation by the Ampere process but the implicit nature of the general solution (Table I, Ex. 20) is all too evident. This is the rule rather than the exception. Mention should also be made of procedures due to Boole and Darboux which may also be found in Forsyth's treatise [19, Vol. 6].

2.8 TABLE OF GENERAL SOLUTIONS

For easy reference we herein tabulate a number of general solutions. Recall that $p = z_x$, $q = z_y$, $r = z_{xx}$, $s = z_{xy}$, and $t = z_{yy}$. The first group in Table I has the general solution in explicit form followed by the implicit group.

TABLE I

Equation	General solution	Reference[a]
	(ϕ and ψ are arbitrary functions)	
1. $s + zp = 0$	$z = \dfrac{2\phi'(y)}{\psi(x) + \phi(y)} - \dfrac{\phi''(y)}{\phi'(y)}$	F
2. $s + e^z = 0$	$e^z = -\dfrac{2\psi'(x)\,\phi'(y)}{[\psi(x) + \phi(y)]^2}$	F
3. $r + t + \lambda^2 h e^{2hz} = 0$ (Liouville's equation)	$\lambda^2 h^2 e^{2hz}(\phi^2 + \psi^2 + 1)^2$ $= \left\{\left(\dfrac{\partial \phi}{\partial x}\right)^2 + \left(\dfrac{\partial \phi}{\partial y}\right)^2\right\}$ where ϕ, ψ are real functions of x and y defined via $\phi + i\psi = F(x+iy)$, F arbitrary.	F
4. $s + \dfrac{\partial}{\partial x}(e^z g(x)) = 0$	$e^z = \dfrac{\phi'(y)}{\psi(x) + \phi(y)\,g(x)}$	F
5. $s + \dfrac{\partial}{\partial x}(e^z H) = 0$ $H = g(x)a(y) + h(x)b(y)$	$e^z = \dfrac{(\partial/\partial y)\{H(\phi'(y) + \rho\phi)\}}{H\psi(x) + H^2(\phi'(y) + \rho\phi)}$ $\rho = \dfrac{\partial}{\partial y}\left\{\log\left(\dfrac{a'b - ab'}{H}\right)\right\}$	F
6. $s + \dfrac{\partial}{\partial x}(\alpha e^z) - \dfrac{\partial}{\partial y}(\beta e^{-z}) = 0$ $\alpha = \dfrac{\rho'(x)}{\rho(x) + \sigma(y)}\dfrac{h(x)}{k(y)}$ $\beta = \dfrac{\sigma'(y)}{\rho(x) + \sigma(y)}\dfrac{k(y)}{h(x)}$	$e^z = \dfrac{\phi'(y) + \beta\psi(x) + \dfrac{1}{\alpha}\dfrac{\partial \alpha}{\partial y}\phi}{\psi' + \dfrac{1}{\beta}\dfrac{\partial \beta}{\partial x}\psi + \alpha\phi}$	F

[a] A: F. Ayres, Jr., "Theory and Problems of Differential Equations." Schaum Publ. Co., New York, 1952.

F: A. R. Forsyth, "Theory of Differential Equations," Vols. 5 and 6, Partial Differential Equations. Dover, New York, 1959.

M: M. H. Martin, *Pacific J. Math.* 3, 165 (1953).

MB: M. Morris and O. E. Brown, "Differential Equations." Prentice-Hall, Englewood Cliffs, New Jersey, 1942.

P: A. V. Pogorelov, "On Monge-Ampere Equations of Elliptic Type." Noordhoff, Groningen, The Netherlands, 1963.

S: I. N. Sneddon, "Elements of Partial Differential Equations." McGraw-Hill, New York, 1957.

2.8 TABLE OF GENERAL SOLUTIONS

TABLE I *(continued)*

Equation	General solution	Reference
7. $s^2 = 4\lambda(x,y)\,pq$	$z = \int u^2\,dx + \dfrac{1}{\lambda}\left(\dfrac{\partial u}{\partial y}\right)^2 dy$ $\dfrac{\partial^2 u}{\partial x\,\partial y} - \dfrac{1}{2\lambda}\dfrac{\partial \lambda}{\partial x}\dfrac{\partial u}{\partial y} - \lambda u = 0$	F
8. $qr - ps = 0$	$z = \phi\{x + \psi(y)\}$	S
9. $qs - pt = 0$	$z = \phi\{y + \psi(x)\}$	S
10. $s + (rt - s^2) = 0$	$z = uv + u\phi'(u) + v\psi'(v) - \phi - \psi$ $x = \phi'(u) + v$ $y = \psi'(v) + u$	S
11. $qs - pt = q^3$	$y + xz = \phi(z) + \psi(x)$	A
12. $q^2 r - 2pqs + p^2 t = 0$	$y + x\phi(z) = \psi(z)$	S
13. $q^2 r - 2pqs + p^2 t = pq^2$	$y = e^x \phi(z) + \psi(z)$	A
14. $q(1+q)r - (1+2q)(1+p)$ $\times s + (1+p)^2 t = 0$	$x = \phi(x + y + z) + \psi(x + z)$	A
15. $qr - (1 + p + q)s$ $+ (1+p)t = 0$	$z = \phi(x + z) + \psi(x + y)$	A
16. $(b+q)^2 r - 2(a+p)(b+q)$ $\times s + (a+p)^2 t = 0$	$z = x\phi(ax + by + z)$ $+ \psi(ax + by + z)$	F
17. $(e^x - 1)(qr - ps) = pqe^x$	$z = \phi[x - e^x - \psi(y)]$	A
18. $q(1+q)r - (1+p+q$ $+ 2pq)s + p(1+p)t = 0$	$z = \phi(x + z) + \psi(y + z)$	F
19. $(r - pt)^2 = q^2 rt$	$z = \tfrac{1}{3}(2v - u^3)\phi'(v) - \tfrac{2}{3}\phi(v)$ $\qquad + \int u^4 \psi'(u)\,du$ $x = \dfrac{1}{u}\phi'(v) + \psi(u)$ $y = u\phi'(v) - \int u^2 \psi'(u)\,du$	F

TABLE I *(continued)*

Equation	General solution	Reference
20. $(1 + q^2)r - 2pqs$ $\quad + (1 + p^2)t = 0$ (Plateau)	$z = i \int \{1 + \psi'^2(u)\}^{1/2} du$ $\quad + i \int \{1 + \phi'^2(v)\}^{1/2} dv$ $x = u + v \quad$ (Monge form) $y = \psi(u) + \phi(v)$	F, P
	[a better form for discussion of minimal surfaces is $z = 2v\phi'' - 2\phi' + 2u\psi'' - 2\psi'$ $x = (1 - v^2)\phi'' + 2v\phi' - 2\phi$ $\quad + (1 - u^2)\psi'' + 2u\psi' - 2\psi$ $y = i\{(1 + v^2)\phi'' - 2v\phi' + 2\phi$ $\quad - (1 + u^2)\psi'' + 2u\psi' - 2\psi\}]$	
21. $r + pt - qs = 0$	$z = u^2\psi''(u) - 2u\psi'(u) + 2\psi(u)$ $\quad + v^2\phi''(v) - 2v\phi'(v) + 2\phi(v)$ $-x = \psi''(u) + \phi''(v)$ $-y = u\psi''(u) - \psi'(u) + v\phi''(v) - \phi'(v)$	F
22. $(q + yt)(r + 1)$ $\quad = (ys - p - x)s$	$z = ux - \frac{1}{2}x^2 + \phi(u) - \psi(u/y)$ $0 = x + \phi'(u) - (1/y)\psi'(u/y)$	F
23. $z(qs - pt) = pq^2$	$y = \phi(z) + z\psi(x)$	

Often general integrals cannot be found in the sense that a full complement (as many as the order) of arbitrary functions cannot be introduced. Nevertheless there is some utility in presenting integrals with fewer arbitrary functions.

Equation	Integral	Reference
24. $xqr - (x + y)s + ypt$ $\quad + xy(rt - s^2) = 1 - pq$	$z = a \log x + bx + \frac{1}{b} y + \phi(x^b y)$	MB
25. $5r + 2s + 5t - 4(rt - s^2) = 6$	$8z = 5x^2 - 2xy + 5y^2 + 2ax$ $\quad + \phi(y + bx)$	MB

2.8 TABLE OF GENERAL SOLUTIONS

TABLE I (continued)

Equation	Integral	Reference
26. $yr - ps + t + y(rt - s^2) = -1$	$6\alpha^2 z = 2y^3 - 3\alpha^2 y^2 + 6\alpha xy + 6\beta y + \phi(\alpha x + \tfrac{1}{2}y^2)$	A

27. $rt - s^2 + \lambda^2 = 0$	Value of λ	Intermediate integral
(see refs. M and P)	0	$q = \phi(p)$
$\lambda = \Gamma(x)P(y)$	1	$q \pm x = \phi(p \mp y)$
$\Gamma = \sqrt{n}\exp\left[\dfrac{S - S_0}{2c_p}\right]$	$P(y) \not\equiv 0$	$p \pm f = \text{const.}, f = \int P\,dy.$
$P(y) = y^{-(n+1)/2}$	y^{-2}	$px + qy - z \pm \Gamma/y = \phi(p \pm 1/y)$
	$x^{m-1}/y^{m+1}\ (m \neq 0)$	$px + qy - z \pm \dfrac{1}{m}\left(\dfrac{x}{y}\right)^m = \text{const}$
	x^{-1}/y	$px + qy - z \pm \ln\left(\dfrac{x}{y}\right) = \text{const}$
	$e^x e^y$	$p - q \pm e^x e^y = \text{const}$

Martin proves that only in these cases does the equation have an intermediate integral and in those cases which arise from these under translations, reflections in the line $y = x$ and dilatations in the x- and y-directions, in the x, y plane.

REFERENCES

1. van Dusen, M. S., *J. Res. Natl. Bur. Std.* **4**, 753 (1930).
2. Forsythe, G. E., and Wasow, W. R., "Finite Difference Methods for Partial Differential Equations," p. 141. Wiley, New York, 1960.
3. Burgers, J. M., *Advan. Appl. Mech.* **1**, 171 (1948).
4. Lagerstrom, P. A., Cole, J. D., and Trilling, L., Problems in the theory of viscous compressible fluids, *Rept. Guggenheim Aeron. Lab., Calif. Inst. Technol.* (1949).
5. Hopf, E., *Commun. Pure Appl. Math.* **3**, 201 (1950).
6. Cole, J. D., *Quart. Appl. Math.* **9**, 225 (1951).
7. Kalman, G., *Ann. Phys. (N.Y.)* **10**, 1 (1960).
8. Boltzmann, L., *Ann. Physik* [3] **53**, 959 (1894).
9. Crank, J., "Mathematics of Diffusion." Oxford Univ. Press, London and New York, 1956.
10. Hill, R., "The Mathematical Theory of Plasticity." Oxford Univ. Press, London and New York, 1950.

11. von Kármán, T., *Bull. Am. Math. Soc.* **46**, 615 (1940).
12. Tricomi, F. G., *in* "Partial Differential Equations and Continuum Mechanics" (R. E. Langer, ed.), p. 207. Univ. of Wisconsin Press, Madison, Wisconsin, 1961.
13. Cherry, T. M., *in* "Partial Differential Equations and Continuum Mechanics" (R. E. Langer, ed.), p. 217. Univ. of Wisconsin Press, Madison, Wisconsin, 1961.
14. Lighthill, M. J., *Proc. Roy. Soc.* **A191**, 323 (1947).
15. Goldstein, S., *Proc. Cambridge Phil. Soc.* **35**, 338 (1939).
16. Mangler, W., *Z. Angew. Math. Mech.* **23**, 243 (1943).
17. Falkner, V. M., and Skan, S. W., *Phil. Mag.* [7] **12**, 865 (1931).
18. Sneddon, I. N., "Elements of Partial Differential Equations." McGraw-Hill, New York, 1957.
19. Forsyth, A. R., "Theory of Differential Equations," Vols. 5 and 6: Partial Differential Equations. Dover, New York, 1959.
20. Courant, R., and Hilbert, D., "Methods of Mathematical Physics," Vols. I and II. Wiley (Interscience), New York, 1953 and 1962.
21. Lax, P. D., *in* "Nonlinear Problems" (R. E. Langer, ed.), p. 3. Univ. of Wisconsin Press, Madison, Wisconsin, 1963.
22. Noh, W. F., and Protter, M., *J. Math. Mech.* **12**, No. 2, 149 (1963).
23. Kamke, E., "Differentialgleichungen, Lösungsmethoden und Lösungen," Vol. 2. Akad. Verlagsges. Leipzig, 1959.

CHAPTER 3

Exact Methods of Solution

3.0 INTRODUCTION

The loss of the superposition principle is a severe blow to one attempting to obtain solutions to a system governed by nonlinear partial differential equations. However, a number of exact methods have been developed for special cases and hopefully these can and will be modified, extended, and generalized as the subject develops.

By exact methods we shall mean any procedure which enables the solution of the equations to be obtained by one or more quadratures of an ordinary differential equation. Thus the similarity approach, previously discussed, is an exact method since one must integrate an ordinary differential equation to obtain the solution. As opposed to these exact methods approximate and numerical methods attack the equation directly. We discuss some of these methods in later chapters.

In linear systems, the analyst may have developed a habit of formulating the problem, complete with all necessary initial and boundary conditions, before proceeding to the solution development. This is generally the correct approach. In a nonlinear problem, he may proceed thusly but is often advised to delay the imposition of the auxiliary conditions—his method of attack may be amenable to some conditions but not to others. The wanton discard of almost any new and novel approach may be harmful to this subject. The analyst should carefully weigh the utility of his method, even though his immediate problem is not solved, and publish it if it has promise.

3.1 THE QUASI-LINEAR SYSTEM

The concept of a quasi-linear system of partial differential equations was introduced in Chapter 2, Section 2.1d. Such a system is not restricted to two equations but has the general form

$$\sum_{i=1}^{n} [a_{ji}u_x{}^i + b_{ji}u_y{}^i] + d_j = 0, \qquad j = 1, 2, ..., n, \tag{3.1}$$

that is to say, n coupled equations in the n unknowns u^i, $i = 1, 2, ..., n$, where the coefficients a_{ji}, b_{ji}, and d_j are generally functions of x, y, and the u^i. If the coefficients are independent of the u^i, the system is linear. While the method we discuss also applies to the linear case we shall not give any amplifying examples for the linear system.

The assumption of a quasi-linear system of the form of Eq. (3.1) is no essential restriction. For a great many nonlinear initial value problems can be transformed into quasi-linear form containing a large number of equations and unknowns. The longitudinal vibration discussed in Chapter 2 and governed by the equation $y_{tt} = [F(y_x)]^2 y_{xx}$ was of this type.

We now adopt the "happier" matrix notation for Eq. (3.1) by setting

$$A = [a_{ji}], \qquad B = [b_{ji}], \qquad d = [d_j], \qquad u = [u^i],$$

where A and B are $n \times n$ matrices, d and u are column vectors. The system then takes the matrix form[†]

$$Au_x + Bu_y + d = 0. \tag{3.2}$$

We shall subject this system to a linear transformation

$$v = Tu \tag{3.3}$$

[†] The notation u_x is used to mean the vector formed from the x derivatives of the u^i, $i = 1, 2, ..., n$.

with determinant of $T \neq 0$; $T = [t_{ji}]$ and the t_{ji} may depend upon x, y, and the u^i but not upon the derivatives of the u^i. Such a transformation has the form

$$v^j = \sum_{i=1}^{n} t_{ji} u^i, \quad j = 1, 2, \ldots, n, \tag{3.4}$$

and under such a transformation Eq. (3.2) takes a new but similar form. This new system is equivalent to the original one in the sense that every solution of one is also a solution of the other.

The linear transformation of Eq. (3.2)

$$TAu_x + TBu_y + Td = 0 \tag{3.5}$$

is used to develop some suitable canonical form. A particularly convenient one is such that

$$TA = ETB, \tag{3.6}$$

when E is a diagonal matrix, say

$$E = \begin{bmatrix} e_1 & 0 & \cdots & 0 \\ 0 & e_2 & 0 & 0 \\ \cdots & \cdots & & \\ 0 & \cdots & \cdots & e_n \end{bmatrix} = [e_j]. \tag{3.7}$$

Under the assumption of Eq. (3.6) we may rewrite Eq. (3.5) as

$$ETBu_x + TBu_y + Td = 0. \tag{3.8}$$

Let us examine the form of these equations by further simplification, setting $TB = A^* = [a_{ji}^*]$, $Td = d^* = [d_j^*]$, so that Eq. (3.8) becomes

$$EA^*u_x + A^*u_y + d^* = 0 \tag{3.9}$$

with the jth equation as

$$\sum_{i=1}^{n} a_{ji}^*(e_j u_x^i + u_y^i) + d_j^* = 0. \tag{3.10}$$

3. EXACT METHODS OF SOLUTION

Now if $\alpha \mathbf{i} + \beta \mathbf{j}$ is the unit vector for which

$$e_j = \frac{\alpha}{\beta} = \cot \theta$$

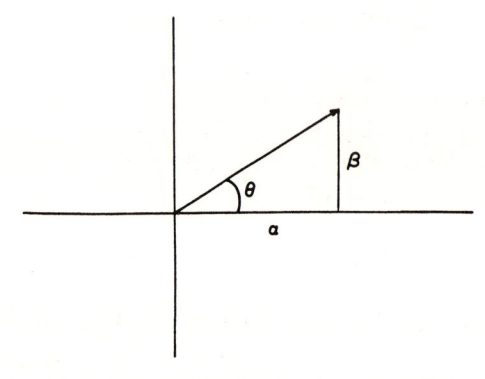

FIG. 3-1. Notation used in the development of characteristics.

we can write (see Fig. 3-1)

$$\begin{aligned} e_j u_x{}^i + u_y{}^i &= \frac{\alpha}{\beta} u_x{}^i + u_y{}^i \\ &= \frac{1}{\beta} [u_x{}^i \alpha + u_y{}^i \beta] \\ &= \frac{1}{\beta} [u_x{}^i \cos \theta + u_y{}^i \sin \theta] \end{aligned} \quad (3.11)$$

which except for the factor $1/\beta$ is the directional derivative in the direction defined by the vector $\alpha \mathbf{i} + \beta \mathbf{j}$ (in Fig. 3-1). This of course depends upon e_j and hence on j. Thus the above remarks mean that every equation of the transformed system (3.9) *contains differentiations in one direction only*. The removal of the complication of having more than one differential operator in each equation brings the theory closer to that of ordinary differential equations.

Calculation of the diagonal matrix E is accomplished from its definition $TA = ETB$. This is equivalent to the system of equations

$$\sum_{k=1}^{n} t_{jk} a_{ki} = \sum_{k=1}^{n} e_j t_{jk} b_{ki}, \quad i = 1, 2, \ldots, n,$$

or

$$\sum_{k=1}^{n}(a_{ki}-e_j b_{ki})t_{jk}=0 \qquad (3.12)$$

which is a system of n homogeneous equations for the t_{jk}, $k = 1, 2, ..., n$. For a nontrivial solution to exist the necessary and sufficient condition is known to be

$$\det(A - e_j B) = 0 \qquad (3.13)$$

which is an algebraic equation generally involving x, y, and u^i since A and B depend upon these.

We now ask for[†]

$$\det(A - \lambda B) = 0$$

to have n *distinct real roots* for if the roots are real and distinct the system is *hyperbolic*. In this case we choose the n roots for $e_j, j = 1, 2, ..., n$, and determine the elements of the T matrix from Eq. (3.12). We are tacitly assuming that

$$\det B \neq 0, \quad \det T \neq 0, \quad \text{and} \quad \det A^* = \det TB \neq 0$$

in all of this discussion.

Equation (3.9) is called the *normal form*. The direction $\alpha_k \mathbf{i} + \beta_k \mathbf{j}$ for which $e_k = \alpha_k/\beta_k$ is called the kth characteristic direction. The n differential equations

$$\cot \theta_k = \frac{dx}{dy} = c_k \qquad (3.14)$$

are called the *characteristics* of the system for their direction, at every point, is precisely the characteristic direction there. As we shall see in Chapter 6 the "method of characteristics" has the above material as its theoretical foundation.

Generally, the characteristic directions are not known until the solution of the differential problem is obtained. However, for a *system of two equations* the concept of characteristics can nevertheless be used to transform the given system into a simpler system. The simpler system, called the *canonical equations*, consists of four rather than two equations—a fair price inasmuch as the new system is easy to compute.

[†] This case is of great importance for the engineer. Physical systems may have multiple roots. The details may be found in Courant and Hilbert [1].

The starting point for the development of the canonical system (for two equations) is the observation that the characteristics form two one parameter families of curves, that may be considered as (not necessarily orthogonal) natural coordinates. Through every point there passes one curve of each family and these curves have different directions at each point since the e_j are different [Fig. 3-2].

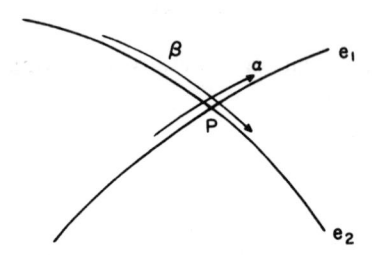

FIG. 3-2. The characteristic net with the parameters α and β illustrated as the new coordinates.

If these curves are to be used as the coordinate axes for a new curvilinear coordinate system, which of the infinitely many transformations

$$x = x(\alpha, \beta)$$
$$y = y(\alpha, \beta) \tag{3.15}$$

shall we use? A particularly useful one is to identify α with the e_1 family and β with the e_2 family so that

$$x_\alpha = \frac{dx}{dy} y_\alpha = e_1 y_\alpha$$
$$x_\beta = e_2 y_\beta . \tag{3.16}$$

Then the quantities $e_j u_x^i + u_y^i$, $j = 1, 2$, become

$$e_1 u_x^i + u_y^i = \frac{x_\alpha}{y_\alpha} u_x^i + u_y^i$$
$$= \frac{1}{y_\alpha} \left[\frac{\partial u^i}{\partial x} \frac{\partial x}{\partial \alpha} + \frac{\partial u^i}{\partial y} \frac{\partial y}{\partial \alpha} \right]$$
$$= u_\alpha^i / y_\alpha \tag{3.17}$$

and

$$e_2 u_x^i + u_y^i = u_\beta^i / y_\beta. \tag{3.18}$$

3.1 THE QUASI-LINEAR SYSTEM

Whereupon, the system $EA^*u_x + A^*u_y + d = 0$, which is equivalent to

$$a_{11}^*(e_1 u_x^1 + u_y^1) + a_{12}^*(e_1 u_x^2 + u_y^2) + d_1 = 0$$
$$a_{21}^*(e_2 u_x^1 + u_y^1) + a_{22}^*(e_2 u_x^2 + u_y^2) + d_2 = 0,$$

becomes

$$a_{11}^* u_\alpha^1 + a_{12}^* u_\alpha^2 + d_1 y_\alpha = 0 \tag{3.19}$$
$$a_{21}^* u_\beta^1 + a_{22}^* u_\beta^2 + d_2 y_\beta = 0.$$

Equations (3.19) together with

$$x_\alpha = e_1 y_\alpha, \qquad x_\beta = e_2 y_\beta \tag{3.20}$$

are the *canonical equations* of the system and have the dependent variables u^1, u^2, x, and y. They can be constructed from the original system without solving it beforehand.

The canonical hyperbolic system shares with the normal form the property that each equation has differentiations in one direction only. Moreover, in the canonical case these directions coincide with the coordinate directions. This considerable improvement is supplemented by the fact that the system does not contain the independent variables, α and β, explicitly.

According to our derivation, every solution of the original set satisfies this canonical characteristic system. The converse—every solution of the canonical system satisfies the original system provided the Jacobian $x_\alpha y_\beta - x_\beta y_\alpha = (e_1 - e_2) y_\alpha y_\beta \neq 0$—is easily verified.

When the two differential equations are linear, then e_1 and e_2 are known functions of x and y. Thus Eqs. (3.20) are not coupled with Eqs. (3.19) and therefore Eqs. (3.20) determine two families of characteristic curves independent of the solution.

In the reducible case (see Section 2.3) when d_1^*, d_2^* are both zero and the other coefficients depend on u^1, u^2 only, the situation is similar to the linear case. Then e_1 and e_2 are known functions of u and v and the differential equations (3.19) are independent of x and y and can thus be solved separately. We might add that if d_1^* and d_2^* do not vanish but depend on u^1 and u^2 then these remarks are also valid.

For reducible equations, the characteristic curves in the (u, v) plane (the images of the characteristics in the (x, y) plane) are independent of the special solution u, v considered. They are obtainable directly from Eqs. (3.19) as

$$a_{11}^* \frac{du^1}{du^2} = -a_{12}^*$$
$$a_{21}^* \frac{du^1}{du^2} = -a_{22}^*. \tag{3.21}$$

Discussion and extension of these ideas may be found in the literature [2, 3, 4, 5] and applications to numerical computation is given in Chapter 7.

3.2 AN EXAMPLE OF THE QUASI-LINEAR THEORY

As an illustration of the above theory we consider the problem of describing the motion of a gas when a sound wave of finite amplitude is being propagated through it. Assuming we have one dimensional isentropic flow of a perfect gas, the governing equations [2] are those of momentum, continuity and the gas laws

$$\rho u_t + \rho u u_x + p_x = 0$$
$$\rho_t + \rho u_x + u \rho_x = 0 \tag{3.22}$$
$$p = (\text{constant})\rho^\gamma, \quad c^2 = \frac{dp}{d\rho}$$

where x is displacement, t is time, $u(x, t)$, $p(x, t)$ and $\rho(x, t)$ are the velocity, pressure, and density, respectively, c is the velocity of sound in the gas, and γ the ratio of specific heats is constant. Let us assume the gas is initially at rest and the pressure, density, and velocity of sound in the gas are p_0, ρ_0, and c_0 throughout.

The pressure is eliminated by using the gas laws which yield $p_x = (\text{constant})\gamma \rho^{\gamma-1} \rho_x$ and $dp/d\rho = (\text{constant})\gamma \rho^{\gamma-1}$ so that

$$p_x = c^2 \rho_x, \tag{3.23}$$

and

$$c^2 = c_0^2 \left[\frac{\rho}{\rho_0}\right]^{\gamma-1}. \tag{3.24}$$

We may then rewrite Eqs. (3.22) as

$$\rho u_t + \rho u u_x + c^2 \rho_x = 0$$
$$\rho_t + \rho u_x + u \rho_x = 0 \tag{3.25}$$
$$c^2 = c_0^2 \left[\frac{\rho}{\rho_0}\right]^{\gamma-1}$$

3.2 AN EXAMPLE OF THE QUASI-LINEAR THEORY

Finally we complete our preparation for the theory of Section 3.2 by a dimensionless formulation obtained by setting

$$u' = u/c_0, \quad c' = c/c_0, \quad \rho' = \rho/\rho_0, \quad x' = x/L, \quad t' = t/L/c_0$$

and in terms of these dimensionless variables (we have dropped the primes already) the mathematical model becomes

$$\begin{aligned} \rho u_t + \rho u u_x + c^2 \rho_x &= 0 \\ \rho_t + u\rho_x + \rho u_x &= 0 \\ c &= \rho^{(\gamma-1)/2}. \end{aligned} \tag{3.26}$$

In order to use the notation of Section 3.2 we set $u = u^1$, $\rho = u^2$, and $t = y$ so that Eqs. (3.26) (we drop the third equation but keep it in mind) become

$$\begin{aligned} u^2 u_x{}^1 + u^1 u_x{}^2 + u_y{}^2 &= 0 \\ u^1 u^2 u_x{}^1 + c^2 u_x{}^2 + u^2 u_y{}^1 &= 0. \end{aligned} \tag{3.27}$$

In matrix form we rewrite these as

$$\begin{bmatrix} u^2 & u^1 \\ u^1 u^2 & c^2 \end{bmatrix} \begin{bmatrix} u_x{}^1 \\ u_x{}^2 \end{bmatrix} + \begin{bmatrix} 0 & 1 \\ u^2 & 0 \end{bmatrix} \begin{bmatrix} u_y{}^1 \\ u_y{}^2 \end{bmatrix} + \begin{bmatrix} 0 \\ 0 \end{bmatrix} = 0 \tag{3.28}$$

so that the matrices A, B, d are

$$A = \begin{bmatrix} u^2 & u^1 \\ u^1 u^2 & c^2 \end{bmatrix}, \quad B = \begin{bmatrix} 0 & 1 \\ u^2 & 0 \end{bmatrix}, \quad d = \begin{bmatrix} 0 \\ 0 \end{bmatrix}.$$

To compute the e_i we form the det $(A - \lambda B) = 0$ and obtain

$$\begin{vmatrix} u^2 & u^1 - \lambda \\ u^1 u^2 - \lambda u^2 & c^2 \end{vmatrix} = 0$$

or expanding and dividing by u^2 ($\neq 0$) we obtain

$$\lambda^2 - 2u^1 \lambda + (u^1)^2 - c^2 = 0. \tag{3.29}$$

The roots of this quadratic are

$$\begin{aligned} e_1 &= u^1 + c \\ e_2 &= u^1 - c, \end{aligned} \tag{3.30}$$

and therefore the characteristics are given by

$$\left.\frac{dx}{dy}\right]_1 = u^1 + c$$
$$\left.\frac{dx}{dy}\right]_2 = u^1 - c.$$
(3.31)

From these results the normal form is easily obtained. First

$$A - e_1 B = \begin{bmatrix} u^2 & -c \\ -cu^2 & c^2 \end{bmatrix}$$

$$A - e_2 B = \begin{bmatrix} u^2 & c \\ cu^2 & c^2 \end{bmatrix}.$$

Second, from $TA = ETB$ we have

$$\begin{bmatrix} t_{11} & t_{12} \\ t_{21} & t_{22} \end{bmatrix} \begin{bmatrix} u^2 & u^1 \\ u^1 u^2 & c^2 \end{bmatrix} = \begin{bmatrix} u^1 + c & 0 \\ 0 & u^1 - c \end{bmatrix} \begin{bmatrix} t_{11} & t_{12} \\ t_{21} & t_{22} \end{bmatrix} \begin{bmatrix} 0 & 1 \\ u^2 & 0 \end{bmatrix}$$

which upon multiplication and equating corresponding components becomes the homogeneous set of equations for the t_{ji}:

$$\begin{aligned} u^2 t_{11} + u^1 u^2 t_{12} &= u^2(u^1 + c) t_{12} \\ u^1 t_{11} + (c)^2 t_{12} &= (u^1 + c) t_{11} \\ u^2 t_{21} + u^1 u^2 t_{22} &= u^2(u^1 - c) t_{22} \\ u^1 t_{21} + (c)^2 t_{22} &= (u^1 - c) t_{21}. \end{aligned}$$
(3.32)

The first two of these equations are

$$u^2 t_{11} - cu^2 t_{12} = 0$$

and

$$-c t_{11} + (c)^2 t_{12} = 0$$

which are clearly identical. Thus one of the coefficients, say t_{11}, may be chosen arbitrarily, thereby fixing the other. If we choose $t_{11} = c$ then $t_{12} = 1$. In a similar fashion the second set of two lead to the result $t_{21} = -c$ and $t_{22} = 1$. Thus our (nonunique) choice of T is

$$T = \begin{bmatrix} c & 1 \\ -c & 1 \end{bmatrix}$$
(3.33)

3.2 AN EXAMPLE OF THE QUASI-LINEAR THEORY

so that the normal form $TAu_x + TBu_y = 0$ becomes, in our original notation,

$$\begin{bmatrix} (u+c)\rho & c(u+c) \\ (u-c)\rho & -c(u-c) \end{bmatrix} \begin{bmatrix} u_x \\ \rho_x \end{bmatrix} + \begin{bmatrix} \rho & c \\ \rho & -c \end{bmatrix} \begin{bmatrix} u_t \\ \rho_t \end{bmatrix} = 0$$

or

$$[(u+c)\rho]u_x + [c(u+c)]\rho_x + \rho u_t + c\rho_t = 0$$
$$[(u-c)\rho]u_x - [c(u-c)]\rho_x + \rho u_t - c\rho_t = 0. \quad (3.34)$$

For the purpose of developing the canonical form we rewrite Eqs. (3.34) as

$$u^2 e_1 u_x^1 + c e_1 u_x^2 + u^2 u_y^1 + c u_y^2 = 0$$
$$u^2 e_2 u_x^1 - c e_2 u_x^2 + u^2 u_y^1 - c u_y^2 = 0$$

or

$$u^2(e_1 u_x^1 + u_y^1) + c(e_1 u_x^2 + u_y^2) = 0$$
$$u^2(e_2 u_x^1 + u_y^1) - c(e_2 u_x^2 + u_y^2) = 0. \quad (3.35)$$

From Eqs. (3.35) it is clear that the canonical characteristic equations are

$$u^2 u_\alpha^1 + c u_\alpha^2 = 0$$
$$u^2 u_\beta^1 - c u_\beta^2 = 0$$

and

$$x_\alpha = (u^1 + c) y_\alpha$$
$$x_\beta = (u^1 - c) y_\beta.$$

Returning to our original variables the canonical system is

$$\rho u_\alpha + c \rho_\alpha = 0$$
$$\rho u_\beta - c \rho_\beta = 0$$
$$x_\alpha = (u+c) t_\alpha$$
$$x_\beta = (u-c) t_\beta. \quad (3.36)$$

We now proceed to introduce the *Riemann invariants* discussed previously (see Section 2.1d) in a different context. We rewrite the first two of Eqs. (3.36) as

$$u_\alpha = -\frac{c}{\rho} \rho_\alpha$$

$$u_\beta = \frac{c}{\rho} \rho_\beta$$

which become upon integration

$$u + l(\rho) = 2r(\beta)$$
$$u - l(\rho) = -2s(\alpha) \tag{3.37}$$

where $l(\rho) = \int_0^\rho c \, d\rho/\rho$. For polytropic gases we had $c = \rho^{(\gamma-1)/2}$ and when this is inserted into the equation for $l(\rho)$ and integrated we easily find

$$l(\rho) = \frac{2c}{\gamma - 1}. \tag{3.38}$$

Thus the Riemann invariants r and s become

$$r = \tfrac{1}{2}u + \frac{c}{\gamma - 1}, \qquad s = -\tfrac{1}{2}u + \frac{c}{\gamma - 1} \tag{3.39}$$

or

$$c = \tfrac{1}{2}(\gamma - 1)(r + s), \qquad u = r - s. \tag{3.40}$$

The following basic statements can be inferred from these results: We had $\alpha \leftrightarrow e_1$ family, $\beta \leftrightarrow e_2$ family. Since $r = r(\beta)$, $s = s(\alpha)$ we observe that

$$r = \frac{u}{2} + \frac{c}{\gamma - 1} \quad \text{is constant along} \quad \frac{dx}{dt} = e_1 = u + c$$
$$-s = \frac{u}{2} - \frac{c}{\gamma - 1} \quad \text{is constant along} \quad \frac{dx}{dt} = e_2 = u - c \tag{3.41}$$

and for this reason are called *invariants*. In the special case $\gamma = 3$ the characteristic velocities are $dx/dt = u + c = 2r$ and $u - c = -2s$, hence these velocities are constant along the characteristics. That is to say the characteristics in the (x, t) plane are straight lines when $\gamma = 3$.

The characteristics in the (u, ρ) plane are the fixed curves

$$u + \frac{1}{\gamma - 1} \rho^{(\gamma-1)/2} \quad \text{constant along the image of } e_1$$
$$u - \frac{1}{\gamma - 1} \rho^{(\gamma-1)/2} \quad \text{constant along the image of } e_2. \tag{3.42}$$

In the (u, c) plane the characteristics are the straight lines

$$\frac{u}{2} + \frac{c}{\gamma - 1} = \text{constant on } \Gamma_1$$
$$\frac{u}{2} - \frac{c}{\gamma - 1} = \text{constant on } \Gamma_2 \tag{3.43}$$

with $c \geqslant 0$.

3.2 AN EXAMPLE OF THE QUASI-LINEAR THEORY

These are shown in Fig. 3-3.

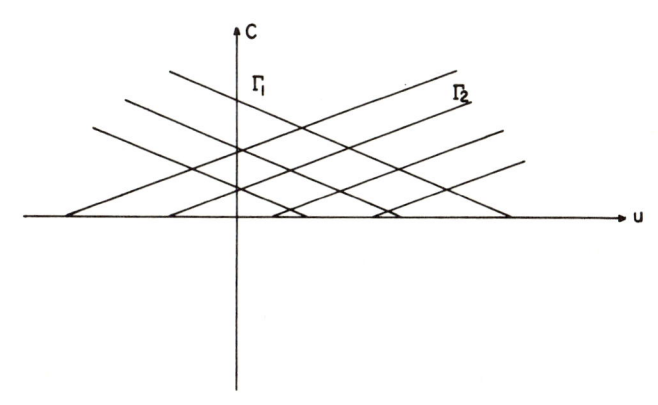

FIG. 3-3. The characteristic net in the velocity-sound speed (u, c) domain for the isentropic flow of a perfect gas.

From Eqs. (3.37) it is clear that $u + c$ and $u - c$ are known functions of r and s. Thus we may write the last two canonical equations as

$$x_\alpha = x_s s_\alpha = (u + c) t_s s_\alpha$$

or

$$x_s = (u + c) t_s \tag{3.44}$$

and similarly

$$x_r = (u - c) t_r . \tag{3.45}$$

These may be regarded as a system of two linear differential equations for x and t as functions of r and s.

Eliminating x between Eqs. (3.44) and (3.45), we obtain the second order linear partial differential equation for t:

$$2c t_{rs} + (u + c)_r t_s - (u - c)_s t_r = 0. \tag{3.46}$$

After t has been obtained x is easily found from either of the preceding equations. Finally we substitute $u = r - s$, $c = \tfrac{1}{2}(\gamma - 1)(r + s)$ and obtain

$$t_{rs} + \frac{\mu}{(r + s)} (t_r + t_s) = 0 \tag{3.47}$$

with $\mu = \tfrac{1}{2}((\gamma + 1)/(\gamma - 1))$. This equation was first obtained by Riemann but has more recently been called the *Poisson — Euler — Dar-*

boux (PED) equation. Because of this equation's importance we shall devote the next section to it and tabulate a few of its properties.

3.3 THE POISSON-EULER-DARBOUX EQUATION [6, 7, 8, 9]

For the special values

$$\gamma = \frac{2N+1}{2N-1}, \qquad N = 0, 1, 2, ..., \qquad (3.48)$$

i.e., $\gamma = -1, 3, 5/3, 7/5, 11/9, ..., \mu = N$, the PED equation may be integrated by means of elementary functions. The value $\gamma = 1.4 = 7/5$ for air occurs among these special values as does the value $\gamma = 11/9 \sim 1.2$ often used for the gases of combustion, CO and CO_2.

With $N = 0$, $\gamma = -1$ so $\mu = 0$ and the PED equation becomes the canonical form of the linear wave equation

$$t_{rs} = 0 \qquad (3.49)$$

so that integration yields the general solution

$$t = f(r) + g(s) \qquad (3.50)$$

in terms of arbitrary functions f and g.

For $N = 1$, $\gamma = 3$, $\mu = 1$ and the PED equation is

$$(r+s)t_{rs} + t_r + t_s = 0 \qquad (3.51)$$

or

$$[(r+s)t]_{rs} = 0. \qquad (3.52)$$

The general solution of Eq. (3.52) is

$$t = \frac{1}{r+s}[f(r) + g(s)] \qquad (3.53)$$

with arbitrary functions f and g as before. Noh and Protter [10] have observed the special character of the case $\gamma = 3$ as far as their functional approach is concerned. This was expected since the characteristics are straight lines.

These special cases suggest consideration of the expression

$$\phi = \frac{\partial^{N-1}}{\partial r^{N-1}}\left[\frac{f(r)}{(r+s)^N}\right] \qquad (3.54)$$

3.3 THE POISSON-EULER-DARBOUX EQUATION

enroute to the solution of the general case, when N is an integer. Upon differentiating we can readily show that

$$\phi_{rs} + \frac{N}{r+s}(\phi_r + \phi_s)$$
$$= -N\frac{\partial^N}{\partial r^N}\frac{f(r)}{(r+s)^{N+1}} + \frac{N}{r+s}\left\{\frac{\partial^N}{\partial r^N}\frac{f(r)}{(r+s)^N} - N\frac{\partial^{N-1}}{\partial r^{N-1}}\frac{f(r)}{(r+s)^{N+1}}\right\}. \quad (3.55)$$

Since

$$\frac{f(r)}{(r+s)^N} = \frac{f(r)}{(r+s)^{N+1}}(r+s)$$

and

$$\frac{\partial^N}{\partial r^N}\left[\frac{f(r)}{(r+s)^N}\right] = (r+s)\frac{\partial^N}{\partial r^N}\left[\frac{f(r)}{(r+s)^{N+1}}\right] + N\frac{\partial^{N-1}}{\partial r^{N-1}}\left[\frac{f(r)}{(r+s)^{N+1}}\right], \quad (3.56)$$

the right-hand side of Eq. (3.55) is zero and therefore Eq. (3.54) furnishes us with a solution of the PED equation. By interchanging r and s a similar solution can be developed. We therefore have

$$t(r, s) = k + \frac{\partial^{N-1}}{\partial r^{N-1}}\left[\frac{f(r)}{(r+s)^N}\right] + \frac{\partial^{N-1}}{\partial s^{N-1}}\left[\frac{g(s)}{(r+s)^N}\right] \quad (3.57)$$

with k constant, f and g arbitrary, and $N \geq 1$, as the general solution of the PED equation for integral N. With the proper choice of f, g, and k the initial conditions can be satisfied.

For the general values of the constant γ, N is not an integer and recourse has to be made to an alternate method such as that of Riemann-Volterra [see 1] or Copson [11]. Copson's procedure is an ingenious use of complex analysis but suffers from a lack of well tabulated functions. The alternate Riemann-Volterra approach, which uses hypergeometric functions, will be briefly discussed. The work of Courant [1] and Magnus [12] should be consulted for additional details.

Given the general linear partial differential equation

$$L(t) = t_{rs} + at_r + bt_s + ct = 0 \quad (3.58)$$

the *adjoint M of L* is defined by the equation

$$M(G) = G_{rs} - (aG)_r - (bG)_s + cG = 0. \quad (3.59)$$

The Green's function $G(r, s; \xi, \eta)$ satisfying Eq. (3.59) is customarily called the *Riemann function*. We observe that a, b, and c may be functions

86   3. EXACT METHODS OF SOLUTION

of r and s and $r = \xi$, $s = \eta$ are two lines along which G satisfies certain conditions as shown in Fig. 3-4.

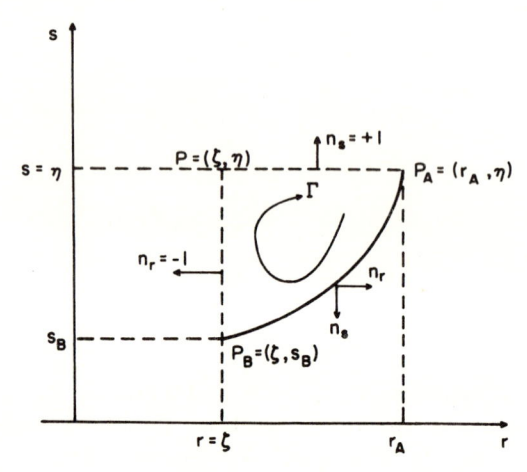

FIG. 3-4. Notation and integration path for Riemann-Volterra integration.

One easily verifies that

$$GL(t) - tM(G) = R_r + S_s \tag{3.60}$$

where

$$R = \tfrac{1}{2}(Gt_s - tG_s) + aGt \tag{3.61}$$

$$S = \tfrac{1}{2}(Gt_r - tG_r) + bGt. \tag{3.62}$$

We now suppose the functions t and G are such that Eqs. (3.58) and (3.59) hold in a region Ω bounded by a sufficiently smooth curve Γ (see Fig. 3-4). Applying Green's theorem to Eq. (3.60) we have

$$0 = \iint_\Omega (R_r + S_s)\, dr\, ds = \int_\Gamma (n_r R + n_s S)\, dl \tag{3.63}$$

where l is the arc length variable on Γ and n_r, n_s are unit outward normals. Using the Γ shown in Fig. 3-4 we can decompose Eq. (3.63) into its component parts and have

$$\int_{P_A}^{P_B} (n_r R + n_s S)\, dl - \int_{P_B}^{P} R\, ds + \int_{P}^{P_A} S\, dr = 0. \tag{3.64}$$

3.3 THE POISSON-EULER-DARBOUX EQUATION

The choice of certain auxiliary conditions for G on $r = \xi$ and $s = \eta$ are motivated by the following observation. The second integral of Eq. (3.64),

$$\int_{P_B}^{P} R \, ds = \int_{P_B}^{P} \{\tfrac{1}{2}(Gt_s - tG_s) + aGt\} \, ds, \qquad (3.65)$$

becomes integrable if $G_s - aG = 0$ on $r = \xi$, i.e., upon the line of integration of Eq. (3.65). Thus we shall ask for G to satisfy the conditions

$$
\begin{aligned}
G_s - aG &= 0 & \text{on} & \quad r = \xi \\
G_r - bG &= 0 & \text{on} & \quad s = \eta \\
G &= 1 & \text{at} & \quad P = (\xi; \eta).
\end{aligned}
\qquad (3.66)
$$

With such conditions on G we find that

$$
\begin{aligned}
\int_{P_B}^{P} R \, ds &= \int_{P_B}^{P} \{\tfrac{1}{2}(Gt_s - tG_s + G_s t)\} \, ds \\
&= \tfrac{1}{2} \int_{P_B}^{P} (Gt)_s \, ds = \tfrac{1}{2} Gt \Big]_{P_B}^{P} \\
&= \tfrac{1}{2} G(P)t(P) - \tfrac{1}{2} G(P_B)t(P_B)
\end{aligned}
\qquad (3.67)
$$

where $G(P)$ means the value of G at the point $P = (\xi, \eta)$.
Similarly,

$$\int_{P}^{P_A} S \, dr = \tfrac{1}{2} G(P_A)t(P_A) - \tfrac{1}{2} G(P)t(P) \qquad (3.68)$$

and thus since $G(P) = 1$ (an arbitrary choice)

$$t(P) = t(\xi, \eta) = \tfrac{1}{2}[G(P_A)t(P_A) + G(P_B)t(P_B)] + \int_{P_A}^{P_B} (n_r R + n_s S) \, dl. \qquad (3.69)$$

This classical result states that the value of t at any point P in the region bounded by $P_A P_B$, by the horizontal line $s = \eta$ and the vertical line $r = \xi$ can be expressed as an integral. We can perform the integration provided that t and the partial derivative t_r or t_s are given along $P_A P_B$.

So much for the general case, Eq. (3.58), which has had considerable utility in work by Ludford [9, 13], Ludford and Martin [7], Zabusky [14], and others.

Corresponding to the PED, Eq. (3.47) is the adjoint equation for the Riemann function

$$G_{rs} - N\left(\frac{G}{r+s}\right)_r - N\left(\frac{G}{r+s}\right)_s = 0 \qquad (3.70)$$

which is subject to

$$G_r - \frac{N}{r+s}G = 0 \quad \text{when} \quad s = \eta$$

$$G_s - \frac{N}{r+s}G = 0 \quad \text{when} \quad r = \xi$$

$$G = 1 \quad \text{when} \quad r = \xi, \quad s = \eta.$$

It can be shown [8] that

$$G(r,s;\xi,\eta) = \left[\frac{\xi+\eta}{r+s}\right]^N F(1-N, N, 1; \psi) \qquad (3.71)$$

$$\psi = -\frac{(r-\xi)(s-\eta)}{(r+s)(\xi+\eta)}.$$

In Eqs. (3.71) we have used the standard notation F for the hypergeometric function $F(\alpha, \beta, \gamma; z)$ defined as

$$F(\alpha, \beta, \gamma; z) = 1 + \sum_{j=1}^{\infty} \frac{\alpha(\alpha+1)(\alpha+2)\cdots(\alpha+j-1)\beta(\beta+1)\cdots(\beta+j-1)}{j!\,\gamma(\gamma+1)\cdots(\gamma+j-1)} z^j. \qquad (3.72)$$

The properties of $F(\alpha, \beta, \gamma; z)$ may be found, for example, in the Bateman Manuscript Project [15] and tables in Jahnke and Emde [16].

3.4 REMARKS ON THE PED EQUATION

We collect here a group of results concerning the PED equation and variations of the equation as developed by Darboux [6], Ludford [9], Ludford and Martin [7], and others. Setting

$$L_m(u) = u_{rs} - \frac{m}{r-s}(u_r - u_s) = 0 \qquad (3.73)$$

we have the following:

3.4 REMARKS ON THE PED EQUATION

I. For positive integral m the general solution (containing two arbitrary functions) of $L_m(u) = 0$ is

$$u = \frac{\partial^{2m-2}}{\partial r^{m-1} \partial s^{m-1}} \left[\frac{f(r) + g(s)}{r - s} \right] \tag{3.74}$$

where f and g are arbitrary functions of r and s, respectively.

II. If u is a solution of $L_m(u) = 0$, then $v = (r - s)^{2m-1} u$ is a solution of $L_{1-m}(v) = 0$ for all values of m.

III. For positive integral m the general solution of $L_{-m}(v) = 0$ is

$$v = (r - s)^{2m+1} \frac{\partial^{2m}}{\partial r^m \partial s^m} \left[\frac{f(r) + g(s)}{r - s} \right]. \tag{3.75}$$

IV. For positive integral m the general solution of the system

$$\begin{aligned} v_r &= \alpha (r - s)^{2m} u_r \\ v_s &= -\alpha (r - s)^{2m} u_s, \end{aligned} \tag{3.76}$$

with $\alpha = $ constant, is

$$\begin{aligned} u &= \frac{m}{\alpha} \frac{\partial^{2m-2}}{\partial r^{m-1} \partial s^{m-1}} \left[\frac{f'(r) + g'(s)}{r - s} \right] \\ v &= (r - s)^{2m+1} \frac{\partial^{2m}}{\partial r^m \partial s^m} \left[\frac{f(r) + g(s)}{r - s} \right]. \end{aligned} \tag{3.77}$$

V. The resolvent of $L_{-m}(u) = 0$ is the solution $v(r, s; \xi, \eta)$ of $L_m(v) = 0$ which satisfies the characteristic conditions $v = r - \xi$ on $s = \eta$ and $v = \eta - s$ on $r = \xi$. Once the resolvent of $L_{-m}(u) = 0$ has been found the overall solution is reduced to a quadrature.

For positive integral m the resolvent of $L_{-m}(u) = 0$ is given by

$$v = \frac{\partial^{2m-2}}{\partial r^{m-1} \partial s^{m-1}} \left[\frac{f(r) + g(s)}{r - s} \right]$$

with

$$f(r) = \frac{[(r - \xi)(r - \eta)]^m}{(m-1)! \, m!} \quad \text{and} \quad g(s) = \frac{[(s - \xi)(s - \eta)]^m}{(m-1)! \, m!}. \tag{3.78}$$

The resolvent of $L_m(u) = 0$ is given by

$$v = -\frac{\partial^{2m-2}}{\partial \xi^{m-1}\partial \eta^{m-1}}\left[\frac{\bar{f}+\bar{g}}{\xi - \eta}\right], \quad \bar{f} = \frac{[(\xi - r)(\xi - s)]^m}{(m-1)!m!}$$

$$\bar{g} = \frac{[(\eta - r)(\eta - s)]^m}{(m-1)!m!}.$$

(3.79)

3.5 ONE-DIMENSIONAL ANISENTROPIC FLOWS

The adiabatic flow behind a decaying shock, such as arises when an expansion wave overtakes a shock, is anisentropic; i.e., the specific entropy varies from one fluid particle to another. The unsteady, one-dimensional, anisentropic flow of a polytropic gas, neglecting the effects of viscosity, heat conduction, and radiation, is represented by the system of equations

$$\rho u_t + \rho u u_x + p_x = 0 \quad (3.80a)$$

$$\rho_t + (\rho u)_x = 0 \quad (3.80b)$$

$$S_t + u S_x = 0 \quad (3.80c)$$

$$S - S_0 = c_v \ln[p\rho^{-\gamma}] \quad (3.80d)$$

where ρ, p, u, and S are the density, pressure, velocity, and entropy, respectively, of the flow. c_v is the specific heat at constant volume, γ is the ratio of the specific heat, and $\tau = 1/\rho$ is the specific volume.

We first illustrate the introduction of an auxiliary function ϕ which allows a reduction of the system (3.80) to one equation. Upon multiplication of Eq. (3.80) by u and adding to Eq. (3.80a) we have the following replacement for Eq. (3.80a):

$$(\rho u)_t + (p + \rho u^2)_x = 0. \quad (3.81)$$

Replacing Eq. (3.80a) by Eq. (3.81) we now introduce the auxiliary function ϕ. To assure satisfaction of continuity (3.80b) set[†]

$$\rho = \phi_{xx}, \quad \rho u = -\phi_{xt}. \quad (3.82)$$

[†] The reader may like to verify the impossibility of using $\rho = \phi_x$, $\rho u = -\phi_t$. A sufficient number of degrees of freedom are available only at the second derivative level.

3.5 ONE-DIMENSIONAL ANISENTROPIC FLOWS

Then Eq. (3.81) becomes $-\phi_{xtt} + (p + \rho u^2)_x = 0$ which suggests setting

$$(p + \rho u^2) = \phi_{tt} \tag{3.83}$$

i.e., Eq. (3.81) is satisfied. We now have

$$u = -\phi_{xt}/\phi_{xx} \tag{3.84}$$

$$p = \phi_{tt} - \rho u^2 = (\phi_{tt}\phi_{xx} - \phi_{xt}^2)/\phi_{xx} \tag{3.85}$$

$$S = c_v \ln\left[\frac{\phi_{tt}\phi_{xx} - \phi_{xt}^2}{\phi_{xx}^{1+\gamma}}\right]. \tag{3.86}$$

There remains the energy equation (3.80c) which is conveniently written, because of Eq. (3.84), as

$$\phi_{xx} S_t - \phi_{xt} S_x = 0.$$

If we take $S = f(\phi_x)$ where f is an arbitrary function, having the necessary differentiability properties, then

$$\phi_{xx} f' \phi_{xt} - \phi_{xt} f' \phi_{xx} \equiv 0$$

and the energy equation is identically satisfied. Returning to Eq. (3.86) and taking exponentials, the equation for ϕ becomes

$$\phi_{xx}\phi_{tt} - \phi_{xt}^2 = F(\phi_x)\phi_{xx}^{1+\gamma} \tag{3.87}$$

where $F(\phi_x) = \exp[(1/c_v)S] = \exp[(1/c_v)f(\phi_x)]$. If the flow is isentropic, then $F(\phi_x)$ is a constant. Equations of this general form have been discussed in Chapter 2.

On occasion specific flows can be constructed from this equation directly. One case concerns a flow when

$$F(\phi_x) = k\phi_x^n \tag{3.88}$$

considered by Smith [17]. Let us attempt a solution of the form[†]

$$\phi(x, t) = x^m H(t) \tag{3.89}$$

for $m \neq 1$, m, k, and n constant. Upon substituting Eq. (3.89) into Eq.

[†] Really a separation of variables assumption. See also Section 3.10.

(3.87) with the given specific form of F, Eq. (3.88), we find that $H(t)$ satisfies

$$m(m-1)HH'' - m^2H'^2 = km^{n+\gamma+1}(m-1)^{1+\gamma}H^{n+\gamma+1} \qquad (3.90)$$

provided

$$n(m-1) = 2\gamma - (\gamma - 1)m. \qquad (3.91)$$

Equation (3.91) is merely the relationship between n and m which ensures that the degree of x is the same on both sides of Eq. (3.90). A first integral of Eq. (3.90) is

$$H'^2 H^{-2m/(m-1)} = \alpha(H^r + \beta) \qquad (3.92)$$

where

$$\alpha = 2km^{(m+\gamma)/(m-1)}(m-1)^{\gamma+1/\gamma-1}$$

$$r = (\gamma - 1)/(m - 1).$$

A second integral of Eq. (3.92) yields (with $\alpha > 0$)

$$t\sqrt{\alpha} = (1-m)\int (g^{1-\gamma} + \beta)^{-\frac{1}{2}}\,dg \qquad (3.93)$$

where $g = H^{-1/(m-1)}$ and the constant of integration, which yields only a constant displacement of the time scale, is ignored.

Explicit integration of Eq. (3.93) in terms of elementary functions will only be possible for certain values of γ. Fortunately one of these is $\gamma = 7/5$ the adiabatic index for dry air. Selecting this value of γ we find two cases.

Case 1. $\beta = 0$. The integral gives

$$g = \{(\gamma + 1)\sqrt{\alpha}\,t/2(1-m)\}^{2/(\gamma+1)} \qquad (3.94)$$

so that

$$\phi = x^m H(t) = x^m g^{1-m}$$

is readily found. From Eqs. (3.84) we can calculate

$$u = 2x/(\gamma + 1)t$$

and the particle trajectories ($\psi = $ constant, $\psi = \phi_x$) are

$$\psi = xt^{-2/(\gamma+1)}. \qquad (3.95)$$

3.5 ONE-DIMENSIONAL ANISENTROPIC FLOWS

Case 2. $\beta \leqslant 0$. The integral yields

$$3\beta^3 \sqrt{\alpha}\, t = (1 - m)(\gamma - 4\beta g^{2/5} + 3\beta^2 g^{4/5})(1 + \beta g^{2/5}). \qquad (3.96)$$

Thus the flow variables may be expressed in terms of x and H as

$$p = F(\phi_x)\phi_{xx}^\gamma = k(m-1)^\gamma x^m (mH)^{(m+\gamma)/(m-1)}$$
$$\rho = m(m-1)x^{m-2}H$$
$$S = c_v \ln[k(mx^{m-1}H)^n] \qquad (3.97)$$
$$u = xH'/(1-m)H = \sqrt{\alpha}\, xH^{1/(m-1)}(H^{2/5(m-1)} + \beta)^{\frac{1}{2}}/(1-m)$$

We now examine a particular case of flow given by Eqs. (3.97) and (3.96). For simplicity we take $\sqrt{\alpha} = 2/15$, $\beta = 1$, and $m = 3/5$. The flow is then given by

$$t = (8 - 4H + 3H^2)\sqrt{1 + H}$$
$$\psi = H/x^{2/5}$$
$$u = x(1 + H)^{1/2}/3H^3 \qquad (3.98)$$
$$p = -2x^{3/5}/225H^5$$
$$\rho = -\frac{6}{25}x^{-7/5}H.$$

Flow properties are obtained by examining this system. Some of these are:

(a) From the t equation, $H \geqslant -1$ so the flow is restricted to $t \geqslant 0$;

(b) For a realistic flow we require $p > 0$, $\rho > 0$. These are true only within

$$x > 0, \quad -1 \leqslant H < 0, \quad 0 \leqslant t < 8$$
$$x < 0, \quad H > 0, \quad t > 8.$$

(c) Stagnation occurs when $x = 0$ and $H = -1$ ($t = 0$) while fluid velocity $\to \infty$ as $H \to 0$ ($t \to 8$);

(d) Initially at $t = 0$ the fluid is at rest with nonuniform pressure and density distributions. The pressure density behavior is $p \to \infty$, $\rho \to 0$ as $x \to \infty$ and $p \to 0$, $\rho \to \infty$ as $x \to 0$.

94   3. EXACT METHODS OF SOLUTION

The method of construction of solutions illustrated by this example has been utilized by other researchers. We shall return to a further discussion later.

3.6 AN ALTERNATE APPROACH TO ANISENTROPIC FLOW

Martin and his co-workers [7, 18, 19] have taken an alternate approach to that discussed above. It is more general than that of Smith's with possibilities of generalization. They obtain a parametric representation for the flow in terms of the independent variables pressure p and particle trajectory $\psi = \phi_x$ in the following way.

The fundamental equations, as discussed in Section 3.5, are (3.80b) and (3.81). From these

$$\rho_t + (\rho u)_x = 0, \qquad (\rho u)_t + (p + \rho u^2)_x = 0$$

we can infer the existence[†] of two functions $\bar{\xi}$, ψ of x and t defined by

$$d\bar{\xi} = \rho u \, dx - (p + \rho u^2) \, dt \tag{3.83'}$$

$$d\psi = \rho \, dx - \rho u \, dt. \tag{3.84'}$$

Along a trajectory $x = x(t)$ of a gas particle in the (x, t) plane we have $dx/dt = u$ and therefore $d\psi \equiv 0$ or $\psi = $ constant along such a trajectory, from Eq. (3.84'). The curves $\psi(x, t) = $ constant are the trajectories of the gas particles, so ψ may be termed a trajectory function.

Close examination of Eq. (3.83') shows that it may be rewritten as

$$\begin{aligned} d\bar{\xi} &= u(\rho \, dx - \rho u \, dt) - p \, dt \\ &= u \, d\psi - p \, dt \\ &= u \, d\psi + t \, dp - d(tp) \end{aligned} \tag{3.85'}$$

or, finally setting $\xi = \bar{\xi} + pt$,

$$d\xi = u \, d\psi + t \, dp. \tag{3.86'}$$

The form of Eqs. (3.84') and (3.86') suggests that ψ and p can be taken as independent variables in place of x and t. This amounts to replacing x, t by the trajectories and isobars which form a more natural system of

---

[†] That is $\bar{\xi}_x = \rho u$, $\bar{\xi}_t = -(p + \rho u^2)$; $\psi_x = \rho$, $\psi_t = -\rho u$ which amounts to the introduction of two auxiliary functions, $\bar{\xi}$ and ψ.

3.6 AN ALTERNATE APPROACH TO ANISENTROPIC FLOW

curvilinear coordinates than the artificially imposed x and t. Of course we cannot do this if the isobars and trajectories coincide. Hence we exclude flows in which each gas particle retains a constant pressure.

The fundamental relations among the variables are easily discerned from the differentials. From Eq. (3.86') we have

$$u = \frac{\partial \xi}{\partial \psi} = \xi_\psi, \qquad t = \xi_p. \tag{3.87'}$$

From Eq. (3.84) ($\tau = 1/\rho$)

$$dx = \tau \, d\psi + u \, dt$$
$$= \tau \, d\psi + \xi_\psi(\xi_{p\psi} \, d\psi + \xi_{pp} \, dp)$$
$$= (\tau + \xi_\psi \xi_{p\psi}) \, d\psi + \xi_\psi \xi_{pp} \, dp \tag{3.88'}$$

and hence

$$\begin{aligned} x_\psi &= \xi_\psi \xi_{p\psi} + \tau \\ x_p &= \xi_\psi \xi_{pp}. \end{aligned} \tag{3.89'}$$

When x is eliminated by differentiation from the last two equations we find that ξ is a solution of the Monge-Ampère equation

$$\xi_{\psi\psi}\xi_{pp} - \xi_{\psi p}^2 = \tau_p. \tag{3.90'}$$

The function $\tau = \tau(\psi, p)$ is an arbitrary function representing the undeterminedness of the system which results from the absence of the specification of the entropy function. Once τ has been specified the procedure discussed above furnishes a solution for the unsteady flow in the parametric form

$$x = x(\psi, p) = \int \{(\tau + \xi_\psi \xi_{p\psi}) \, d\psi + \xi_\psi \xi_{pp} \, dp\} \tag{3.91'}$$
$$t = \xi_p(\psi, p), \qquad u = \xi_\psi(\psi, p)$$

representing the trajectory of the gas particle in the (x, t) plane in terms of the pressure and the velocity at each point of the trajectory. The density is calculated from $\tau = \tau(\psi, p)$ which was prescribed.

For a polytropic gas, we had in Eq. (3.80d) that τ assumes the form

$$\tau = p^{-1/\gamma} \exp[(S - S_0)/c_v] \qquad (S_0 \text{ constant}), \tag{3.92'}$$

where prescription of the entropy function $S = S(\psi)$ is required to specify the variation from trajectory to trajectory.

From the discussion of Section 2.5b the general solution of Eq. (3.90') results from the solutions of one or the other (or both) of the differential systems

$$d\xi - u\,d\psi - \tau\,dp = 0, \qquad d\xi - u\,d\psi - \tau\,dp = 0$$
$$du + \lambda\,dp = 0, \qquad\qquad du - \lambda\,dp = 0 \qquad (3.93')$$
$$dt - \lambda\,d\psi = 0, \qquad\qquad dt + \lambda\,d\psi = 0$$

(compare the first with Eq. (3.86') and the others with Eq. (2.234)) where $\lambda^2 = -\tau_p$, or $\lambda = \tau/c$, $c^2 = (\partial p/\partial \rho)_s$ is the local speed of sound. τ does not appear explicitly in Eqs. (3.93') and there are only two families of characteristics instead of the expected three. The effect of the third family (trajectories $\psi = $ constant) is felt in λ, which may be prescribed to vary discontinuously with ψ. The last two equations of (3.93') are simple in form and are well suited to computation.

Calculation of the Jacobian

$$J = \frac{\partial(x, t)}{\partial(\psi, p)} = \tau t_p \qquad (3.94')$$

so the mapping of the (ψ, p) plane on the (x, t) plane is locally one to one if the specific volume τ and t_p remain finite and nonzero along a trajectory.

Since $dx = \tau\,d\psi + u\,dt$ we have that along a trajectory ($\psi = $ constant) $dx/dt = u$. More generally a curve $p = p(\psi)$ in the (ψ, p) plane maps into a curve $x = x(t)$ in the (x, t) plane upon which

$$\frac{dx}{dt} = u + \tau \frac{d\psi}{dt} = u + \frac{\tau}{dt/d\psi}$$
$$= u + \frac{\tau}{t_\psi + t_p p'} \qquad (3.95')$$

where $p' = dp/d\psi$. Consequently, if some physical quantity q is constant along a curve $p = p(\psi)$ then this quantity is propagated at a rate, relative to the gas, of

$$r_q = \frac{\tau}{t_\psi + t_p p'}. \qquad (3.96')$$

3.6 AN ALTERNATE APPROACH TO ANISENTROPIC FLOW

We can select q as p, u, τ, and $p\tau$ and therefore the rates at which *constant* pressure, velocity, specific volume, and temperature are propagated with respect to the gas, are

$$r_p = \tau/t_\psi, \qquad r_u = -\frac{\tau t_\psi}{\tau_p}, \qquad r_\tau = \frac{\tau \tau_p}{\tau_p t_\psi - \tau_\psi t_p}$$

$$r_T = \frac{\tau(p\tau_p + \tau)}{p(\tau_p t_\psi - \tau_\psi t_p) + \tau t_\psi}.$$
(3.97')

The first relation of system (3.97') follows immediately from Eq. (3.96') since $p = $ constant, $dp \equiv 0$. We shall derive the second relation which is typical. If $u = $ constant, $u = \xi_\psi$, $du = \xi_{\psi\psi} d\psi + \xi_{\psi p} dp \equiv 0$ so

$$p' = \frac{dp}{d\psi} = -\frac{\xi_{\psi\psi}}{\xi_{\psi p}} = -\frac{\tau_p + t_\psi^2}{t_p t_\psi}$$
(3.98')

utilizing the Monge-Ampère equation for ξ and the relation $t = \xi_p$. From Eq. (3.98') and some algebra the relation for r_u follows. For a polytropic gas the absolute temperature $T = p\tau/R$ so the result r_T is obtained by setting $q = p\tau$ and then $d(p\tau) = 0$.

Lastly, from the relations $\lambda^2 = -\tau_p$, $\lambda = \tau/c$ we have

$$c^2 = -\tau^2/\tau_p = r_p r_u,$$
(3.99)

that is to say, the local speed of sound is the geometric mean of the rates at which pressure and velocity are propagated with respect to the gas. If the gas is assumed to be polytropic then the relation Eq. (3.92') is used to eliminate τ_p from r_T, r_p and r_τ are used to eliminate t_ψ and $(\tau_p t_\psi - \tau_\psi t_p)$ resulting in

$$\frac{\gamma - 1}{r_T} = \frac{\gamma}{r_p} - \frac{1}{r_\tau}.$$
(3.100)

Up to this point the results have not been specifically for anisentropic gases. Now, isentropic flows are characterized by $S = $ constant or $\tau = \tau(p)$ so that $\lambda = \lambda(p)$. The differential systems (3.93') possess the integrals

$$u + l(p) = 2r, \qquad u - l(p) = 2s$$
(3.101)

where r, s are constants (the Riemann invariants) and $l(p) = \int \lambda(p) dp$. The results are of the same form as those previously described but in view of the simplicity of the form we develop them again. From the relationships between t, u, and ξ we can write Eq. (3.101) as

$$\xi_\psi + l = 2r, \qquad \xi_\psi - l = 2s \tag{3.102}$$

which provide intermediate integrals of the Monge-Ampere equation (3.90′) for ξ. Integrating with respect to ψ the general solutions are

$$\xi = P(p) + (2r - l)\psi, \qquad \xi = P(p) + (2s + l)\psi \tag{3.103}$$

where P is an arbitrary function. Since these are solutions of Eq. (3.90′), we have immediately

$$\begin{aligned} x &= \int uP'' \, dp - (u - c)\lambda\psi, & t &= P' - \lambda\psi, & u &= 2r - l \\ x &= \int uP'' \, dp + (u + c)\lambda\psi, & t &= P' + \lambda\psi, & u &= 2s + l. \end{aligned} \tag{3.104}$$

Equations (3.104) can be rewritten as

$$\begin{aligned} x - (u - c)t &= \int uP'' - (u - c)P', & u &= 2r - l(p) \\ x - (u + c)t &= \int uP'' + (u + c)P', & u &= 2s + l(p). \end{aligned} \tag{3.105}$$

From these it is clear that along each isobar ($p = $ constant) u is constant, depending on one parameter. Thus the flow is a "*simple wave*" type, i.e., this defines a simple wave in isentropic flow.[†]

The preceding discussion on isentropic flow suggests the definition: *The flow given by Eqs. (3.91′) when ξ is a solution of an intermediate integral of Eq. (3.90′) will be called a simple wave.* We saw in Chapter 2 that the Monge-Ampère equation

$$\xi_{\psi\psi}\xi_{pp} - \xi_{\psi p}^2 + \lambda^2 = 0, \qquad \lambda = \Gamma(\psi)P(p) \tag{3.106}$$

has intermediate integrals only in seven cases. If we concern ourselves only with polytropic gases

$$\Gamma(\psi) = \sqrt{n} \exp[S - S_0/2c_v], \qquad P(p) = p^{-(n+1)/2}, \qquad n = 1/\gamma \tag{3.107}$$

[†] We discuss this concept in greater detail in Chapter 7.

3.6 AN ALTERNATE APPROACH TO ANISENTROPIC FLOW

then the following observations hold:

(1) if $\lambda = 0$, Eqs. (3.107) are not possible;
(2) if $\lambda = 1$, we have isentropic flow of a Kármán-Tsien gas ($\gamma = -1$);
(3) if $P(p) \neq 0$, isentropic flow of an arbitrary gas, already discussed, occurs;
(4) if $\lambda = p^{-2}$, corresponds to isentropic flow with $\gamma = 1/3$;
(5) if $\lambda = \psi^{-1}/p$, corresponds to anisentropic flow with $\gamma = 1$;
(6) if $\lambda = e^\psi e^p$, Eqs. (3.107) are not possible.

This leaves

$$\lambda = \psi^{m-1}/p^{m+1} \qquad (m \neq 0) \tag{3.108}$$

and its translations and reflections in the line $p = \psi$ as the only case of interest. The translations are

$$\lambda = (\psi - \psi_0)^{m-1}/(p - p_0)^{m+1} \tag{3.109}$$

and the reflections are

$$\lambda = (p - p_0)^{m-1}/(\psi - \psi_0)^{m+1} \tag{3.110}$$

with the second case (3.110) included in Eq. (3.109) by the proper choice of m. For polytropic gases we have $p_0 = 0$ so without loss of generality we may choose $\psi_0 = 0$. This returns us to Eq. (3.108) with the intermediate integral

$$u\psi + tp - \xi \pm \frac{1}{m}\left(\frac{\psi}{p}\right)^m = \text{constant.} \tag{3.111}$$

However, the dilatations where ψ is replaced by $Ak^{1/2m}\psi$ and p by $Ak^{-1/2m}p$ ($m \neq 0$) provide a slight generalization with intermediate integrals

$$u\psi + tp - \xi - \frac{k}{m}\left(\frac{\psi}{p}\right)^m = r, \qquad u\psi + tp - \xi + \frac{k}{m}\left(\frac{\psi}{p}\right)^m = s \tag{3.112}$$

for Eq. (3.106) with

$$\lambda = k\psi^{m-1}/p^{m+1} \qquad (m \neq 0). \tag{3.113}$$

Here r and s are constant, $m = \frac{1}{2}(\gamma^{-1} - 1)$,

$$(S - S_0) = c_v \ln\left\{\frac{k^2\psi^{2m-2}}{2m+1}\right\}; \tag{3.114}$$

and

$$\tau = \left(\frac{k^2}{2m+1}\right)\psi^{2m-2}/p^{2m+1}. \tag{3.115}$$

To summarize: For $\gamma \neq \frac{1}{3}, \pm 1$ the only anisentropic simple waves in a polytropic gas arise from the solutions of the intermediate integrals of the Monge-Ampère equation (3.106) with $m = \frac{1}{2}(\gamma^{-1} - 1)$.

If one continues with the isentropic flow and introduces the Riemann invariants as the new independent variables, previously derived results, obtained via the PED equation, are obtained. We omit these previously discussed details. On the other hand the procedure for anisentropic flow are of interest.

3.7 GENERAL SOLUTION FOR ANISENTROPIC FLOW

For anisentropic flow the characteristic system (3.93') becomes

$$\begin{aligned}
\xi_s - u\psi_s - tp_s &= 0, & \xi_r - u\psi_r - tp_r &= 0 \\
u_s + \lambda p_s &= 0, & u_r - \lambda p_r &= 0 \\
t_s - \lambda \psi_s &= 0, & t_r + \lambda \psi_r &= 0 \\
\lambda &= k\frac{\psi^{m-1}}{p^{m+1}},
\end{aligned} \tag{3.116}$$

i.e., the invariants r and s defined in the intermediate integrals (3.112) are new independent variables. Thus differentiation in each of the sets of Eq. (3.116) are in one direction only. To show how this is accomplished we examine $du + \lambda \, dp = 0$ of the first set. If $u = u(r, s)$ then $du = u_r \, dr + u_s \, ds$ but in the s direction (i.e., $r = $ constant) $du = u_s \, ds$ so that $du + \lambda \, dp = 0$ becomes $u_s \, ds + \lambda p_s \, ds = 0$.

From the intermediate integrals (3.112) we have by subtraction of the first from the second $(2k/m)(\psi/p)^m = s - r$ so that

$$\frac{\psi}{p} = \left[\frac{m}{2k}(s - r)\right]^{1/m} \tag{3.117}$$

Alternatively, by addition of the two intermediate integrals it follows that

$$u\psi + tp - \xi = \tfrac{1}{2}(r + s). \tag{3.118}$$

3.7 GENERAL SOLUTION FOR ANISENTROPIC FLOW

Thus we may write the last two pairs of Eqs. (3.116) in a more useful form. The details for one of these follows:

$$u_r = \lambda p_r = k \frac{\psi^{m-1}}{p^{m-1}} \frac{p_r}{p^2} = k \left[\frac{m}{2k}(s-r) \right]^{1-(1/m)} \pi_r$$

$$\left. \begin{array}{l} u_r = \beta[r-s]^{2M}\pi_r \\ u_s = -\beta[r-s]^{2M}\pi_s \end{array} \right\} \qquad (3.119)$$

where

$$\pi = p^{-1}, \qquad \beta = -k[2k(2M-1)]^{-2M}$$

and

$$\left. \begin{array}{l} w_r = \alpha(r-s)^{2N}t_r, \qquad w_s = -\alpha(r-s)^{2N}t_s \\ w = \psi^{-1}, \qquad \alpha = k^{-1}[2k(2N+1)]^{-2N} \end{array} \right\} \qquad (3.120)$$

where

$$M = \frac{1}{2}\left(1 - \frac{1}{m}\right) = \frac{3\gamma-1}{2(\gamma-1)}, \qquad N = M-1.$$

By partial differentiation we can show that π, u, t, and w are solutions of PED equations (same notation as Section 3.4)

$$L_M(\pi) = \pi_{rs} - \frac{M}{r-s}(\pi_r - \pi_s) = 0 \qquad (3.121a)$$

$$L_{-M}(u) = u_{rs} + \frac{M}{r-s}(u_r - u_s) = 0 \qquad (3.121b)$$

$$L_N(t) = t_{rs} - \frac{N}{r-s}(t_r - t_s) = 0 \qquad (3.121c)$$

$$L_{-N}(w) = w_{rs} + \frac{N}{r-s}(w_r - w_s) = 0. \qquad (3.121d)$$

As in the isentropic case the solution of an initial value problem reduces to quadrature, once the resolvent is known. For N a positive integer the resolvents of $L_N(t) = 0$ and $L_{-N}(w) = 0$ are given in Section 3.4.

Starting with a solution $t(r, s)$ of $L_N(t) = 0$ we obtain in succession

$$\psi^{-1} = w = w(r,s) = \alpha \int (r-s)^{2N}\{t_r\,dr - t_s\,ds\} \quad \text{from (3.120);} \quad (3.122a)$$

$$p^{-1} = \pi = \pi(r,s) = \psi^{-1}\left[\frac{m}{2k}(s-r)\right]^{1/m} = \left[\frac{2k(2M-1)}{r-s}\right]^{2M-1} w$$
$$\text{from Eq. (3.117);} \quad (3.122b)$$

$$u = u(r, s) = \beta \int (r - s)^{2M} \{\pi_r \, dr - \pi_s \, ds\} \quad \text{from Eq. (3.120)}; \quad (3.122\text{c})$$

$$\tau = \tau(r, s) = \frac{k^2 \pi^{2m+1}}{(2m+1)w^{2m-2}} \quad \text{from Eq. (3.115)}; \quad (3.122\text{d})$$

$$c = c(r, s) = \frac{k\pi^m}{(2m+1)w^{m-1}} \quad \text{from } \lambda = \tau/c \text{ and Eq. (3.106)}; \quad (3.122\text{e})$$

and finally

$$x = x(r, s) = \int \{(u - c)t_r \, dr + (u + c)t_s \, ds\}. \quad (3.122\text{f})$$

The transformation $x = x(r, s)$, $t(r, s)$ maps the characteristic (r, s) plane upon the (x, t) plane and carries the curves $\psi(r, s) = $ constant, $p(r, s) = $ constant, $u(r, s) = $ constant, $\tau(r, s) = $ constant, $c(r, s) = $ constant in the characteristic (r, s) plane into trajectories, isobars, isovels, isopycnics and isothermals, respectively, in the (x, t) plane. The coordinate curves $r = $ constant, $s = $ constant map into the mach curves.

For N positive and integral, corresponding to $\gamma = (2N + 1)/(2N - 1)$ (as in the isentropic case) the functions π, u; t, w; τ; and c (but not x) may be expressed explicitly in terms of two arbitrary functions $R(r)$ and $S(s)$.

From Eqs. (3.76), Section 3.4, the solutions of π and u involving two arbitrary functions $R(r)$ and $S(s)$ are

$$\pi = \frac{2(2M - 1)}{\alpha} \frac{\partial^{2M-2}}{\partial r^{M-1} \partial s^{M-1}} \left[\frac{R' + S'}{r - s}\right] \quad (3.123\text{a})$$

$$u = -\frac{(r - s)^{2M+1}}{2M(2M - 1)} \frac{\partial^{2M}}{\partial r^M \partial s^M} \left[\frac{R + S}{r - s}\right] \quad (3.123\text{b})$$

where $-R/2M(2M - 1)$ and $-S/2M(2M - 1)$ have been taken as the arbitrary functions and $\alpha/\beta = -4(2M - 1)^2$. Also using Section 3.4 we have

$$t = \frac{N}{\alpha} \frac{\partial^{2N-2}}{\partial r^{N-1} \partial s^{N-1}} \left[\frac{R'' + S''}{r - s}\right] \quad (3.123\text{c})$$

$$w = (r - s)^{2N+1} \frac{\partial^{2N}}{\partial r^N \partial s^N} \left[\frac{R' + S'}{r - s}\right] \quad (3.123\text{d})$$

and thus from Eqs. (3.122d) and (3.122e)

$$\tau = \frac{(r - s)^{4(N+1)}}{2\alpha(2N - 1)} \left[\frac{\partial^{2N}}{\partial r^N \partial s^N} \left(\frac{R' + S'}{r - s}\right)\right]^3 \quad (3.123\text{e})$$

$$c = \frac{(r - s)^{2N+2}}{2(2N - 1)} \frac{\partial^{2N}}{\partial r^N \partial s^N} \left(\frac{R' + S'}{r - s}\right). \quad (3.123\text{f})$$

3.8 VIBRATION OF A NONLINEAR STRING

A relation for x is most easily obtained by utilizing Eq. (3.122f) knowing first the functions $u(r, s)$, $c(r, s)$, and $t(r, s)$.

An example is easily calculated by selecting values of the parameters. Ludford and Martin [7] discuss one of these and show the flow characteristics. Since the example is somewhat artificial it will not be given here.

3.8 VIBRATION OF A NONLINEAR STRING

Zabusky [14] considered the nonlinear longitudinal oscillations of a string whose governing equation is

$$y_{tt} = [F(y_x)]^2 y_{xx}. \tag{3.124}$$

The reduction of this equation to a set of first order equations was carried out in Chapter 2, Section 2.1d, the results being given in Eqs. (2.60) as

$$r_t - F[B^{-1}\{\tfrac{1}{2}(r+s)\}] r_x = 0$$
$$s_t + F[B^{-1}\{\tfrac{1}{2}(r+s)\}] s_x = 0 \tag{3.125}$$

where r and s are the Riemann invariants

$$r = v + \int_{u_0}^{u} F(\eta)\, d\eta = v + B(u)$$

$$s = -v + \int_{u_0}^{u} F(\eta)\, d\eta = -v + B(u)$$

and $u = y_x$, $v = y_t$.

We now recognize from the form of Eqs. (3.125) that they can be linearized by the hodograh transformation, that is to say we interchange the variables r, s for x, t. This is done, as discussed in Chapter 2, by means of the transformation

$$r_x = jt_s, \qquad r_t = -jx_s, \qquad s_x = -jt_r, \qquad s_t = jx_r$$

where the Jacobian from the (x, t) plane to the (r, s) plane is given by

$$j = r_x s_t - s_x r_t$$

and $j \neq 0$ for a one-to-one transformation. The resulting linearized equations, in the hodograph plane, are

$$x_s + F[B^{-1}\{\tfrac{1}{2}(r+s)\}]t_s = 0$$
$$x_r - F[B^{-1}\{\tfrac{1}{2}(r+s)\}]t_r = 0. \qquad (3.126)$$

Let us be more specific now and set

$$F^2[y_x] = [1 + \epsilon y_x]^\alpha$$

or

$$F^2[u] = [1 + \epsilon u]^\alpha. \qquad (3.127)$$

Then, as can easily be computed,

$$\left.\begin{matrix}r\\s\end{matrix}\right\} = \pm v + [\tfrac{1}{2}\epsilon(\alpha+2)]^{-1}(1+\epsilon u)^{1+(\alpha/2)} \qquad (3.128)$$

and by adding these equations and taking the inverse we have

and
$$u = \epsilon^{-1}\{-1 + [\tfrac{1}{4}\epsilon(\alpha+2)(r+s)]^{2/(2+\alpha)}\}$$
$$v = \tfrac{1}{2}(r-s). \qquad (3.129)$$

Upon utilizing Eqs. (3.129) an explicit representation for $F(u)$ in terms of r and s is

$$F(u) = (1+\epsilon u)^{\alpha/2} = [\beta(r+s)]^{\alpha/(2+\alpha)} \qquad (3.130)$$

where

$$\beta = \tfrac{1}{4}\epsilon(\alpha+2).$$

Thus Eqs. (3.126) become

$$x_s + [\beta(r+s)]^{\alpha/(2+\alpha)}t_s = 0$$
$$x_r - [\beta(r+s)]^{\alpha/(2+\alpha)}t_r = 0 \qquad (3.131)$$

and when x is eliminated by differentiation the equation for t is the familiar PED equation

$$t_{rs} + \frac{n}{r+s}(t_r + t_s) = 0 \qquad (3.132)$$

3.8 VIBRATION OF A NONLINEAR STRING

where $n = \frac{1}{2}\alpha(2 + \alpha)^{-1}$. The equation for x is the PED equation

$$x_{rs} - \frac{n}{r+s}(x_r + x_s) = 0. \tag{3.133}$$

Unfortunately, the values of n, that concern us here, are not integers and therefore the solutions are not available in terms of elementary functions (see Section 3.3). However, in that section we developed the general Riemann-Volterra approach for the solution of the initial value problem, commencing with Eq. (3.58) and ending with Eq. (3.72). The longitudinal oscillation is therefore obtainable by way of this procedure.

Set $\alpha = 1$, so that $n = \frac{1}{6}$ and $r = v + (2/3\epsilon)(1 + \epsilon u)^{3/2}$, $s = -v + (2/3\epsilon)(1 + \epsilon u)^{3/2}$. As boundary conditions (on y) let us select

$$y(0, t) = y(1, t) = 0 \tag{3.134}$$

and as initial conditions

$$y(x, 0) = a \sin \pi x$$
$$y_t(x, 0) = 0. \tag{3.135}$$

Since the Riemann-Volterra method is simplest for initial value problems we *convert to an initial value problem*. To accomplish this introduce a new variable x', $-\infty < x' < \infty$ (note: our original range on dimensionless x was $0 \leqslant x \leqslant 1$) and require the solution to be periodic over the spatial interval $\Delta x' = 2$ and odd about the origin. These requirements amount to insisting that

$$y(-x', t) = -y(x', t) = 0$$
$$y(x' + 2, t) = y(x', t). \tag{3.136}$$

That the boundary conditions (3.134) are satisfied is seen as follows: The oddness condition of (3.136) requires that $y(0, t) = -y(0, t)$ and this can be true if and only if $y(0, t) = 0$. Further, the same odd character of y requires that $y(1, t) = -y(-1, t) = -y(1, t)$ where the final relation follows from the periodicity condition. Thus $y(1, t) = 0$ and the boundary conditions are satisfied. As we have seen this reduction to an initial value problem was accomplished by a certain periodic extension of our x domain. The initial conditions are now

$$y(x', 0) = a \sin \pi x' \quad \text{for} \quad -\infty < x' < \infty. \tag{3.137}$$
$$y_t(x', 0) = 0$$

106   3. EXACT METHODS OF SOLUTION

The transformed initial line ($t = 0$) in the (r, s) plane is obtained by noting that at $t = 0$, $y_t(x', 0) = v = 0$. Thus the initial condition line is the straight line $r = s$, as shown in Fig. 3-5. Along the initial line, since $y(x', 0) = a \sin \pi x'$, the mapping is given by

$$r|_{t=0} = \rho(x') = s|_{t=0} = \sigma(x') = \frac{2}{3\epsilon}(1 + \epsilon u|_{t=0})^{\frac{3}{2}}$$

$$= \frac{2}{3\epsilon}(1 + \epsilon y_x|_{t=0})^{\frac{3}{2}} = \frac{2}{3\epsilon}(1 + \epsilon a\pi \cos \pi x')^{\frac{3}{2}}$$

$$= \frac{2}{3\epsilon} f^{\frac{3}{2}}(x') \tag{3.138}$$

where $f(x') = (1 + \epsilon a\pi \cos \pi x')$ and the symbols ρ and σ have been used to designate the values of r and s at $t = 0$.

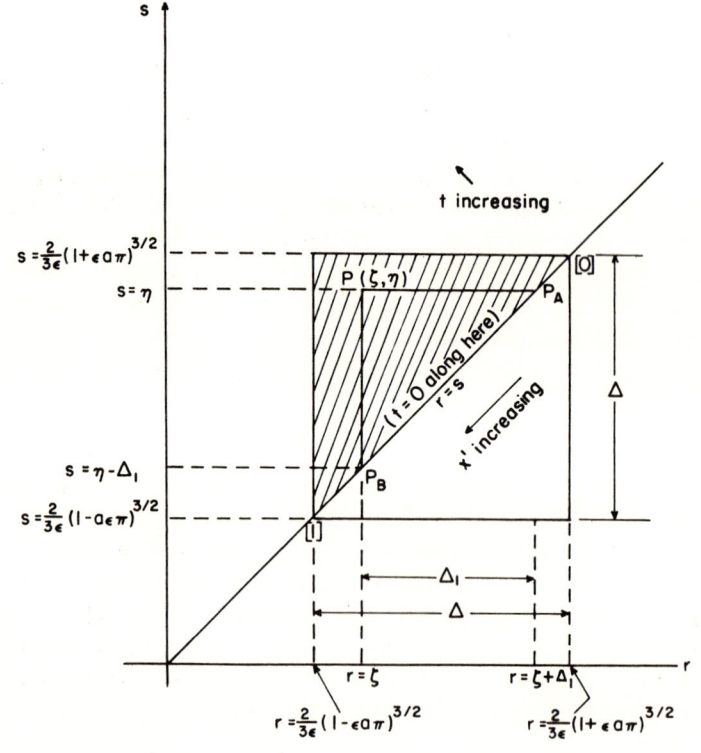

Fig. 3-5. Integration domain in Riemann invariant coordinates for a nonlinear vibration problem.

3.8 VIBRATION OF A NONLINEAR STRING

Referring to Fig. 3-5 note that $P(\xi, \eta)$, chosen in the region $t > 0$, describes an arbitrary point in the (r, s) plane at which the solution is to be evaluated. As r and s are unchanged along vertical and horizontal lines, respectively, then

$$\xi = \left(\frac{2}{3\epsilon}\right) f^{\frac{3}{2}}(x_A), \qquad \eta = \left(\frac{2}{3\epsilon}\right) f^{\frac{3}{2}}(x_B). \tag{3.139}$$

At $t = 0$, each point in $0 < x' < 1$, of the physical plane, is mapped uniquely into a point along the initial line of the (r, s) plane by the mapping (3.138). The solution at $P(\xi, \eta)$ is influenced by the initial conditions along $P_A P_B$. If at $t = 0$ we proceed outside the "main" interval $0 \leqslant x' \leqslant 1$, then r and s will oscillate over the initial line of the figure since $u = y_x$, at $t = 0$, is a periodic function. This suggests that the (r, s) plane is many sheeted.

Consider now any point in the shaded region of Fig. 3-5. We obtain the solution for x and t by Riemann's method. Along $P_A P_B$, $t = 0$, $dt = 0$, so that

$$r_{x'}|_{t=0} = \rho_{x'} = (x_\rho')^{-1} = s_x|_{t=0} = \sigma_{x'} = (x_\sigma')^{-1}. \tag{3.140}$$

From the Hodograph relations $r_x = jt_s$, $r_t = -jx_s$, $s_x = -jt_r$, $s_t = jx_r$, $j = r_x s_t - s_x r_t = J^{-1} = [x_r t_s - x_s t_r]^{-1}$ we see that relations (3.140) imply that

$$j = -2 r_x s_x [\beta(r + s)]^{\frac{1}{3}}. \tag{3.141}$$

Thus along $P_A P_B$ (i.e., $t = 0$)

$$t_r = t_\rho = -s_x|_{t=0} j^{-1} = \{2\rho_{x'}[\beta(\rho + \sigma)]^{\frac{1}{3}}\}^{-1} = -t_\sigma \tag{3.142}$$

where again we recall that ρ and σ are generic terms for r and s at $t = 0$. The solution for t follows immediately from Eq. (3.69)

$$t(\xi, \eta) = \tfrac{1}{2}[G(P_B)t(P_B) + G(P_A)t(P_A)] + \int_{P_A}^{P_B} (n_r R + n_s S) \, dl$$

by noting that $t(P_A) = t(P_B) = 0$, $R = \tfrac{1}{2}(Gt_\sigma - tG_s) + aGt = \tfrac{1}{2}Gt_\sigma$, $S = \tfrac{1}{2}Gt_\rho$, and with $n_r = (\tfrac{1}{2})^{\frac{1}{2}} = -n_s$

$$t(\xi, \eta) = -\sqrt{1/2} \int_{P_A}^{P_B} Gt_\rho \, dl. \tag{3.143}$$

The G of Eq. (3.143) is the Riemann function associated with the t

equation. In order to express the integral over x' it is easily seen from the right triangle of Fig. 3-5 that

$$dl = -\sqrt{2}\, d\rho = -\sqrt{2}\, \rho_{x'}\, dx' = -\sqrt{2}\, \sigma_{x'}\, dx' = \sqrt{2}\, a\pi^2 \sin \pi x' f^{\frac{1}{2}}(x')\, dx'.$$

In addition, t_ρ is available from Eq. (3.142) so

$$t(\xi, \eta) = \tfrac{1}{2} \int_{x_A}^{x_B} G f^{-\frac{1}{2}}(x')\, dx' \qquad (3.144)$$

where $x_A = x(P_A)$, $x_B = x(P_B)$.

In similar fashion the solution for x is expressible as

$$x(\xi, \eta) = \tfrac{1}{2}[H(P_A)x(P_A) + H(P_B)x(P_B)] + \tfrac{1}{2} \int_{x_A}^{x_B} (H_\sigma - H_\rho) x' \rho_{x'}\, dx' \qquad (3.145)$$

and H is the Riemann function associated with the x equation.

The only difference between the x and t equations is the sign of n. Thus we write $G = G^{(+)}$, $H = G^{(-)}$ where $G^{(\pm)}$ comes from Eq. (3.71) as

$$G^{(\pm)}(\xi, \eta; \rho, \sigma) = \left[\frac{\rho + \sigma}{\xi + \eta}\right]^{\pm n} F[1 \mp n, \pm n, 1, \psi]$$

or alternatively the form of Shaw [2]

$$G^{(\pm)}(\xi, \eta; \rho, \sigma) = \left[\frac{\rho + \sigma}{\xi + \eta}\right]^{\pm n} P_{(\pm n - 1)}(q) \qquad (3.146)$$

where $q = 1 + 2(\rho - \xi)(\sigma - \eta)/[(\xi + \eta)(\rho + \sigma)]$ and P_μ is the Legendre function of order μ.

Utilizing the above results we can write the solutions for t and x in the more explicit forms

$$t(\xi, \eta) = \frac{1}{2}\left[\frac{4}{3\epsilon(\xi + \eta)}\right]^{\frac{1}{6}} \int_{x_A}^{x_B} P_{-5/6}(q) f^{-\frac{1}{2}}(x')\, dx' \qquad (3.147)$$

$$x(\xi, \eta) = \frac{1}{2}\left[\frac{3\epsilon}{4}(\xi + \eta)\right]^{\frac{1}{6}} \{x_A f^{-\frac{1}{4}}(x_A) + x_B f^{-\frac{1}{4}}(x_B)\}$$
$$- a\left(\frac{3\pi\epsilon}{4}\right)^2 (\xi - \eta)\left[\left(\frac{3\epsilon}{4}\right)(\xi + \eta)\right]^{-5/6} \int_{x_A}^{x_B} \frac{dP_{1/6}}{dq} x' \sin \pi x' f^{-5/4}(x')\, dx'.$$

Various approximate and numerical schemes are used to evaluate these integrals as functions of ξ, η in the (r, s) plane thus obtaining t and x as functions of r and s or alternatively r and s as functions of t and x. u and v are then determined from r and s. Further details, many computational in nature, may be found in the referenced paper. The reader may well wish to examine a numerical or approximate procedure before adopting this approach for nonintegral n.

3.9 OTHER EXAMPLES OF THE QUASI-LINEAR THEORY

In addition to the examples discussed above the following applications of the quasi-linear theory are illustrative.

Supersonic flow theory to 1948 and applications are discussed, with detailed references, by Courant and Friedrichs [2]. Two dimensional nozzle flow has been examined by Frankl [20], Falkovich [21] and reviewed in von Mises [22]. More recently the quasi-linear theory has been utilized by Friedrichs [23], Pai [24], Weir [25] and Gunderson [26, 27] in magnetohydrodynamic flow. Three dimensional supersonic flows have been considered by Coburn [28] and Dolph and Coburn [29] using extensions of the quasi-linear theory.

A number of interesting mathematical properties of the equations of gas dynamics are contained in Bers [30]. Application of the quasi-linear theory to problems of plasticity is available in Hoffmann and Sachs [31].

We shall refer again to this theory when we discuss the numerical method of characteristics which arises from it.

3.10 DIRECT SEPARATION OF VARIABLES

Some nonlinear problems are solvable, under special conditions, by familiar "linear" methods but a word of caution must be interjected here. One of these familiar linear methods is that of separation of variables. In illustration we consider first a vibration problem due to Oplinger [32].

The Carrier equations (1.39) and (1.40) describe the transverse and longitudinal oscillation of stretched string. We consider small transverse oscillations only for which the governing equation is

$$\frac{\partial(T \sin \theta)}{\partial x} = m \frac{\partial^2 y}{\partial t^2} \qquad (3.148)$$

110   3. EXACT METHODS OF SOLUTION

where T, m, y, x, and t are tension, mass per unit length, transverse displacement, distance, and time, respectively.

The assumption of small displacement implies that

$$\text{(i)} \quad \frac{\partial y}{\partial x} = \tan\theta \approx \theta \approx \sin\theta; \tag{3.149a}$$

$$\text{(ii)} \quad T = T(t), \quad \text{i.e.,} \quad \frac{\partial T}{\partial x} = 0. \tag{3.149b}$$

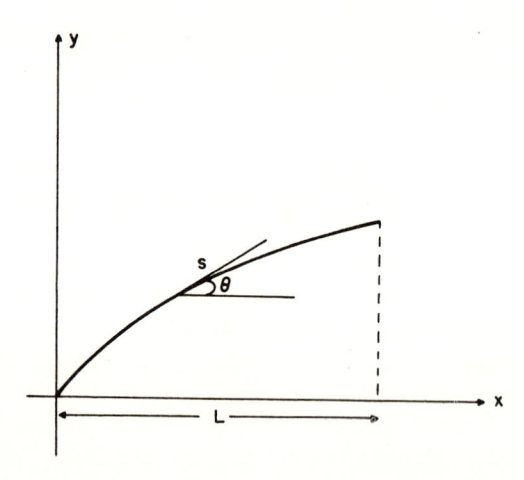

FIG. 3-6. Forced nonlinear vibration solved by direct separation of variables.

Referring to Fig. 3-6 we assume the tension relation

$$T - T_0 = E\epsilon = Ea\left(\frac{S-L}{L}\right) \tag{3.150}$$

where T_0, E, a, S, L are, respectively, tension when the string is on the axis, Young's modulus, cross-sectional area, arc length, and length. Relation (3.149b) implies that Eq. (3.148) may be written

$$\left(\frac{T}{m}\right)\frac{\partial^2 y}{\partial x^2} = \frac{\partial^2 y}{\partial t^2}. \tag{3.151}$$

A relation for T may be obtained by noting that

$$S = \int_0^L [1 + (y_x)^2]^{\frac{1}{2}}\, dx$$

3.10 DIRECT SEPARATION OF VARIABLES

but for small amplitudes

$$[1 + (y_x)^2]^{\frac{1}{2}} = 1 + \tfrac{1}{2}y_x^2 + O(y_x^4)$$

and thus

$$S \cong L + \tfrac{1}{2}\int_0^L y_x^2\, dx. \tag{3.152}$$

When we substitute Eq. (3.152) into Eq. (3.150) the expression for T is

$$T = T_0 + \frac{1}{2}\frac{Ea}{L}\int_0^L y_x^2\, dx \tag{3.153}$$

clearly demonstrating the dependence of T on t alone. Upon setting $c^2 = T_0/m$ Eq. (3.151) becomes

$$y_{tt} - c^2\left(1 + \frac{Ea}{2LT_0}\int_0^L y_x^2\, dx\right)y_{xx} = 0 \tag{3.154}$$

where the nonlinearity might be called "weak" because of its form.

The actual physical problem is one in which the end of the string (at $x = L$) is subjected to a periodic oscillation. Before we specify the form of this boundary condition examination of the form of the solutions obtainable by separation techniques is required. Since superposition is not possible the solutions obtainable by separation, or by any other procedure, are not amenable to all forms of boundary conditions.[†]

Upon trying to find solutions of the form $y = F(x)G(t)$ it is evident that

$$\int_0^L y_x^2\, dx = G^2(t)\int_0^L \left(\frac{dF}{dx}\right)^2 dx = IG^2(t)$$

and the separation equations for Eq. (3.154) become

$$F'' + \nu^2 F = 0$$

and
$$G'' + \nu^2 c^2\left[1 + \frac{Ea}{2LT_0}IG^2\right]G = 0 \tag{3.155}$$

where ν^2 is a "separation" constant.

[†] This procedure has been facetiously described as the "find a solution and then look for a problem technique." Some also dress it up with the name "inverse method."

3. EXACT METHODS OF SOLUTION

The solutions of these equations are

$$F(x) = A \sin \nu x + B \cos \nu x$$
$$G(t) = D \operatorname{cn}(ft, K)$$

with $\operatorname{cn}(ft, K)$ a Jacobian periodic elliptic function and f, K are combinations of the physical constants.

Possible boundary conditions are severely restricted by this form. Since the string is held fixed at $x = 0$ we have $y(0, t) = F(0)G(t) = 0$, so $F(0) = 0$ which requires that $B \equiv 0$. The solution now has the form

$$y(x, t) = \sin \nu x [D \operatorname{cn}(ft, K)].$$

With a driven right-hand end of periodicity t' and maximum amplitude α we ask that

$$y(L, t) = F(L)G(t) = \alpha G(t)$$

and $G(t + t') = G(t)$. These requirements lead to the solution

$$y(x, t) = \alpha \frac{\sin \nu x}{\sin \nu L} \operatorname{cn}(ft, K) \tag{3.156}$$

where

$$f = 1/t', \qquad K = \int_0^{\pi/2} [1 - k^2 \sin^2 \psi]^{-\frac{1}{2}} d\psi, \quad \text{and} \quad k^2 = \frac{b}{2(d+b)},$$

$$b = \frac{c^2 \nu^2 E a \alpha^2}{2T_0} \left\{ \frac{\nu^2 L + (\nu \sin 2\nu L)/2}{2 \sin^2(\nu L)} \right\} \quad \text{and} \quad d = c^2 \nu^2.$$

A second example will further demonstrate the restrictions created by the *ad hoc assumption that the solution be separable*.

Tomotika and Tamada [33] studied certain idealized two-dimensional transonic flows of compressible fluids by means of separation and other assumed forms of solution (these will be examined in a later section). Later Sedov [34], Lidov [35], and Keller [36] have used the separation technique to obtain some exact solutions of gas flows. We examine Keller's problem in some detail. His approach like Ludford and Martin [7] is to investigate the flow differential equations without regard to initial or boundary conditions and in this way obtain a class of solutions depending upon an arbitrary function. The arbitrary function is adjusted to satisfy particular initial or boundary conditions.

Consider the spherical, cylindrical, or one-dimensional gas flows which are inviscid, nonheat-conducting and have a polytropic equation

3.10 DIRECT SEPARATION OF VARIABLES

of state. Such flows are most often investigated in terms of the Eulerian equations involving a radius r and time t. In this case we do not use the Eulerian form but a Lagrangian formulation. The equations in Lagrangian form describe the motion in terms of the paths of the individual particles of the gas, i.e., the coordinates x, y, z of the particles as functions of time t and three parameters a, b, c which characterize the individual particles; a, b, c are often chose as the coordinates of the particles at $t = 0$. Courant and Friedrichs [2] give the equations of fluid motion as:

Conservation of mass: $\quad (\rho \Delta) = 0$

Conservation of momentum: $\quad \rho x_{tt} + p_x = 0$

$$\rho y_{tt} + p_y = 0$$
$$\rho z_{tt} + p_z = 0 \quad (3.157)$$

Changes of state are adiabatic: $\quad \dot{S} = 0$

Equation of state: $\quad p = f(\rho, S).$

In this system $(\ )_t$ means time differentiation and $\Delta = \partial(x, y, z)/\partial(a, b, c)$ the Jacobian of $x(a, b, c, t)$, $y(a, b, c, t)$, and $z(a, b, c, t)$. Because of the appearance of Δ in Eqs. (3.157) the Lagrangian form is often cumbersome to use. This objection does not hold for motions symmetrical enough to be characterized by merely one space coordinate. In such cases, which concern us here, the Lagrange representation is often advantageous.

In one-dimensional flow (flow with spherical symmetry) the Lagrange concept requires us to attach a number h to each plane section (spherical shell) of particles normal to the x-axis and thus the changing position of each section is describable by a function $x(h, t)$. The quantities p, ρ, S are then functions of h and t. The quantity h may be chosen in many arbitrary ways. A natural choice, based upon the law of conservation of mass is

$$h = \int_{x(0,t)}^{x(h,t)} \rho \, dx. \quad (3.158)$$

In explanation of this implicit choice we think of the flow taking place, in a tube of unit cross section, along the x-axis. Attach the reference value $h = 0$ to any section, moving with the flow, and for any other section let h be defined as the magnitude of the mass of the medium in the tube between that section and the zero section. The sign of h is positive or negative according as the zero section is to the left or the right of the section considered. The quantity $\rho = \rho(x, t)$ is the density regarded from the Eulerian view as a function of x and t.

In order to obtain the differential equation for x as a function of h and t we need to convert the Lagrange form to the h, t system of coordinates. From the Leibniz rule[†] we find by differentiating Eq. (3.158) with respect to h that

$$1 = \rho[x(h, t), t]x_h(h, t) = \rho(h, t)x_h(h, t) \tag{3.159a}$$

or

$$x_h(h, t) = \tau(h, t) \tag{3.159b}$$

where τ is the specific volume. The Lagrange form (Eqs. (3.157)) for one-dimensional flow take the form (recall $a = h$)

$$(\rho x_h)_t = 0 \tag{3.160a}$$

$$\rho x_{tt} = -p_x = -p_h/x_h \tag{3.160b}$$

$$S_t = 0 \tag{3.160c}$$

$$p = f(\rho, S) = g(\tau, S). \tag{3.160d}$$

It is clear from Eq. (3.159a) that Eq. (3.160a) is superfluous and that Eq. (3.160b) may be written as

$$x_{tt} = -p_h. \tag{3.161}$$

From Eq. (3.160c) it is apparent that $S = S(h)$ is independent of time t.

The preparations for the elimination of ρ, τ, and p are now completed. Starting from Eq. (3.161),

$$-x_{tt} = p_h = g_\tau \tau_h + g_S S_h$$

$$= g_\tau x_{hh} + g_S S_h$$

or

$$x_{tt} = -g_\tau x_{hh} - g_S S_h \tag{3.162}$$

is the equation for x as a function of h and τ. If the flow is isentropic, $S(h) = S_0$ (constant), so Eq. (3.162) becomes[‡]

$$x_{tt} = -g_\tau x_{hh}. \tag{3.163}$$

[†] To recall the Leibniz rule:

$$\frac{\partial}{\partial t} \int_{a(t)}^{b(t)} f(t, x)\, dx = \int_{a(t)}^{b(t)} \frac{\partial f}{\partial t}\, dx + f[t, b(t)]\frac{db}{dt} - f[t, a(t)]\frac{da}{dt}$$

[‡] The quantity $\sqrt{-g_\tau(\tau, S)} = k = \rho c$ is often called the *acoustic impedance* of the medium.

3.10 DIRECT SEPARATION OF VARIABLES

The cases of spherical or cylindrical flow require a generalization of Eq. (3.158) to

$$h = \int_{(y_0, t)}^{y(h,t)} r^{n-1} \rho(r, t) \, dr \tag{3.164}$$

where $y(h, t)$ represents the radius at time t of the particle with the Lagrangian coordinate h. Here the symbol n is the spatial dimension. Clearly Eq. (3.164) reduces to Eq. (3.158) when $n = 1$. As before we assume that all flow variables depend upon y, t only and flow occurs only in the y-direction.

The particle velocity $u = y_t$ and by differentiating Eq. (3.164) with respect to h we find (analogous to (3.159)) that

$$\tau(h, t) = \rho^{-1} = y^{n-1} y_h. \tag{3.165}$$

The Lagrange form for spherical flow ($n = 3$) and cylindrical flow ($n = 2$) are the same as Eqs. (3.160) with y replacing x. Thus we have again

$$y_{tt} = -p_h/\rho y_h = -y^{n-1} p_h \tag{3.166}$$

and eliminating p, since $S = S(h)$ only, the equation for y is

$$y_{tt} = -y^{n-1}[g_\tau(y^{n-1} y_h)_h + g_S S_h]. \tag{3.167}$$

If the gas is polytropic

$$g(\tau, S) = g_0 + A(S)\tau^{-\gamma}$$

where $A(S)$, γ, and g_0 are assumed known. Equation (3.167) then takes the form

$$y_{tt} = \gamma A(S)(y^{n-1} y_h)^{-\gamma-1}(y^{n-1} y_h)_h y^{n-1} - A_h(y^{n-1} y_h)^{-\gamma} y^{n-1}. \tag{3.168}$$

We find solutions of Eq. (3.168) by separation of variables. One this is completed the flow variables u, ρ, S, and p may be calculated from $u = y_t$, Eq. (3.165), $S = S(h)$ (assumed known), and $p = g(\tau, S)$. Upon seeking product solutions of the form

$$y(h, t) = f(h)j(t) \tag{3.169}$$

and introducing a separation constant λ the separation equations for Eq. (3.168) are

$$j'' - \lambda j^{n(1-\gamma)-1} = 0 \tag{3.170}$$

$$-A[(f^{n-1} f')^{-\gamma}]' f^{n-2} - A'(f^{n-1} f')^{-\gamma} f^{n-2} = \lambda. \tag{3.171}$$

3. EXACT METHODS OF SOLUTION

To solve Eq. (3.170) multiply by j' and integrate; the result is

$$(j')^2 = \frac{2\lambda}{n(1-\gamma)} j^{n(1-\gamma)} + a, \qquad \gamma \neq 1$$
$$(j')^2 = 2\lambda \log j + a, \qquad \gamma = 1 \tag{3.172}$$

with an integration constant a. Providing $j \neq$ constant another integration gives the explicit representation (for t)

$$\int_{j_0}^{j} \left[\frac{2\lambda}{n(1-\gamma)} j^{n(1-\gamma)} + a \right]^{-\frac{1}{2}} dj = t, \qquad \gamma \neq 1$$
$$\int_{j_0}^{j} [2\lambda \log j + a]^{-\frac{1}{2}} dj = t, \qquad \gamma = 1. \tag{3.173}$$

The solution of Eq. (3.171) is more difficult. To obtain it we differentiate and obtain

$$\gamma A[f''f^{n-1} + (n-1)f^{n-2}(f')^2](f^{n-1}f')^{-\gamma-1}f^{n-2} - f^{n-2}(f^{n-1}f')^{-\gamma}A' = \lambda. \tag{3.174}$$

At this point rather than working with $f = f(h)$ it is more convenient to introduce the inverse function $h = h(f)$ and denote $h'(f)$ by $q(f)$. Recalling that $f' = 1/h' = q^{-1}$ and $f'' = -q'/(q)^3$ we find that Eq. (3.174) becomes

$$\gamma A[-q'q^{-3}f^{n-1} + (n-1)f^{n-2}q^{-2}][f^{n-1}q^{-1}]^{-\gamma-1}f^{n-2}$$
$$-f^{n-2}[f^{n-1}q^{-1}]^{-\gamma}A'(h) = \lambda. \tag{3.175}$$

For $\gamma \neq 1$, two auxiliary function $z(f)$, $B(f)$ defined by $z = q^{\gamma-1}$, $B(f) = A(h(f))$ are introduced into Eq. (3.175). With these definitions, and with primes denoting differentiations with respect to f, Eq. (3.175) becomes

$$z' + z\left[-(n-1)(\gamma-1)f^{-1} + \frac{\gamma-1}{\gamma} (\log B)' \right]$$
$$+ \frac{\lambda(\gamma-1)}{\gamma B} f^{(n-1)(\gamma-1)+1} = 0. \tag{3.176}$$

Setting $G =$ constant, the solution of Eq. (3.176) is obtained as that of an elementary first order equation

$$z = f^{(n-1)(\gamma-1)} B^{(1-\gamma)/\gamma} \left[G - \frac{\lambda(\gamma-1)}{\gamma} \int f B^{-1/\gamma} df \right]. \tag{3.177}$$

Thus we have $q = z^{1/(\gamma-1)}$ and since $h' = q$

$$h = \int_{f_0}^{f} f^{n-1}B^{-1/\gamma}\left[G - \frac{\lambda(\gamma-1)}{\gamma}\int fB^{-1/\gamma}\,df\right]^{1/(\gamma-1)} df$$

yielding $f = f(h)$ implicitly. If we define $F(f)$ by means of

$$F(f) = \left[G - \frac{\lambda(\gamma-1)}{\gamma}\int fB^{-1/\gamma}\,df\right]^{1/(\gamma-1)} \tag{3.178}$$

and $\lambda \neq 0$, this may be solved for $B(f)$ and yields

$$B(f) = (-\lambda f)^{\gamma}(F')^{-\gamma}F. \tag{3.179}$$

The solution for the flow variables can now be easily established. For $\gamma \neq 1$, $\lambda \neq 0$,

$$u(y, t) = y_t = f(h)j_t = yj'j^{-1} \tag{3.180}$$

$$\tau(y, t) = -\lambda y j^{n-1}/F'(yj^{-1}) \tag{3.181}$$

$$p(y, t) = g_0 + j^{-n\gamma}F(yj^{-1}) \tag{3.182}$$

are the flow quantities; in fact these expressions yield a solution of the Eulerian equations of motion for an arbitrary function F provided that $j(t)$ is given by Eq. (3.173).

3.11 OTHER SOLUTIONS OBTAINED BY AD HOC ASSUMPTIONS

Alternate types of gas motion can be described by choosing solutions of other forms. Solutions of the form $y = H(\alpha t + \beta h)$ with α, β constants may be found for Eq. (3.168). Solutions of the progressing wave type have been investigated by numerous authors. These can be obtained if y has the form

$$y = t^{\alpha}f[K(h)t]$$

where α is constant; f and K are functions to be determined.

In a completely different study Tomotika and Tamada [33] have studied the two-dimensional transonic flow of compressible fluids by *ad hoc* methods. This is done by introducing a hypothetical gas which approximates the real gas in the vicinity of the critical state when the fluid velocity becomes just equal to the velocity of sound.

The two-dimensional steady irrotational flow of a compressible fluid has been previously discussed in Chapter 2, Section 2.3c. In the discussion the flow was described by the differential equations

$$\phi_x = u = q \cos \theta, \qquad \phi_y = v = q \sin \theta$$
$$\psi_x = -\rho q \sin \theta, \qquad \psi_y = \rho q \cos \theta. \qquad (3.183)$$

Here the usual notations are used: (x, y) are the cartesian coordinates in in the plane of fluid motion; (u, v) are the velocity components, θ denotes the direction of the velocity vector, and ϕ, ψ are the potential and stream functions, respectively. The velocity magnitude q and density ρ are normalized to be unity at the critical velocity.

To develop the equation, we shall use as the basic model of the hypothetical gas, we consider x and y to be functions of ϕ and ψ. Such interchange of coordinates has been previously used with some success. Beginning with

$$\phi = \phi(x, y), \qquad \psi = \psi(x, y)$$

we differentiate both with respect to ϕ and solve for x_ϕ and y_ϕ. Then differentiate both with respect to ψ and solve for x_ψ and y_ψ. The results, upon substituting Eq. (3.183) are

$$x_\phi = q^{-1} \cos \theta, \qquad x_\psi = -(\rho q)^{-1} \sin \theta$$
$$y_\phi = q^{-1} \sin \theta, \qquad y_\psi = (\rho q)^{-1} \cos \theta. \qquad (3.184)$$

Upon eliminating both x and y the equations of motion in the (ϕ, ψ) plane become

$$\theta_\phi = \frac{\rho}{q} q_\psi \qquad (3.185)$$

$$\theta_\psi = \frac{q}{\rho} \left[\frac{1}{c^2} - \frac{1}{q^2} \right] q_\phi. \qquad (3.186)$$

In this deduction $d\rho/dq$ has been replaced, using Bernoulli's equation in differential form, by

$$d\rho/dq = -\rho q/c^2$$

with c^2 normalized to unity at the critical state.

The form of Eq. (3.185) suggests the introduction of a new dependent variable $w = w(\rho, q)$ as

$$w_\psi = \frac{\rho}{q} q_\psi = w_q q_\psi. \qquad (3.187)$$

3.11 OTHER SOLUTIONS OBTAINED BY AD HOC ASSUMPTIONS

The last form suggests that $w_q = \rho/q$ or

$$w = \int_1^q \frac{\rho}{q} \, dq \tag{3.188}$$

where we observe that $q = 1$ corresponds to $w = 0$. The basic equations (3.185) and (3.186) are now obviously transformed into

$$\begin{aligned} \theta_\phi &= w_\psi \\ \theta_\psi &= -Xw_\phi \end{aligned} \tag{3.189}$$

where

$$X = (q\rho^{-1})^2(q^{-2} - c^{-2}). \tag{3.190}$$

In order to complete the approximating equation we assume the gas is polytropic which assumption implies, after some algebra, that

$$\rho = \left[\frac{\gamma+1}{2} - \frac{\gamma-1}{2} q^2\right]^{1/(\gamma-1)}$$

$$c = \left[\frac{\gamma+1}{2} - \frac{\gamma-1}{2} q^2\right]^{\frac{1}{2}}$$

where $\gamma =$ is the ratio of specific heats. Utilizing these expressions it is clear from the transformation (3.188) that w is a function of q only and $X = X(w)$ from Eq. (3.190). Thus we can expand $X(w)$ in powers of w, about 0:

$$X(w) = X(0) + X'(w)\big|_{w=0} w + O(w^2).$$

But since $w = 0 \leftrightarrow q = 1$ ($c = 1$) we see that $X(0) = 0$ and

$$X(w) = \left(\frac{dX}{dq}\right)\left(\frac{dq}{dw}\right)_{q=1} w + O(w^2)$$

$$X(w) = -(\gamma + 1)w + O(w^2). \tag{3.191}$$

In the case of a nearly uniform transonic gas, the value of w is small. Neglecting the term $O(w^2)$ the basic equations for the "hypothetical gas" become

$$\left.\begin{aligned} \theta_\phi &= w_\psi \\ \theta_\psi &= k(w^2)_\phi \\ k &= \frac{\gamma+1}{2} . \end{aligned}\right\} \tag{3.192}$$

Eliminating θ yields the equation

$$(kw)_{\psi\psi} = [(kw)^2]_{\phi\phi} \tag{3.193}$$

which only approximates the nearly uniform transonic flow of the real gas obeying the adiabatic law. However, they may be taken as the exact fundamental equations of a hypothetical gas obeying an appropriate law of change of state. The relation which specifies the hypothetical gas is

$$X = -(\gamma + 1)w$$

which by virtue of Eqs. (3.190) and (3.188) takes the form

$$\left\{ \frac{q^2}{\rho} \left(\frac{1}{\rho q} \right)' \right\}' = (\gamma + 1)(\rho/q) \tag{3.194}$$

(differentiation with respect to q). Solving this subject to $\rho_{q=1} = 1$, $\rho'_{q=1} = -1$, numerically gives us the picture of $\rho(q)$ for the gas. Near the critical point the approximation to the real flow is excellent.

Exact solutions for Eq. (3.193) can be obtained by making the *ad hoc* assumptions for kw as follows:

$$\begin{aligned}
&\text{(a)} \quad kw = f(\phi + \lambda\psi) \\
&\text{(b)} \quad kw = \Omega(\phi) + \Gamma(\psi) \\
&\text{(c)} \quad kw = \Omega(\phi)\Gamma(\psi) \\
&\text{(d)} \quad kw = \Gamma_0(\psi) + \Gamma_1(\psi)\phi^2 \\
&\text{(e)} \quad kw = \Gamma(\phi + \psi^2) + 2\psi^2.
\end{aligned} \tag{3.195}$$

The details of these separate cases may be found in Tomotika and Tamada [33] and Tamada [37]. Here we examine case (e) briefly.

If the flow has the form (3.195) expressed as

$$kw = \Gamma(u) + 2\psi^2, \qquad u = \phi + \psi^2$$

then upon inserting this expression into Eq. (3.193) the ordinary differential equation for $\Gamma(u)$ becomes

$$\frac{d}{du}\left[\Gamma \frac{d\Gamma}{du}\right] - \frac{d\Gamma}{du} - 2 = 0. \tag{3.196}$$

One integration is immediate to give

$$\Gamma \frac{d\Gamma}{du} - \Gamma - 2u = 0 \tag{3.197}$$

where a constant of integration is included in u. If we set $d\Gamma/du = \lambda$ then Eq. (3.197) is expressible as

$$u = \tfrac{1}{2}(\lambda - 1)\Gamma. \tag{3.198}$$

But if $\Gamma' = \lambda$ then $d^2\Gamma/du^2 = \lambda\, d\lambda/d\Gamma$ so Eq. (3.196) becomes

$$\Gamma\lambda \frac{d\lambda}{d\Gamma} = -(\lambda - 2)(\lambda + 1)$$

which immediately integrates to give

$$\Gamma = 2^{\tfrac{3}{2}}a(\lambda - 2)^{-\tfrac{2}{3}}(\lambda + 1)^{-\tfrac{1}{3}} \tag{3.199}$$

where $2^{\tfrac{3}{2}}a$ is a constant of integration.

Thus Eq. (3.198) describes u as a function of λ and Γ, and Eq. (3.199) describes Γ as a function of λ. Elimination of λ between these two expressions gives

$$(\Gamma - 2u)^2(\Gamma + u) = 2a^3 \tag{3.200}$$

and finally from Eq. (3.192)

$$k\theta = 2\psi[\Gamma + 2\phi + \tfrac{2}{3}\psi^2]. \tag{3.201}$$

A physical realization of this flow may be obtained from the following observations. The symmetric character of the flow about the streamline $\psi = 0$ is evident from the choice of kw and the results following that choice. Thus we may replace two streamlines $\psi = \psi_0$ and $\psi = -\psi_0$ by two solid walls. This is a Laval nozzle having the line $\psi = 0$ as its axis. The transition from Taylor flow to Meyer flow is easily investigated by means of this solution [38, 39].

REFERENCES

1. Courant, R., and Hilbert, D., "Methods of Mathematical Physics," Vol. 2. Wiley (Interscience), New York, 1962.
2. Courant, R., and Friedrichs, K. O., "Supersonic Flow and Shock Waves." Wiley (Interscience), New York, 1948.
3. Courant, R., and Lax, P., *Commun. Pure Appl. Math.* **2**, 255 (1949).
4. Friedrichs, K. O., *Am. J. Math.* **70**, 555 (1948).
5. Gelfand, I. M., *Usp. Mat. Nauk.* **14**, No. 2, 87 (1959).
6. Darboux, G., "Leçons sur la théorie générale des surfaces," Part 2, p. 54. Gauthier-Villars, Paris, 1915.

7. Ludford, G. S. S., and Martin, M. H., *Commun. Pure Appl. Math.* **7**, 45 (1954).
8. Riemann, B., "Gesammelte mathematische Werke," (R. Dedekind and H. Weber, eds.), Vol. I, Chapter 8, p. 144 (para 2). Teubner, Leipzig, 1876.
9. Ludford, G. S. S., *Proc. Cambridge Phil. Soc.* **48**, 499 (1952).
10. Noh, W. F., and Protter, M., *J. Math. Mech.* **12**, No. 2, 149 (1963).
11. Copson, E. T., *Proc. Roy. Soc.* **A216**, 539 (1953).
12. Magnus, W., and Oberhetinger, F., "Formulae and Theorems for the Special Functions of Mathematical Physics." Chelsea, New York, 1949.
13. Ludford, G. S. S., *J. Rational Mech. Anal.* **3**, 77 (1954).
14. Zabusky, N. J., *J. Math. Phys.* **3**, 1028 (1962).
15. Erdelyi, A. (ed.), "Bateman Manuscript Project." McGraw-Hill, New York, 1955.
16. Jahnke, E., and Emde, F., "Tables of Functions." Teubner, Leipzig, 1938.
17. Smith, P., *Appl. Sci. Res.* **A12**, No. 1, 66 (1963).
18. Martin, M. H., *Can. J. Math.* **5**, 37 (1953).
19. Martin, M. H., *Quart. Appl. Math.* **8**, 137 (1950).
20. Frankl, F. I., *Izv. Akad. Nauk SSSR, Ser. Mat.* **9**, 387 (1945).
21. Falkovich, S. V., *Natl. Advisory Comm. Aeron. Rept.* No. NACA-TM-1212 (1947).
22. von Mises, R., "Mathematical Theory of Compressible Flow." Academic Press, New York, 1958.
23. Friedrichs, K., *Los Alamos Lab. Rept.* No. 2105 (1957).
24. Pai, S. I., "Magnetogas-Dynamics and Plasma Dynamics." Springer, Vienna (Prentice-Hall, Englewood Cliffs, New Jersey), 1962.
25. Weir, D., *Proc. Cambridge Phil. Soc.* **57**, 890 (1961).
26. Gundersen, R. M., *Am. Inst. Aeron. Astronautics Journal* **1**, 1191 (1963).
27. Gundersen, R. M., *J. Math. Anal. Appl.* **6**, 86 (1962).
28. Coburn, N., *Quart. Appl. Math.* **15**, 237 (1957).
29. Dolph, C. L., and Coburn, N., *Proc. 1st Symp. Appl. Math., Providence, Rhode Island, 1947* p. 55. Am. Math. Soc., Providence, Rhode Island, 1949.
30. Bers, L., "Mathematical Aspects of Subsonic and Transsonic Gas Dynamics." Wiley, New York, 1958.
31. Hoffman, O., and Sachs, G., "Introduction to the Theory of Plasticity for Engineers." McGraw-Hill, New York, 1953.
32. Oplinger, D. W., *J. Acoust. Soc.* **32**, 1529 (1960).
33. Tomotika, S., and Tamada, K., *Quart. Appl. Math.* **7**, 381 (1949).
34. Sedov, L. I., *Dokl. Akad. Nauk SSSR* **90**, 735 (1953).
35. Lidov, M. L., *Dokl. Akad. Nauk SSSR* **97**, 409 (1954).
36. Keller, J. B., *Quart. Appl. Math.* **14**, 171 (1956).
37. Tamada, K., Studies on the two-dimensional flow of a gas, with special reference to the flow through various nozzles, Ph.D. Thesis, University of Kyoto, Japan (1950).
38. Taylor, G. I., *Brit. Aeron. Res. Council Repts. and Mem.* No. 1381 (1931).
39. Goertler, H., *Z. Angew. Math. Mech.* **19**, 325 (1939).

CHAPTER **4**

Further Analytic Methods

4.0 INTRODUCTION

In this chapter we continue our discussion of exact methods for obtaining the solutions of equations. Further *ad hoc* methods are discussed, similarity concepts are used, some useful source solutions are derived and combined methods are discussed.

4.1 AN AD HOC SOLUTION FROM MAGNETO-GAS DYNAMICS

The governing equations of an ideal gas which is viscous (viscosity = μ), heat conducting (thermal conductivity = k), electrically conducting (electrical conductivity = σ), and one dimensional (see Pai [1, 2]) are

$$p = \rho RT \quad (4.1)$$

$$\rho_t + (\rho u)_x = 0 \quad (4.2)$$

$$\rho u_t + \rho u u_x + p_x + \mu_e H H_x - \tfrac{4}{3}(\mu u_x)_x = 0 \quad (4.3)$$

$$H_t + (uH)_x - (\nu_H H_x)_x = 0 \quad (4.4)$$

$$\rho h_t + \rho u h_x - p_t - \tfrac{4}{3}(\mu u u_x)_x - (kT_x)_x + \mu_e H_x(uH - \nu_H H_x) = 0. \quad (4.5)$$

In this system ρ, u, p, H, h, and T are density, velocity, pressure, magnetic field (planar and perpendicular to u), stagnation enthalpy ($h = c_p T + u^2/2$), and temperature, respectively. The physical constants μ_e, ν_H, c_p, and R are magnetic permeability, $\nu_H = 1/\mu_e \sigma$, specific heat at constant pressure, and gas constant, respectively.

In the following discussion we shall treat the ideal gas which is inviscid ($\mu \equiv 0$) nonheat-conducting ($k \equiv 0$) and has infinite electrical conductivity ($\nu_H \equiv 0$). Equations (4.2)–(4.5) thus become

$$\rho_t + (\rho u)_x = 0 \tag{4.2a}$$

$$\rho u_t + \rho u u_x + p_x + \mu_e H H_x = 0 \tag{4.3a}$$

$$H_t + (uH)_x = 0 \tag{4.4a}$$

$$\rho h_t + \rho u h_x - p_t + \mu_e u H H_x = 0. \tag{4.5a}$$

Since Eqs. (4.2a) and (4.4a) are identical we may write

$$H = A\rho = \frac{H_0}{\rho_0}\rho \tag{4.6}$$

where A is a constant and H_0, ρ_0 refers to values at the stagnation point. From this result it is clear that H and ρ have the same functional form.

For anisentropic flow, the equation of state of a gas may be written as

$$p/\rho^\gamma = b \exp[S/c_v] = b\theta$$

where $b = (p_0/\rho_0^\gamma) \exp(-S_0/c_v)$ and S is the entropy. As in ordinary gas dynamics of inviscid fluids we can show from Eqs. (4.1), (4.3a), and (4.5a) that

$$\left[\frac{\partial}{\partial t} + u\frac{\partial}{\partial x}\right] \ln \theta = \frac{D}{Dt} \ln \theta \equiv 0 \tag{4.7}$$

i.e., there is no change in entropy along any line of flow. One of the particular solutions of Eq. (4.7) is $\theta = $ constant throughout the flow (isentropic) but in general a shock wave might occur and the entropy changes in the flow field.

Some critical speeds are useful. First, the local speed of sound,

$$c^2 = \left(\frac{\partial p}{\partial \rho}\right)_\theta = \frac{\gamma p}{\rho} = b\gamma\theta\rho^{\gamma-1}; \tag{4.8}$$

4.1 AN AD HOC SOLUTION FROM MAGNETO-GAS DYNAMICS

Second, the local speed of Alfven's wave,

$$V_H^2 = \frac{\mu_e H^2}{\rho} = \mu_e A^2 \rho = \mu_e \frac{H_0^2}{\rho_0^2} \rho = V_{H_0}^2 \frac{\rho}{\rho_0} ; \qquad (4.9)$$

and finally the effective speed of sound in a one-dimensional flow c_e,

$$c_e^2 = c^2 + V_H^2. \qquad (4.10)$$

Generally c_e is a function of both θ and ρ except that for isentropic flow c_e depends *only* on ρ.

After these preliminaries we illustrate a method of finding an exact solution of Eqs. (4.2a)–(4.5a) originally due to Pai [2]. When the flow is isentropic ($\theta \equiv$ constant) we may find solutions of the equations such that density is a function of velocity alone. That is we may write

$$\rho = f(u) \qquad (4.11)$$

where f is to be determined. Under this assumption equation (4.2a) becomes

$$f'(u)u_t + f(u)u_x + uf'(u)u_x = 0$$

or

$$u_t + uu_x = -\frac{f}{f'} u_x. \qquad (4.12)$$

Upon recalling Eq. (4.6) we see that the assumption $\rho = f(u)$ implies that $H = Af(u)$. Thus Eq. (4.3a) becomes

$$fu_t + fuu_x + (\gamma b \theta \rho^{\gamma-1}) f u_x + \mu_e A^2 \rho f' u_x = 0. \qquad (4.13)$$

But, since the effective speed c_e is defined as

$$c_e^2 = \gamma b \theta \rho^{\gamma-1} + \mu_e A^2 \rho,$$

Eq. (4.13) can be written in the form

$$u_t + uu_x = -\frac{c_e^2 f'}{f} u_x . \qquad (4.14)$$

Now, if $\rho = f(u)$ is the solution required then Eqs. (4.12) and (4.14) must be the same. Therefore,

$$c_e^2 \frac{f'}{f} = \frac{f}{f'}$$

or
$$\frac{f'}{f} = \frac{1}{\rho}\frac{d\rho}{du} = \pm \frac{1}{c_e} \tag{4.15}$$
which upon integration give
$$u = \pm \int_{\rho_0}^{\rho} \frac{c_e(\rho)}{\rho} d\rho. \tag{4.16}$$

Actually, Eq. (4.15) is sufficient to continue to the exact solution since upon substituting it into Eqs. (4.12) we obtain
$$u_t + (u \pm c_e)u_x = 0 \tag{4.17}$$
and
$$\rho_t + (u \pm c_e)\rho_x = 0. \tag{4.18}$$
Since both of these equations are first order the methods of first order partial differential equations (see Section 2.5a) apply. Direct integration by means of the Lagrange equations yields the solutions
$$\begin{aligned} u &= F[x - (u \pm c_e)t] \\ \rho &= G[x - (u \pm c_e)t] \end{aligned} \tag{4.19}$$
where F and G are arbitrary functions determined from initial conditions.

The solution $u = F[x - (u + c_e)t]$ means that the variation of u with respect to a point moving with velocity $u + c_e$ is zero, i.e., the disturbance is propagated with an instantaneous velocity $u + c_e$. If $u \ll c_e$ then $u = F[x - c_e t]$ approximately, so c_e is the velocity of propagation of an infinitesimal disturbance and therefore is called the effective speed of sound. For waves whose amplitude is finite, the velocity of wave propagation is different at various positions of the flow field and the shape of the wave is distorted as the wave propagates. For waves of the type $u = F[x - (u + c_e)t]$ the velocity of propagation at the crest is greater than at the trough. Therefore the crest will overtake the trough, a shock wave will form and the assumption of isentropic flow breaks down.

4.2 THE UTILITY OF LAGRANGIAN COORDINATES

The equations of an inviscid, non-heat-conducting gas have been described and utilized in special flows in Section 3.10, Eqs. (3.157). Only one dimensional and spherically symmetric problems were

4.2 THE UTILITY OF LAGRANGIAN COORDINATES

considered there. Noh and Protter [3] have found the Lagrange formulation to be a convenient vehicle for obtaining "soft" solutions of the equations of gas dynamics. In Section 2.5 general solutions for partial differential equations of first order were discussed. In particular the solution of $u_t + uu_x = 0$ was found to be $u = f[x - ut]$. A function $u(x, t)$ is said to be a soft[†] solution of $u_t + uu_x = 0$, subject to the initial condition $u(x, 0) = f(x)$, if u satisfies the functional relation $u = f[x - tu(x, t)]$. Sufficient differentiability of f will insure that a soft solution is also a strict solution. The development of soft solutions in two and three dimensions is most easily carried out, not by extending the techniques for one dimension (see Eqs. (2.173)–(2.180)), but by utilizing Lagrangian coordinates. We shall discuss the results in two dimensions although the methods are perfectly general.

The two-dimensional gas-dynamic equation *with constant pressure* in Euler form are

$$u_t + uu_x + vu_y = 0 \tag{4.20a}$$

$$v_t + uv_x + vv_y = 0 \tag{4.20b}$$

$$\rho_t + (\rho u)_x + (\rho v)_y = 0 \tag{4.20c}$$

$$\epsilon_t + (\epsilon u)_x + (\epsilon v)_y = 0 \tag{4.20d}$$

$$\epsilon = e + p$$

with initial conditions

$$\begin{aligned} u(x, y, 0) &= f(x, y), & v(x, y, 0) &= g(x, y) \\ \rho(x, y, 0) &= h(x, y), & \epsilon(x, y, 0) &= k(x, y). \end{aligned} \tag{4.21}$$

In the usual notation u, v are velocity components, ρ is density, e is internal energy per unit volume, and $\epsilon = e + p$, p the pressure.

In Lagrange coordinates each particle of the fluid is identified by two parameters a, b and time t and the path of each particle is given by

$$\begin{aligned} x &= x(a, b, t) \\ y &= y(a, b, t) \end{aligned} \tag{4.22}$$

[†] The term soft solution is used to distinguish these from the "weak" solutions of Lax [4]. Weak solutions of nonlinear equations automatically satisfy certain jump conditions across discontinuities, such as occur in shocks. Soft solutions have no jump conditions. The relation between the two types is an open question.

where a, b are chosen as the coordinates of the particles at $t = 0$. We recall that for any function $F(x, y, t)$

$$\frac{dF}{dt} = \frac{\partial F}{\partial t} + \frac{\partial F}{\partial x}\frac{dx}{dt} + \frac{\partial F}{\partial y}\frac{dy}{dt} = F_t + uF_x + vF_y \qquad (4.23)$$

and that the Jacobian $J = \partial(x, y)/\partial(a, b) = x_a y_b - x_b y_a$ characterize the essentials for a transformation from (x, y) to (a, b) space. Selecting u and v for F in Eq. (4.23) provides us with the equation of motion as

$$\frac{du}{dt} = 0 \qquad (4.24\text{a})$$

$$\frac{dv}{dt} = 0 \qquad (4.24\text{b})$$

and by definition

$$\frac{\partial x}{\partial t} = u \qquad (4.24\text{c})$$

$$\frac{\partial y}{\partial t} = v \qquad (4.24\text{d})$$

$$J = x_a y_b - x_b y_a.$$

The equation of continuity

$$\rho_t + u\rho_x + v\rho_y = -\rho(u_x + v_y)$$

becomes

$$\frac{d\rho}{dt} = -\rho \frac{J_t}{J}$$

or

$$\frac{1}{\rho}\frac{d\rho}{dt} + \frac{1}{J}\frac{dJ}{dt} = 0$$

which integrates to

$$\rho J = \text{constant} = \rho(a, b, 0) J(a, b, 0). \qquad (4.24\text{e})$$

Similarly, since the equation for ϵ is exactly the form of that for ρ,

$$\epsilon J = \epsilon(a, b, 0) J(a, b, 0). \qquad (4.24\text{f})$$

4.2 THE UTILITY OF LAGRANGIAN COORDINATES

These six equations (4.24a–f) together with the initial conditions

$$x(a, b, 0) = a, \qquad y(a, b, 0) = b$$
$$u(a, b, 0) = f(a, b), \qquad v(a, b, 0) = g(a, b) \qquad (4.25)$$
$$\rho(a, b, 0) = h(a, b), \qquad \epsilon(a, b, 0) = k(a, b)$$

constitute the Lagrange system.

We now proceed to demonstrate the following result. The soft solution of Eqs. (4.20a)–(4.20d) and satisfying the initial conditions (4.21) is given by the set

$$u = f(\omega, \sigma), \qquad v = g(\omega, \sigma)$$
$$\rho = h(\omega, \sigma)/\Delta, \qquad \epsilon = k(\omega, \sigma)/\Delta \qquad (4.26)$$

where

$$\omega = x - tu \qquad (4.27)$$
$$\sigma = y - tv \qquad (4.28)$$
$$\Delta = (1 + tf_\omega)(1 + tg_\sigma) - t^2 f_\sigma g_\omega. \qquad (4.29)$$

The symbol f_ω means take the partial derivative of f with respect to the grouping $\omega = x - tu$.

The equations $u_t = 0$, $v_t = 0$ assert that along the particle paths

$$u(a, b, t) = \text{constant} = f(a, b)$$
$$v(a, b, t) = \text{constant} = g(a, b) \qquad (4.30)$$

where the constant in each case has been determined from the initial conditions (4.25). Going next to $x_t = u$ and integrating we get

$$x(a, b, t) = \text{constant} + \int_0^t u(a, b, t)\, dt$$
$$= x(a, b, 0) + \int_0^t f(a, b)\, dt$$
$$= x(a, b, 0) + tf(a, b)$$
$$= a + tu.$$

Similarly $y(a, b, t) = b + tv$. Upon solving for a and b,

$$a = x - tu, \qquad b = y - tv$$

4. FURTHER ANALYTIC METHODS

which when substituted in Eq. (4.30) completes the derivation for u and v.

The derivation for ρ and ϵ are the same so only one will be given. When we differentiate $x = a + tu$, $y = b + tv$ with respect to a and b we get

$$x_a = 1 + tu_a = 1 + tf_a = 1 + tf_\omega, \qquad x_b = tf_\sigma$$
$$y_a = tg_\omega, \qquad\qquad\qquad\qquad\qquad y_b = 1 + tg_\sigma.$$

Thus

$$J = x_a y_b - x_b y_a = (1 + tf_\omega)(1 + tg_\sigma) - t^2 f_\sigma g_\omega = \Delta$$

and $J(a, b, 0) = 1$. From Eq. (4.24e)

$$\rho = \frac{\rho(a, b, 0)}{J} J(a, b, 0)$$
$$= \frac{h(a\ b)}{\Delta} = \frac{h(\omega\ \sigma)}{\Delta}$$

as given in Eq. (4.26).

The soft solutions for the three-dimensional equations

$$u_t + uu_x + vu_y + wu_z = 0, \qquad \rho_t + (\rho u)_x + (\rho v)_y + (\rho w)_z = 0$$
$$v_t + uv_x + vv_y + wv_z = 0, \qquad \epsilon_t + (\epsilon u)_x + (\epsilon v)_y + (\epsilon w)_z = 0$$
$$w_t + uw_x + vw_y + ww_z = 0, \qquad\qquad\qquad\qquad \epsilon = e + p$$

with initial conditions

$$u(x, y, z, 0) = f(x, y, z), \qquad \rho(x, y, z, 0) = h(x, y, z)$$
$$v(x, y, z, 0) = g(x, y, z), \qquad \epsilon(x, y, z, 0) = k(x, y, z)$$
$$w(x, y, z, 0) = m(x, y, z),$$

are

$$u = f(\omega, \sigma, \gamma), \quad v = g(\omega, \sigma, \gamma), \quad w = m(\omega, \sigma, \gamma), \quad \rho = h(\omega, \sigma, \gamma)/\Delta,$$
$$\epsilon = k(\omega, \sigma, \gamma)/\Delta$$

where

$$\omega = x - tu, \quad \sigma = y - tv, \quad \gamma = z - tw, \quad \text{and} \quad \Delta = \frac{\partial(x, y, z)}{\partial(a, b, c)}.$$

4.2 THE UTILITY OF LAGRANGIAN COORDINATES

These methods are applicable to the more general system

$$u_t + \varphi(u)u_x + \psi(v)u_y + \Gamma(w)u_z = 0$$
$$v_t + \varphi(u)v_x + \psi(v)v_y + \Gamma(w)v_z = 0$$
$$w_t + \varphi(u)w_x + \psi(v)w_y + \Gamma(w)w_z = 0$$

which have the soft solutions

$$u = f[x - t\varphi(u), y - t\psi(v), z - t\Gamma(w)]$$
$$v = g[x - t\varphi(u), y - t\psi(v), z - t\Gamma(w)]$$
$$w = h[x - t\varphi(u), y - t\psi(v), z - t\Gamma(w)]$$

satisfying the initial conditions

$$u(x, y, z, 0) = f(x, y, z), \quad v(x, y, z, 0) = g(x, y, z)$$
$$\text{and} \quad w(x, y, z, 0) = h(x, y, z).$$

The case of variable pressure in one dimension will now be considered. We suppose $p = p(x, t)$ is prescribed and that *e is the internal energy per unit mass* so that $\epsilon = \rho e$. The basic equations are

$$u_t + uu_x + \frac{1}{\rho} p_x = 0 \tag{4.31a}$$

$$\rho_t + (\rho u)_x = 0 \tag{4.31b}$$

$$\epsilon_t + (\epsilon u)_x + pu_x = 0 \tag{4.31c}$$

in Eulerian form. The assumed initial conditions are

$$u(x, 0) = f(x), \quad \rho(x, 0) = g(x), \quad \epsilon(x, 0) = h(x). \tag{4.32}$$

The form of Eqs. (4.31) in Lagrangian coordinates is obtained as before as

$$\frac{du}{dt} + \frac{1}{\rho} p_x = 0$$

$$\frac{\partial x}{\partial t} = u$$

$$\rho x_a = \text{constant}$$

$$\frac{de}{dt} - \frac{p}{\rho^2} \frac{d\rho}{dt} = 0$$

(4.33)

with the new initial conditions

$$u(a, 0) = f(a), \quad \rho(a, 0) = g(a), \quad x(a, 0) = a, \quad e(a, 0) = \frac{h(a)}{g(a)}.$$

Since $x(a, 0) = a$, $\partial x/\partial a(a, 0) = 1$, we find that

$$\rho x_a = \rho(a, 0) x_a(a, 0) = g(a). \tag{4.34}$$

The first equation of Eq. (4.33) may therefore be written as

$$\frac{du}{dt} + p_x \frac{x_a}{g(a)} = 0$$

but

$$\frac{\partial p}{\partial x}\frac{\partial x}{\partial a} = \frac{\partial p}{\partial a} = p_a$$

so the momentum equation becomes

$$\frac{du}{dt} + \frac{p_a}{g(a)} = 0. \tag{4.35}$$

But both $g(a)$ and p_a are known functions so integration results in

$$u(a, t) = f(a) - \frac{1}{g(a)} \int_0^t p_a \, dt \tag{4.36}$$

and the x integration gives

$$x(a, t) = a + \int_0^t u(a, t) \, dt. \tag{4.37}$$

Knowledge of x and the third equation of the system (4.33) allows the solution for ρ to be obtained as

$$\rho = \frac{g(a)}{x_a} = \frac{g(a)}{1 + \int_0^t u_a \, dt}. \tag{4.38}$$

Lastly,

$$e(a, t) = \int_0^t \frac{p \rho_t}{\rho^2} \, dt + \frac{h(a)}{g(a)}. \tag{4.39}$$

In principle the solutions (4.36)–(4.39) constitute the completion of the problem. For general initial conditions this process may be difficult to carry out explicitly. The case where it can always be done is that one where the pressure is a prescribed function of t alone. When $p = p(t)$, $p_x \equiv 0$ so Eqs. (4.31) become

$$u_t + u u_x = 0$$
$$\rho_t + (\rho u)_x = 0 \qquad (4.40)$$
$$\epsilon_t + (\epsilon u)_x + p u_x = 0.$$

The first two of these are the same as those previously considered so the solutions are

$$u(x, t) = f(\omega)$$
$$\rho(x, t) = \frac{g(\omega)}{1 + t f'(\omega)}$$
$$\epsilon(x, t) = \frac{g(\omega)}{1 + t f'(\omega)} e$$
$$\omega = x - tu.$$

From Eq. (4.39) and the result that $\rho_t/\rho = -f'(\omega)/g(\omega)$ we find that

$$e(a, t) = \frac{h(a)}{g(a)} - \frac{f'(a)}{g(a)} \int_0^t p(t)\, dt$$

or finally

$$\epsilon(x, t) = \frac{h(\omega) - f'(\omega) \int_0^t p(t)\, dt}{1 + t f'(\omega)}.$$

This procedure is easily extended to the corresponding higher dimensional case.

Zwick [5] has found the Lagrangian form most useful in his investigation of the behavior of small gas bubbles in a liquid.

4.3 SIMILARITY VARIABLES

The concept of similarity solutions and their limitations was introduced in Sections 2.2 with the Boltzmann transformation and Section 2.3d where similarity solutions for the boundary layer equations were discussed. In this and subsequent sections we shall clarify these ideas and enlarge them to include a wider range of examples.

We recall that the physical meaning of the term "similarity" relates to internal—or self similitude in the problem. As an example consider the laminar flow over a flat plate—"similar" solutions are those for which the x-component of velocity, u, has the property that two velocity profiles $u(x, y)$ located at different coordinates x differ only by a scale factor. The mathematical interpretation of the term "similarity" is a transformation of variables, so carried out, that a reduction in the number of independent variables is achieved. This similarity transformation will reduce a problem in two independent variables from a partial differential equation and the auxiliary conditions, to an ordinary differential equation and appropriate boundary conditions. The method can be extended to reducing a set of partial differential equations in two independent variables to a set of ordinary differential equations. It may also be applied repeatedly (or all at once) to reduce a partial differential equation in more than two independent variables, first to fewer independent variables and finally to an ordinary differential equation.

The mathematical models of physical problems can be classified into two mathematical categories: those which are "well posed" and those which are not. A well posed problem is characterized by two basic properties: (i) the differential equation and sufficient auxiliary conditions (initial and boundary) are known to insure a unique solution; and (ii) the solution is continuously dependent upon the auxiliary conditions. Using the "well posed" or not dichotomy we may classify similarity solutions under *assumed transformations*[†] (see Abbott and Kline [6]):

(1) Differential equations and some (but not all) auxiliary conditions are given;
   (a) no similarity variables exist;
   (b) one similarity variable exists;
   (c) many similarity variables exist.

(2) Well-posed problems;
   (a) no similarity variables exist;
   (b) one similarity variable exists.

Again we emphasize the phrase "similarity solutions under assumed transformations."

There are two[‡] well-developed methods for developing similarity variables, the use of transformation groups and the "separation of

---

[†] We underscore this phrase to emphasize that other similarity variables may exist. Those who claim establishment of all similarity variables in reality have established variables only for a particular class of transformations.

[‡] Hansen [73] discusses two other procedures, the *free parameter method* and *dimensional analysis*.

4.4 SIMILARITY VIA ONE-PARAMETER GROUPS

variables" approach. Before discussing these let us fix the idea by considering ("horrors") a linear example to which these methods also apply.

Consider the parabolic equation $u_t = \nu u_{xx}$ together with the auxiliary conditions $u(x, 0) = f(x)$ (all $x \geq 0$), $u(0, t) = u_0(t)$ (all $t > 0$) and $u(\infty, t) = u_\infty(t)$. We now ask (for this well-posed problem), "does a transformation of variables exist which reduces the number of independent variables from two to one," i.e., which reduces the equation to an ordinary differential equation. If so, the dependent variable u would transform to a function of a new independent variable η, i.e., $u(x, t) \to u(\eta)$ where $\eta = \eta(x, t)$. The resulting equation would be a second-order ordinary differential equation in η. It would therefore require *two* boundary conditions (on η), unlike the original equation in which *three* are required.

This reduction in number of auxiliary conditions requires that two of the original conditions coalesce, i.e., are related by

$$u[\eta(x, \alpha)] = u[\eta(\beta, t)]. \tag{a}$$

If such a condition were not true it would be impossible to satisfy all three of the original boundary conditions in terms of only two boundary conditions on η. The particular boundary conditions that are so reduced are found by comparison of the original conditions with that above (a). If such a condition cannot be found a similarity solution will generally not exist in terms of the variable η.

For the specific problem we see that condition (a) can be satisfied in two possible ways that is choose η so that $u(x, 0) = u(\infty, t)$ or $u(x, 0) = u(0, t)$. The first requires $f(x) = u_\infty(t)$, i.e., both must be constant (say λ) and the second requires $f(x) = u_0(t) =$ constant (say Ω). In the first case we have thus selected $\alpha = 0$ and $\beta = \infty$. One transformation consistent with this condition $\eta(x, 0) = \eta(\infty, t)$ is $\eta = ax^n/t^m = a(x/t^{m/n})^n$ where a is constant and n, m are to be determined. A transformation consistent with the condition ($\alpha = 0$, $\beta = 0$) $\eta(x, 0) = \eta(0, t)$ is $\eta = ax^m t^n$ ($m, n > 0$).

4.4 SIMILARITY VIA ONE-PARAMETER GROUPS

Birkhoff [7] was probably the first to apply a *general* method of a one parameter group of transformation to develop similarity solutions in some areas of fluid mechanics. Birkhoff's technique called "the method of search for symmetric solutions" is based upon the invariance under a group G of transformations of a system Σ of partial differential equations. A simple example is provided by the equation $\phi_t + \frac{1}{2}\phi_x^2 = \phi_{xx}$. If we make the transformation $x = a\eta$, $t = a^2\tau$ the form of the partial differential equation is $\phi_\tau + \frac{1}{2}(\phi_\eta)^2 = \phi_{\eta\eta}$ which is the same as before. Thus the equation is invariant under this transformation whose (one) parameter is a.

Birkhoff's first method consists in searching for a solution which is also invariant under the group G of transformations. This procedure is based upon a general theory due to Morgan [8] for obtaining similarity solutions of partial differential equations using one-parameter group theory. According to Morgan, seeking of similarity solutions of a system of partial differential equations is equivalent to the determination of the invariant solutions of these equations under the appropriate one-parameter group of transformations. This procedure as elaborated by Manohar [9] includes in a general way results obtained by Schuh [10] for two dimensional, unsteady, laminar, incompressible boundary layer flows. The method also applies to the three-dimensional boundary layer equations as discussed by Geis [11] in the axisymmetric case and Hansen [12] and Morgan [13] in the general case.

The method of one parameter groups of transformations is simple and and straight forward. A set of similarity variables (invariants of the group) are developed by solving a set of simultaneous equations. These equations also suggest other possible sets of transformations and corresponding similarity variables.

a. General Theory

Let Σ be a system of partial differential equations given by $\Phi_j = 0$ ($j = 1, 2, ..., n$) in which x_i ($i = 1, ..., m$) and y_j ($j = 1, 2, ..., n$) are the independent and dependent variables, respectively. Let a group Γ_1 consisting of a set of transformations be defined as

$$\Gamma_1: \begin{array}{ll} \bar{x}_1 = a^{\alpha_1}x_1, \quad \bar{x}_r = a^{\alpha_r}x_r, ... & (r = 2, ..., m) \\ \bar{y}_j = a^{\gamma_j}y_j & (j = 1, ..., n) \end{array} \qquad (4.41)$$

where the parameter $a \neq 0$ is real and α_r, γ_j are to be determined from the condition that the system Σ be constant conformally invariant[†]

[†] A set of functions $\phi_j(x_i)$ is said to be "conformally invariant" under a one-parameter (a) group $\bar{x}_i \to x_i$ if $\phi_j(x_i) = f_j(\bar{x}_i; a) \phi_j(\bar{x}_i), i = 1, 2, ..., m, j = 1, 2, ..., n$, where the $\phi_j(\bar{x}_i)$ are exactly the same functions of the $\bar{x}_i$ as the ϕ_j are of the x_i. These functions will be said to be "constant conformally invariant" if $f_j(\bar{x}_i; a)$ are independent of the $\bar{x}_i$ and "absolutely invariant" if $f_j \equiv 1$ for all j.

(including absolutely) under Γ_1. This requirement gives rise to a set of simultaneous equations, nontrivial solutions of which generate similarity variables which are invariants of Γ_1.

We suppose x_1 is the independent variable to be eliminated. There are two cases: If $\alpha_1 \neq 0$ the invariants of Γ_1 are

$$\eta_r = \frac{x_r}{x_1^{\beta_r}}, \qquad \beta_r = \alpha_r/\alpha_1 \qquad (r = 2, \ldots, m) \tag{4.42}$$

and

$$f_j(\eta_2, \eta_3, \ldots, \eta_m) = \frac{y_j(x_1, x_2, \ldots, x_m)}{x_1^{\gamma_j/\alpha_1}} \qquad (j = 1, 2, \ldots, n).$$

If $\alpha_1 = 0$ and the simultaneous equations have a nontrivial solution, then we may choose a group Γ_2 consisting of

$$\bar{x}_1 = x_1 + \ln a, \qquad \bar{x}_r = a^{\alpha_r} x_r, \qquad \bar{y}_j = a^{\gamma_j} y_j \tag{4.43}$$

and the invariants of the group are

$$\eta_r = \frac{x_r}{\exp(\alpha_r x_1)} \qquad (r = 2, \ldots, m)$$

$$f_j(\eta_2, \ldots, \eta_m) = \frac{y_j(x_1, \ldots, x_m)}{\exp[\gamma_j x_1]} \qquad (j = 1, 2, \ldots, n). \tag{4.44}$$

The ingenuity of the first authors to utilize similarity variables is striking. With this theory much of this "educated guessing" is removed. However, the facility of the innovator cannot be eliminated.

b. An Example

The first example, considered in some detail, will be the equations for unsteady, incompressible, two-dimensional boundary layer flow. The governing equations, in the usual notation are

$$u_t + uu_x + vu_y = -\frac{1}{\rho} p_x + \nu u_{yy}$$

$$u_x + v_y = 0 \tag{4.45}$$

with the boundary conditions

$$u = v = 0 \quad \text{for} \quad y = 0; \quad u \to u_e(x, t) \quad \text{for} \quad y \to \infty. \tag{4.46}$$

Introducing the stream function $\psi(u = \psi_y, \ v = -\psi_x)$ and applying Bernoulli's equation for flow outside the boundary layer, so that $-(1/\rho)p_x = u_{et} + u_e u_{ex}$, we get the equation for ψ as

$$\psi_{yt} + \psi_y \psi_{xy} - \psi_x \psi_{yy} = u_{et} + u_e u_{ex} + v \psi_{yyy} \qquad (4.47)$$

with boundary conditions

$$\psi_x = \psi_y = 0 \quad \text{for} \quad y = 0; \quad \psi_y \to u_e(x, t) \quad \text{for} \quad y \to \infty. \qquad (4.48)$$

We use the general theory on Eq. (4.47). As in Eq. (4.41) let the group Γ_1 be

$$\bar{x} = a^m x, \quad \bar{y} = a^n y, \quad \bar{t} = a^r t, \quad \bar{\psi} = a^s \psi, \quad \bar{u}_e = a^p u_e$$

with $a \neq 0$ and m, n, r, s, p real numbers to be determined. Upon substitution into Eq. (4.47) we obtain

$$a^{n+r-s}\bar{\psi}_{\bar{y}\bar{t}} + a^{m+2n-2s}[\bar{\psi}_{\bar{y}}\bar{\psi}_{\bar{x}\bar{y}} - \bar{\psi}_{\bar{x}}\bar{\psi}_{\bar{y}\bar{y}}]$$
$$= a^{r-p}\bar{u}_{e\bar{t}} + a^{m-2p}\bar{u}_e\bar{u}_{e\bar{x}} + va^{3n-s}\bar{\psi}_{\bar{y}\bar{y}\bar{y}}. \qquad (4.49)$$

The requirement of constant conformally (absolute) invariance requires that these coefficients all be the same, i.e., the system of equations for m, n, r, s, and p is

$$n + r - s = m + 2n - 2s = r - p = m - 2p = 3n - s. \qquad (4.50)$$

We now choose to eliminate t (we could just as easily eliminate x or y first) so that the solutions of Eq. (4.50) for $r \neq 0$ are equivalent to the solutions[†] of

$$\frac{n}{r} + 1 - \frac{s}{r} = \frac{m}{r} + 2\frac{n}{r} - 2\frac{s}{r} = 1 - \frac{p}{r} = \frac{m}{r} - 2\frac{p}{r} = 3\frac{n}{r} - \frac{s}{r}. \qquad (4.51)$$

The solution of this system can proceed along a variety of lines. We attempt to solve it (nontrivially) by setting $m/r = A$ where A is an arbitrary constant. Then $p/r = (m/r) - 1 = A - 1$; and successively $s/r = A - \frac{1}{2}$, and $n/r = \frac{1}{2}$. Other alternates are possible but will result in the same system multiplied by an arbitrary constant. From

---

[†] Division by r is convenient since the invariants have exponents of the form $s/r, n/r$, etc., as in Eqs. (4.42).

4.4 SIMILARITY VIA ONE-PARAMETER GROUPS

Eq. (4.42) the corresponding invariants of the group, which are called similarity variables, are given by

$$\xi = x/t^A, \quad \eta = y/t^{\frac{1}{2}}, \quad f(\xi, \eta) = \frac{\psi(x, y, t)}{t^{A-\frac{1}{2}}}, \quad V(\xi) = \frac{u_e(x, t)}{t^{A-1}}. \tag{4.52}$$

Thus we have immediately that for similarity in t, the velocity of the outer flow should be of the form

$$u_e(x, t) = V(\xi) t^{A-1} = V\left(\frac{x}{t^A}\right) t^{A-1}. \tag{4.53}$$

The partial differential equation for f is obtained by substituting Eq. (4.52) into Eq. (4.47) and boundary conditions (4.48). Carrying out the differentiations gives

$$\nu f_{\eta\eta\eta} + \tfrac{1}{2}\eta f_{\eta\eta} - (A-1)f_\eta + A\xi f_{\xi\eta} + f_\xi f_{\eta\eta} - f_\eta f_{\xi\eta} + (A-1)V(\xi)$$
$$- A\xi \frac{dV}{d\xi} + V \frac{dV}{d\xi} = 0 \tag{4.54}$$

with†

$$\frac{\partial f}{\partial \xi} = \frac{\partial f}{\partial \eta} = 0 \quad \text{for} \quad \eta = 0; \quad \frac{\partial f}{\partial \eta} \to V(\xi) \quad \text{for} \quad \eta \to \infty. \tag{4.55}$$

There is no reason why a further reduction cannot be accomplished. In order to eliminate ξ from Eqs. (4.54) and (4.55) we proceed exactly as above and consider a new group Γ_2 of transformations

$$\bar{\xi} = b^p \xi, \quad \bar{\eta} = b^r \eta, \quad \bar{f} = b^s f, \quad \bar{V} = b^m V. \tag{4.56}$$

If Eq. (4.54) is to be absolute conformally invariant under Eq. (4.56) we must have the system

$$s - 3r = s - r = 2s - 2r - p = m = 2m - p \tag{4.57}$$

satisfied. With $p \neq 0$ the solution of the system

$$\frac{s}{p} - 3\frac{r}{p} = \frac{s}{p} - \frac{r}{p} = \frac{2s}{p} - 2\frac{r}{p} - 1 = \frac{m}{p} = 2\frac{m}{p} - 1$$

---

† The reader may observe that $V(\xi)$ in the boundary condition could have been eliminated by choosing $f(\xi, \eta) = V(\xi) f_1(\xi, \eta)$. Further the coefficient ν of Eq. (4.54) could have been made 1 by selecting $\xi = bx/t^A$, $\eta = cy/t^{\frac{1}{2}}$ and selecting b and c properly. However, these are niceties and really extraneous to the method.

is $r/p = 0$, $s/p = 1$, $m/p = 1$. The invariants of the group defined by Eq. (4.56) are therefore

$$\lambda = \eta, \qquad F(\lambda) = \frac{f(\xi, \eta)}{\xi}, \qquad \frac{V(\xi)}{\xi} = K \quad \text{(constant)}. \qquad (4.58)$$

Therefore the elimination of both x and t can be accomplished by introduction of the variables

$$\lambda = y/t^{\frac{1}{2}}, \qquad F(\lambda) = \frac{\psi(x, y, t) t^{\frac{1}{2}}}{x} \qquad (4.59)$$

provided the outer velocity u_e is expressed by

$$u_e(x, t) = Kx/t \qquad (4.60)$$

as discovered by Schuh [10]. The equation for F is obtained either by applying Eq. (4.59) to the original Eq. (4.47) or by applying Eq. (4.58) to Eq. (4.54). In either event we obtain the F equation

$$\nu F''' + \tfrac{1}{2}\lambda F'' + F' + FF'' - (F')^2 + K(K-1) = 0$$
$$F = F' = 0 \quad \text{for} \quad \lambda = 0; \qquad F' \to K \quad \text{for} \quad \lambda \to \infty. \qquad (4.61)$$

The previously excluded case $r = 0$ for Eqs. (4.50) gives rise to the system

$$n - s = m + 2n - 2s = -p = m - 2p = 3n - s$$

which has the nontrivial solution

$$n = 0 \qquad m = s = p = B \qquad (4.62)$$

where B is an arbitrary constant. From Eq. (4.44) the invariants (similarity variables) are

$$\xi_1 = x/e^{Bt}, \qquad \eta_1 = y$$
$$f_1(\xi_1, \eta_1) = \frac{\psi(x, y, t)}{e^{Bt}}, \qquad V_1(\xi_1) = \frac{u_e(x, t)}{e^{Bt}}. \qquad (4.63)$$

Therefore for similarity in t, the velocity u_e may have the form

$$u_e(x, t) = e^{Bt} V_1(xe^{-Bt}). \qquad (4.64)$$

The differential equation for f_1 is

$$\nu \frac{\partial^3 f_1}{\partial \eta_1^3} - B\frac{\partial f_1}{\partial \eta_1} + B\xi_1 \frac{\partial^2 f_1}{\partial \xi_1 \partial \eta_1} - \frac{\partial f_1}{\partial \eta_1}\frac{\partial^2 f_1}{\partial \xi_1 \partial \eta_1} + \frac{\partial f_1}{\partial \xi_1}\frac{\partial^2 f_1}{\partial \eta_1^2}$$
$$+ B\left(V_1 - \xi_1 \frac{dV_1}{d\xi_1}\right) + V_1 \frac{dV_1}{d\xi_1} = 0 \quad (4.65)$$

with boundary conditions

$$\frac{\partial f_1}{\partial \xi_1} = \frac{\partial f_1}{\partial \eta_1} = 0 \quad \text{for} \quad \eta_1 = 0; \quad \frac{\partial f_1}{\partial \eta_1} \to V_1(\xi_1) \quad \text{as} \quad \eta_1 \to \infty.$$

By now the procedure is clear so just the results will be stated. In Eq. (4.65) we find that ξ_1 can be eliminated by

$$\lambda_1 = \eta_1, \quad F_1(\lambda_1) = \frac{f_1(\xi_1, \eta_1)}{\xi_1}, \frac{V_1(\xi_1)}{\xi_1} = \text{constant} = K_1$$

that is both x and t can be eliminated from Eq. (4.47) by the similarity variables

$$\lambda_1 = y, \quad F_1(\lambda_1) = \frac{\psi(x, y, t)}{x}$$

provided $u_e(x, t) = K_1 x$.

Other special cases may be considered. When Eq. (4.47) is independent of t, the boundary layer flow is steady and an application of this theory leads to the familiar similarity variables of such flows as given in Pai [14], Curle [15], and Meksyn [16].

4.5 EXTENSIONS OF THE SIMILARITY PROCEDURE

An extension of the preceding theory, when two or more variables are to be eliminated, follows almost immediately. We suppose two variables are to be eliminated, say x_1 and x_2. Let Γ_1 be the same as that of Eq. (4.41), i.e., a one parameter group. If $\alpha_1 = \alpha_2 \neq 0$ the invariants of Γ_1 are

$$\eta_r = \frac{x_r}{(bx_1 + cx_2)^{\alpha_r/\alpha_1}} \quad (r = 3, ..., m)$$

and
$$f_j(\eta_3, ..., \eta_m) = \frac{y_j(x_1, ..., x_m)}{(bx_1 + cx_2)^{\gamma_j/\alpha_1}}$$
(4.66)

where b and c are any arbitrary constants.

4. FURTHER ANALYTIC METHODS

If $\alpha_1 = \alpha_2 = 0$, the invariants (similarity variables) are

$$\eta_r = \frac{x_r}{\exp[(bx_1 + cx_2 + d)a_r]}$$

$$f_j(\eta_3, ..., \eta_m) = \frac{y_j(x_1, ..., x_m)}{\exp[(bx_1 + cx_2 + d)\gamma_j]}.$$

(4.67)

Instead of the group Γ_1 we may use the two parameter group Γ_2 consisting of

$$\Gamma_2: \quad \bar{x}_1 = a^{\alpha_1}x_1, \quad \bar{x}_2 = b^{\beta_2}x_2, \quad \bar{x}_r = a^{\alpha_r}b^{\beta_r}x_r \quad (r = 3, ..., m)$$

and

$$\bar{y}_j = a^{\gamma_j}b^{\delta_j}y_j \quad (j = 1, 2, ..., n).$$

(4.68)

The invariants of Γ_2, under the requirement that the system of partial differential equations be absolute conformally invariant under Γ_2, are the similarity variables. Four cases arise:

(i) $\alpha_1 \neq 0$, $\beta_2 \neq 0$

$$\eta_r = \frac{x_r}{x_1^{\alpha_r/\alpha_1} x_2^{\beta_r/\beta_2}} \quad (r = 3, ..., m)$$

$$f_j(\eta_3, ..., \eta_m) = \frac{y_j(x_1, ..., x_m)}{x_1^{\gamma_j/\alpha_1} x_2^{\delta_j/\beta_2}} \quad (j = 1, ..., n)$$

(4.69)

(ii) $\alpha_1 \neq 0$, $\beta_2 = 0$

$$\eta_r = \frac{x_r}{x_1^{\alpha_r/\alpha_1} \exp(\beta_r x_2)}$$

$$f_j = \frac{y_j}{x_1^{\gamma_j/\alpha_1} \exp[\delta_j x_2]}$$

(4.70)

(iii) $\alpha_1 = 0$, $\beta_2 \neq 0$

$$\eta_r = \frac{x_r}{\exp[\alpha_r x_1] x_2^{\beta_r/\beta_2}}$$

$$f_j = \frac{y_j}{\exp[\gamma_j x_1] x_2^{\delta_j/\beta_2}}$$

(4.71)

(iv) $\alpha_1 = \beta_2 = 0$

$$\eta_r = \frac{x_r}{\exp[\alpha_r x_1 + \beta_r x_2]}$$

$$f_j = \frac{y_j}{\exp[\gamma_j x_1 + \delta_j x_2]}.$$

(4.72)

4.5 EXTENSIONS OF THE SIMILARITY PROCEDURE

The details of application of these extensions are similar to those already discussed. Examples are given by Hansen [12] and Manohar [9] for the general boundary layer equations for steady laminar incompressible flows.

We remark here that we may replace $bx_1 + cx_2$ of Eq. (4.66) by $bx_1 + cx_2 + d$, for the boundary layer equations, since they are invariant under a transformation $\bar{x}_1 = x_1 + d$.

The techniques developed in Sections 4.4 and 4.5 are applicable to other problems besides those of the boundary layer equations. As an example we consider briefly the nonlinear diffusion equation

$$\frac{\partial}{\partial x}\left[c^p \frac{\partial c}{\partial x}\right] + \frac{\partial}{\partial y}\left[c^p \frac{\partial c}{\partial y}\right] = \frac{\partial c}{\partial t} \tag{4.73}$$

written out as

$$c^p\{c_{xx} + c_{yy}\} + pc^{p-1}\{c_x^2 + c_y^2\} - c_t = 0. \tag{4.74}$$

Application of the one parameter group (cf. Ames [74])

$$\Gamma_1: \quad \bar{x} = a^m x, \quad \bar{y} = a^n y, \quad \bar{t} = a^r t, \quad \text{and} \quad \bar{c} = a^s c \tag{4.75}$$

to Eq. (4.74) requires that

$$2m - (p+1)s = 2n - (p+1)s = r - s \tag{4.76}$$

for invariance of Eq. (4.74). If we wish to eliminate t, select $r \neq 0$, so that Eqs. (4.76) are satisfied by

$$\frac{s}{r} = A, \quad \frac{m}{r} = \frac{n}{r} = \frac{1 + pA}{2} = \alpha. \tag{4.77}$$

The invariants of Γ_1 are

$$\xi = x/t^\alpha, \quad \eta = y/t^\alpha, \quad f(\xi, \eta) = \frac{c(x, y, t)}{t^A} \tag{4.78}$$

and the differential equation for f is

$$f^p[f_{\xi\xi} + f_{\eta\eta}] + pf^{p-1}[f_\xi^2 + f_\eta^2] - Af + \alpha[\xi f_\xi + \eta f_\eta] = 0. \tag{4.79}$$

Similarly we may eliminate ξ (or η). The invariants are

$$\theta = \eta/\xi, \quad \psi(\theta) = \frac{f(\xi, \eta)}{\xi^{2/p}}$$

and the ordinary differential equation for ψ is

$$\psi^{p-1}\left\{(1+\theta^2)\psi'' + 2\theta\left(1-\frac{2}{p}\right)\psi' + \frac{2}{p}\left(\frac{2}{p}-1\right)\psi\right\}$$

$$+ p\psi^{p-2}\left\{(\psi')^2 + \left(\frac{2}{p}\psi - \theta\psi'\right)^2\right\} + \frac{1}{p} = 0. \quad (4.80)$$

Summarizing, we can eliminate both x and t by means of the similarity variables

$$\theta = y/x, \qquad \psi(\theta) = \frac{c(x, y, t)}{x^{2/p} t^{A(1-2/p)}} \quad (4.81)$$

which form clearly indicates the special role of $p = 2$. When $p = 2$ the differential equation (4.80) is considerably simplified.

Often the boundary conditions stipulate further restrictions on the existence of similarity solutions as we have observed in the boundary layer problems. Since the development here results from a limited use of Morgan's results claim cannot be made that no other similarity solutions exist.

Abbot and Kline [6] have a substantial bibliography of problems prior to 1960 and Polskii and Shvets [17] have obtained several similarity solutions in magnetohydrodynamics.

4.6 SIMILARITY VIA SEPARATION OF VARIABLES

Birkhoff [7] has suggested an alternate method which he calls "separation of variables". Essentially, the method depends upon the introduction of some unknown transformation function(s), suggested by possible similarity. Then the condition that the transformed partial differential equation should be solvable by the method of separation of variables and also satisfy the boundary conditions determines the unknown function(s). This method and the one (or more) parameter group method are often equivalent, except in certain degenerate cases. When the method is applied one has to assume the general form of similarity variable to begin with and in addition make all substitutions into the equation. The resulting differential equations and the boundary conditions must be examined before one obtains the specific similarity variables. Abbott and Kline [6] have elaborated this approach and applied it to some problems. We illustrate the "separation of variables" approach with

4.6 SIMILARITY VIA SEPARATION OF VARIABLES

two examples, the first is typical of a well-posed problem, the second where not all auxiliary conditions are specified.

The viscous boundary layer equations along a flat plate have previously served as excellent examples. Utilizing them again will serve to illustrate the (often) subtle techniques that may be required for the construction of similarity solutions by this technique.

The (steady) Navier-Stokes equations are of elliptic type while the boundary layer equations are of parabolic type. This transition from an elliptic set to a parabolic equation is one of the fundamental consequences of Prandtl's asymptotic transformation. But because of this a difficulty enters—centering around the specification of an adequate "initial condition."[†] For the complete mathematical determination an initial condition on the velocity profile at some specified value of x is required in addition to the boundary conditions on y.

Case A. *A well-posed problem.* Suppose a steady flow with constant free stream velocity u_0 impinges on a flat plate which is parallel to the x-axis, with leading edge at $x = 0$. Equations (1.21) describe the flow. These boundary layer equations $u_x + v_y = 0$, $uu_x + vu_y = vu_{yy}$ subject to the boundary conditions[‡]

$$u(x > 0, y = 0) = 0, \quad v(x > 0, y = 0) = 0, \quad u(x, \infty) = u_0 \quad \text{(a)}$$

can be transformed to an equation in the single stream function ψ defined by means of $\psi_y = u$, $\psi_x = -v$. The governing equation for ψ becomes

$$\psi_y \psi_{xy} - \psi_x \psi_{yy} = v \psi_{yyy}. \quad \text{(b)}$$

For the purpose of illustrating the method of attack on a "well posed" problem we add to boundary conditions (a) the "initial" condition

$$u = u_0 \quad \text{when} \quad x = 0 \quad (y \neq 0) \quad \text{(c)}$$

which is questionable for small nonzero y. Its use will be avoided in the not well posed case.

The problem now concerns Eq. (b) subject to the auxiliary conditions

$$\psi_y(x > 0, y = 0) = 0, \quad \psi_y(x \geqslant 0, y = \infty) = u_0$$
$$\psi_x(x > 0, y = 0) = 0, \quad \psi_y(0, y \neq 0) = u_0. \quad \text{(d)}$$

[†] Parabolic equations define initial value problems.
[‡] The equations are not numbered in this section since most of them repeat previous equations.

The problem is well posed although the questionable nature of the last condition in (d) is again pointed out.

We seek a transformation which will reduce (b) to a third order ordinary differential equation and the new independent variable must reduce the number of boundary conditions from four to three. Thus we might ask that

$$\frac{\partial \psi}{\partial y}[\eta(\alpha, y)] = \frac{\partial \psi}{\partial y}[\eta(x, \beta)], \tag{e}$$

although alternate conditions could be set on ψ_x or any combination of ψ and its derivatives as long as the number of auxiliary conditions is reduced to 3. If $\alpha = 0$ and $\beta = \infty$ one possible form for η is $\eta = a \cdot y^n/x^m$. Now the value of n can be chosen arbitrarily, because if there is a solution in the form $u(\eta)$, there must be a solution in the form $u = u(\eta^n)$. The "best" value of n is that value which leads to the ordinary differential equation which is easiest to solve. Actual solution may be delayed until the form of the equation is determined. In this example we set $n = 1$, choose a second variable $\xi = x$ and require the resulting equation, in the new variables ξ and η to be separable.[†] The necessary transformations are

$$\psi_x = \psi_\xi - m\frac{\eta}{\xi}\psi_\eta$$

$$\psi_{xy} = a\xi^{-m}\psi_{\eta\xi} - ma\xi^{-(m+1)}[\psi_\eta + \eta\psi_{\eta\eta}]$$

$$\psi_y = a\xi^{-m}\psi_\eta$$

$$\psi_{yy} = a^2\xi^{-2m}\psi_{\eta\eta}$$

$$\psi_{yyy} = a^3\xi^{-2m}\psi_{\eta\eta\eta}$$

whereupon Eq. (b) becomes

$$\psi_{\xi\eta}\psi_\eta - m\xi^{-1}(\psi_\eta)^2 - \psi_\xi\psi_{\eta\eta} = va\xi^{-m}\psi_{\eta\eta\eta}. \tag{f}$$

We now ask for Eq. (f) to be separable so that when $\psi = b \cdot g(\xi)f(\eta)$ Eq. (f) becomes, after division by $(f')^2$,

$$\xi^m g' - m\xi^{m-1}g = \xi^m g'\frac{ff''}{(f')^2} + \frac{va}{b}\frac{f'''}{(f')^2}. \tag{g}$$

[†] The original equation (b) does separate but the only solution satisfying the boundary conditions is $\psi \equiv$ constant, which is trivial, for it leads to the velocity components $u = v = 0$ everywhere.

4.6 SIMILARITY VIA SEPARATION OF VARIABLES

The boundary conditions on η are

$$abgf'(0)\xi^{-m} = 0 \quad \text{at} \quad \eta = 0$$
$$-bg'f(0) = 0 \quad \text{at} \quad \eta = 0 \qquad \text{(h)}$$
$$abgf'(\infty)\xi^{-m} = u_0 \quad \text{at} \quad \eta = \infty.$$

Inspecting Eq. (g) we see that it will separate (we require each coefficient of f, which involves g to be constant) if $\xi^m g'(\xi) = \text{constant} = c_1$. The solution of this equation for g is

$$g = c_1 \ln \xi + c_2 \qquad \text{for} \quad m = 1$$
$$g = c_1(1-m)^{-1}\xi^{1-m} + c_2 \qquad \text{for} \quad m \neq 1.$$

Consequently, Eq. (g) has an infinite number of separation variables η, one for each value of m, $-\infty < m < \infty$. If there is a unique value of m (or none) it will be determined from the boundary conditions (Eq. (h)) by requiring $g(\xi)$ to be compatible with the boundary conditions.

Thus if $m = 1$ the condition at $\eta = \infty$ gives

$$ab\xi^{-1}[c_1 \ln \xi + c_2]f'(\infty) = u_0$$

which is satisfied only for $\xi = x = \text{constant}$. For $m \neq 1$ the condition at $\eta = \infty$ gives

$$ab[c_1(1-m)^{-1}\xi^{1-2m} + c_2\xi^{-m}]f'(\infty) = u_0$$

which is satisfied if $m = \tfrac{1}{2}$, $c_2 = 0$ or $m = 0$, $c_1 = 0$. Both $m = 1$ and $m = 0$, $c_1 = 0$ imply that ψ is a function of y alone so that the velocity $v = -\psi_x$ is identically zero. The physics of the problem denies this possibility. Thus $m = \tfrac{1}{2}$, $c_2 = 0$ leads to a unique similarity variable.

The function g is then $g = 2c_1\xi^{\tfrac{1}{2}}$ so that the left-hand side of Eq. (g) is identically zero. The separation constant is thus zero and the differential equation for f is

$$f''' + \frac{bc_1}{v a} ff'' = 0. \qquad \text{(i)}$$

To complete the problem we absorp the parameters into the variables and introduce the Abbott-Kline "natural coordinates."[†] These are

---

[†] A dimensionless formulation of the original problem would avoid this requirement.

motivated by a desire to have the Eq. (i) parameter free as well as the boundary conditions. The $\eta = \infty$ condition suggests taking (for example) $ab = \frac{1}{2}$, $c_1 = u_0$ and $c_1 b/\nu a = \frac{1}{2}$. Thus $a^2 = u_0/\nu$. Summarizing:

$$2f''' + ff'' = 0$$
$$f(0) = 0, \quad f'(0) = 0, \quad f'(\infty) = 1$$

and $\eta = y(u_0/\nu x)^{\frac{1}{2}}$, the well-known Blasuis problem.

Case B. *The initial condition unspecified.* In this case we consider Eq. (b) subject to first three boundary conditions (d), neglecting the initial condition at $x = 0$. Consequently, we cannot use a relation analogous to Eq. (e) but a similarity solution may nevertheless exist. As in case A, we define $\xi = x$ and $\eta = ay/\gamma(x)$ where we seek a $\gamma(x)$ such that the resulting equation in ξ and η can be separated. To change variables we need the following transformations:

$$\psi_x = \psi_\xi - \gamma^{-1}\gamma'\eta\psi_\eta, \qquad \psi_y = a\gamma^{-1}\psi_\eta$$
$$\psi_{xy} = a\gamma^{-1}\psi_{\xi\eta} - a\gamma'\gamma^{-2}[\psi_\eta + \eta\psi_{\eta\eta}]$$
$$\psi_{yy} = a^2\gamma^{-2}\psi_{\eta\eta}, \qquad \psi_{yyy} = a^3\gamma^{-3}\psi_{\eta\eta\eta}.$$

When these transformations are applied to Eq. (b) there results

$$\gamma\psi_\eta\psi_{\eta\xi} - \gamma'(\psi_\eta)^2 - \gamma\psi_\xi\psi_{\eta\eta} = \nu a\psi_{\eta\eta\eta} \tag{j}$$

We now attempt to separate Eq. (j) with $\psi = b \cdot g(\xi)f(\eta)$ and obtain after division by $(f')^2$

$$\gamma g' - \gamma'g = \gamma g' \frac{ff''}{(f')^2} + \frac{\nu a}{b}\frac{f'''}{(f')^2}. \tag{k}$$

Equation (k) separates if $\gamma g' = $ constant $= c_1$. This constitutes one condition on the two unknown functions γ and g. Another relation is required—this may be furnished by the boundary conditions or it may arise from outside physical considerations such as conservation requirements. Usually the conservation of energy, mass, etc., are expressible in the form

$$\int_{y_1}^{y_2} E(x, y)\, dy = \text{constant} = A$$

where y_1 and y_2 are constants. If it is possible to write the function

4.6 SIMILARITY VIA SEPARATION OF VARIABLES

$E(x, y)$ as a function of ξ alone times a function of η alone (ξ and η are the similarity variables) then the conservation law becomes

$$\int_{y_1}^{y_2} F(\xi) G(\eta) \, d\eta = F(\xi) \int_{y_1}^{y_2} G(\eta) \, d\eta = A$$

so that

$$F(\xi) = \frac{A}{\int_{y_1}^{y_2} G(\eta) \, d\eta} = B = \text{constant}.$$

The function $F(\xi)$ is usually a relation between g and γ and its constancy provides the second relation between g and γ. An example will be found in the next section.

To return to the specific problem of finding a relation between γ and g for the boundary layer flow we examine the boundary condition at $\eta = \infty$, i.e.,

$$\frac{\partial \psi}{\partial y}(x, \infty) = \frac{a}{\gamma} \frac{\partial \psi(x, \infty)}{\partial \eta} = ab \frac{g}{\gamma} f'(\infty) = u_0. \tag{l}$$

Since a, b, $f'(\infty)$, and u_0 are all constants the required relation is $g/\gamma = c_2 = \text{constant}$. When the relation $g = c_2 \gamma$ and $\gamma g' = c_1$ are solved simultaneously the result is

$$g = [2c_1 c_2 (\xi + \xi_0)]^{\frac{1}{2}}, \quad \gamma = [2(c_1/c_2)(\xi + \xi_0)]^{\frac{1}{2}}. \tag{m}$$

Returning to Eq. (k) it is seen that the left-hand side is zero, therefore the separation constant is zero and the equation for f is

$$f''' + \frac{c_1 b}{\nu a} f'' f = 0.$$

To make η a natural coordinate we select $c_1 = u_0$, $ab = \frac{1}{2}$, $c_1 b/\nu a = \frac{1}{2}$, and $c_2 = 2c_1$ so that $a^2 = u_0/\nu$ and $f'(\infty) = 1$. Thus the problem is again the Blasius problem but the similarity variable is no longer $y(u_0/\nu x)^{\frac{1}{2}}$ but now is[†]

$$\eta = y \left[\frac{u_0}{\nu(x + x_0)} \right]^{\frac{1}{2}}.$$

[†] This relation is also obtainable from the one parameter group approach by noting the invariance of the boundary layer equations under $x \to \bar{x} + \bar{x}_0$.

Determination of x_0 is not possible by means of the mathematical model for this case. The magnitude of x_0 is fixed by conditions at the leading edge of the plate but the present data seems to be too inaccurate to make meaningful estimates [16].

4.7 SIMILARITY AND CONSERVATION LAWS

In 1959 Pattle [18] published, almost without discussion, a class of simple exact solutions for the equation of diffusion or thermal conduction:

$$\frac{\partial}{\partial x}\left[T^n \frac{\partial T}{\partial x}\right] = \frac{\partial T}{\partial t}, \qquad n > 0. \tag{4.82}$$

Boyer [19] used the method of "separation of variables" to developed Pattle's solution and more generally that for the q-dimensional equation with spherical symmetry (see also Ames [74])

$$r^{1-q} \frac{\partial}{\partial r}\left[r^{q-1} T^n \frac{\partial T}{\partial r}\right] = \frac{\partial T}{\partial t}. \tag{4.83}$$

When the *general* form of the similarity variable is assumed to be $r/R(t)$ a trial solution, in separated form

$$T(r, t) = U(t) Y(r/R(t)) \tag{4.84}$$

with U, R, Y unspecified is attempted. Setting $x = r/R(t)$, for convenience, and substituting Eq. (4.84) into Eq. (4.83) gives[†]

$$U^{n+1}(t) x^{1-q} \frac{d}{dx}[x^{q-1} Y^n Y'] = R(t)[RYU' - xUY'R']. \tag{4.85}$$

[†] An elementary result can be applied here for what follows. A necessary and sufficient condition that

$$\begin{vmatrix} a_1(x) & a_2(x) \\ b_1(t) & b_2(t) \end{vmatrix} = V(t) Z(x)$$

is that

$$\begin{vmatrix} a_1(x) & a_2(x) \\ a_1'(x) & a_2'(x) \end{vmatrix} \cdot \begin{vmatrix} b_1(t) & b_2(t) \\ b_1'(t) & b_2'(t) \end{vmatrix} \equiv 0.$$

4.7 SIMILARITY AND CONSERVATION LAWS

The right-hand side of Eq. (4.85) may be expressed as

$$\begin{vmatrix} Y(x) & xY'(x) \\ UR' & U'R \end{vmatrix}$$

which is "separable" (see footnote result) if and only if

$$\begin{vmatrix} Y & xY' \\ Y' & (xY')' \end{vmatrix} \cdot \begin{vmatrix} UR' & U'R \\ (UR')' & (U'R)' \end{vmatrix} \equiv 0. \qquad (4.86)$$

Since neither U nor R are specified we select a relation between them so that Eq. (4.86) is always satisfied. Recalling that a determinant has the value zero if one column (or row) is a constant multiple of another column (row) we ask that

$$UR' = -AU'R$$

so that

$$R(t) = [U(t)]^{-A} \qquad (4.87)$$

with A constant. The right-hand side of Eq. (4.85) takes the form

$$R^2 U'[Y + AxY']$$

so that the equation separates into

$$\frac{x^{1-q}(d/dx)[x^{q-1}Y^n Y']}{Y + AxY'} = \frac{R^2 U'}{U^{n+1}} = -B \qquad (4.88)$$

with B constant.

The form of the similarity variables is still not clear. At this point we invoke boundary conditions to determine the constant A. Several cases will be considered.

One special case of interest is that of constant total heat (energy or matter in general) corresponding to diffusion of a fixed quantity of thermal energy, initially at the origin.[†] The "constant quantity" condition can be stated mathematically as requiring that E, given by

$$E = s(q) \int_0^\infty T(r, t) r^{q-1}\, dr \qquad (4.89)[‡]$$

[†] Bomb blasts, bursts of radiation, transmission line pulses and seepage of liquids into porous bodies are physical examples of such phenomena.

[‡] $s(q)$ is the area of the unit sphere in q dimensions.

be constant. Substituting Eq. (4.84) into Eq. (4.89) gives

$$E = R^q(t)U(t)s(q) \int_0^\infty Y(x)x^{q-1}\,dx$$

thereby requiring that

$$R^q(t)U(t) = U^{1-Aq}$$

be constant. Hence $A = 1/q$ and

$$R = U^{-1/q}. \tag{4.90}$$

A second case is that of a point source producing heat (by say chemical or radioactive means) at a constant rate, i.e., the flux at $r = 0$ is

$$K(T)\frac{\partial T}{\partial r} = \text{constant} = \gamma. \tag{4.91}$$

This requirement becomes

$$T^n \frac{\partial T}{\partial r}\bigg|_{r=0} = U^n(t)R^{-1}(t)Y^n(x)Y'(x)\big|_{x=0} = \text{constant}$$

or that

$$R^{-1}U^n = U^{n+A}$$

be independent of t. Therefore, $A = -n$ and

$$R(t) = U^n. \tag{4.92}$$

In either case A has been determined from auxiliary conditions and Eqs. (4.88) can now be attacked. For the first case with $R = U^{-1/q}$ the integral of $R^2 U'/U^{n+1} = -B$ is

$$U(t) = [B(n+2/q)t]^{-q/(nq+2)}. \tag{4.93}$$

so that

$$R(t) = [B(n+2/q)t]^{1/(nq+2)}. \tag{4.94}$$

With $A = 1/q$ the spatial part may be written as

$$\frac{d}{dx}[x^{q-1}Y^n Y'] = -\frac{B}{q}\frac{d}{dx}[x^q Y] \tag{4.95}$$

4.7 SIMILARITY AND CONSERVATION LAWS

which integrates once to

$$x^{q-1}Y^n Y' = -\frac{B}{q} x^q Y + \alpha.$$

If $Y(0) = C$ and $x^{q-1}Y'(x) \to 0$ as $x \to 0$, a second integration and constant evaluation yields

$$Y(x) = C\left[1 - \frac{nBx^2}{2qC^n}\right]^{1/n}. \tag{4.96}$$

Without losing an arbitrary constant we can set $nB = 2qC^n$ and evaluate C from the constant quantity condition

$$E = s(q)\int_0^\infty Y(x)x^{q-1}\,dx = s(q)C\int_0^1 [1 - x^2]^{1/n} x^{q-1}\,dx.$$

After some algebra the solution is expressible as

$$T(r, t) = \begin{cases} \frac{E}{gR^q}\left[1 - \left(\frac{r}{R}\right)^2\right]^{1/n}, & 0 \leqslant r \leqslant R(t) \\ 0, & r > R(t) \end{cases} \tag{4.97}$$

$$R(t) = \left[\frac{2}{n}(nq + 2)\left(\frac{E}{q}\right)^n t\right]^{1/(nq+2)}$$

$$g(n, q) = \frac{2}{(nq + 2)}\pi^{q/2}\frac{\Gamma(1/n)}{\Gamma((1/n) + q/2)}. \tag{4.98}$$

The solution (4.97) describes a burst appearing initially as a heat pole and then expanding behind a sharp front of radius $R(t)$, with zero temperature outside. The complete solution was obtainable in this case. For other values of A numerical or approximate methods may be required. In addition the similarity variable $x = r/R(t)$ has been constructed by means of separation. Other applications are presented in Boyer's paper.

The engineering literature is very rich in similarity solutions obtained by Birkhoff's method of "separation of variables." Unfortunately, since the choice of appropriate variables is difficult to explain, the student often feels "lost" in trying to understand how to proceed. The use of the group invariant approach eliminates this high degree of uncertainty. We conclude this section by examining another example of Birkhoff's method.

4. FURTHER ANALYTIC METHODS

Riley [20, 21] develops a similarity solution of the incompressible laminar boundary layer equations for radial flow, in the absence of a pressure gradient, when there is an axially symmetric swirling component of velocity w. With the usual notation of x as radial distance from the axis of symmetry, y as the coordinate parallel to the axis, u, v as the velocity components in these directions and $\nu = \mu/\rho$ the governing equations are

$$uu_x + vu_y - w^2/x = \nu u_{yy} \tag{4.99}$$

$$uw_x + vw_y + uw/x = \nu w_{yy} \tag{4.100}$$

$$(xu)_x + (xv)_y = 0. \tag{4.101}$$

The momentum equation (4.99) and continuity equation (4.101) are derived on the basis that u and w are of the same order of magnitude, but v is of lower order. Equation (4.100) is the momentum equation in the λ-direction where λ is defined so that (x, λ, y) form an orthogonal coordinate system.

The system of equations (4.99)–(4.101) is simplified by the introduction of a stream function ψ such that

$$xu = \psi_y, \qquad xv = -\psi_x \tag{4.102}$$

whereupon Eq. (4.99) and (4.100) become

$$x^{-1}\psi_{xy}[\psi_x + \psi_y] - (x^{-1}\psi_y)^2 - w^2 = \nu \psi_{yyy} \tag{4.103}$$

and

$$\psi_y w_x - \psi_x w_y + \psi_y w/x = \nu x w_{yy}. \tag{4.104}$$

A similarity solution of Eqs. (4.103) and (4.104) is now sought in the form

$$\psi = A x^\gamma g_1(x) f(\eta)$$

$$\eta = \frac{\gamma A}{\nu} x^{\gamma-2} g_2(x) y \tag{4.105}$$

$$w = \frac{\gamma A^2}{\nu} \epsilon x^{2\gamma-4} g_3(x) h(\eta).$$

The author gives little discussion of the reasons for such an ingenious choice which illustrates the confusion about these transformations. However, some idea of where they arise, can be inferred from the following observations.

4.7 SIMILARITY AND CONSERVATION LAWS

(1) The boundary layer equations for radial laminar incompressible flow with axial symmetry

$$uu_x + vu_y = vu_{yy}$$
$$(xu)_x + (xv)_y = 0 \quad (4.106)$$

have the similarity solution

$$\psi = A(x^3 + l^3)^{\gamma/3} f(\eta) \quad (4.107)$$
$$\eta = \frac{\gamma A}{\nu} x(x^3 + l^3)^{(\gamma-3)/3} y$$

as obtained by Watson and later by Riley [22]. Here $f(\eta)$ satisfies

$$f''' + ff'' + \alpha(f')^2 = 0 \quad (4.108)$$

and

$$\alpha = (3 - 2\gamma)/\gamma.$$

(2) Since the equation for f has been the subject of considerable investigation it would be advantageous to attempt to keep the same form in the more general problem.

(3) Additional freedom is supplied by the introduction of the undetermined functions $g_1(x)$, $g_2(x)$, $g_3(x)$, and $h(\eta)$.

These observations, after some trial and error, lead to the similarity variables of Eqs. (4.105). In this system, f still satisfies Eq. (4.108), ϵ is a constant and A and γ are the constants from Eq. (4.107). Substitution of system (4.105) into Eqs. (4.103) and (4.104) shows that when

$$G(x) = \left(1 - \frac{\epsilon^2}{x^2}\right)^{\frac{3}{2}} + \frac{l^3}{x^3}$$

that

$$g_1(x) = G^{\gamma/3}, \quad g_2(x) = \left(1 - \frac{\epsilon^2}{x^2}\right)^{\frac{1}{2}} G^{(\gamma-3)/3}, \quad g_3(x) = G^{(2\gamma-3)/3},$$

and $h(\eta) = f'(\eta)$. Thus the similarity solution of Eqs. (4.103) and (4.104) is given by

$$\psi = A[(x^2 - \epsilon^2)^{\frac{3}{2}} + l^3]^{\gamma/3} f(\eta)$$
$$\eta = \frac{\gamma A}{\nu}(x^2 - \epsilon^2)^{\frac{1}{2}}[(x^2 - \epsilon^2)^{\frac{3}{2}} + l^3]^{(\gamma-3)/3} y$$
$$w = \frac{\gamma A^2}{\nu} \frac{\epsilon}{x}[(x^2 - \epsilon^2)^{\frac{3}{2}} + l^3]^{(2\gamma-3)/3} f'(\eta).$$

4.8 GENERAL COMMENTS ON TRANSFORMATION GROUPS

We have previously had several opportunities to refer to Birkhoff's classic "Hydrodynamics" [7]. This work is especially important because of its clarification of the relationship between modeling, dimensional analysis and similarity and the mathematical theory of transformation groups. We shall abstract some of these ideas here.

The procedures of dimensional analysis as clearly stated in the monograph of Bridgman [23] are based upon the following general mathematical assumptions.

I. There are certain independent "fundamental units" q_i such that for any positive real numbers λ_i, $i = 1, 2, ..., n$, we can transform units by means of

$$T(q_i) = \lambda_i q_i \qquad (4.109)$$

(in mechanics $n = 3$, and q_i are length, mass, and time).

II. There are "derived quantities" Q_j (e.g., viscosity, density) which are homogeneous in the sense that under Eq. (4.109) each Q_j is multiplied by a "conversion factor" given by

$$T(Q_j) = Q_j \lambda_1^{a_{j1}} \lambda_2^{a_{j2}} \cdots \lambda_n^{a_{jn}}, \qquad j = 1, ..., n. \qquad (4.110)$$

The exponents a_{jk} are called the "dimensions" of Q_j; if they are all zero, then Q_j is called dimensionless.

III. Any other quantity Q_0 is determined by $Q_1, ..., Q_n$ through a functional relationship

$$Q_0 = f(Q_1, Q_2, ..., Q_n). \qquad (4.111)$$

IV. The relation $Q_0 = f(Q_1, ..., Q_n)$ is "unit free," in the sense of being preserved by any transformation.

V. The quantities $Q_1, ..., Q_n$ involve all n fundamental units.

The validity of these assumptions is and has been hotly debated. Birkhoff discusses some of these arguments. Our concern is not with these discussions but rather with mathematical means of demonstrating the proof of a principle. To illustrate the idea we consider the principle: "In Newtonian fluid mechanics, length, time, and mass can be taken as independent fundamental units". To convince oneself of the validity

4.8 GENERAL COMMENTS ON TRANSFORMATION GROUPS

of this statement (a special case of assumption IV above) one must resort to the "inspectional analysis" of Ruark [24] and Birkhoff. Inspectional analysis can be defined as "testing for invariance under the transformation (4.109), every fundamental mathematical equation on which a theory is based." This test has occurred most commonly in connection with "dynamic similitude" of fluid flows.

Two fluid motions ϕ and ϕ' are "dynamically similar" if they can be described by coordinate systems which are related by transformations of space-time-mass, of the form

$$x_i' = \alpha x_i, \qquad t' = \beta t, \qquad m' = \gamma m. \tag{4.112}$$

The equations of a hydrodynamical theory are easily tested for invariance under Eq. (4.112).

We must not make the mistake of supposing that the applications of "inspectional analysis" are limited to transformations of the form (4.112). In fact inspectional analysis is applicable to any group[†] of transformations. A group of transformations is a set which contains an identity, the inverse of any member and the "product" of any two members. The term *"product"* is used here to describe the group operation which may be any operation. Ehrenfest-Afanassjewa [25] bases this assertion on the axiom: "if the hypotheses of a theory are invariant under a group G, then so are its conclusions. Conversely, the set of all one-to-one transformations leaving any set of equation invariant forms a group."

After the group of transformation (4.112) the ten parameter *Galilei-Newton* group is most important for mechanics. This group is generated by the three parameter subgroup S of space translations

$$x_i' = x_i + c_i \qquad (i = 1, 2, 3); \tag{4.113}$$

the one-parameter subgroup T of time translations

$$t' = t + c; \tag{4.114}$$

the three-parameter subgroup R of rigid rotations

$$x_i' = \sum_{k=1}^{3} a_{ik} x_k \qquad (i = 1, 2, 3) \tag{4.115}$$

[†] The group concept and properties are sketched in Appendix I.

where $A = (a_{ik})$ is a three by three orthogonal[†] matrix; and the three parameter subgroup M of moving coordinates, being translated with constant velocity,

$$x_i' = x_i - b_i t. \tag{4.116}$$

We can easily verify that Newton's three laws of motion are invariant under the Galilei-Newton group and that these transformations do not change the definitions of such material constants as density, viscosity and so forth. These remarks imply that Newtonian Mechanics is invariant under the Galilei-Newton group as well as under the group (4.112) of dynamic similitude.

In view of these remarks about the invariant character of Newtonian mechanics under at least two groups it is surprising that relatively little use of the similarity concept has been made in solid mechanics. The probable reasons for this are the restrictions imposed by the boundary conditions. As an example the damped wave equation

$$u_{xx} = u_{tt} + (u_t)^2$$

can easily be shown by means of one-parameter group methods, to have the similarity solution $f(\eta) = \psi(x, t)$, $\eta = t/x$ where $f(\eta)$ satisfies

$$(\eta^2 - 1)f'' + \eta f' - (f')^2 = 0.$$

Greater utilization of these methods should help in obtaining solutions in solid nonlinear continuum mechanics.

4.9 SIMILARITY APPLIED TO MOVING BOUNDARY PROBLEMS

One dimensional melting-freezing problems are often called Stefan problems although similarity solutions were probably first given by Neumann. Most of the studies discussed by Carslaw and Jaeger [26], Ruoff [27], Knuth [28], and Horvay [29] concerned problems with moving boundaries using the classical similarity variable $xt^{-\frac{1}{2}}$. An alternate approach introduced by Landau [30] involved the transformation variable $x - a/(\delta - a)$, where a is the original slab thickness. Hamill and Bankoff [31] utilize a related idea which has the great advan-

---

[†] A matrix is orthogonal if $A^T = A^{-1}$, i.e., $AA^T = A^TA = I$.

4.9 MOVING BOUNDARY PROBLEMS

tage in that it immobilizes the moving boundary. The computation may then be performed for an initial value problem instead of a boundary value problem with two fixed end points and one "floating" intermediate point.

Let a semi-infinite slab of phase 1 be originally at temperature T_1 (for time $t < 0$). At $t = 0$, corresponding to the appearance of phase 2, a constant temperatature T_W is applied at the face $x = 0$ as shown in Fig. 4-1. Let the thickness of this new phase be $\delta(t)$ and the moving

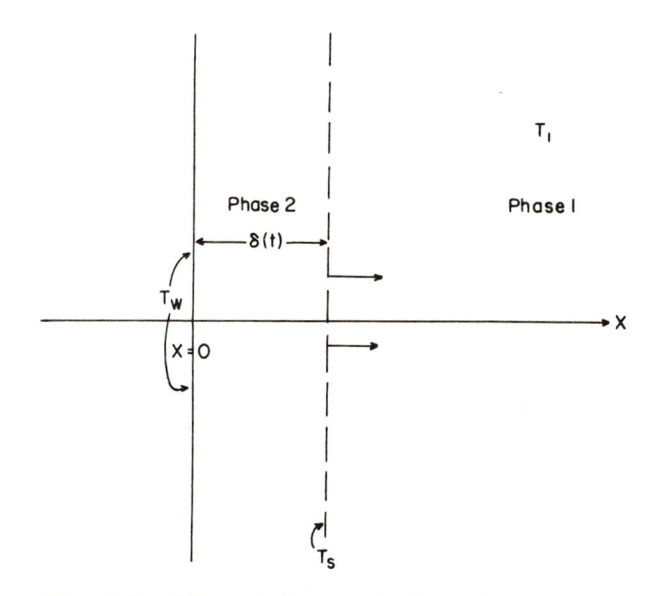

FIG. 4-1. Schematic for a moving boundary problem.

boundary be at the transition temperature T_s. If the densities ρ_i of the two phases remain constant but unequal and the specific heats c_i and conductivities k_i depend upon the temperature the equations for the temperature field in each phase are:

Phase 1:

$$\rho_1 c_1 \left\{ \frac{\partial T_1}{\partial t} + u_1 \frac{\partial T_1}{\partial x_1} \right\} = \frac{\partial}{\partial x_1} \left(k_1 \frac{\partial T_1}{\partial x_1} \right) \tag{4.117}$$

Phase 2:

$$\rho_2 c_2 \frac{\partial T_2}{\partial t} = \frac{\partial}{\partial x_2} \left(k_2 \frac{\partial T_2}{\partial x_2} \right). \tag{4.118}$$

The quantity u_1 is the velocity of phase 1 resulting from the difference in density of the two phases. The boundary conditions are

$$T_1(\delta, t) = T_2(\delta, t) = T_s \quad \text{(melting point)}$$
$$T_2(0, t) = T_W \tag{4.119}$$

and the initial conditions are

$$\delta(0) = 0$$
$$T_1(x, 0) = f(x). \tag{4.120}$$

Lastly, one further condition must be described to couple the two subsystems at the free boundary.

Since the temperature at the interface remains constant, the surface of separation is a constant energy surface. Neglecting kinetic and potential energy effects, the conservation of energy reduces to an enthalpy balance:

$$k_1 \frac{\partial T_1}{\partial x_1}\bigg|_{+\epsilon} = \pm \rho_2 \Omega \frac{d\delta}{dt} + k_2 \frac{\partial T_2}{\partial x_2}\bigg|_{-\epsilon} \tag{4.121}$$

where Ω is the latent heat and $\pm\epsilon$ denotes that the gradients are measured an infinitesimal distance ϵ on both sides of the surface. For constant densities we can write

$$u_1 = -\left(\frac{\rho_2 - \rho_1}{\rho_1}\right) \frac{d\delta}{dt} \tag{4.122}$$

so that Eq. (4.117) can be replaced by

$$\rho_1 c_1 \left[\frac{\partial T_1}{\partial t} - \left(\frac{\rho_2 - \rho_1}{\rho_1}\right) \frac{d\delta}{dt} \frac{\partial T_1}{\partial x_1}\right] = \frac{\partial}{\partial x_1}\left(k_1 \frac{\partial T_1}{\partial x_1}\right). \tag{4.117a}$$

A change in independent variable is now made by defining

$$\lambda_i = x_i/\delta(t) \tag{4.123}$$

and under this transformation Eqs. (4.117a) and (4.118) become

$$\rho_1 c_1 \left\{\frac{\partial T_1}{\partial t} - \left(\lambda_1 + \frac{\rho_2 - \rho_1}{\rho_1}\right) \frac{1}{\delta} \frac{d\delta}{dt} \frac{\partial T_1}{\partial \lambda_1}\right\} = \frac{1}{\delta^2} \frac{\partial}{\partial \lambda_1}\left(k_1 \frac{\partial T_1}{\partial \lambda_1}\right)$$

$$\rho_2 c_2 \left\{\frac{\partial T_2}{\partial t} - \frac{\lambda_2}{\delta} \frac{d\delta}{dt} \frac{\partial T_2}{\partial \lambda_2}\right\} = \frac{1}{\delta^2} \frac{\partial}{\partial \lambda_2}\left(k_2 \frac{\partial T_2}{\partial \lambda_2}\right)$$

where $T_i = T_i(\lambda_i, t)$, $i = 1, 2$, and $\lambda_1 \geq 1, 0 \leq \lambda_2 \leq 1$.

4.9 MOVING BOUNDARY PROBLEMS

The energy balance at the interface becomes

$$\frac{k_1}{\delta}\frac{\partial T_1}{\partial \lambda_1}\bigg|_{+\epsilon} = \pm \rho_2 \Omega \frac{d\delta}{dt} + \frac{k_2}{\delta}\frac{\partial T_2}{\partial \lambda_2}\bigg|_{-\epsilon}.$$

To explain what follows later it is convenient to rewrite the last three equations in the alternate form

$$\rho_1 c_1 \frac{\partial T_1}{\partial t} = \frac{1}{\delta^2}\left\{\frac{\partial}{\partial \lambda_1}\left(k_1 \frac{\partial T_1}{\partial \lambda_1}\right) + \frac{c_1}{2}(\rho_2 - \rho_1 + \lambda_1 \rho_1)\frac{d\delta^2}{dt}\frac{\partial T_1}{\partial \lambda_1}\right\} \quad (4.124)$$

$$\rho_2 c_2 \frac{\partial T_2}{\partial t} = \frac{1}{\delta^2}\left\{\frac{\partial}{\partial \lambda_2}\left(k_2 \frac{\partial T_2}{\partial \lambda_2}\right) + \frac{c_2}{2}\rho_2 \lambda_2 \frac{d\delta^2}{dt}\frac{\partial T_2}{\partial \lambda_2}\right\} \quad (4.125)$$

and

$$k_1 \frac{\partial T_1}{\partial \lambda_1}\bigg|_{+\epsilon} = \pm \frac{1}{2}\rho_2 \Omega \frac{d\delta^2}{dt} + k_2 \frac{\partial T_2}{\partial \lambda_2}\bigg|_{-\epsilon}. \quad (4.126)$$

The boundary conditions are

$$T_1(1, t) = T_s = T_2(1, t)$$
$$T_2(0, t) = T_W. \quad (4.127)$$

As yet the initial temperature distribution $T_1(\lambda_1, 0) = f(\lambda_1)$ has not been specified. Neither have we specified a second boundary condition for phase 1. These will be forced upon us as a consequence of the desire for a similarity solution.

Here we consider the possibility of finding particular solutions to the system which are independent of time and still meet all the requirements imposed above. Thus if $T_i = T_i(\lambda_i)$ we see that Eq. (4.126) requires that $d\delta^2/dt = \text{constant} = 2\alpha$ or that $\delta(t)$ be proportional to $t^{\frac{1}{2}}$. Thus Eqs. (4.124) and (4.125) become

$$\frac{d}{d\lambda_1}\left(k_1 \frac{dT_1}{d\lambda_1}\right) + c_1(\rho_2 - \rho_1 + \lambda_1 \rho_1)\alpha \frac{dT_1}{d\lambda_1} = 0 \quad (4.128)$$

and

$$\frac{d}{d\lambda_2}\left(k_2 \frac{dT_2}{d\lambda_2}\right) + c_2 \rho_2 \lambda_2 \alpha \frac{dT_2}{d\lambda_2} = 0. \quad (4.129)$$

Apart from the requirement that $\delta(t) \sim t^{\frac{1}{2}}$, the consistency requirements are that the boundary conditions (4.127) as well as all other boundary conditions be time-independent. Let us specify $T_1(\infty) = T_\infty$

(constant). Since $\lambda \to \infty$ as $x \to \infty$ and as $t \to 0$ this also represents the uniform initial condition for the liquid temperature. Thus the problem is characterized by the two ordinary differential equations (4.128) and (4.129), the auxiliary conditions

$$\begin{aligned} T_1(1) = T_s, & \quad T_2(1) = T_s \\ T_1(\infty) = T_\infty, & \quad T_2(0) = T_W, \end{aligned} \quad (4.130)$$

and at $\lambda_1 = \lambda_2 = 1$ the enthalpy condition

$$k_1 \frac{dT_1}{d\lambda_1} - k_2 \frac{dT_2}{d\lambda_2} = \pm \rho_2 \Omega \alpha.$$

4.10 SIMILARITY CONSIDERATIONS IN THREE DIMENSIONS

As a last example of the application of this powerful procedure we consider three dimensional boundary layer flows of non-Newtonian fluids by means of the "separation of variables". Several authors have solved the steady boundary layer flows near a stagnation point, near a rotating disk or over a plane, e.g., Srivastava [32], Jain [33], and Jones [34]. Unsteady boundary layer flows have been examined by Datta [35] and Wu [36]. The particular case of interest to us here is due to Datta [37].

We consider the problem of the three-dimensional boundary layer over a spinning cone, of a non-Newtonian fluid that is incompressible, inelastic and of Reiner-Rivlin type.[†] The constitutive equations for such a fluid are

$$\begin{aligned} \tau_{ij} &= -p\delta_{ij} + \tau'_{ij} \\ \tau'_{ij} &= 2\mu e_{ij} + 4\mu_c e_{ik} e_{kj} \end{aligned} \quad (4.131)$$

$i, j = 1, 2, 3$, where τ_{ij}, e_{ij}, p, μ, μ_c, and δ_{ij} are the stress components, rate of strain components, pressure, coefficient of viscosity, coefficient of cross viscosity, and $\delta_{ij} = 1$ if $i = j$ and 0 otherwise, respectively. In addition we know that

$$e_{ij} = \tfrac{1}{2}(v_{i,j} + v_{j,i}) \quad (4.132)$$

where the comma subscript notation $(\,,j)$ means differentiation and the v_i are velocity components.

[†] See Eringen [38] for these concepts.

4.10 SIMILARITY CONSIDERATIONS IN THREE DIMENSIONS

Let α be the half angle opening of the cone (see Fig. 4-2) and choose cartesian coordinates (x, y, z) with origin 0 at the cone's vertex and

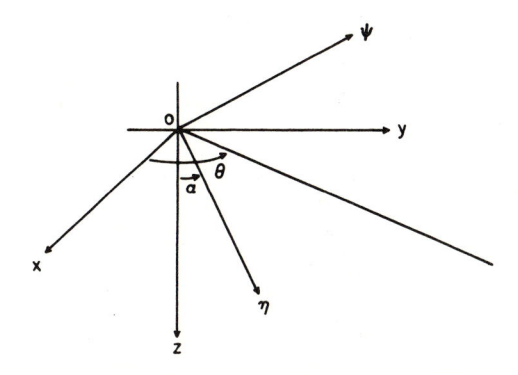

FIG. 4-2. Coordinate systems for steady boundary layer flows in three dimensions.

z-axis along the cone axis. An alternate set of orthogonal coordinates (η, ψ, θ) is convenient, with η along the cone generatrix,

$$x = (\eta \sin \alpha + \psi \cos \alpha) \cos \theta \quad (4.133)$$

$$y = (\eta \sin \alpha + \psi \cos \alpha) \sin \theta$$
$$z = \eta \cos \alpha - \psi \sin \alpha. \quad (4.134)$$

In this new coordinate system the physical components of rate of strain are easily determined to be

$$e_{\eta\eta} = \frac{\partial v_\eta}{\partial \eta}, \quad e_{\psi\psi} = \frac{\partial v_\psi}{\partial \psi}, \quad e_{\theta\theta} = \frac{v_\eta \sin \alpha + v_\psi \cos \alpha}{\eta \sin \alpha + \psi \cos \alpha}$$

$$e_{\eta\psi} = \frac{1}{2}\left[\frac{\partial v_\eta}{\partial \psi} + \frac{\partial v_\psi}{\partial \eta}\right], \quad e_{\eta\theta} = \frac{1}{2}\left[\frac{\partial v_\theta}{\partial \eta} - \frac{v_\theta \sin \alpha}{\eta \sin \alpha + \psi \cos \alpha}\right] \quad (4.135)$$

$$e_{\psi\theta} = \frac{1}{2}\left[\frac{\partial v_\theta}{\partial \psi} - \frac{v_\theta \cos \alpha}{\eta \sin \alpha + \psi \cos \alpha}\right].$$

The continuity equation is

$$\frac{\partial v_\eta}{\partial \eta} + \frac{\partial v_\psi}{\partial \psi} + \frac{v_\eta \sin \alpha + v_\psi \cos \alpha}{\eta \sin \alpha + \psi \cos \alpha} = 0; \quad (4.136)$$

and the equations of momentum are

$$\rho \left[v_\eta \frac{\partial v_\eta}{\partial \eta} + v_\psi \frac{\partial v_\eta}{\partial \psi} - \frac{v_\theta^2 \sin \alpha}{\eta \sin \alpha + \psi \cos \alpha} \right]$$

$$= -\frac{\partial p}{\partial \eta} + \frac{\partial \tau'_{\eta\eta}}{\partial \eta} + \frac{\partial \tau'_{\eta\psi}}{\partial \psi} + \frac{(\tau'_{\eta\eta} - \tau'_{\theta\theta}) \sin \alpha + \tau'_{\eta\psi} \cos \alpha}{\eta \sin \alpha + \psi \cos \alpha}$$

$$\rho \left[v_\eta \frac{\partial v_\psi}{\partial \eta} + v_\psi \frac{\partial v_\psi}{\partial \psi} - \frac{v_\theta^2 \cos \alpha}{\eta \sin \alpha + \psi \cos \alpha} \right]$$
$$= -\frac{\partial p}{\partial \psi} + \frac{\partial \tau'_{\psi\psi}}{\partial \psi} + \frac{\partial \tau'_{\eta\psi}}{\partial \eta} + \frac{(\tau'_{\psi\psi} - \tau'_{\theta\theta}) \cos \alpha + \tau'_{\eta\psi} \sin \alpha}{\eta \sin \alpha + \psi \cos \alpha} \quad (4.137)$$

$$\rho \left[v_\eta \frac{\partial v_\theta}{\partial \eta} + v_\psi \frac{\partial v_\theta}{\partial \psi} + \frac{v_\eta \sin \alpha + v_\psi \cos \alpha}{\eta \sin \alpha + \psi \cos \alpha} v_\theta \right]$$
$$= \frac{\partial \tau'_{\eta\theta}}{\partial \eta} + \frac{\partial \tau'_{\psi\theta}}{\partial \psi} + \frac{2(\tau'_{\eta\theta} \sin \alpha + \tau'_{\psi\theta} \cos \alpha)}{\eta \sin \alpha + \psi \cos \alpha}.$$

In all these equations conical symmetry has been taken into consideration and the τ_{ij} are given by the second equation of (4.131) and by Eq. (4.135).

The boundary layer assumptions, which are compatible with the continuity Eqs. (4.136), are

$$v_\eta = O(1), \qquad v_\psi = O(\delta/\eta_0), \qquad v_\theta = O(1)$$

which are generally valid at a large distance from the cone's vertex. Now with $\nu = \mu/\rho$, $\nu_c = \mu_c/\rho$, and assuming

$$\nu_c = O(\nu), \qquad \delta/\eta_0 = O\left(\frac{1}{\eta_0} \sqrt{\frac{\nu}{\omega \sin \alpha}}\right),$$

where ω is the angular velocity of rotation, we deduce the appropriate boundary layer equations by retaining all terms that are $O(1)$. Thus Eqs. (4.136) and (4.137) become

$$\frac{\partial v_\eta}{\partial \eta} + \frac{\partial v_\psi}{\partial \psi} + \frac{v_\eta}{\eta} = 0 \quad (4.138)$$

$$v_\eta \frac{\partial v_\eta}{\partial \eta} + v_\psi \frac{\partial v_\eta}{\partial \psi} - \frac{v_\theta^2}{\eta} = -\frac{1}{\rho} \frac{\partial p}{\partial \eta} + \nu \frac{\partial^2 v_\eta}{\partial \psi^2}$$
$$+ \nu_c \left[\frac{\partial}{\partial \eta} \left(\frac{\partial v_\eta}{\partial \psi} \right)^2 + \frac{\partial}{\partial \psi} \left(\frac{\partial v_\theta}{\partial \eta} \cdot \frac{\partial v_\theta}{\partial \psi} - \frac{v_\theta}{\eta} \frac{\partial v_\theta}{\partial \psi} - 2 \frac{v_\eta}{\eta} \frac{\partial v_\eta}{\partial \psi} \right) \right.$$
$$\left. + \frac{1}{\eta} \left\{ \left(\frac{\partial v_\eta}{\partial \psi} \right)^2 - \left(\frac{\partial v_\theta}{\partial \psi} \right)^2 \right\} \right];$$

4.10 SIMILARITY CONSIDERATIONS IN THREE DIMENSIONS

$$v_\eta \frac{\partial v_\theta}{\partial \eta} + v_\psi \frac{\partial v_\theta}{\partial \psi} + \frac{v_\eta v_\theta}{\eta} = \nu \frac{\partial^2 v_\theta}{\partial \psi^2}$$

$$+ \nu_c \left[\frac{\partial}{\partial \eta} \left(\frac{\partial v_\theta}{\partial \psi} \frac{\partial v_\eta}{\partial \psi} \right) + \frac{\partial}{\partial \psi} \left(\frac{\partial v_\theta}{\partial \eta} \frac{\partial v_\eta}{\partial \psi} - \frac{v_\theta}{\eta} \frac{\partial v_\eta}{\partial \psi} - 2 \frac{\partial v_\theta}{\partial \psi} \frac{\partial v_\eta}{\partial \eta} \right) \right. \quad (4.139)$$

$$\left. + \frac{2}{\eta} \frac{\partial v_\theta}{\partial \psi} \frac{\partial v_\eta}{\partial \psi} \right];$$

$$- \frac{v_\theta^2}{\eta} \cot \alpha = - \frac{1}{\rho} \frac{\partial p}{\partial \psi}$$

$$+ \nu_c \left[\frac{\partial}{\partial \psi} \left\{ \left(\frac{\partial v_\eta}{\partial \psi} \right)^2 + \left(\frac{\partial v_\theta}{\partial \psi} \right)^2 \right\} + (\cot \alpha) \frac{1}{\eta} \left(\frac{\partial v_\eta}{\partial \psi} \right)^2 \right].$$

These equations are to be solved subject to the boundary conditions

$$v_\eta(\eta, 0) = 0, \quad v_\psi(\eta, 0) = 0, \quad v_\theta(\eta, 0) = \eta \omega \sin \alpha$$

$$\lim_{\psi \to \infty} v_\eta(\eta, \psi) = 0, \quad \lim_{\psi \to \infty} v_\theta(\eta, \psi) = 0. \quad (4.140)$$

In the spirit of the separation of variables apprach we assume Eqs. (4.138) and (4.139) have solutions of the form

$$v_\eta = f(\eta)F(\psi), \quad v_\psi = g(\eta)G(\psi), \quad v_\theta = h(\eta)H(\psi) \quad (4.141)$$

with, as yet, none of the functions specified. From the equation of continuity Eq. (4.138) the relation

$$\frac{F(\psi)}{G'(\psi)} = - \frac{\eta g(\eta)}{f(\eta) + \eta f'(\eta)} = -\lambda/2 \quad (4.141a)$$

is obtained thus reducing the number of degrees of freedom by 2. The quantity λ is a constant.

To satisfy the third condition of Eq. (4.140) we choose $h(\eta) = \eta$. After these preliminaries the equations of momentum take the following form upon inserting Eqs. (4.141) and (4.141a):

$$f(\eta)f'(\eta)F^2(\psi) + g(\eta)f(\eta)F'(\psi)G(\psi) - \eta H^2(\psi) = - \frac{1}{\rho} \frac{\partial p}{\partial \eta}$$

$$+ \nu f(\eta)F''(\psi) + \nu_c \left[2f(\eta)f'(\eta)\{F'(\psi)\}^2 - 2 \frac{\partial}{\partial \psi} \left\{ \frac{(f(\eta))^2}{\eta} F(\psi)F'(\psi) \right\} \right.$$

$$\left. + \frac{1}{\eta} \{(f(\eta))^2(F'(\psi))^2 - \eta^2(H'(\psi))^2\} \right]$$

$$f(\eta)F(\psi)H(\psi) + g(\eta)G(\psi)H'(\psi) + f(\eta)F(\psi)H(\psi) = \nu\eta H''(\psi) \tag{4.142}$$

$$+ \nu_c \left[\frac{\partial}{\partial \eta}(\eta f(\eta)F'(\psi)H'(\psi)) + \frac{\partial}{\partial \psi}\{-2\eta f'(\eta)H'(\psi)F(\psi)\} + 2H'(\psi)f(\eta)F'(\psi) \right]$$

$$- \eta(H(\psi))^2 \cot \alpha = -\frac{1}{\rho}\frac{\partial p}{\partial \psi} + \nu_c \left[\frac{\partial}{\partial \psi}\{(f(\eta))^2(F'(\psi))^2 + \eta^2(H'(\psi))^2\} \right.$$

$$\left. + \frac{\cot \alpha}{\eta}(f(\eta))^2(F'(\psi))^2 \right].$$

From these it appears that a similarity solution is possible (not unique ?) if

$$f(\eta) = \eta, \quad p(\eta, \psi) = \eta^2 p_1(\psi) + \eta p_2(\psi) \cot \alpha, \tag{4.143}$$

$p_1 \sim O(1)$, $p_2 \sim O(\delta/\eta_0)$. Thus we obtain $g(\eta) = \lambda$ from Eq. (4.141). Substituting these relations into Eq. (4.142) we obtain, upon equating coefficients of η and η^2 (all relations in ψ),

$$F^2 + \lambda F'G - H^2 = \nu F'' - \nu_c[(F')^2 + 2FF'' + 3(H')^2] \tag{4.144}$$

$$\frac{1}{\rho}p_1(\psi) = \nu_c[(F')^2 + (H')^2] \tag{4.145}$$

$$\frac{1}{\rho}p_2'(\psi) = \nu_c(F')^2 + H^2 \tag{4.146}$$

$$2FH + \lambda GH' = \nu H'' + 2\nu_c[F'H' - FH'']. \tag{4.147}$$

Equations (4.144), (4.147), and (4.141a) give us the functions F, G, and H. After this calculation Eqs. (4.145) and (4.146) give the pressure distribution within the boundary layer. These calculations in dimensionless form have been carried out by Srivastava [32] and Datta [37] and are summarized in the latter paper.

The utilization of similarity concepts allows the solution of complex problems of which the above is one example. This analytic device, while "boundary condition" limited does considerably expand our horizon of understanding. When applicable it is far superior to a direct numerical approach.

4.11 GENERAL DISCUSSION OF SIMILARITY

Essentially two methods have been developed and shown to apply to development of similarity variables. Both the group invariants and the "separation of variables" leads to some general conclusions which can act as a simple guide in using these ideas.

(a) The existence of a similarity variable usually implies the lack of a characteristic length in one or more coordinates of the problem.

(b) In some cases it is possible to determine a unique similarity variable directly from the partial differential equation even though some (or all) boundary conditions are missing. In other cases it is necessary to find other conditions to uniquely define a similarity variable. These other conditions may arise from the boundary conditions or from additional information such as conservation theorems.

(c) We emphasize again that in both methods it is necessary to employ or infer a particular class of transformations. If a similarity variable is not found for that class it cannot be concluded that none exist, but only that there is no similarity variable under that class of transformations. Establishment of one (or an infinite number) of separation variables for a given partial differential equation never demonstrates that all possible separation variables have been found. Others may arise from other transformations.

(d) Similarity solutions are often asymptotic solutions to a given problem and therefore may have great utility in this area of limiting solutions.

(e) The group invariants method is probably the easier, more straight forward method. Both methods have been found to agree except in certain degenerate cases.

(f) Reduction of an equation to an ordinary differential equation is sometimes possible by repeated application of these procedures (or by one application of the linear combination). This ordinary differential equation is nonlinear—usually requiring the use of numerical or approximate methods.

4.12 INTEGRAL EQUATION METHODS

Conversion of an equation to integral equation form is a natural way for the development of iterative methods. The difficulties that may arise have to do with problems of integration and questions of convergence of the iteration process.

In illustration of the idea we consider an application to a transmission line, originally due to Boyer [19]. Let the transmission line of infinite extent have a nonlinear series resistance and linear shunt capacitance so that the governing equations are

$$\frac{\partial i}{\partial x} = -\alpha \frac{\partial \phi}{\partial t}, \qquad E = -\frac{\partial \phi}{\partial x}, \qquad E = r(E)i \qquad (4.148)$$

where $r(E)$, α, i, E, and ϕ represent resistance per unit length, capacitance per unit length, current, electric field, and capacitance voltage drop, respectively. If we eliminated both i and ϕ from Eqs. (4.148) we get

$$\frac{\partial^2}{\partial x^2}\left[\frac{E}{r(E)}\right] = \frac{\partial}{\partial x}\left[\frac{d}{dE}\left(\frac{E}{r(E)}\right)\frac{\partial E}{\partial x}\right] = \alpha \frac{\partial E}{\partial t}. \qquad (4.149)$$

Equation (4.149) is clearly of the same form as the nonlinear diffusion equation (1.10) when we set $D(E) = (1/\alpha)(d/dE)(E/r(E))$ and identify E with C.

If instead of eliminating both i and ϕ from Eq. (4.148) we eliminate only ϕ the resulting system is

$$E = r(E)i, \qquad i_{xx} = \alpha E_t. \qquad (4.150)$$

From the first of these equations we can solve for E and write

$$E(x, t) = F[i(x, t)] \qquad (4.151)$$

by solving for E. From the second equation of Eq. (4.150) one finds upon integrating from 0 to x (once) that

$$\frac{\partial i}{\partial x} = \alpha \int_0^x E_t(y, t)\, dy + A(t).$$

A second integration gives

$$i(x, t) = \alpha \int_0^x \int_0^x E_t(y, t)\, dy^2 + A(t)x + B(t)$$

$$= \alpha \int_0^x (x - y)E_t(y, t)\, dy + A(t)x + B(t). \qquad (4.152)$$

To achieve the final result of Eq. (4.152) we have utilized the easily proved result (see e.g., Hildebrand [39])

$$\underbrace{\int_a^x \cdots \int_a^x f(x)\, dx \cdots dx}_{n \text{ times}} = \frac{1}{(n-1)!}\int_a^x (x-y)^{n-1} f(y)\, dy.$$

4.12 INTEGRAL EQUATION METHODS

An iteration method may now be applied, based upon Eqs. (4.151) and (4.152). We choose $i_1(x, t)$ by means of any information available (or arbitrarily) and iterate by means of

$$E_n = F(i_n), \qquad n = 1, 2, \ldots$$
$$i_{n+1} = \alpha \int_0^x (x - y)(E_n)_t \, dy.$$
(4.153)

As an illustration of this procedure we suppose the "constant quantity" condition holds, that is,

$$\int_0^\infty E(y, t) \, dy = \text{constant}$$

and that $r(E) = r_0(E_0/E)^m$ $(m > 0)$[†]. Further we assume that $A(t) = B(t) \equiv 0$ and $E > 0$ for $0 \leqslant x \leqslant R(t)$. Returning to Eq. (4.152) we can write

$$\frac{\partial i}{\partial x} = \alpha \int_0^\infty E_t(y, t) \, dy - \alpha \int_x^\infty E_t(y, t) \, dy$$

or

$$\frac{\partial i}{\partial x} = -\alpha \int_x^\infty E_t(y, t) \, dy$$

by virtue of the constant quantity condition. Integrating once again we finally have the iteration

$$E_n = [r_0 E_0{}^m i_n]^{1/(m+1)}$$

and

$$i_{n+1} = \alpha \int_x^\infty (y - x) \frac{\partial E_n}{\partial t} \, dy.$$
(4.154)

Suppose the pulse front velocity is constant, that is $R(t) = vt$. For arbitrary I_1 and λ_1 take i_1 as

$$i_1(x, t) = \begin{cases} I_1(vt - x)^{\lambda_1}, & 0 \leqslant x \leqslant vt \\ 0, & x > vt \end{cases}$$

then

$$E_1(x, t) = \begin{cases} [I_1 r_0 E_0{}^m (vt - x)^{\lambda_1}]^{1/(m+1)}, & 0 \leqslant x \leqslant vt \\ 0, & x > vt \end{cases}$$

[†] This case has good physical meaning for $m \sim 5$ according to Boyer [19].

and applying the iteration procedure, Eq. (4.154) yields after integration

$$i_{n+1}(x, t) = \begin{cases} I_{n+1}(vt - x)^{\lambda_{n+1}}, & 0 \leq x \leq vt \\ 0, & x > vt \end{cases}$$

where

$$\lambda_{n+1} = (m + 1)^{-1}\lambda_n + 1$$

$$I_{n+1} = \alpha v E_0 \left[\frac{r_0 I_n}{E_0}\right]^{1/(m+1)} \lambda_{n+1}^{-1}.$$

(4.155)

The question of convergence can be examined relatively easily for this special case. As $n \to \infty$, i.e., as the number of iteration steps increases indefinitely, does the procedure converge and if so what is the dependency upon I_1 and λ_1? By way of answering this question we solve the first order difference equation (4.155) by recursion[†] and get

$$\lambda_{n+1} = \left(\frac{1}{m+1}\right)^n \lambda_1 + \sum_{j=1}^{n} \left(\frac{1}{m+1}\right)^{j-1}, \quad m > 0,$$

whose limit as $n \to \infty$ is

$$\lim_{n \to \infty} \lambda_{n+1} = 1 + 1/m$$

which is independent of λ_1. In a similar manner we can show that

$$\lim_{n \to \infty} I_{n+1} = E_0 r_0^{1/m} \left[\frac{\alpha v m}{m+1}\right]^{1+1/m}$$

independent of I_1. Thus the iteration does in fact converge to

$$i(x, t) = \lim_{n \to \infty} i_n(x, t) = \begin{cases} E_0 r_0^{1/m} \left[\frac{\alpha v m}{m+1}(vt - x)\right]^{1+1/m}, & 0 \leq x \leq vt \\ 0, & x > vt \end{cases}$$

which is easily verified to be a solution of our original system. From the expression

$$E_n = [r_0 E_0^m i_n]^{1/(m+1)},$$

$E = \lim_{n \to \infty} E_n$ is easily obtained.

In principle the iteration given via Eqs. (4.153) can be used for any differentiable functions $r(E)$ and $R(t)$. Necessary and sufficient conditions

---

[†] The theory of difference equations is discussed, for example, in Hildebrand [39], Levy and Lessman [40], and Milne-Thomson [41].

for convergence of the iteration, in the general case, appear to be difficult to obtain without becoming overrestrictive. Such questions are usually most easily answered for specific cases. The iterative nature of the solution suggests that this procedure may be of great value in generating approximate solutions to equations. These integral equation methods will be utilized in Chapter 5.

4.13 THE HODOGRAPH

The general concept of the hodograph transformation, complete with several examples, was discussed in Section 2.3. The transformation was used, in an auxiliary way, in Section 3.8 where the vibration of a nonlinear string was discussed. In this section we shall discuss the general utility of the hodograph method and give some examples and references.

Hamilton, in 1846, coined the word "hodograph" for a velocity-locus associated with a moving particle. If the components of velocity are $u_1(t)$, $u_2(t)$, and $u_3(t)$, the hodograph (literally "speed" graph) is the locus of a point whose position coordinates in an auxiliary space are $u_i(t)$, $i = 1, 2, 3$. This amounts to using the velocity components u_i as the independent variables, in terms of which everything else, including the original position coordinates x_i, is to be expressed.

For a fluid in steady motion, the string of fluid-particles on any one streamline all give the same hodograph locus (neglecting the time parameter), so there is a set of hodographs which are maps of the streamlines onto the velocity space. In fluid dynamics the hodograph method is so named because it emphasizes these loci. If, for a specific problem, they have been found the actual streamlines are obtained by mapping back from the velocity space to the physical space.

The hodograph procedure is feasible in any system where the field equations are relations connecting the u_i and $\partial u_i/\partial x_j$. The basic equations for converting to velocity space are given in Section 2.3. If the hodograph field equations should be easier to solve, subject to such boundary conditions as may be prescribed, than the original equations the hodograph method will be advantageous. One such important case where this occurs is that of the steady irrotational motion of an ideal, inviscid gas in two dimensions (see Eqs. (2.98) *et. seq.* and Section 2.3b) whose Eulerian equations

$$(c^2 - u^2)u_x - uv(u_y + v_x) + (c^2 - v^2)v_y = 0, \qquad u_y - v_x = 0 \qquad (4.156)$$

are nonlinear but whose hodograph equations

$$(c^2 - u^2)y_v + uv(x_v + y_u) + (c^2 - v^2)x_u = 0, \qquad x_v - y_u = 0 \qquad (4.157)$$

are linear. The hodograph relations for the stream function ψ are

$$\psi_u = (1 - q^2)^\beta (vx_u - uy_u), \qquad \psi_v = (1 - q^2)^\beta (vx_v - uy_v) \qquad (4.158)$$

where $q^2 = u^2 + v^2$, $\beta = 1/(\gamma - 1)$, and $q^2 + 2\beta c^2 = 1$.

A solution of Eq. (4.157) is a pair of relations

$$x = x(u, v), \qquad y = y(u, v)$$

whose inversion u, v must satisfy Eq. (4.156). The related function ψ, given in Eq. (4.158), can be expressed in terms of x and y. In practice we can sometimes find explicit analytic solutions for Eqs. (4.157) and (4.158)—which is practicable because these equations are linear and relatively simple in form. This linearity also allows for the use of superposition. The inversion, usually, can only be done numerically. We tabulate $\psi(u, v)$, determine thence a set of points (u, v) for which ψ is an assigned constant and then from $x = x(u, v)$, $y = y(u, v)$ determine the corresponding set (x, y) and thus plot as many streamlines as desired.

For the inversion to be possible it is necessary that the Jacobian J should be analytic and nonzero, so that the mapping between the uv and xy planes is locally one to one. In extended regions there may be points where J is zero or infinite. These give rise to branch points of u, v, or x, y as functions of the other pair and the extended mapping is one to one between Riemann surfaces that are usually multiple-sheeted. This phenomena actually arises in physically interesting cases as discussed by Lighthill [42] and Cherry [43, 44]. They show that near the axial sonic point there is a one to one correspondence between the (x, y) plane and a three sheeted hodograph or uv surface. Thus the hodograph solutions $x(u, v)$, $y(u, v)$ must have a branch point (sonic point) where $v = 0$ and $u = c$. This may be in part confirmed from Eq. (4.157) for which the boundary conditions are $y = 0$, $x = f^{-1}(u)$ on $v = 0$ and the coefficient $c^2 - u^2$ (of y_v) is identically zero for $u = c$. A lengthy discussion of the hodograph and its singularities is found in Courant and Friedrichs [45, pp. 62–69 and 256].

The hodograph procedure enables us to plot the streamlines, with uniform accuracy, out to a natural boundary of the flow field. This would not be practicable by direct numerical solution of the nonlinear equations (4.156).

4.14 SIMPLE EXAMPLES OF HODOGRAPH APPLICATION

Special solutions of nonlinear problems have been obtained by numerous authors using the hodograph or modifications. Among these are the classic papers of Molenbroek and Chaplygin (see Chapter 2). For subsonic flows see Tsien [46], Ringleb [47], Bers [48], Bergman [49], Lin [50], and von Mises [51]. Laitone [52] does an excellent job of summarizing, both for subsonic and supersonic flow. Transsonic flows are discussed by Cherry [43, 44], Tsien [53], Courant and Friedrichs [45], Bers [54], Frankl [55], Lighthill [42], and others. Von Kármán obtains the similarity law of transsonic flow in [56].

For our simple examples we recall Eq. (2.100)

$$(c^2 - u^2)X_{vv} + 2uvX_{uv} + (c^2 - v^2)X_{uu} = 0 \qquad (2.100)$$

for the stream function X. As we have previously mentioned this equation, obtained from the hodograph field equations, is linear. Every solution $X(u, v)$ defined in a certain region of the u, v plane leads to a flow given by u and v as functions of x and y provided the Jacobian

$$J = x_u y_v - x_v y_u = X_{uu} X_{vv} - X_{uv}^2 \qquad (4.159)$$

is not zero. Equation (2.100) is elliptic for subsonic flow and hyperbolic for supersonic flow.

For subsonic flow J never vanishes unless $\psi_{uu} = \psi_{uv} = \psi_{vv} = 0$. This follows for $q^2 = u^2 + v^2 < c^2$ since $-(c^2 - u^2)[X_{uu}X_{vv} - X_{uv}^2] = (c^2 - u^2)X_{uv}^2 + 2uvX_{uv}X_{uu} + (c^2 - v^2)X_{uu}^2$ and the right-hand member is a positive definite quadratic form. A similar argument for $j = u_x v_y - u_y v_x$ shows that j never vanishes for nonconstant subsonic flow. The advantage of linearity gained by this hodograph is paid for by boundary condition complication; that is, the boundaries in the (u, v) plane, corresponding to given boundaries in the (x, y) plane depend on the problem solution.

On the other hand the problem of supersonic flow is complicated by the fact that the Jacobians J and j may change sign. The hodograph of the flow is not simple in such cases. If j changes sign the image of the flow in the (u, v) plane covers certain parts of the flow two or three times. If J changes sign the functions $x(u, v)$, $y(u, v)$ do not represent a flow globally (throughout the x, y plane) since the image in the (x, y) plane possesses a fold covering parts of the plane two or three times. The edge of such a fold is called a limiting line or natural boundary. In spite of such singularities special significant flow fields are obtainable by hodograph means for supersonic as well as subsonic flow.

To develop such flows we appeal to a more amenable form of Eq. (2.100). From Section (2.3b) we have the relations

$$\rho\phi_\theta = q\psi_q, \qquad \rho q\phi_q = (q^2/c^2 - 1)\psi_\theta \qquad (4.160)$$

for the potential function $\phi(x, y)$ and stream function $\psi(x, y)$. By means of a similar development (to that used to get (4.160)) the Legendre transforms $X(u, v)$, $Y(u, v)$ defined via Eqs. (2.102) and (2.104) can be shown to satisfy relations similar to Eqs. (4.160), namely

$$\begin{aligned} q\rho^{-1}Y_q &= (1 - q^2/c^2)X_\theta \\ \rho^{-1}Y_\theta &= -qX_q. \end{aligned} \qquad (4.161)$$

The relation $X = ux + vy - \phi$ in polar coordinates $q^2 = u^2 + v^2$, $u = q\cos\theta$, $v = q\sin\theta$, takes the form

$$\phi = qX_q - X, \qquad \phi_q = qX_{qq}. \qquad (4.162)$$

Now Eq. (2.100) can either be converted directly into q, θ variables or Y can be eliminated from Eq. (4.161). In either event there results

$$c^2 X_{qq} - (q^2 - c^2)(q^{-1}X_q + q^{-2}X_{\theta\theta}) = 0 \qquad (4.163)$$

for X. The quantities ψ and Y, related through

$$Y(\rho u, \rho v) = \rho u y - \rho v x - \psi(x, y)$$

take the form

$$\psi = \rho q Y_{\rho q} - Y = (1 - q^2/c^2)^{-1} q Y_q - Y \qquad (4.164)$$

in polar coordinates.

Once a solution $X(q, \theta)$ has been obtained it is sometimes convenient to use the relations

$$\begin{aligned} q\psi_q &= \rho(qX_q - X)_\theta \\ \psi_\theta &= \rho(qX_q + X_{\theta\theta}) \end{aligned} \qquad (4.165)$$

to determine the stream function directly.

We now examine some special flows by means of Eq. (4.161) or (4.163). Several cases are detailed and others are given with little comment.

a. Circulatory Flow

$X(q, \theta) = k\theta = k \tan^{-1}(v/u)$ is a simple solution of Eq. (4.163). From $x = X_u$, $y = X_v$ we have

$$x = -kv(u^2 + v^2)^{-1}, \qquad y = ku(u^2 + v^2)^{-1}.$$

To carry out the necessary inversion (to obtain u and v) we note that $x^2 + y^2 = k^2(u^2 + v^2)^{-1} = k^2q^{-2}$. Thus $q = k(x^2 + y^2)^{-\frac{1}{2}} = kr^{-1}$, $u = kyr^{-2}$, $v = -kxr^{-2}$. These relations represent a circulatory flow having the angular velocity kr^{-2}. The corresponding stream function is obtainable from Eq. (4.165). Clearly ψ is independent of θ, since $\psi_\theta = 0$, and $\psi_q = -k\rho q^{-1}$ so that

$$\psi = -k \int_{q_0}^{q} \rho q^{-1} \, dq.$$

b. Radial Flow

$\psi = k\theta$ is a simple solution for then a function $\phi = \phi(q)$ satisfying Eq. (4.160) is thus determinable. If X depends only upon q, $X = X(q)$, we find from Eq. (4.165)

$$X = k \int_{q_0}^{q} \rho^{-1} q^{-1} \, dq.$$

From $x = X_u$, $y = X_v$ it therefore follows that $x = k\rho^{-1}q^{-2}u$, $y = k\rho^{-1}q^{-2}v$, and $r = k\rho^{-1}q^{-1}$. With $k > 0$ we can obtain q as a functions of $k^{-1}r$ which is written in the general form $q = Q(k^{-1}r)$ and the velocity components become

$$u = xr^{-1}Q(k^{-1}r), \qquad v = yr^{-1}Q(k^{-1}r).$$

These relations represent purely radial flow which is either purely subsonic or purely supersonic (see Courant and Friedrichs [45 p. 253], and Chapter 5 for further discussion).

c. Spiral Flow

Spiral flow can be synthesized by superimposing the function X for the purely radial and the purely cirulatory flow to obtain

$$X = k_0 \theta + k \int_{q_0}^{q} \rho^{-1} q^{-1} \, dq$$

with the stream function

$$\psi = -k_0 \int_{q_0}^{q} \rho q^{-1} \, dq + k\theta.$$

Further, we obtain

$$x = (-k_0 v + k\rho^{-1} u) q^{-2}, \qquad y = (k_0 u + k\rho^{-1} v) q^{-2}$$

and

$$r = (k_0^2 + k^2 \rho^{-2})^{\frac{1}{2}} q^{-1}.$$

Computing the Jacobian J from (4.159) gives

$$J = k^2(\rho^{-1} q^{-1})_q \rho^{-1} q^{-2} - k_0^2 q^{-4} = k^2 \rho^{-2} q^{-2}(c^{-2} - q^{-2}) - k_0^2 q^{-4}. \qquad (4.166)$$

This expression for J takes both positive and negative values and is zero for a certain value q_e of q. Thus the solution represents two branches of a flow, one being purely subsonic, the other being partly subsonic and partly supersonic.

d. Mixed Flow (Ringleb [47])

A simple example of a flow being first subsonic, then supersonic, and finally subsonic again, is obtainable by seeking solutions X, Y or ϕ, ψ which are products of functions of q with functions of θ (i.e., by direct separation of variables). Since the technique is somewhat different we illustrate it. Recalling that $c^2 = c_0^2 - \frac{1}{2}(\gamma - 1)q^2$, set $\psi = F(\theta)G(q)$, $\phi = f(\theta)g(q)$ into the simultaneous equations (4.160). Thus

$$\begin{aligned} q\rho f'g &= q^2 F G' \\ q\rho f g' &= (q^2/c^2 - 1) F' G \end{aligned} \qquad (4.167)$$

are the equations for F, G, f, and g. Dividing the first equation into the second and separating gives

$$\frac{fF}{f'F'} = (c^{-2} - q^{-2}) \frac{Gg}{G'g'} = \text{(constant)}.$$

Selecting the constant as -1, the equation in θ requires that $f'F' = -fF$ which is satisfied, for example,[†] by choosing $f(\theta) = \cos \theta$, $F(\theta) = \sin \theta$.

[†] Other solutions are of course possible. One way of proceeding is to select (say) f and then determine F from the differential equation $f'F' = kfF$.

From the first equation of Eq. (4.167) we then find that $-\rho g = qG'$ and the second equation yields the relation $q\rho g' = ((q^2/c^2) - 1)G$. Both of these equations are satisfied by the choice $g = k\rho^{-1}q^{-1}$ and $G = kq^{-1}$ as well (possibly) by many other choices.

The simplest case is thus characterized by

$$\phi = k\rho^{-1}q^{-1}\cos\theta \quad \text{and} \quad \psi = kq^{-1}\sin\theta.$$

From these relations it follows via Eq. (4.165) that

$$X = kq\left\{\int_{q_0}^{q} \rho^{-1}q^{-3}\,dq\right\}\cos\theta$$

and then

$$x = k\int_{q_0}^{q} \rho^{-1}q^{-3}\,dq + k\rho^{-1}q^{-2}\cos^2\theta$$

$$y = k\rho^{-1}q^{-2}\cos\theta\sin\theta.$$

4.15 THE HODOGRAPH IN MORE COMPLICATED PROBLEMS

Numerous studies have been carried out in which the hodograph mapping is of fundamental importance. The majority of these require numerical computation for their successful completion. Fluid mechanics will again be used to describe the basic approach—specifically the article of Fisher [57] is outstanding in description.

We consider the problem of determining the cavity ratios for compressible flow past two symmetrically placed parallel flat plates as shown in Fig. 4-3a. Because of symmetry we need consider only the upper half plane. As has been previously described (Chapter 2) the problem is

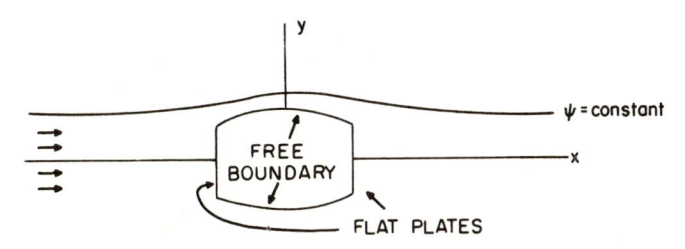

FIG. 4-3. (a) The physical plane for the cavity ratio problem in compressible flow.

then characterized by the nonlinear elliptic partial differential equation

$$[(\rho c)^2 - \psi_y^2]\psi_{xx} + 2\psi_x\psi_y\psi_{xy} + [(\rho c)^2 - \psi_x^2]\psi_{yy} = 0 \quad (4.168)$$

in D, for the stream function ψ of the subsonic, irrotational isentropic, steady, inviscid compressible flow. The boundary condition is $\psi \equiv 0$ on Γ (see Fig. 4-3b). The inherent difficulties of this formulation are the

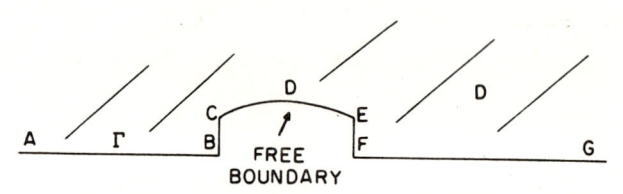

FIG. 4-3. (b) The integration domain in the original variables.

nonlinearity of the partial differential equation and the prescription of the boundary data on an unknown arc which is to be determined as part of the solution. Both of these difficulties may be overcome by using the polar variables $q^2 = u^2 + v^2$, $\theta = \tan^{-1}(v/u)$ to map the physical plane onto the hodograph plane as was previously done (recall $\rho u = \psi_x$, $\rho v = -\psi_y$). A further transformation, due to Christianovisch [58], to the "pseudo-logarithmic" hodograph (θ, λ) plane given by

$$\frac{d\lambda}{dq} = \frac{\sqrt{1-M^2}}{q} \quad (4.169)$$

places both the differential equation and the region of interest into forms which are more amenable to treatment. Here $M = q/c$ is the Mach number. This transformation has also been used (in more available references) by Lighthill [59] and Bergman and Epstein [60]. Its utility is summarized by Laitone [52].

When Eq. (4.169) is applied to the (q, θ) equation (2.121) the resulting equation is

$$\psi_{\lambda\lambda} + \psi_{\theta\theta} + g(\lambda)\psi_\lambda = 0$$

$$g(\lambda) = -\frac{\gamma+1}{2}\frac{M^4}{(1-M^2)^{\frac{3}{2}}} \quad (4.170)$$

and the integration domain is depicted in Fig. 4-3c. Equation (4.170) is linear and is simplified to the form of a Laplacian plus a lower order term. The integration domain is a semi-infinite slit rectangular domain

4.15 MORE COMPLICATED PROBLEMS

with the original free boundary CDE of Fig. 4-3b mapped into the segment of the θ-axis between $-\pi/2$ and $\pi/2$ (and hence known!). In obtaining the analytic or numerical solution of Eq. (4.170) one must use care at the sonic singularity. Cherry [44] and Laitone [52] summarize the difficulties and results of this linear problem.

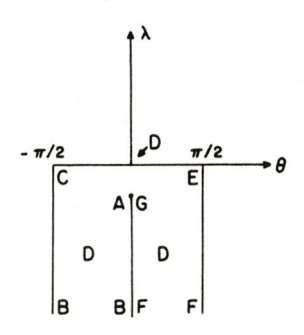

FIG. 4-3. (c) The integration domain in the pseudo-logarithmic plane.

After obtaining a solution of the system (4.170) in the (θ, λ) plane one obtains the shape of the extremal body in the physical plane from the integrations

$$x(\theta, \lambda) = \int x_\theta \, d\theta + x_\lambda \, d\lambda = \int x_\theta \, d\theta + x_q q_\lambda \, d\lambda$$

$$y(\theta, \lambda) = \int y_\theta \, d\theta + y_q q_\lambda \, d\lambda$$

where

$$x_q = -\frac{1}{\rho q} \left[\psi_\theta \left(\frac{1 - M^2}{q} \right) \cos \theta + \psi_q \sin \theta \right]$$

$$x_\theta = \frac{1}{\rho q} [q \psi_q \cos \theta - \psi_\theta \sin \theta]$$

and

$$y_q = \frac{1}{\rho q} \left[-\psi_\theta \left(\frac{1 - M^2}{q} \right) \cos \theta + \psi_q \sin \theta \right]$$

$$y_\theta = \frac{1}{\rho q} [q \psi_q \sin \theta + \psi_\theta \cos \theta].$$

4.16 UTILIZATION OF THE GENERAL SOLUTIONS OF CHAPTER 2

Chapter 2 was concluded by a tabulation of a number of general solutions to equations. In this section we illustrate how one of these general solutions may be used to develop solutions to specific problems. The utility of this method has received little recognition.

If Q is the monomolecular heat of reaction, k the thermal conductivity, and W the reaction velocity then the appropriate equation for the thermal balance between the heat generated by a chemical reaction and that conducted away is

$$k \nabla^2 T = -QW. \tag{4.171}$$

The usual expression for W is the Arrhenius relation

$$W = ca \exp[-E/RT]$$

where c is concentration, a is a frequency factor, and E is the energy of activation. As a first approximation write

$$-\frac{E}{RT} = -\frac{E}{RT_0}\left[\frac{1}{1 + (T - T_0)/T_0}\right] \approx -\frac{E}{RT_0}\left[1 - \frac{T - T_0}{T_0}\right] \tag{4.172}$$

for T sufficiently close to T_0. Then with $\theta = (E/RT_0^2)(T - T_0)$ the equation for θ is

$$\nabla^2 \theta + A \exp \theta = 0 \tag{4.173}$$

where

$$A = \frac{Q}{k} \frac{E}{RT_0^2} \exp\left(-\frac{E}{RT_0}\right).$$

Chambre [61] has obtained the solution for Eq. (4.173) in the case of spherical symmetry (one independent variable).

Two-dimensional steady vortex motion of an incompressible fluid is governed by an equation of the form

$$\nabla^2 \psi + g(\psi)g'(\psi) = 0. \tag{4.173a}$$

One means of derivation of Eq. (4.173a) is by means of a variational principle given, for example, in Bateman [62]. With the stream function

4.16 GENERAL SOLUTIONS OF CHAPTER 2

ψ defined by $u = -\psi_y$, $v = \psi_x$ so that continuity $u_x + v_y = 0$ is satisfied, consider the variation of the integral

$$I = \frac{1}{2} \iint_R \left[\left(\frac{\partial \psi}{\partial x}\right)^2 + \left(\frac{\partial \psi}{\partial y}\right)^2 + (\dot{s})^2 \right] dx\, dy \qquad (4.174)$$

where we have set

$$\dot{s} = \frac{\partial(\psi, s)}{\partial(x, y)} = u\frac{\partial s}{\partial x} + v\frac{\partial s}{\partial y}. \qquad (4.175)$$

Letting ψ and s have variations which vanish at the boundary of the region R, we obtain the two equations, from the variation of I,

$$\frac{\partial(\dot{s}, \psi)}{\partial(x, y)} = \dot{s}_x \psi_y - \dot{s}_y \psi_x = 0 \qquad (4.176)$$

and

$$\psi_{xx} + \psi_{yy} + \frac{\partial}{\partial x}\left(\dot{s}\frac{\partial s}{\partial y}\right) - \frac{\partial}{\partial y}\left(\dot{s}\frac{\partial s}{\partial x}\right) = 0. \qquad (4.177)$$

The form of Eq. (4.176) implies that $\dot{s} = g(\psi)$ whereupon Eq. (4.177) becomes

$$\psi_{xx} + \psi_{yy} + g(\psi)g'(\psi) = 0 \qquad (4.178)$$

where $g(\psi)$ is an arbitrary function which, when R is infinite, must be such that the integral I (Eq. (4.174)) has meaning. This usually means that u, v, and s must vanish at infinity. In the special case where $g(\psi) = \lambda e^{\beta \psi}$, the vortex equation becomes

$$\psi_{xx} + \psi_{yy} + \lambda^2 \beta e^{2\beta \psi} = 0. \qquad (4.179)$$

Equations of the form (4.179) also occurs in the theory of the space charge of electricity around a glowing wire [63], and in the nebular theory for the distribution of mass of gaseous interstellar material under the influence of its own gravitational field. Walker [64] considers this problem in some detail.

The general solution to problems of the form (4.179) is quite old probably dating back to Liouville (1853). We consider a modified form of Pockel's method for obtaining the general solution to

$$\psi_{xx} + \psi_{yy} + \mu \exp(a\psi) = 0. \qquad (4.180)$$

Upon setting $\psi = F(\theta)$, Eq. (4.180) takes the form

$$F'[\theta_{xx} + \theta_{yy}] + F''[\theta_x^2 + \theta_y^2] = -\mu \exp(a\psi). \qquad (4.181)$$

Recalling that $\theta_{xx} + \theta_{yy} = 0$ has the general solution

$$\theta = f_1(x+iy) + f_2(x-iy) \qquad (4.182)$$

where f_1, f_2 are sufficiently differentiable arbitrary functions, the choice of $\theta(x, y)$ as Eq. (4.182) reduces Eq. (4.181) to

$$\exp a\psi = -\frac{1}{\mu} F''(\theta) \left[\left(\frac{\partial \theta}{\partial x}\right)^2 + \left(\frac{\partial \theta}{\partial y}\right)^2 \right]. \qquad (4.183)$$

We want this to be put into a form which gives real positive values to the right-hand side. This is easily done because of the three degrees of freedom, F, f_1, and f_2. By making the choice of $f_2(x-iy) = [f_1(x-iy)]^{-1}$ and denoting g, h as the conjugate (real) functions given by

$$g(x, y) + ih(x, y) = f_1(x+iy),$$

which is always possible, Eq. (4.182) becomes

$$\theta(x, y) = g + ih + (g - ih)^{-1} = (g^2 + h^2 + 1)(g - ih)^{-1}.$$

Carrying out the differentiations and then selecting $F(\theta) = \ln \theta$ finally yields the solution

$$\exp[a\psi] = \frac{2}{\mu a} \frac{\{(\partial g/\partial x)^2 + (\partial g/\partial y)^2\}}{(g^2 + h^2 + 1)^2}. \qquad (4.184)$$

This remarkable form gives us the general solution of Eq. (4.179) and the method of attack seems capable of generalization. Comparison with the general solution, Ex. 3, of Table I (Chapter 2) is easily made.

Typical application of the general solution usually proceeds via the *inverse method*. In this procedure we usually examine promising solutions for possible application to specific physical problems. In the case of the nebular theory the relation (4.184) is essentially a density so that it can supply possible surfaces of equal density when any known conjugate functions are assigned.

Thus suppose that $r^2 = x^2 + y^2$ with the conjugate functions $g = (r/b)^n \cos n\theta$, $h = (r/b)^n \sin n\theta$.

Thus

$$\exp[a\psi] = \frac{2}{\mu a} \frac{\{g_r^2 + (1/r^2)g_\theta^2\}}{(g^2 + h^2 + 1)}$$

$$= \frac{2n^2}{\mu a r^2} \frac{(r/b)^{2n}}{[(r/b)^{2n} + 1]^2}$$

or

$$H = \exp[a\psi] = \frac{2n^2}{\mu a r^2} \left\{ \left(\frac{r}{b}\right)^n + \left(\frac{b}{r}\right)^n \right\}^{-2}. \quad (4.185)$$

When $n < 1$, H becomes infinite as $r \to 0$, while elsewhere it is finite and vanishes at infinity. When $n = 1$, H is finite at the origin and elsewhere finite, vanishing at infinity. For $n > 1$, H is zero at the origin, rising to a finite maximum and then going to zero at infinity.

If we interpret the solution (4.185) in the sense of the vorticity problem described by Eq. (4.179) and set $a = 2\beta$, $\mu = \lambda^2\beta$, then

$$e^{2\beta\psi} = \frac{n^2}{\lambda^2\beta^2 r^2} \left\{ \left(\frac{r}{b}\right)^n + \left(\frac{b}{r}\right)^n \right\}^{-2}$$

or

$$e^{-\beta\psi} = \frac{\lambda\beta r}{n} \left\{ \left(\frac{r}{b}\right)^n + \left(\frac{b}{r}\right)^n \right\}.$$

From this expression one can easily calculate the velocity components. When $n = 1$ they are

$$u = \frac{2y}{\beta(r^2 + b^2)}, \qquad v = -\frac{2x}{\beta(r^2 + b^2)}$$

which are very like those components for a line vortex but have the advantage that they do not become infinite at the origin.

The choice of other conjugate functions will generate other solutions via Eq. (4.184).

4.17 SIMILAR SOLUTIONS IN HEAT AND MASS TRANSFER

The use of the similarity developments of Sections 4.3–4.11 are often preceded by attempts to transform an equation into the same form as that for another variable. We then say that the solutions are similar but not in the sense used previously. The exact meaning will be clarified below.

Physical problems often involve the interactions of fluid, chemical reaction and heat transfer factors. The following examples will be of this general type.

If a material with concentration c is diffusing through a moving viscous incompressible fluid the equations are

$$uu_x + vu_y = -p_x + \mu(u_{xx} + u_{yy}) \tag{4.186a}$$

$$uv_x + vv_y = -p_y + \mu(v_{xx} + v_{yy}) \tag{4.186b}$$

$$uc_x + vc_y = D(c_{xx} + c_{yy}) \tag{4.186c}$$

$$u_x + v_y = 0. \tag{4.186d}$$

Meksyn [16] gives the similar solution $c = au + b$, valid for the specific case of boundary layer flow on a flat plate where the concentration c is constant at the plate's surface. The reaction rate must therefore also be constant at the plate's surface. Fainzil'ber [65, 66, 67] has considered more complicated problems in some detail. An expansion of his analysis is given here.

Upon differentiating Eq. (4.186a) with respect to y and subtracting from it the x differentiated form of Eq. (4.186b) we get

$$u(u_y - v_x)_x + v(u_y - v_x)_x = \mu[(u_y - v_x)_{xx} + (u_y - v_x)_{yy}]. \tag{4.187}$$

In the development of Eq. (4.187) the continuity relation (4.186d) has been used. Upon setting $\omega = u_y - v_x$, Eq. (4.187) takes the form

$$u\omega_x + v\omega_y = \mu[\omega_{xx} + \omega_{yy}] \tag{4.188}$$

which is exactly the same form as that for the concentration equation (4.186c). That is to say, the new variable $\Gamma = c - A\omega$ has the equation

$$u\Gamma_x + v\Gamma_y = \mu(\Gamma_{xx} + \Gamma_{yy}). \tag{4.189}$$

The solution $\Gamma = B =$ constant follows immediately so that

$$c = A\omega + B \tag{4.190}$$

represents the similar integral of vorticity (ω) and concentration fields for flow past a profile of arbitrary shape.

For the specific case of flow past a plate a more general integral is

4.17 SOLUTIONS IN HEAT AND MASS TRANSFER

obtainable using the boundary layer theory. With $\partial p/\partial x = 0$ the boundary layer from of Eqs. (4.186) is

$$uu_x + vu_y = \mu u_{yy}$$
$$uc_x + vc_y = Dc_{yy} \tag{4.191}$$
$$u_x + v_y = 0.$$

From these it is clear that $c = au + b$ as given by Meksyn. Second the equation for Γ becomes

$$u\Gamma_x + v\Gamma_y = \mu\Gamma_{yy}$$

so that comparison with the first equation of Eq. (4.191) gives

$$\Gamma = ku + d$$
or $$\tag{4.192}$$
$$c = A\omega + ku + B'.$$

To evaluate the coefficients we note that at the edge of the boundary layer $u = \bar{u}$, $\omega = 0$, $c = \bar{c}$ so that $B' = \bar{c} - k\bar{u}$. Then at the surface of the plate $(y = 0)$, $(\partial \omega/\partial y)_0 = 1/\mu(dp/dx) = 0$, $(\partial c/\partial y)_0 = r(x)$ where the latter term is the concentration gradient for the case of a surface reaction and $r(x)$ is the law of surface solubility. From Eq. (4.192)

$$r(x) = k\left(\frac{\partial u}{\partial y}\right)_0$$

and u is given by the known Blasius ([14], [15], [16]) solution of the boundary layer flow in the form

$$u = \bar{u}f[y\bar{u}(/\mu x)^{\frac{1}{2}}] \tag{4.193}$$

where f is tabulated. Thus the required form of $r(x)$ is

$$r(x) = \alpha x^{-\frac{1}{2}} \tag{4.194}$$

where $\alpha = kf'(0)\bar{u}(\bar{u}/\mu)^{\frac{1}{2}}$ for this similar solution.

By an analogous approach one can construct similarity integrals of vortex and temperature fields. A number of these cases are considered by Fainzil'ber [67]. In the boundary layer form the equations of motion, continuity, and energy for a viscous gas flow past a profile of arbitrary form may be written

$$uu_x + vu_y = vu_{yy} - \frac{1}{\rho}\frac{dp}{dx}$$

$$u_x + v_y = 0 \tag{4.195}$$

$$c_p(uT_x + vT_y) = \frac{\lambda}{\rho} T_{yy} + vu_y^2 + \frac{u}{\rho}\frac{dp}{dx}$$

where T, c_p, λ, ρ, ν are temperature, specific heat, thermal conductivity, density, and kinematic viscosity (μ/ρ), respectively. By methods of differentiation and consolidation like those in the previous example one finds, for unity Prandtl number,

$$uS_x + vS_y = \nu S_{yy} \tag{4.196}$$

and

$$S = T + u^2/2c_p - A\omega. \tag{4.197}$$

In Eq. (4.197) we have let $\omega = \partial u/\partial y$ which is the vorticity in the boundary layer. Clearly Eq. (4.196) is satisfied by taking $S = B = $ constant so that from Eq. (4.197) we find

$$T = A\omega - B - u^2/2c_p. \tag{4.198}$$

4.18 SIMILARITY INTEGRALS IN COMPRESSIBLE GASES

The elementary concepts discussed in Section 4.17 will be used herein on two more difficult problems. First we consider the development of similarity integrals of velocity, temperature and concentration fields in the boundary layer of a compressible gas. With the Prandtl number (Pr $= \mu c_p/\lambda$) equal to 1, the hydrodynamic and thermodynamic equations for the boundary layer of a compressible gas on a flat plate are

$$\rho u u_x + \rho v u_y = (\mu u_y)_y \tag{4.199a}$$

$$\rho u = \psi_y, \qquad \rho v = -\psi_x \tag{4.199b}$$

$$\rho u t_x + \rho v t_y = (\mu t_y)_y \tag{4.199c}$$

where ψ is the stream function and $t = T + u^2/2c_p$ is the stagnation

4.18 SIMILARITY INTEGRALS IN COMPRESSIBLE GASES

temperature. To this system we apply the Dorodnitsyn transformation [68] which changes the independent variables from x and y to

$$x \quad \text{and} \quad \eta = \int_0^y \rho(x, y)\, dy. \tag{4.200}$$

In these new variables the system (4.199) becomes

$$\psi_\eta \psi_{x\eta} - \psi_x \psi_{\eta\eta} = (\mu\rho\psi_{\eta\eta})_\eta \tag{4.201}$$

$$\psi_\eta t_x - \psi_x t_\eta = (\mu\rho t_\eta)_\eta. \tag{4.202}$$

We now make the usual initial assumption of a linear dependence of viscosity on temperature $\mu = \alpha T$ and use the equation of state $\rho = p/RT$ so that

$$\mu\rho = \frac{\alpha p}{R} = \text{constant} \tag{4.203}$$

by virtue of the constant pressure assumption. Using relation (4.203) we now reduce the complexity of Eqs. (4.201) and (4.202) by introducing the new dependent variable

$$\psi_1 = \psi/\mu\rho$$

whereupon those equations become

$$(\psi_1)_\eta (\psi_1)_{x\eta} - (\psi_1)_x (\psi_1)_{\eta\eta} = (\psi_1)_{\eta\eta\eta} \tag{4.204}$$

and

$$(\psi_1)_\eta t_x - (\psi_1)_x t_\eta = t_{\eta\eta}. \tag{4.205}$$

Upon examination these equations suggest another differentiation of Eq. (4.204) with respect to η, giving

$$(\psi_1)_\eta (\psi_1)_{x\eta\eta} - (\psi_1)_x (\psi_1)_{\eta\eta\eta} = (\psi_1)_{\eta\eta\eta\eta}. \tag{4.204a}$$

Now upon comparing Eq. (4.204a) with (4.205) they have exactly the same form if the new dependent variable

$$\tau = (\mu\rho)^2 (\psi_1)_{\eta\eta}. \tag{4.206}$$

is introduced. Under this new variable Eq. (4.204a) becomes

$$(\psi_1)_\eta \tau_x - (\psi_1)_x \tau_\eta = \tau_{\eta\eta}. \tag{4.207}$$

4. FURTHER ANALYTIC METHODS

Before proceeding we clarify the meaning of τ. Since $u = (1/\rho)\,\partial\psi/\partial y = \partial\psi/\partial\eta$ we have

$$\tau = (\mu\rho)^2 \frac{\partial^2 \psi_1}{\partial \eta^2} = \mu\rho \frac{\partial^2 \psi}{\partial \eta^2} = \mu\rho \frac{\partial u}{\partial \eta} = \mu \frac{\partial u}{\partial y},$$

i.e., τ represents the *shear stress*.

Since Eqs. (4.207) and (4.205) have the same form we infer the integral

$$\tau = At + B \tag{4.208}$$

which represents the condition of similarity of the shear stress and temperature fields in the boundary layer of a compressible gas.

At the edge of the boundary layer $\tau = 0$ and $t = \bar{t} =$ constant, so that $B = A\bar{t}$ and Eq. (4.208) may be written as

$$\frac{\tau}{t - \bar{t}} = A = \text{constant}, \tag{4.208a}$$

Since $u = v = 0$ at $y = 0$ we have from Eq. (4.199a) that $(\partial \tau/\partial y)_0 = 0$. Consequently, Eq. (4.208) gives

$$\left.\frac{\partial t}{\partial y}\right|_0 = \left.\frac{\partial T}{\partial y}\right|_0 = 0,$$

i.e., the integral (4.208) corresponds to the absence of a heat flow through the surface of a profile in the flow.

From Eq. (4.208) we obtain the temperature distribution in the flow as

$$T = \frac{\tau}{A} + \bar{t} - u^2/2c_p$$

$$= (\mu\rho)^2 \left[\frac{1}{A} \frac{\partial^2 \psi_1}{\partial \eta^2} - \frac{1}{2c_p}\left(\frac{\partial \psi_1}{\partial \eta}\right)^2\right] + \bar{t} \tag{4.209}$$

where $\psi_1(x, \eta) = x^{\frac{1}{2}} F(\xi)$ and $F(\xi)$ is the Blasius function, $\xi = \eta/\sqrt{x}$ (see Pai [14], Curle [15], and Meksyn [16]). Equation (4.209) may be written in terms of ξ, by elementary operations, as

$$T = (\mu\rho)^2 \left[\frac{1}{A\sqrt{x}} F''(\xi) - \frac{1}{2c_p}(F')^2\right] + \bar{t}. \tag{4.209a}$$

4.18 SIMILARITY INTEGRALS IN COMPRESSIBLE GASES

When $y = 0$, $\eta = 0$ and since $F'(0) = 0$ we get the temperature of the surface as

$$T_0 = \bar{t} + m/\sqrt{x}.$$

Thus the similarity integral (4.208) corresponds to a specific variable surface temperature.

Analogous integrals for similarity of concentration and shear stress may be obtained.

As a last example we take into account the effect of concentration on the viscosity and diffusion coefficients. Confining our attention to the case of a surface reaction in flow over a flat plate the equations are

$$uc_x + vc_y = [D(c)c_y]_y \tag{4.210a}$$

$$uu_x + vu_y = [\mu(c)u_y]_y \tag{4.210b}$$

$$u = \psi_y, \qquad v = -\psi_x.$$

If $\sigma = \mu/\rho D = 1$, then the system of equations (4.210) has the integral $c = au + b$ because of the obvious relation between the two equations. Changing the dependent variable to $\tau = \mu(c)\, \partial u/\partial y$ and independent variables to x and $u(x, y)$, Eq. (4.210b) becomes

$$\frac{\partial^2 \tau}{\partial u^2} = -u \frac{\partial}{\partial x}(\mu/\tau). \tag{4.211}$$

With $\mu(c) = Ac^n$, Eq. (4.211) takes the final form

$$\tau^2 \frac{\partial^2 \tau}{\partial u^2} = Au(au + b)^n \frac{\partial \tau}{\partial x}. \tag{4.212}$$

The boundary conditions on (4.212) are

$$\left.\frac{\partial \tau}{\partial u}\right|_0 = 0$$
$$\tau(x, \bar{u}) = 0. \tag{4.213}$$

A solution by means of one parameter group theory may be found for this system.

4.19 SOME DISJOINT REMARKS

An ingenious general method of exact solution of the concentration-dependent diffusion equation has been introduced by Philip [69, 70].

Consider the nonlinear diffusion equation

$$C_t = [D(C)C_x]_x \qquad (4.214)$$

subject to the conditions

$$C(0, t) = 1, \quad C(x, 0) = 0, \quad C(\infty, t) = 0. \qquad (4.215)$$

The following analysis also applies to the conditions

$$\int_0^1 x\, dC = 0, \quad t > 0; \quad C = 1, \quad x < 0, \quad t = 0;$$
$$C = 0, \quad x > 0, \quad t = 0, \qquad (4.216)$$

but we confine our attention to Eq. (4.215). The similarity variable $\phi = xt^{-\frac{1}{2}}$ has already been considered. When this transformation is applied to Eq. (4.214) the resulting ordinary differential equation is

$$2 \frac{d}{d\phi}\left(D \frac{dC}{d\phi}\right) + \phi \frac{dC}{d\phi} = 0 \qquad (4.217)$$

subject to $C(0) = 1$, $C(\infty) = 0$. The variation of C between zero and one suggests that after one integration C may be used as the independent variable. By integration, Eq. (4.217) becomes

$$\int_0^C \phi\, dC = -2D \bigg/ \frac{d\phi}{dC}$$

or

$$D(C) = -\frac{1}{2} \frac{d\phi}{dC} \int_0^C \phi\, dC. \qquad (4.218)$$

It therefore follows that the solution of Eq. (4.214) subject to Eq. (4.215), $\phi(C)$, exists in exact form so long as $D(C)$ has the form of Eq. (4.218). We ask that ϕ be a single valued function of C, $0 \leqslant C \leqslant 1$ and since $C(\phi = 0) = 1$, $C(\infty) = 0$ it follows that

$$\phi(1) = 0. \qquad (4.219)$$

For D to exist in $0 \leqslant C \leqslant 1$ it is necessary that $d\phi/dC$ and $\int_0^C \phi \, dC$ both exist or that the product exists in the range. However, if a finite number of discontinuities in D are allowable, or if D is permitted to be infinite at a finite set of points in $0 \leqslant C \leqslant 1$, $d\phi/dC$ may either not exist or be infinite at the corresponding points.

The usual physical phenomena has $D \geqslant 0$. The flux of diffusing substances in the positive x-direction is

$$Q = -D \frac{\partial C}{\partial x} = -t^{-\frac{1}{2}} D \frac{dC}{d\phi}. \tag{4.220}$$

For $D \geqslant 0$, $Q \geqslant 0$, $dC/d\phi \leqslant 0$. Accordingly we have the further condition that

$$\frac{d\phi}{dC} \leqslant 0, \qquad 0 \leqslant C \leqslant 1.$$

Our problem is now characterized by Eq. (4.218) subject to

$$\phi(1) = 0, \qquad \frac{d\phi}{dC} \leqslant 0 \quad \text{on} \quad 0 \leqslant C \leqslant 1.$$

Philip tabulates thirty-three solutions obtained by this process. All previously known solutions as discussed by Crank [71] are obtainable by this procedure. For example if for $n > 0$

$$D(C) = \tfrac{1}{2} n C^n [1 - C^n/(1+n)]$$

then $\phi(C) = 1 - C^n$. Application of these methods to practical problems is discussed by Philip [69].

Nehari [72] and others have investigated the solution of the equation

$$\nabla^2 u = f(u) \tag{4.221}$$

with the main purpose of obtaining bounds for the solution. His results (and others) are discussed in Chapter 8. As a byproduct of this work the following special solutions are obtained for Eqs. (4.221):

$$f(u) = e^u; \qquad u = 2 \log \frac{2\sqrt{2}\, R}{R^2 - r^2}$$

$$f(u) = u^{(n+2)/(n-2)}, \qquad n \geqslant 3; \qquad u = \left[\frac{R\sqrt{n(n-2)}}{R^2 - r^2} \right]^{8/(n-2)}.$$

References

1. Pai, S. I., "Magnetogas-Dynamics and Plasma Dynamics." Springer, Vienna (Prentice-Hall, Englewood Cliffs, New Jersey), 1962.
2. Pai, S. I., *Proc. 5th Midwestern Conf. Fluid Mech.*, *1956* p. 251. Univ. of Michigan Press, Ann Arbor, Michigan, 1957.
3. Noh, W. F., and Protter, M. H., *J. Math. Mech.* **12**, No. 2, 149 (1963).
4. Lax, P. D., *Commun. Pure. Appl. Math.* **7**, 159 (1954).
5. Zwick, S. A., *J. Math. Phys.* **37**, 246 (1958).
6. Abbot, D. E., and Kline, S. J., Simple methods for classification and construction of similarity solutions of partial differential equations, *Air Force Office Sci. Res. Rept.* No. AFOSR-TN-60-1163 (1960).
7. Birkhoff, G., "Hydrodynamics," Chapter 4 and 5. Princeton Univ. Press, Princeton, New Jersey, 1960.
8. Morgan, A. J. A., *Quart. J. Math.* (*Oxford*) **2**, 250 (1952).
9. Manohar, R., Some similarity solutions of partial differential equations of boundary layer equations, *Math. Res. Center* (*Univ. of Wis.*) *Tech. Summary Rept.* No. 375 (1963).
10. Schuh, H., Über die ähnlichen Lösungen der instationaren Laminaren Grenzschichtgleichungen in inkompressiblen Strömungen, *in* "50 Jahre Grenzschichtforschung" (H. Görtler and W. Tollmien, eds.), p. 149. Vieweg, Braunschweig, 1955.
11. Geis, T., Ähnliche Grenzschichten an Rotationskörpern, *in* "50 Jahre Grenzschichtforschung" (H. Görtler and W. Tollmien, eds.), p. 294. Vieweg, Braunschweig, 1955.
12. Hansen, A. G., *Trans. ASME* **80**, 1553 (1958).
13. Morgan, A. J. A., *Trans. ASME* **80**, 1559 (1958).
14. Pai, S. I., "Viscous Flow Theory," Vol. 1, Laminar Flow. Van Nostrand, Princeton, New Jersey (1956).
15. Curle, N., "The Laminar Boundary Layer Equations." Oxford Univ. Press, London and New York, 1962.
16. Meksyn, D., "New Methods in Laminar Boundary Layer Theory." Macmillan (Pergamon), New York, 1961.
17. Polskii, N. I., and Shvets, I. T., *Soviet Phys.—Doklady* **6**, 121 (1961).
18. Pattle, R. E., *Quart. J. Mech. Appl. Math.* **12**, 407 (1959).
19. Boyer, R. H., *J. Math. Phys.* **41**, 41 (1962).
20. Riley, N., *Quart. J. Mech. Appl. Math.* **15**, 439 (1962).
21. Riley, N., *Quart. J. Mech. Appl. Math.* **15**, 459 (1962).
22. Riley, N., *Quart. J. Mech. Appl. Math.* **14**, 197 (1961).
23. Bridgman, P. W., "Dimensional Analyses," 2nd ed. Yale Univ. Press, New Haven, Connecticut, 1931.
24. Ruark, A. E., *J. Elisha Mitchell Sci. Soc.* **51**, 127 (1935).
25. Ehrenfest-Afanassjewa, Mrs. T., *Phil. Mag.* **1**, 257 (1926).
26. Carslaw, H. S., and J. C. Jaeger, "Conduction of Heat in Solids," 2nd ed. Oxford Univ. Press, London and New York, 1959.
27. Ruoff, A. L., *Quart. Appl. Math.* **16**, 197 (1958).
28. Knuth, E. L., *Phys. Fluids* **2**, 84 (1959).

REFERENCES

29. Horvay, G. J., *J. Heat Transfer* **82**, 37 (1960).
30. Landau, H. G., *Quart. Appl. Math.* **8**, 81 (1950).
31. Hamill, T. D., and S. G. Bankoff, Personal communication (1963).
32. Srivastava, A. C., *Bull. Calcutta Math. Soc.* **50**, 57 (1958).
33. Jain, M. K., *Appl. Sci. Res.* **A10**, 410 (1961).
34. Jones, J. R., *Z. Angew. Math. Phys.* **12**, 328 (1961).
35. Datta, S. K., *Appl. Phys. Quart.* **7**, 2 (1961).
36. Wu, C. S., *Appl. Sci Res.* **A8**, 140 (1959).
37. Datta, S. K., *Univ. Wisconsin Math. Res. Center Tech. Rept.* No. 358 (1959).
38. Eringen, C. E., "Nonlinear Theory of Continuous Media." McGraw-Hill, New York, 1962.
39. Hildebrand, F. B., "Methods of Applied Mathematics." Prentice-Hall, Englewood Cliffs, New Jersey, 1952.
40. Levy, H., and Lessmann, F., "Finite Difference Equations," Macmillan, New York, 1961.
41. Milne-Thomson, L. M., "The Calculus of Finite Differences." Macmillan, New York, 1933.
42. Lighthill, M. J., *Proc. Roy. Soc.* **A191**, 323 (1947).
43. Cherry, T. M., *J. Australian Math. Soc.* **1**, 80 (1959).
44. Cherry, T. M., Trans-sonic nozzle flows found by the hodograph method, *in* "Partial Differential Equations and Continuum Mechanics" (R. E. Langer, ed.), p. 217. Univ. of Wisconsin Press, Madison, Wisconsin, 1961.
45. Courant, R., and Friedrichs, K. O., Supersonic Flow and Shock Waves." Wiley (Interscience), New York, 1948.
46. Tsien, H. S., *J. Aeron. Sci.* **6**, 339 (1939).
47. Ringleb, F., *Z. Angew. Math. Mech.* **20**, 185 (1940).
48. Bers, L., On the circulatory subsonic flow of compressible fluid past a circular cylinder, *Natl. Advisory Comm. Aeron. Rept.* No. NACA-TN-970 (1945).
49. Bergman, S., On the two-dimensional flows of compressible fluids, *Natl. Advisory Comm. Aeron. Rept.* No. NACA-TN-972 (1945).
50. Lin, C. C., *Quart. Appl. Math.* **4**, 291 (1946).
51. von Mises, R., "Mathematical Theory of Compressible Fluid Flow." Academic Press, New York, 1958.
52. Laitone, E. V., Mathematical theory of compressible fluids, *in* "Nonlinear Problems of Engineering" (W. F. Ames, ed.), Chapter 12. Academic Press, New York, 1964.
53. Tsien, H. S., *Calif. Inst. Technol. (Pasadena) Galcit Publ.* No. 3 (1946).
54. Bers, L., "Mathematical Aspects of Subsonic and Transonic Gas Dynamics." Wiley, New York, 1958.
55. Frankl, F. I., *Izv. Akad. Nauk SSSR, Ser. Mat.* **9**, 121 (1945); also *Natl. Advisory Comm. Aeron. Rept.* No. NACA-TM-1155.
56. von Kármán, T., *J. Math. Phys.* **26**, 182 (1947).
57. Fisher, D. D., *J. Math. Phys.* **42**, 14 (1963).
58. Christianovich, S. A., Gas flow over bodies at high subsonic velocities, *Trudy Centr. Aeron. Inst.* No. 481 (1940). [In Russian.]
59. Lighthill, M. J., *Proc. Roy. Soc.* **A191**, 352 (1947).
60. Bergman, S., and Epstein, B., *J. Math. Phys.* **26**, 195 (1948).

61. Chambré, P. L., *J. Chem. Phys.* **20**, 1795 (1952).
62. Bateman, H., "Partial Differential Equations of Mathematical Physics," p. 166. Cambridge Univ. Press. London and New York, 1959.
63. Richardson, O. W., "The Emission of Electricity From Hot Bodies," p. 50. Longmans Green, New York, 1921.
64. Walker, G. W., *Proc. Roy. Soc.* **A91**, 410 (1915).
65. Fainzil'ber, A. M., *Dokl. Akad. Nauk SSSR* **112**, No. 4, p. 607 (1957).
66. Fainzil'ber, A. M., *Dokl. Akad. Nauk SSSR* **106**, No. 5, p. 793 (1956).
67. Fainzil'ber, A. M., *Intern. J. Heat Mass Transfer* **5**, 1069 (1962).
68. Dorodnitsyn, A. A., *Appl. Math. Mech.* **6**, 449 (1942).
69. Philip, J. R., *Australian J. Phys.* **13**, 1 (1960).
70. Philip, J. R., *Australian J. Phys.* **13**, 13 (1960).
71. Crank, J., "Mathematics of Diffusion." Oxford Univ. Press, London and New York, 1956.
72. Nehari, Z., *Proc. Am. Math. Soc.* **14**, 829 (1963).
73. Hansen, A. G., "Similarity Analyses of Boundary Value Problems in Engineering." Prentice-Hall, Englewood Cliffs, New Jersey, 1964.
74. Ames, W. F., *Ind. Eng. Chem. Fundamentals* **4**, 72 (1965).

CHAPTER **5**

Approximate Methods

5.0 INTRODUCTION

By approximate methods we shall mean analytical procedures for obtaining solutions in the form of functions which are close, in some sense, to the exact solution of the nonlinear problem. Thus numerical methods fall into a separate category (see Chapter 7) since they result in tables of values as opposed to functional forms.

Approximate methods may be classified into three broad categories; "asymptotic", "weighted residual" and "iterative." Combinations of these methods may (and have) been used to develop alternate *ad hoc* procedures. Asymptotic methods have at their foundation a desire to obtain solutions that are approximately valid when a physical parameter (or variable) of the problem is very small (or very large). Typical of these methods, whose construction is intimately related to the parameter and variables of the problem are the perturbation procedures.

The weighted residual methods, probably originating in the calculus of variation, ask for the approximate solution to be close to the exact solution in the sense that the difference between them (residual) is minimized in some sense. The collocation procedure requires the residual to vanish at a set of points while Galerkin's procedure requires weighted integrals of the residual to vanish. The property of solution continuity is then invoked to (hopefully) maintain small error over the whole integration field. Weighted residual methods are often called *direct methods* of the calculus of variations although they need not be related to a variational problem.

The iterative methods are analogous to the Picard method of ordinary differential equations in that repetitive calculation via some operation F, whose character is $u_{n+1} = F(u_n, u_{n-1}, ...)$ successively improves the approximation. Transformation of the equations to an integral equation leads to iteration. An example of this type was discussed in Section 4.12.

5.1 PERTURBATION CONCEPTS

The perturbation or small parameter method, often attributed to Poincaré [1] is a common analytic tool for the approximate solutions of nonlinear problems. In essence it consists in developing the solution of a nonlinear boundary or initial value problem in (usually ascending) powers of a parameter which either appear explicitly in the original problems or are introduced in some artificial manner. A perturbed system is one which differs slightly from a known standard system. The expansion in terms of the perturbation parameter provides a means for obtaining solutions to the perturbation system by utilizing the known properties of the standard system. The method has been most often used to investigate the behavior of slightly nonlinear systems. Extensive treatments of the perturbation concept as they apply to ordinary differential equations may be found in Stoker [2], Minorsky [3] and in a multiparameter way Nowinski and Ismail [4]. The brief but illuminating monograph of Bellman [5] is very useful although he restricts his attention to at most ordinary differential equations.

Suppose that a certain equation and possibly the boundary conditions depend upon a parameter ϵ (say). The general perturbation problem is that of finding the solution for small values of ϵ, given the solution for $\epsilon = 0$ (the standard). If the order of the equations and the number of boundary conditions remains fixed in the procedure the problem is called a *regular perturbation* problem. However, if the order of the equation is lowered when $\epsilon = 0$, and if one or more boundary conditions have to be discarded the perturbation problem is called *singular*. Examples of singular perturbation problems include the boundary layer theory and the theory of the edge effect in elasticity and plasticity.

Modifications of the basic idea are quite common. For example one may think of the determination of a small departure from equilibrium as a perturbation problem not directly involving another parameter. This is quite common usage by physicists as we shall see in a later section.

We begin our investigation with a regular perturbation.

5.2 REGULAR PERTURBATIONS IN VIBRATION THEORY

Consider the nonlinear wave motion described by the equation

$$u_{xx} = u_{tt} + \lambda u_t + \epsilon u^3; \quad \lambda \neq 0 \tag{5.1}$$

with ϵ small, $0 \leqslant t < \infty$, $0 \leqslant x \leqslant \pi$, boundary conditions $u(0, t) = u(\pi, t) = 0$ and initial conditions $u(x, 0) = f(x)$, $u_t(x, 0) = g(x)$.

If the parameter ϵ were to vanish, the system would be linear and solutions could readily be obtained. The parameter ϵ can then be thought of as a measure of how much the system is perturbed from the standard which in this case is the linear system.[†] The basic idea of the method is to seek a solution in the form of a series in ascending powers of ϵ. Thus

$$u(x, t) = u_0(x, t) + \epsilon u_1(x, t) + \epsilon^2 u_2(x, t) + \cdots \tag{5.2}$$

where the u_j are to be determined. We evaluate them by requiring that $u_0(x, t)$ must satisfy *all* the initial and boundary conditions and $u_j(x, t)$, $j \geqslant 1$ to satisfy *only* homogeneous auxiliary conditions. Upon inserting Eq. (5.2) into Eq. (5.1) we obtain

$$u_{0xx} + \epsilon u_{1xx} + \epsilon^2 u_{2xx} + \cdots = (u_{0tt} + \lambda u_{0t}) + \epsilon(u_{1tt} + \lambda u_{1t}) + \epsilon^2(u_{2tt} + \lambda u_{2t})$$
$$+ \cdots + \epsilon[u_0^3 + 3\epsilon u_0^2 u_1 + 3\epsilon^2(u_0^2 u_2 + u_1^2 u_0) + \cdots]$$

or

$$(u_{0xx} - u_{0tt} - \lambda u_{0t}) + \epsilon(u_{1xx} - u_{1tt} - \lambda u_{1t} - u_0^3)$$
$$+ \epsilon^2(u_{2xx} - u_{2tt} - \lambda u_{2t} - 3u_0^2 u_1) + \cdots = 0. \tag{5.3}$$

If Eq. (5.3) is to be satisfied identically in ϵ, then the coefficient of each power of ϵ must separately vanish. Thus we obtain the set of equations for u_j:

$$\begin{aligned} u_{0xx} - u_{0tt} - \lambda u_{0t} &= 0 \\ u_{1xx} - u_{1tt} - \lambda u_{1t} &= u_0^3 \\ u_{2xx} - u_{2tt} - \lambda u_{2t} &= 3u_0^2 u_1 \\ &\vdots \end{aligned} \tag{5.4}$$

[†] While the standard is usually the solution of a linear problem there is no *a priori* reason for this. One can select the standard from the solution of a non-linear problem if appropriate. We shall illustrate this idea later.

with the auxiliary conditions

$$u_0(0, t) = u_0(\pi, t) = 0, \quad u_0(x, 0) = f(x), \quad u_{0t}(x, 0) = g(x);$$
$$u_j(0, t) = u_j(\pi, t) = u_j(x, 0) = u_{jt}(x, 0) = 0, \quad j \geqslant 1.$$

The first of Eqs. (5.4) represents the standard or unperturbed system and each succeeding equation has exactly the same form except they are inhomogeneous with different forcing terms. The fact that all equations after the first have zero boundary and initial conditions requires only the determination of particular solutions. The methods of solution are standard linear methods. For ϵ small, compared to one, a useful approximation is obtained by selecting only the first correction

$$u(x, t) = u_0(x, t) + \epsilon u_1(x, t).$$

Care must be exercised to avoid secular terms. Benney and Niell [6] note that when $\lambda = 0$ in Eq. (5.1) the usual perturbation scheme fails because of secular terms.

5.3 PERTURBATION AND PLASMA OSCILLATIONS

Most theoretical effort devoted to the study of plasmas in nonequilibrium states is based on the Boltzmann-Landau-Vlasov equations given in a review article by Dolph [7] as

$$\frac{\partial f_\sigma}{\partial t} + \mathbf{v} \cdot \nabla f_\sigma + \frac{\eta e}{m}\left[\mathbf{E} + \frac{1}{c}\mathbf{v} \times \mathbf{H}\right] \cdot \nabla_v f_\sigma = \frac{\partial f_\sigma}{\partial t}\bigg]_{\text{coll}}$$

$$\nabla \cdot \mathbf{E} = 4\pi e(n - n_0)$$

$$\nabla \cdot \mathbf{B} = 0$$

$$c\nabla \times \mathbf{B} = 4\pi \mathbf{J} + \frac{\partial \mathbf{E}}{\partial t}$$

$$c\nabla \times \mathbf{E} = \frac{\partial \mathbf{B}}{\partial t}$$

$$n = \sum_\sigma \eta \iiint f_\sigma \, d^3v \quad (5.5)$$

$$\mathbf{J} = \sum_\sigma \eta \iiint \mathbf{v} f_\sigma \, d^3v$$

$$\mathbf{D} = \epsilon \mathbf{E}$$

$$\mathbf{B} = \mu \mathbf{H}.$$

5.3 PERTURBATION AND PLASMA OSCILLATIONS

In these equations we have assumed the plasma can be fully described by a statistical distribution function $f_\sigma(\mathbf{r}, \mathbf{v}, t)$ for each species (usually σ is taken to be 1 and 2 with $\eta = 1$ for ions and -1 for electrons) which determines the number of particles of type σ in a differential element of the $\mathbf{r}, \mathbf{v}$ dimension-velocity space at time t. The symbols e, m, c, $\mathbf{E}$, $\mathbf{H}$, $\mathbf{B}$, $\mathbf{J}$, $\mathbf{D}$, ϵ, and μ have their usual meaning as in electromagnetic theory while the zero moment $n = \Sigma_\sigma \eta e \int f_\sigma \, d^3v$ defines the plasma density and the first moment the current $\mathbf{J}$ in the plasma. Our concern here will be with a somewhat restricted version of Eq. (5.5) which is described subsequently.

For a given distribution function $f(\mathbf{r}, \mathbf{v}, t)$ and for some quantity $Q(\mathbf{v}, \mathbf{r}, t)$ an *average* quantity $\langle Q(\mathbf{r}, t) \rangle$ is calculated by means of the triple integral over all of velocity space

$$\langle Q(\mathbf{r}, t) \rangle = \frac{1}{n(\mathbf{r}, t)} \iiint Q(\mathbf{r}, \mathbf{v}, t) f \, dv_1 \, dv_2 \, dv_3. \quad (5.6)$$

The first equation of system (5.5) for f is

$$\frac{\partial f}{\partial t} + \sum_{i=1}^{3} \frac{\partial f}{\partial x_i} \frac{dx_i}{dt} + \sum_{i=1}^{3} \frac{\partial f}{\partial v_i} \frac{dv_i}{dt} = \frac{\partial f}{\partial t}\bigg]_{\text{coll}} \quad (5.7)$$

i.e., the coefficients of $\nabla_v f_\sigma$ are accelerations. Multiply Eq. (5.7) by any integrable quantity $Q(\mathbf{v})$ and integrate the result over all of velocity space getting ($v_i = dx_i/dt$, $F_i/m = dv_i/dt$)

$$\iiint Q(\mathbf{v}) \frac{\partial f}{\partial t} d\mathbf{v} + \sum_{i=1}^{3} \iiint Q(\mathbf{v}) v_i \frac{\partial f}{\partial x_i} d\mathbf{v}$$

$$+ \sum_{i=1}^{3} \iiint Q(\mathbf{v}) \frac{F_i}{m} \frac{\partial f}{\partial v_i} d\mathbf{v} = \iiint Q(\mathbf{v}) \frac{\partial f}{\partial t}\bigg]_{\text{coll}} d\mathbf{v}. \quad (5.8)$$

From the definition Eq. (5.6) the first of these integrals is obtainable as

$$\iiint Q(\mathbf{v}) \frac{\partial f}{\partial t} d\mathbf{v} = \frac{\partial}{\partial t} \iiint Q(\mathbf{v}) f \, d\mathbf{v} = \frac{\partial}{\partial t} [n\langle Q \rangle]. \quad (5.9)$$

By a similar argument the second integral is easily seen to be

$$\sum_{i=1}^{3} \frac{\partial}{\partial x_i} [n\langle Qv_i \rangle]. \quad (5.10)$$

Since
$$\frac{\partial}{\partial v_i}[QF_i f] = QF_i \frac{\partial f}{\partial v_i} + f \frac{\partial (QF_i)}{\partial v_i}$$

we can write the typical term of the third sum

$$I_3 = \iiint Q(\mathbf{v}) \frac{F_i}{m} \frac{\partial f}{\partial v_i} d\mathbf{v}$$
$$= \frac{1}{m} \iiint \frac{\partial}{\partial v_i}(QF_i f)\, dv_1\, dv_2\, dv_3 - \frac{1}{m} \iiint f \frac{\partial (QF_i)}{\partial v_i}\, dv_1\, dv_2\, dv_3. \quad (5.11)$$

Because of the conservation of energy the assumption that $f \to 0$ as $v_i \to \pm\infty$ is valid. Under this assumption the first term of Eq. (5.11) vanishes for each i and

$$I_3 = -\frac{1}{m} \iiint f \frac{\partial (QF_i)}{\partial v_i}\, d\mathbf{v} = -\frac{n}{m} \left\langle \frac{\partial (QF_i)}{\partial v_i} \right\rangle.$$

Composing all these results we have, for a reasonable distribution function f, that a collisionless plasma gives rise to

$$\frac{\partial}{\partial t}[n\langle Q\rangle] + \sum_{i=1}^{3} \frac{\partial}{\partial x_i}[n\langle Qv_i\rangle] - \sum_{i=1}^{3} \frac{n}{m} \left\langle \frac{\partial (QF_i)}{\partial v_i} \right\rangle = 0. \quad (5.12)$$

For $Q = 1$, $\langle Q \rangle = 1$ and Eq. (5.12) becomes the familiar continuity equation with a production term

$$\frac{\partial n}{\partial t} + \sum_{i=1}^{3} \frac{\partial}{\partial x_i}(n\langle v_i\rangle) = \sum_{i=1}^{3} \frac{n}{m} \left\langle \frac{\partial F_i}{\partial v_i} \right\rangle.$$

In most considered cases the forces are independent of the velocities and therefore the right-hand side vanishes. If we take $Q = m\mathbf{v}$, Eq. (5.12) generates the momentum equation and successively higher moments give the energy equation, etc.

In order to reduce the complexity we consider the one dimensional analog of Eq. (5.5) as treated by Chandrasekhar [8] in the case of electrostatic ($\mathbf{B} = 0$) plasma oscillations where the positive ions are at rest. Equations (5.5) become

$$\frac{\partial f}{\partial t} + v \frac{\partial f}{\partial x} + \frac{eE_x}{m} \frac{\partial f}{\partial v} = 0 \quad (5.13a)$$

$$\frac{\partial E}{\partial x} = 4\pi e(n - n_0) \quad (5.13b)$$

5.3 PERTURBATION AND PLASMA OSCILLATIONS

where n is the density described by the zero moment, $n = \int_{-\infty}^{\infty} f\, dv$. The first moment of Eq. (5.13a) (note that E_x is assumed independent of v) by Eq. (5.12) is

$$\frac{\partial n}{\partial t} + \frac{\partial}{\partial x}[n\langle v \rangle] = 0; \qquad (5.14)$$

the second moment is

$$\frac{\partial}{\partial t}[n\langle v \rangle] + \frac{\partial}{\partial x}[n\langle v^2 \rangle] - \frac{e}{m} E_x n = 0; \qquad (5.15)$$

and the third moment is

$$\frac{\partial}{\partial t}[n\langle v^2 \rangle] + \frac{\partial}{\partial x}[n\langle v^3 \rangle] - \frac{2e}{m} E_x n\langle v \rangle = 0. \qquad (5.16)$$

The person with a knowledge of fluid mechanics will recognize the first equation as the conservation of mass equation, the second as the conservation of momentum equation and the third as the expression for the conservation of energy. To Eqs. (5.14)–(5.16) we add the Poisson equation (5.13b). The unknowns of this system are n, $\langle v \rangle$, $\langle v^2 \rangle$, $\langle v^3 \rangle$ and E, that is five in number, but only four equations appear. Should we take another moment? This will introduce a new equation but also the new dependent variable $\langle v^4 \rangle$. This then does not seem to be the answer especially since the physical interpretation of the new equation(s) is impaired.

One method of circumventing this dilemma is to obtain some information by making a further assumption. We assume that collisions are sufficiently frequent to insure a local Maxwellian velocity distribution

$$f(x, v, t) = n(x, t)(2\pi\theta)^{-\frac{1}{2}} \exp[-(v - u)^2/2\theta] \qquad (5.17)$$

where $u = \langle v \rangle$, $n =$ density, $\theta = KT/m$, K the Boltzmann constant, $T =$ kinetic temperature (defined by $KT = mV^2$) and m is electron mass.

Since $\langle A + B \rangle = \langle A \rangle + \langle B \rangle$ we can easily show by simple application of Eq. (5.6) that

$$\begin{aligned}\langle v^2 \rangle &= \langle v \rangle^2 + \theta = u^2 + \theta \\ \langle v^3 \rangle &= u^3 + 3u\theta.\end{aligned} \qquad (5.18)$$

Using Eq. (5.18) the basic system (5.13b), (5.14)–(5.16) now becomes

$$\frac{\partial n}{\partial t} + \frac{\partial}{\partial x}[nu] = 0 \qquad (5.19a)$$

$$\frac{\partial [nu]}{\partial t} + \frac{\partial}{\partial x}[nu^2 + n\theta] - \frac{e}{m} E_x n = 0 \qquad (5.19b)$$

$$\frac{\partial}{\partial t}[nu^2 + n\theta] + \frac{\partial}{\partial x}[n(u^3 + 3u\theta)] - \frac{2e}{m} E_x nu = 0 \qquad (5.19c)$$

$$\frac{\partial E_x}{\partial x} = 4\pi e(n - n_0) \qquad (5.19d)$$

for the *four unknowns* n, u, $\theta = KT/m$, and E_x. As we see θ is essentially temperature.

If n_0 and θ_0 refer to the equilibrium values let us consider, in the manner of a perturbation, small departures from equilibrium. Thus we set

$$n = n_0 + n_1$$
$$\theta = \theta_0 + \theta_1$$

and substitute into Eqs. (5.19) getting

$$(n_0 + n_1)_t + [(n_0 + n_1)u]_x = 0$$

$$[(n_0 + n_1)u]_t + [(n_0 + n_1)(u^2 + \theta_0 + \theta_1)]_x - \frac{e}{m} E_x(n_0 + n_1) = 0$$

$$[(n_0 + n_1)(u^2 + \theta_0 + \theta_1)]_t$$
$$+ [(n_0 + n_1)(u^3 + 3u\theta_0 + 3u\theta_1)]_x - \frac{2e}{m} E_x(n_0 + n_1)u = 0 \qquad (5.20)$$

$$(E_x)_x = 4\pi e n_1.$$

The terms $n_1 u$, $n_1 u^2$, $n_0 u^2$, $n_1 \theta_1$, $E_x n_1$, etc., are discarded because they are second (or higher order) corrections. When this elimination has been completed the linear system with only first order terms is

$$\frac{\partial n_1}{\partial t} + n_0 \frac{\partial u}{\partial x} = 0 \qquad (5.21a)$$

$$n_0 \frac{\partial u}{\partial t} + \frac{\partial}{\partial x}(n_0 \theta_1 + n_1 \theta_0) = \frac{e}{m} n_0 E_x \qquad (5.21b)$$

$$\frac{\partial}{\partial t}(n_0 \theta_1 + n_1 \theta_0) + 3n_0 \theta_0 \frac{\partial u}{\partial x} = 0 \qquad (5.21c)$$

$$\frac{\partial E_x}{\partial x} = 4\pi e n_1. \qquad (5.21d)$$

The unknowns are now n_1, θ_1, u, and E_x.

5.3 PERTURBATION AND PLASMA OSCILLATIONS

Equations (5.21) are all linear, so strictly speaking our (nonlinear) interest has been transformed away. However, the further work is of sufficient interest to warrant inclusion. Using the first Eq. (5.21a) we can eliminate $\partial u/\partial x$ in Eq. (5.21c) and obtain after one integration and evaluation of the integration "function"

$$n_0 \theta_1 = 2n_1 \theta_0 . \tag{5.22}$$

Utilizing Eq. (5.22) in Eq. (5.21b) results in

$$n_0 \frac{\partial u}{\partial t} + 3\theta_0 \frac{\partial n_1}{\partial x} - \frac{e}{m} E_x n_0 = 0. \tag{5.23}$$

To eliminate E_x we differentiate Eq. (5.23) with respect to x and substitute $4\pi e n_1$ for $\partial E_x/\partial x$ (see Eq. (5.21d)).
Thus

$$n_0 \frac{\partial^2 u}{\partial x \, \partial t} + 3\theta_0 \frac{\partial^2 n_1}{\partial x^2} - \omega_p^2 n_1 = 0 \tag{5.24}$$

where

$$\omega_p^2 = \frac{4\pi n_0 e^2}{m}$$

is the plasma frequency squared. Finally, from Eq. (5.21a) there results

$$-\frac{\partial^2 n_1}{\partial t^2} = n_0 \frac{\partial^2 u}{\partial x \, \partial t}$$

so that Eq. (5.24) becomes the wave equation

$$\frac{\partial^2 n_1}{\partial t^2} - 3\theta_0 \frac{\partial^2 n_1}{\partial x^2} + \omega_p^2 n_1 = 0. \tag{5.25}$$

Equation (5.25) is solvable by classical means. However, since the main interest is the so called *dispersion relation*, i.e., an equation which predicts how the plasma will oscillate in dependence upon the various physical parameters, we shall find this by assuming n_1 has a traveling wave solution of the classical form

$$n_1 = N \exp[i(\omega t - kx)] \tag{5.26}$$

and upon substitution get the (eigen) relation

$$\omega^2 = \omega_p^2 + 3\theta_0 k^2. \tag{5.27}$$

Equation (5.27) is the dispersion relation for one dimension, obtained also by Bohm and Gross [9] for a slightly different case. In the two dimensional case, we obtain by a similar analysis

$$\omega^2 = \omega_p^2 + 2\theta_0 k^2$$

and in three dimensions (Gross [10])

$$\omega^2 = \omega_p^2 + (\tfrac{5}{3})\theta_0 k^2.$$

5.4 PERTURBATION IN ELASTICITY

Numerous authors have utilized the method of perturbation to consider various nonlinear problems in elasticity. Chien [11] examined the large deflections of a circular clamped plate under uniform pressure and Nash and Cooley [12] carried through a similar analysis for elliptic plates. Previous analyses employed expansions into Mathieu functions and the Ritz method both of which turn out to be more cumbersome than the perturbation method coupled with a polynomial solution of the resulting linearized problem.

Let a thin, initially flat elliptic plate have semimajor axis a and seminor axis b. Let the plate be of thickness h and be subjected to uniform pressure of magnitude q. When x is chosen along the major axis, y along the minor axis and z normal to the middle surface the equations for the x, y, z components of displacement, designated by u, v, w are the von Kármán equations

$$u_{xx} + \left(\frac{1-\nu}{2}\right)u_{yy} + \left(\frac{1+\nu}{2}\right)v_{xy} + w_x w_{xx} + \left(\frac{1-\nu}{2}\right)w_x w_{yy}$$
$$+ \left(\frac{1+\nu}{2}\right)w_y \cdot w_{xy} = 0 \quad (5.28\text{a})$$

$$\left(\frac{1+\nu}{2}\right)u_{xy} + \left(\frac{1-\nu}{2}\right)v_{xx} + v_{yy} + w_y w_{yy} + \left(\frac{1+\nu}{2}\right)w_x w_{xy}$$
$$+ \left(\frac{1-\nu}{2}\right)w_{xx} w_y = 0 \quad (5.28\text{b})$$

$$D\nabla^4 w = q + w_{xx}\left\{\frac{Eh}{1-\nu^2}\left[u_x + \tfrac{1}{2}w_x^2 + \nu v_y + \frac{\nu}{2}w_y^2\right]\right\}$$
$$+ w_{yy}\left\{\frac{Eh}{1-\nu^2}\left[v_y + \tfrac{1}{2}w_y^2 + \nu u_x + \frac{\nu}{2}w_x^2\right]\right\}$$
$$+ w_{xy}\left\{\frac{Eh}{1+\nu}\left[u_y + v_x + w_x w_y\right]\right\}. \quad (5.28\text{c})$$

5.4 PERTURBATION IN ELASTICITY

We wish to solve this system for the large deflections of a clamped-edge elliptical plate $x^2/a^2 + y^2/b^2 = 1$ with the boundary conditions

$$\left.\begin{array}{l} u = v = w = 0 \\ w_x = w_y = 0 \end{array}\right\} \quad \text{along} \quad \frac{x^2}{a^2} + \frac{y^2}{b^2} = 1. \tag{5.29}$$

A perturbation procedure is based upon the smallness of the central deflection of the plate, which we designate by λ. Expanding q, w, u, and v in ascending powers of λ yields the following relations:

$$q = \alpha_1 \lambda + \alpha_3 \lambda^3 + \cdots \tag{5.30a}$$

$$u = s_2(x, y)\lambda^2 + s_4(x, y)\lambda^4 + \cdots \tag{5.30b}$$

$$v = t_2(x, y)\lambda^2 + t_4(x, y)\lambda^4 + \cdots \tag{5.30c}$$

$$w = w_1(x, y)\lambda + w_3(x, y)\lambda^3 + \cdots. \tag{5.30d}$$

Here α_1, α_3 are constants and the functions s_2, s_4, t_2, t_4, w_1, and w_3 are to be determined. The loading q (arbitrarily chosen) has an obvious relation to w and symmetry dictates an even character (in λ) for both u and v. Since λ is the central deflection of the plate (at $x = 0$, $y = 0$) we must require that

$$w_1(0, 0) = 1, \quad w_3(0, 0) = w_5(0, 0) = \cdots = 0.$$

Substituting relations (5.30) into Eq. (5.28c) and equating terms involving λ leads to the usual small deflection result for $w_1(x, y)$, to be found in books on elasticity (e.g., Timoshenko's "Plates and Shells"),

$$D = \frac{Eh^3}{12(1 - \nu^2)}, \quad w_1 = q \left[D \left(\frac{24}{a^2} + \frac{16}{a^2 b^2} + \frac{24}{b^2} \right) \right]^{-1} \left[1 - \frac{x^2}{a^2} - \frac{y^2}{b^2} \right]^2. \tag{5.31}$$

Substitution of Eq. (5.30) into Eq. (5.28a) and (5.28b) and equating terms in λ^2 gives the simultaneous linear equations

$$(s_2)_{xx} + \left(\frac{1-\nu}{2}\right)(s_2)_{yy} + \left(\frac{1+\nu}{2}\right)(t_2)_{xy} + (w_1)_x (w_1)_{xx}$$
$$+ \left(\frac{1-\nu}{2}\right)(w_1)_x w_{1yy} + \left(\frac{1+\nu}{2}\right)(w_1)_y (w_1)_{xy} = 0 \tag{5.32}$$

$$\left(\frac{1+\nu}{2}\right)(s_2)_{xy} + \left(\frac{1-\nu}{2}\right)(t_2)_{xx} + (t_2)_{yy} + (w_1)_y (w_1)_{yy}$$
$$+ \left(\frac{1+\nu}{2}\right)(w_1)_x (w_1)_{xy} + \left(\frac{1-\nu}{2}\right)(w_1)_{xx}(w_1)_y = 0. \tag{5.33}$$

From Eqs. (5.29) we infer the boundary conditions

$$s_2(x, y) = t_2(x, y) = 0 \quad \text{along} \quad \frac{x^2}{a^2} + \frac{y^2}{b^2} = 1.$$

The solution of this boundary value problem characterized by the inhomogeneous Eq. (5.32) and (5.33) is solved by attempting a polynomial solution. Polynomial solutions satisfying these equations and possessing the odd characteristics that this problem demands must have the form

$$s_2(x, y) = \left(1 - \frac{x^2}{a^2} - \frac{y^2}{b^2}\right) x(A_2 + B_2 x^2 + C_2 y^2 + D_2 x^4 + E_2 y^4 + F_2 x^2 y^2) \tag{5.34}$$

$$t_2(x, y) = \left(1 - \frac{x^2}{a^2} - \frac{y^2}{b^2}\right) y(a_2 + b_2 x^2 + c_2 y^2 + d_2 x^4 + e_2 y^4 + f_2 x^2 y^2) \tag{5.35}$$

To determine the twelve coefficients $A_2, ..., F_2$, $a_2, ..., f_2$ one substitutes Eqs. (5.31), (5.34), and (5.35) into the partial differential Eqs. (5.32) and (5.33) and equates terms involving corresponding powers of x and y. Twelve linear simultaneous equations are thus obtained.

Lastly the first correction to w, that is $w_3(x, y)$ is obtained by the same process of using Eq. (5.30) in Eq. (5.28c) and equating terms involving λ^3; thus

$$D\nabla^4 w_3 = \alpha_3 + (w_1)_{xx} \left\{ \frac{Eh}{1-\nu^2} \left[(s_2)_x + \tfrac{1}{2}(w_1)_x^2 + \nu(t_2)_y + \frac{\nu}{2}(w_1)_y^2 \right]\right\}$$

$$+ (w_1)_{yy} \left\{ \frac{Eh}{1-\nu^2} \left[(t_2)_y + \tfrac{1}{2}(w_1)_y^2 + \nu(s_2)_x + \frac{\nu}{2}(w_1)_x^2 \right]\right\}$$

$$+ 2(w_1)_{xy} \left\{ \frac{Eh}{1-\nu^2} [(s_2)_y + (t_2)_x + (w_1)_x(w_1)_y]\right\} \tag{5.36}$$

with the boundary conditions

$$w_3 = (w_3)_x = (w_3)_y = 0 \quad \text{on} \quad \frac{x^2}{a^2} + \frac{y^2}{b^2} = 1.$$

A polynomial solution of even character

$$w_3 = \left(1 - \frac{x^2}{a^2} - \frac{y^2}{b^2}\right)^2 (B_3 x^2 + C_3 y^2 + D_3 x^4 + E_3 y^4 + F_3 x^2 y^2)$$

is obtained by inserting in Eq. (5.35) and solving the six linear algebraic equations for α_3, B_3, C_3, D_3, E_3, and F_3. When $a = 2b$ the two term load deflection relation is

$$q = D\left[\frac{24}{a^4} + \frac{16}{a^2b^2} + \frac{24}{b^4}\right]\lambda + \frac{19.2419\,D}{h^2b^4}\lambda^3$$

a result which has been shown to agree well with experiment (see Nash and Cooley [12]).

This problem can also be examined by variational methods.

5.5 OTHER APPLICATIONS

Perturbation methods are particularly appropriate whenever the problem under consideration closely resembles one which is exactly solvable. The presumption is made that these differences are not singular in character but that one may change from the exactly solvable situation to the problem under consideration in a gradual fashion as expressed by requiring that the perturbation be a continuous function of a parameter λ.

Nonlinearities may occur in the boundary conditions as well as in the equations so the method may be useful there. *Surface* perturbations refer to deviations in the boundary surface or boundary conditions, or both, from the exactly solvable form. Problems of this type continually arise in the linear arena—for example we may wish to determine the heat transfer characteristics in a slab with trapezoidal cross section, a surface which is not a coordinate surface in which the scalar Helmholtz equation separates [13, 14].

Alternatively *volume* perturbations refer to deviations within the volume which remove the problem from the exactly solvable class. These are most often of the type we have already met in the previous sections, i.e., the occurrence of viscous effects, chemical reactions or a potential depending upon the solution.

When the deviations from the exactly solvable problem becomes large, the method of perturbation becomes inconvenient and then a variational method is more appropriate. And in all events the problem associated with convergence are difficult ones and often are intractable except in terms of the solution to the original problem which we seek.

Morse and Feshbach [15, pp. 1001–1106] discuss the methods as applied to linear problems. Their main concern is with deviations from an exactly solvable problem whose equation is of the four standard types: Diffusion, Laplace, Schroedinger, or Wave equation. Convergence considerations are more amenable to treatment in these cases and are carried through in some detail for several situations. However, even in this area where more machinery is available, the intractability of the convergence problem for the perturbation series is the main barrier to the utilization of these procedures.

5.6 PERTURBATION ABOUT EXACT SOLUTIONS

There is no *a priori* reason why the perturbation technique must be applied to a system whose exact solution is obtained from a linear equation. In fact there are several advantages if the exact solvable system retains some of the nonlinearities. This is especially true in physical areas, such as fluid mechanics, where the essential features are due to the nonlinear terms. Many phenomena are simply not capable of analysis by linear techniques no matter how hard we try. The literature of elementary nonlinear vibration theory contains a wealth of such examples as relaxation oscillations and subharmonic resonance.

To illustrate the idea of perturbations about known exact solutions of nonlinear problems we consider two examples. The basic ideas discussed here were originally discussed by Ames and Laganelli [16] and have been but slightly utilized by this writting (1964).

As a first example consider the case of two dimensional, constant pressure, gas dynamics in which material and energy are added to the flow so that Eqs. (4.20) are modified to the form

$$u_t + uu_x + vu_y = 0 \tag{5.37a}$$

$$v_t + uv_x + vv_y = 0 \tag{5.37b}$$

$$\rho_t + (\rho u)_x + (\rho v)_y = \lambda F(u) \tag{5.37c}$$

$$\epsilon_t + (\epsilon u)_x + (\epsilon v)_y = \lambda G(u) \tag{5.37d}$$

$$\epsilon = e + p$$

and λ is a small parameter. From Section 4.2 we know that the initial value problem given by Eqs. (5.37), with $\lambda = 0$, and with the initial conditions $u(x, y, 0) = f(x, y)$, $v(x, y, 0) = g(x, y)$, $\rho(x, y, 0) = h(x, y)$, $\epsilon(x, y, 0) = k(x, y)$ is exactly solvable. Therefore we adopt this solution of the nonlinear system as our basic solution and proceed.

5.6 PERTURBATION ABOUT EXACT SOLUTIONS

Let
$$u = u_0 + \lambda u_1 + \lambda^2 u_2 + \cdots$$
$$v = v_0 + \lambda v_1 + \lambda^2 v_2 + \cdots$$
$$\rho = \rho_0 + \lambda \rho_1 + \lambda^2 \rho_2 + \cdots \qquad (5.38)$$
$$\epsilon = \epsilon_0 + \lambda \epsilon_1 + \lambda^2 \epsilon_2 + \cdots$$

and substitute these relations into Eqs. (5.37) (where we take $F(u) = G(u) = u^2$ for definiteness). Upon equating like powers of λ we find that in λ^0

$$u_{0t} + u_0 u_{0x} + v_0 u_{0y} = 0$$
$$v_{0t} + u_0 v_{0x} + v_0 v_{0y} = 0$$
$$\rho_{0t} + (\rho_0 u_0)_x + (\rho_0 v_0)_y = 0 \qquad (5.39)$$
$$\epsilon_{0t} + (\epsilon_0 u_0)_x + (\epsilon_0 v_0)_y = 0$$

and in λ the linear system

$$u_{1t} + u_0 u_{1x} + v_0 u_{1y} + u_{0x} u_1 + u_{0y} v_1 = 0$$
$$v_{1t} + u_0 v_{1x} + v_0 v_{1y} + v_{0x} u_1 + v_{0y} v_1 = 0$$
$$\rho_{1t} + (\rho_1 u_0 + u_1 \rho_0)_x + (\rho_0 v_1 + \rho_1 v_0)_y = u_0^2 \qquad (5.40)$$
$$\epsilon_{1t} + (\epsilon_1 u_0 + u_1 \epsilon_0)_x + (\epsilon_0 v_1 + \epsilon_1 v_0)_y = u_0^2.$$

The solution of the initial value problem described by Eqs. (5.39) subject to the initial conditions

$$u_0(x, y, 0) = f(x, y), \quad v_0(x, y, 0) = g(x, y), \quad \rho_0(x, y, 0) = h(x, y)$$

and $\epsilon_0(x, y, 0) = k(x, y)$ is given in Eqs. (4.26)–(4.29) as the implicit form

$$u_0(x, y, t) = f(x - tu_0, y - tv_0)$$
$$v_0(x, y, t) = g(x - tu_0, y - tv_0)$$
$$\rho_0(x, y, t) = h(x - tu_0, y - tv_0)/\Delta$$
$$\epsilon_0(x, y, t) = k(x - tu_0, y - tv_0)/\Delta$$

where

$$\Delta = (1 + tf_\omega)(1 + tg_\sigma) - t^2 f_\sigma g_\omega$$

210   5. APPROXIMATE METHODS

and the symbol f_ω, g_σ means differentiate with respect to the grouping $\omega = x - tu$, $\sigma = y - tv$ respectively.

Having obtained the solution for the zero order system, the first-order correction is then obtained from the linear system (5.40) subject to the homogeneous boundary conditions $u_1(x, y, 0) = v_1(x, y, 0) = \rho_1(x, y, 0) = \epsilon_1(x, y, 0) = 0$. The complexities of solution depend strongly upon the form of f, g, h, and k since the zero order solutions are implicit.

A second example will be based upon the Hopf transformation discussed in Chapter 2 and also in a more general way by Ames [17]. First we recall that if F is a solution of the diffusion equation

$$F_t = \nu \nabla^2 F \tag{5.41}$$

then

$$\mathbf{u} = -2\nu \nabla (\ln F) \tag{5.42}$$

is a solution of

$$\mathbf{u}_t + (\mathbf{u} \cdot \nabla)\mathbf{u} = \nu \nabla^2 \mathbf{u}. \tag{5.43}$$

Consider the modified Burgers' equation

$$u_t + u u_x = \nu u_{xx} + \lambda G(u) \tag{5.44}$$

where $G(u)$ may have various forms e.g., for a chemical reaction G may be an exponential, power function etc. For definiteness let $G(u) = u^2$ and let λ be a small parameter. With

$$u = u_0 + \lambda u_1 + \cdots$$

the equation for u_0 becomes

$$u_{0t} + u_0 u_{0x} = \nu u_{0xx} \tag{5.45}$$

and for u_1

$$u_{1t} + (u_0 u_1)_x = \nu u_{1xx} + u_0^2. \tag{5.46}$$

The solution of Eq. (5.45) is obtained from

$$u_0 = -2\nu \frac{\partial}{\partial x} (\ln F)$$

where $F_t = \nu F_{xx}$. u_1 is then obtained by solving the linear equation, with variable coefficients (5.46).

5.7 THE SINGULAR PERTURBATION PROBLEM

To introduce the concepts involved in singular perturbations we first consider an example of a linear system first detailed by Carrier [18]. The derivation is omitted.

The essential features of the long time average motion of the fluid in an ocean basin is specified by the linear boundary value problem

$$\epsilon \nabla^4 \psi - \psi_x = \sin y \qquad (5.47)$$

with boundary conditions

$$\begin{aligned} \psi = \psi_x = 0 &\quad \text{on } AB \text{ and } BC \\ \psi = \psi_{yy} = 0 &\quad \text{on } AC \end{aligned} \qquad (5.48)$$

where AB and BC are land boundaries and AC is a curve along which $\psi_{yy} = 0$ as shown in Fig. (5-1). Here $\psi_y = u$ and $\psi_x = -v$. The value of ϵ for the Carrier problem is 0.035.

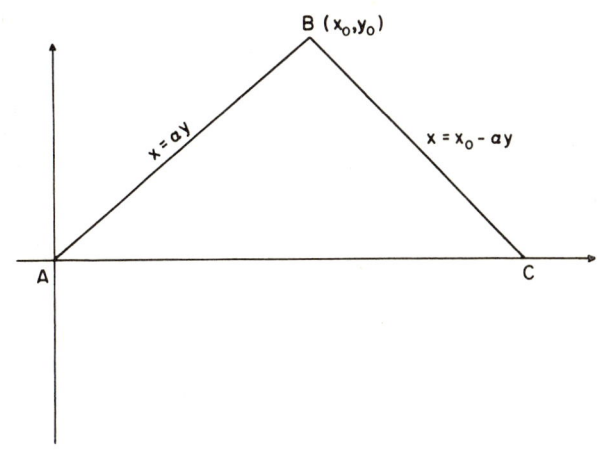

FIG. 5-1. The idealized ocean used in illustrating a singular perturbation problem.

The problem just described is characterized by three essential features: (1) *the coefficient of the terms involving the highest order derivatives is small compared to unity*; (2) *the other important terms have coefficients of order unity*; (3) *in the coordinate system chosen the characteristic lengths describing the size of the integration domain are of order unity.*

Moving ahead boldly, so that we observe the problems first-hand, we note that for $\epsilon \ll 1$ a good approximation to a solution of Eq. (5.47) is obtainable when the term with the coefficient ϵ is ignored. That function is

$$\psi_0 = -[x + f(y)] \sin y. \tag{5.49}$$

Adopting Eq. (5.49) as the approximate solution of Eq. (5.47) leads to the difficulty that only $f(y)$ is at our disposal to satisfy the boundary conditions (5.48). ψ_0 cannot satisfy those conditions and therefore *is not* a good approximate solution to our problem. What has been attempted here is a singular perturbation (see Section 5.1) since the order of the partial differential equation has been reduced.

What ways are there of circumventing the dilemma described above? One way is the so-called "inner-outer solution" method first described by von Kármán [19] in a slightly different context. To motivate this idea consider any function ψ which has the property that the maximum values of ψ, ψ_x, ψ_{xxxx}, ψ_{yyyy}, etc. are all of the same order of magnitude. Such a function can be a solution of Eq. (5.47) only if it is essentially of the form $\psi_0 + \epsilon \psi_1 + \cdots = \psi_0 + O(\epsilon)$. But we have established that a function of this form cannot be the required ψ. Thus ψ will be "steep" somewhere in the domain which means that at least one of the higher derivatives of ψ is large compared to ψ. In particular, one of the contributions of $\nabla^4 \psi$ must be of order $1/\epsilon$ compared to ψ_x if ψ is to be different from $\psi_0 + O(\epsilon)$. Here we *hope* that the steep portion of the function ψ is confined to a region very close to the boundary. This is the *"boundary layer assumption"*. If this turns out to be true then the outer solution (solution away from the boundaries of the domain) is represented very well by ψ_0 but we still need to find the behavior of ψ near the boundary. Of course one cannot always be sure that the steep part of the function is confined in this manner, but the analysis of any particular problem along the lines to be illustrated will lead to anomalies if such is not the case.

Returning to our linear ocean flow problem it is convenient to represent the boundary layer solution (that is the solution for the steep area near the boundary) as the sum of three functions, one for each portion of the boundary (see Fig. 5-1)

$$\psi = \psi_0 + \varphi_{AB} + \varphi_{BC} + \varphi_{CA}. \tag{5.50}$$

Our previous thinking makes us anticipate that φ_{AB} will be steep near AB but diminish rapidly with increasing distance from AB. Similarly

5.7 THE SINGULAR PERTURBATION PROBLEM

φ_{BC} and φ_{CA} are large near BC and CA, respectively, and small elsewhere. To determine φ_{AB} it is convenient to adopt a new coordinate system so chosen that one coordinate essentially measures distance *from* the boundary whereas the other measures distances along the boundary. The scale (units) in which these distances are measured should be such that the maximum values of φ_{AB} and its derivatives all are of the same order of magnitude. We henceforth call such functions *regular*. Thus the *coordinate stretching* must depend upon the parameter ϵ.

For the boundary layer near AB let ξ be the variable measuring distance from the boundary, which of course must vanish at the boundary, i.e., when $x = \alpha y$, and η be the variable measuring distance along the boundary so that $\varphi_{AB} = \varphi_{AB}(\xi, \eta)$. These can be chosen as

$$\xi = \epsilon^n(x - \alpha y), \qquad \eta = \lambda y \tag{5.51}$$

where n and λ must be determined subsequently.

On AB the boundary conditions, from Eq. (5.48), are

$$\psi_0(\alpha y, y) + \varphi_{AB}(0, \eta) = \frac{\partial \psi_0}{\partial x}(\alpha y, y) + \epsilon^n \frac{\partial \varphi_{AB}}{\partial \xi}(0, \eta) = 0. \tag{5.52}$$

The regularity of ψ_0 implies the regularity of φ_{AB} in η if $\lambda = 1$, i.e., $\eta = y$. Thus there is no stretch in the y-coordinate. The number n must be determined by the criterion that $\epsilon \nabla^4 \psi$ contribute an amount which is of the same order in ϵ as the contribution of ψ_x.

Having detailed the strategy, and the reasons for it, we now introduce coordinates (5.51) into Eq. (5.47), with $\psi = \psi_0 + \varphi_{AB}$, and obtain

$$(1 + \alpha^2)^2 \epsilon^{4n+1}(\varphi_{AB})_{\xi\xi\xi\xi} - \epsilon^n(\varphi_{AB})_\xi + 0(1) = 0. \tag{5.53}$$

For the contributions of the two leading terms to be of the same order of magnitude (in ϵ) we must have $4n + 1 = n$, or $n = -\frac{1}{3}$ so that finally the equation in φ_{AB} is

$$(1 + \alpha^2)^2 (\varphi_{AB})_{\xi\xi\xi\xi} - (\varphi_{AB})_\xi + O(\epsilon^{\frac{1}{3}}) = 0. \tag{5.54}$$

We may now decide to represent φ_{AB} formally as a perturbation series in $\epsilon^{\frac{1}{3}}$ or equivalently from a practical point of view discard the term in $O(\epsilon^{\frac{1}{3}})$ from Eq. (5.54). When this is done the first order solutions of Eq. (5.54) are

$$F(\eta), \qquad G_n(\eta) \exp[\gamma e^{2n\pi i/3} \xi]$$

where $\gamma = (1 + \alpha^2)^{-\frac{2}{3}}$, $n = 1, 2, 3$ and F, G_n are to be determined from the boundary conditions. But since φ_{AB} must decay rapidly as we leave the boundary we must discard the solutions $F(\eta)$ and the case $n = 3$, $G_3(\eta) \exp(\gamma \xi e^{2\pi i}) = G_3(\eta) \exp(\gamma \xi)$. Thus only the solution

$$\varphi_{AB} = G_1(\eta) \exp[\gamma \xi e^{2\pi i/3}] + G_2(\eta) \exp[\gamma \xi e^{4\pi i/3}] \tag{5.55}$$

is acceptable.

By now the method is clear so we do not give the arguments for boundaries BC and CA in detail. For BC define $\xi' = (x + x_0 - \alpha y)\epsilon^n$, $\eta' = y$, again obtaining Eq. (5.54). This time near $\xi' = 0$, that is near BC,

$$\varphi_{BC} = G_3(\eta') \exp(\gamma \xi'). \tag{5.56}$$

The parameters $f(y)$, $G_1(y)$, $G_2(y)$, and $G_3(y)$ may now be chosen so that the sum of the functions defined by

$$\psi = -[x + f(y)] \sin y + G_1(y) \exp[\gamma e^{2\pi i/3} \xi]$$
$$+ G_2(y) \exp[\gamma e^{4\pi i/3} \xi] + G_3(y) \exp[\gamma \xi'] \tag{5.57}$$

can satisfy the boundary conditions on AB and BC. Further algebraic details and other physical details may be found in Munk [20, 21].

There is one important detail omitted. That is, at the boundary CA ($y = 0$) conditions (5.48) are not as yet satisfied. The model, Eq. (5.47), does not contain provisions for this and the foregoing boundary layer analysis would not give a decaying boundary layer solution for this edge. We must either convince ourselves that this is of no consequence or else we must modify the mathematics. In this case we have an easy out because of inaccuracies in physical data. For if the function $\sin y$, which is associated with the yearly average wind intensity, were replaced by a carefully chosen wind intensity function which differs very little from $\sin y$, the boundary condition would be satisfied by ψ_0 and boundary layer or other correction would not be necessary.

Corrections at the corners of the region also do not seem warranted for like the wind intensity the ocean boundaries are not defined this accurately.

We can now summarize what is usually meant by a singular perturbation problem:

"A singular perturbation problem is characterized by a small coefficient ϵ multiplying the term involving the highest order derivative. A function

$\psi_0 + \varphi$ is found such that ψ_0 is a regular solution of the differential equation Σ and hence a solution when the ϵ term is ignored. φ is a function which makes $\psi_0 + \varphi$ a solution of an approximation to the original differential equation Σ and which contributes nontrivially only near the boundaries. Finally one can check that the amount by which Σ fails to be satisfied by $\psi_0 + \varphi$ is small compared to the contribution of the other terms." Note that this last statement proves nothing about the accuracy with which the true solution is approximated by $\psi_0 + \varphi$ but is usually accepted as a reliable indication of that accuracy.

5.8 SINGULAR PERTURBATIONS IN VISCOUS FLOW

In previous sections we have had several opportunities to refer to the boundary layer theory. We shall see that the boundary layer theory of Prandtl can be considered as a singular perturbation procedure.

We consider the two dimensional flow of a compressible viscous fluid over a flat plate. We suppose the plate lies in the $x = z$ plane, the coordinate origin is taken at the leading edge of the plate, the free stream flow is in the x-direction and the characteristic length of the plate is L. The boundary layer assumption here is that when the Reynolds number is large, the effects of viscosity are confined to a narrow region (the boundary layer) near the plate. With δ the measure of the boundary layer thickness we remark that $\delta \ll L$.

The basic equations of this problem are

$$u_t + uu_x + vu_y = -\rho^{-1}p_x + \rho^{-1}\frac{\partial}{\partial x}\{\mu[2u_x - \tfrac{2}{3}(u_x + v_y)]\}$$
$$+ \rho^{-1}\frac{\partial}{\partial y}[\mu(u_y + v_x)] \quad (5.58a)$$

$$v_t + uv_x + vv_y = -\rho^{-1}p_y + \rho^{-1}\frac{\partial}{\partial y}\{\mu[2v_y - \tfrac{2}{3}(u_x + v_y)]\}$$
$$+ \rho^{-1}\frac{\partial}{\partial x}[\mu(u_y + v_x)] \quad (5.58b)$$

$$\rho_t + (\rho u)_x + (\rho v)_y = 0 \quad (5.58c)$$

where the body forces are assumed to be zero. It is an easy matter to include the energy equation in this analysis but no additional ideas are introduced by so doing.

From Section 5.7 we saw that all variables should be of the same order of magnitude, say unity. Thus we nondimensionalize Eq. (5.58) by setting

$$\xi = x/L, \quad \eta = y/\delta, \quad \bar{u} = u/U, \quad \bar{v} = v/V, \quad \bar{t} = t/(L/U)$$
$$\bar{\rho} = \rho/\rho_0, \quad \bar{p} = p/p_0, \quad \text{and} \quad \bar{\mu} = \mu/\mu_0. \quad (5.59)$$

The quantities U, V, p_0, ρ_0 and μ_0 are reference values of the corresponding quantities such as those at the edge of the boundary layer. These dimensionless variables are all of the order of unity, $O(1)$. But are u and v (or U and V) of the same order of magnitude? To answer this question we integrate the steady form of Eq. (5.58c) across the boundary layer i.e., with the boundary conditions $y = 0$, $v = 0$; $y = \delta$, $v = V$, $\rho = \rho_0$, and obtain

$$\rho_0 V = -\int_0^\delta (\rho u)_x \, dy$$

or

$$V/U = -\frac{\delta}{L} \left[\int_0^1 (\bar{\rho}\bar{u})_\xi \, d\eta \right]. \quad (5.60)$$

The bracketed quantity is of the order unity so that $VL/U\delta = O(1)$, that is $V \ll U$ since $\delta \ll L$.

The nondimensional form of Eqs. (5.58) is

$$\frac{\partial \bar{u}}{\partial \bar{t}} + \bar{u}\frac{\partial \bar{u}}{\partial \xi} + \bar{v}\frac{\partial \bar{u}}{\partial \eta}\frac{LV}{\delta U}$$
$$O(1) \quad O(1) \quad O(1)$$

$$= -\frac{p_0}{\rho_0 U^2 \bar{\rho}} \frac{\partial \bar{p}}{\partial \xi} + \frac{1}{R_e \bar{\rho}} \frac{\partial}{\partial \xi}\left[\bar{\mu}\left(2\frac{\partial \bar{u}}{\partial \xi} - \frac{2}{3}\left(\frac{\partial \bar{u}}{\partial \xi} + \frac{\partial \bar{v}}{\partial \eta}\frac{VL}{U\delta}\right)\right)\right]$$
$$O(1) \quad O(\delta^2) \quad O(1) \quad O(1) \quad O(1)$$

$$+ \frac{1}{R_e \bar{\rho}}\frac{L}{\delta}\frac{\partial}{\partial \eta}\left[\bar{\mu}\left(\frac{\partial \bar{u}}{\partial \eta}\frac{L}{\delta} + \frac{\partial \bar{v}}{\partial \xi}\frac{V}{U}\right)\right] \quad (5.61a)$$
$$O(\delta) \quad O(1/\delta) \quad O(\delta)$$

$$\frac{\partial \bar{v}}{\partial \bar{t}}\frac{V}{U} + \bar{u}\frac{\partial \bar{v}}{\partial \xi}\frac{V}{U} + \bar{v}\frac{\partial \bar{v}}{\partial \eta}\frac{LV^2}{\delta U^2}$$
$$O(\delta) \quad O(\delta) \quad O(\delta)$$

$$= -\frac{p_0 L}{\rho_0 U^2 \bar{\rho}\delta}\frac{\partial \bar{p}}{\partial \eta} + \frac{1}{R_e \bar{\rho}}\left\{\frac{L}{\delta}\frac{\partial}{\partial \eta}\left[\bar{\mu}\left(2\frac{\partial \bar{v}}{\partial \eta}\frac{VL}{U\delta} - \frac{2}{3}\left(\frac{\partial \bar{u}}{\partial \xi} + \frac{\partial \bar{v}}{\partial \eta}\frac{LV}{\delta U}\right)\right)\right]\right.$$
$$O(1) \quad O(\delta^2) \; O(1/\delta) \quad O(1) \quad O(1) \quad O(1)$$

$$\left.+ \frac{\partial}{\partial \xi}\left[\bar{\mu}\left(\frac{\partial \bar{u}}{\partial \eta}\frac{L}{\delta} + \frac{\partial \bar{v}}{\partial \xi}\frac{V}{U}\right)\right]\right\} \quad (5.61b)$$
$$O(1/\delta) \quad O(\delta)$$

5.8 SINGULAR PERTURBATIONS IN VISCOUS FLOW

$$\frac{\partial \bar{\rho}}{\partial \bar{t}} + \frac{\partial(\bar{\rho}\bar{u})}{\partial \xi} + \frac{LV}{\delta U}\frac{\partial(\bar{\rho}\bar{v})}{\partial \eta} = 0. \qquad (5.61c)$$

$O(1) \quad O(1) \quad\quad O(1)$

Some simple exact solutions of the Navier-Stokes equations (and experiments) indicate that δ is proportional to the square root of $\nu_0 = \mu_0/\rho_0$, thus $\mathrm{Re} = UL/\nu_0 = O(\delta^{-2})$. Thus we may record the order of magnitude for each term in Eqs. (5.61). This is done under the various terms in those equations.

If we neglect the terms of order δ and higher and revert to the unbarred original variables we have the boundary layer equations

$$\frac{\partial u}{\partial t} + u\frac{\partial u}{\partial x} + v\frac{\partial u}{\partial y} = -\frac{1}{\rho}\frac{\partial p}{\partial x} + \frac{1}{\rho}\frac{\partial}{\partial y}\left[\mu\frac{\partial u}{\partial y}\right] + O(\delta^2)$$

$$\frac{\partial p}{\partial y} + O(\delta) = 0 \qquad (5.62)$$

$$\frac{\partial \rho}{\partial t} + \frac{\partial(\rho u)}{\partial x} + \frac{\partial(\rho v)}{\partial y} = 0.$$

The advantages of these equations, though they are still nonlinear, is the reduction in the number of viscous terms and the fact that the pressure p is a known function in the boundary layer equations, i.e., $p = p(x)$, instead of an unknown in the original equations. Of course our discard of terms of $O(\delta)$ and higher has lowered the order of the partial differential equations from four to three and therefore this is a *singular perturbation problem*.

Upon re-examination of the method discussed above we recall that $\mathrm{Re}^{-1} = O(\delta^2)$ or alternatively $\delta = O(\mathrm{Re}^{-\frac{1}{2}})$. Now $V/U = O(\delta) = O(\mathrm{Re}^{-\frac{1}{2}})$ This suggests that an alternate approach may be fruitful. We follow the Lighthill-Kuo method [22, 23]. Consider Eqs. (5.58) in the steady state for a constant viscosity incompressible fluid.

$$uu_x + vu_y = -\rho^{-1}p_x + \nu[u_{xx} + u_{yy}]$$
$$uv_x + vv_y = -\rho^{-1}p_y + \nu[v_{xx} + v_{yy}] \qquad (5.63)$$
$$u_x + v_y = 0.$$

In accordance with the above motivation introduce the dimensionless quantitites

$$\xi = x/L, \quad \eta = \mathrm{Re}^{\frac{1}{2}}y/L, \quad \bar{u} = u/U, \quad \bar{v} = R_e^{\frac{1}{2}}v/U \qquad (5.64)$$

where U, $\bar{p}$ are the x component of velocity and the pressure at infinity respectively and $\mathrm{Re} = UL/\nu$. The resulting equations are

$$\bar{u}\bar{u}_\xi + \bar{v}\bar{u}_\eta = -(\rho U^2)^{-1} p_\xi + \mathrm{Re}^{-1}\bar{u}_{\xi\xi} + \bar{u}_{\eta\eta}$$
$$\mathrm{Re}^{-1}(\bar{u}\bar{v}_\xi + \bar{v}\bar{v}_\eta - \bar{v}_{\eta\eta}) = -(\rho U^2)^{-1}{}_\eta p + \mathrm{Re}^{-2}\bar{v}_{\xi\xi} \quad (5.65)$$
$$\bar{u}_\xi + \bar{v}_\eta = 0.$$

Finally, we set $\bar{p} = (p - p_\infty)/\rho U^2$ and introduce the stream function ψ so that $\psi_\eta = \bar{u}$, $\psi_\xi = -\bar{v}$, resulting in the system

$$\psi_\eta \psi_{\xi\eta} - \psi_\xi \psi_{\eta\eta} - \psi_{\eta\eta\eta} = -\bar{p}_\xi + \mathrm{Re}^{-1}\psi_{\xi\xi\eta}$$
$$\mathrm{Re}^{-1}[\psi_\xi \psi_{\xi\eta} - \psi_\eta \psi_{\xi\xi} + \psi_{\xi\eta\eta}] + \bar{p}_\eta = -\mathrm{Re}^{-2}\psi_{\xi\xi\xi}. \quad (5.66)$$

At this point we pause to notice that in the expressions (5.64) the variable y/L, which is a measure of the distance from the boundary, has been stretched by the parameter $\epsilon = \mathrm{Re}^{\frac{1}{2}}$ as has v/U. We arrived at this by a different argument than that previously used in Section (5.7) but the end result is the same i.e., one term, $\psi_{\eta\eta\eta}$, from the higher order derivatives, has the same order of magnitude as $\psi_\eta\psi_{\xi\eta} - \psi_\xi\psi_{\eta\eta}$ in the "region of steepness" the boundary layer. Since $\psi_{\eta\eta\eta} = \bar{u}_{\eta\eta}$, we have the physical interpretation that the viscous contribution of the gradient of the velocity gradient, in the normal direction to the plate, is the same order of magnitude as the inertia terms.

The flow field can now be obtained as the potential flow or "outer solution" outside the boundary layer, where the effect of viscosity is of lower order than inertial effects, plus the boundary layer or "inner solution" ψ which satisfies Eqs. (5.66). To obtain ψ and p we expand in terms of the small parameter $\epsilon = \mathrm{Re}^{-\frac{1}{2}}$ as

$$\psi = \psi^{(0)} + \epsilon\psi^{(1)} + \epsilon^2\psi^{(2)} + \cdots$$
$$\bar{p} = \bar{p}^{(0)} + \epsilon\bar{p}^{(1)} + \epsilon^2\bar{p}^{(2)} + \cdots \quad (5.67)$$

and find in ϵ^0

$$\psi^{(0)}_\eta \psi^{(0)}_{\xi\eta} - \psi^{(0)}_\xi \psi^{(0)}_{\eta\eta} - \psi^{(0)}_{\eta\eta\eta} = -\bar{p}_\xi; \quad (5.68)$$

in ϵ^1

$$\psi^{(0)}_\eta \psi^{(1)}_{\xi\eta} + \psi^{(0)}_{\xi\eta}\psi^{(1)}_\eta - \psi^{(0)}_\xi \psi^{(1)}_{\eta\eta} - \psi^{(0)}_{\eta\eta}\psi^{(1)}_\xi = -\bar{p}^{(1)}_\xi + \psi^{(1)}_{\eta\eta\eta} \quad (5.69)$$
$$\bar{p}^{(1)}_\eta = 0.$$

All higher order equations for $\psi^{(n)}$, $n > 0$ are linear. Equation (5.68) is immediately recognizable as the ordinary boundary layer equation while the linear correction equations (5.69) *et. seq.* (not shown) yield successive improvements.

The main difficulty with this approach lies in the fact that the solution of the zero order equations is singular in the whole line $x = 0$. Modifications must be made in order to avoid passing on these singularities in accentuated form to the rest of the successive approximations. Lighthill and Kuo [22, 23] present methods for accomplishing this goal.

The boundary layer equations are the first approximation of the Navier-Stokes equations for very large Reynolds number. Because of this central position considerable work has been done by numerous researchers. A central position in these studies has been occupied by the introduction of similarity variables—a topic we have discussed in some detail. Most of these methods ultimately depend upon the numerical approximate solution of (usually) some nonlinear ordinary differential equation. Such graphs constitute a highly acceptable engineering solution. On the other hand closed form or exact solutions, whether explicit or implicit, are highly desirable in understanding the peculiar characteristics (if any) of equations.

In Section 4.1 we considered the *ad hoc* assumption, made by Pai, that $\rho = f(u)$, which lead to a class of solutions in magnetogasdynamics. Our problems are more complicated than those discussed by Pai but a similar assumption should lead to valuable information about the boundary layer equations.

5.9 THE "INNER-OUTER" EXPANSION (A MOTIVATION)

Stokes, in 1851, concerned himself with the problem of incompressible flow past a sphere obtaining a solution by completely neglecting the inertia of the fluid so that the equations become

$$0 = -\operatorname{grad} p + \nu \nabla^2 \mathbf{u}$$
$$\nabla \cdot \mathbf{u} = 0 \tag{5.70}$$

where p, ν, and $\mathbf{u}$ are kinematic pressure, kinematic viscosity and velocity respectively. The linear problem of determining the Stokes stream function satisfying the no slip condition at the sphere boundary and the uniform stream condition at infinity results in the well known form

$$\psi = \frac{Ua^2}{4}\left[2r^2 - 3r + \frac{1}{r}\right]\sin^2\theta.\dagger \tag{5.71}$$

---

† Hereafter we take $U = 1$, $a = 1$ and consider only dimensionless ψ.

5. APPROXIMATE METHODS

Here U is the undisturbed stream velocity, a is the sphere radius, the origin of the spherical coordinates (r, θ, γ) is taken at the center of the sphere and the line $\theta = 0$ is in the direction of the undisturbed stream.

The Stokesian theory breaks down when

$$\text{Re}\, r = \frac{Ua}{\nu} r = O(1) \tag{5.72}$$

and Re is the (small) Reynolds number. Discussions of the reasons for this breakdown are detailed by numerous authors notably Lagerstrom and Cole [24], Proudman and Pearson [25], and others. From Eq. (5.72) we see that Stokes solution (5.71), for sufficiently small Re, does not break down until the region in which the flow is nearly a uniform stream is reached. Thus the solution does provide a uniformly valid approximation to the *total velocity* distribution and therefore a valid approximation to many bulk properties of the flow, such as drag. The solution *is not* a uniform approximation to the distribution of velocity gradients, i.e., the derivatives of the velocity at a great distance from the body are seriously in error. This error is most critical in the problem of obtaining a second approximation to the flow, for the reason that the neglected inertia terms involve velocity gradients. Stokes theory is therefore not self-consistent at a great distance from the body.

In mathematical terms these deficiencies of the theory of Stokes result from the fact that the perturbation represented by a small nonzero Reynolds number has a singularity at infinity. *In all singular perturbation problems a uniformly valid approximation to the neglected terms in the governing equation is a necessary prerequisite for the determination of a second approximation to the solution anywhere in the field.* Additional discussion of these points are given in Lagerstrom [24], Proudman [25], and Kaplun [26].

Oseen [27] in 1910 showed that the determination of a uniformly valid first approximation to the velocity and all of its derivatives is itself a linear problem which may be solved analytically. Oseen's technique revolves about taking inertia forces into account in the region where they are comparable with viscous forces but neglecting them in the Stokes' region of flow. Further, this is the region where uniform stream conditions have almost been attained so the appropriate equation (Oseen Approximation) is

$$\mathbf{U} \cdot \nabla \mathbf{u} = -\text{grad}\, p + \nu \nabla^2 \mathbf{u} \tag{5.73}$$

where $\mathbf{U}$ represents the uniform stream. The asymptotic flow field

5.9 THE "INNER-OUTER" EXPANSION (A MOTIVATION)

obtained from the Oseen approximation occupies an important place in viscous flow theory. Exact solutions of Eq. (5.73) are difficult to obtain as discussed by Goldstein [28] but in this case there is no justification for going to the expenditure of effort to find a solution which satisfies the boundary conditions to a higher order of approximation than that in the governing equation. Thus the Oseen solution, in terms of the dimensionless stream function ψ,

$$\psi = \frac{1}{4}\left[2r^2 + \frac{1}{r}\right]\sin^2\theta - \frac{3}{2\,\text{Re}}(1 + \cos\theta)[1 - \exp(-\tfrac{1}{2}\text{Re}\,r)(1 - \cos\theta)] \tag{5.74}$$

is an adequate approximation.

The function given by Eq. (5.74) is easily shown to satisfy Eq. (5.73), the relevant boundary conditions at infinity and when $r = O(1)$ an expansion of the second term show that

$$\psi = \frac{1}{4}\left[2r^2 + \frac{1}{r}\right]\sin^2\theta - \frac{3}{2\,\text{Re}}(1 + \cos\theta)[\tfrac{1}{2}\text{Re}\,r(1 - \cos\theta) + O(\text{Re}^2)]$$

$$\psi = \frac{1}{4}\left[2r^2 - 3r + \frac{1}{r}\right]\sin^2\theta + O(\text{Re}) \tag{5.75}$$

which agrees with the Stokes solution, and therefore satisfies the boundary conditions on the sphere, to an adequate approximation. That Eq. (5.74) provides a *uniform approximation* to the stream disturbance follows from its method of derivation. Lastly we remark that the Oseen equations (and solution) is only a first approximation to the governing equations and therefore cannot be used to derive *any* second approximation to *any* flow property![†]

In two (see footnote) and three dimensions the solutions of Oseen's equation are a starting point for the determination of higher approximations to the flow. Lagerstrom and Cole [24] have carried out such investigations in their basic paper. In principle, since Oseen's equation contains Re as a free parameter (a necessary consequence of

---

[†] The differences in the two-dimensional case are drastic since for slow streaming motion there is no solution to Stokes' equation (The Stokes' Paradox). We do no more here than remark that a useful discussion of the two dimensional case is included in Proudman and Pearson [25]. The relevant solution of Oseen's equations provide a uniformly valid approximation to the velocity and all its derivatives in the two dimensional flow past an infinite cylinder of finite cross section.

the uniform validity of Oseen's approximation), one could construct an iterative method of using a lower order approximation to calculate the inertia terms. This adaptation of an unsuccessful method of Whitehead [29] (originally attempted as a correction to the Stokes problem) applied to Oseen's equation gives

$$(\mathbf{u}^{(n)} \cdot \nabla)\mathbf{u}^{(n+1)} = -\operatorname{grad} p^{(n+1)} + \nu \nabla^2 \mathbf{u}^{(n+1)} \tag{5.76}$$

where $\mathbf{u}^{(0)} = \mathbf{U}$ the undisturbed flow and $\mathbf{u}^{(1)}$ is the result obtained from the Oseen function (5.75). Equations (5.76), or rather their combination with the continuity equation, are solved to the appropriate degree of accuracy. The expansion generated in this way is perhaps the most economic expansion possible and offers some interesting possibilities.

However, an alternate procedure using both the Stokes (inner) and Oseen (outer) expansions is usually preferable to the iterative method of Whitehead applied to the Oseen system. This alternate procedure is discussed in the next section.

5.10 THE INNER AND OUTER EXPANSIONS

The use of the terms inner and outer solution dates at least as far back as 1934 when von Kármán and Millikan [19] studied boundary layer separation. While this is not the main point of our exposition it is sufficiently of interest to be sketched. The laminar boundary layer equations

$$\begin{aligned} uu_x + vu_y &= u_\infty u_{\infty x} + \nu u_{yy}, \\ u_x + v_y &= 0 \end{aligned} \tag{5.77}$$

become

$$Z_x = \nu u Z_{\psi\psi} \tag{5.78}$$

under the change of variables (von Mises transformation, Section 2.2) $Z = u_\infty^2 - u^2$, x, and ψ (stream function) as independent variables instead of x and y. The difficulty caused by a singularity at $x = \psi = 0$ must be mentioned. If we introduce, instead of x, the potential function ϕ of the external potential flow, $\phi = \int_0^x u_\infty(x)\, dx$ then Eq. (5.78) becomes

$$Z_\phi = \nu \frac{u}{u_\infty} Z_{\psi\psi} = \nu \left[1 - \frac{Z}{u_\infty^2}\right]^{\frac{1}{2}} Z_{\psi\psi}. \tag{5.79}$$

5.10 THE INNER AND OUTER EXPANSIONS

In the outer portions of the flow, $\psi \to \infty$, Z tends to zero (i.e., $u/u_\infty \to 1$)[†] so that Z satisfies, to a first approximation, the diffusion equation

$$Z_\phi = \nu Z_{\psi\psi}. \quad (5.80)$$

Von Kármán and Millikan recognized that the velocity profile in the retarded flow has a point of inflection. Choosing this point as the matching point, which occurs at $\psi = \psi_m$, where

$$\frac{\partial^2 u}{\partial y^2} = \frac{u}{2} \frac{\partial^2 (u^2)}{\partial \psi^2} = 0$$

one constructs an "inner" solution to Eq. (5.79) and joins the two at ψ_m. The inner solution is constructed by approximating to the equation for u^2 in terms of ϕ and ψ. From the definition of Z we have $Z_{\psi\psi} = -(u^2)_{\psi\psi}$ and $Z_\phi = (u_\infty^2)_\phi - (u^2)_\phi = 2u_\infty u_\infty' - u_\phi^2$ so that Eq. (5.79) becomes

$$\nu(u^2)_{\psi\psi} = -\frac{2u_\infty^2 u_\infty'}{u}\left(1 - \frac{1}{2u_\infty u_\infty'}(u^2)_\phi\right) \quad (5.81)$$

(note that $u_\infty' = du_\infty/d\phi$, thus we are assuming u_∞ has a ϕ expansion of the form $u = \Sigma b_n \phi^n$). For the inner solution, Eq. (5.81) is replaced by the approximate form

$$\nu(u^2)_{\psi\psi} = -\frac{2u_\infty^2 u_\infty'}{u}\left(1 - \frac{u}{u_m}\right) \quad (5.82)$$

where u_m is the velocity u which occurs at the inflection point ψ_m. This equation is an ordinary differential equation in ψ, with parameter ϕ, having the boundary conditions

$$u = u_m: \quad (u^2)_\psi{}^I = (u^2)_\psi{}^{\bar{O}}$$
$$u = 0: \quad \psi^I = 0$$

where I = inner and $\bar{O}$ = outer solution. Further details and questions of accuracy are discussed in Von Kármán and Millikan [19] and the historical utility of the method is also given by Pai [30].

We now return to our main line of discussion whose position was motivated in Section 5.9 by the flow about a sphere. The procedure involves simultaneous consideration of locally valid expansions close

---

[†] This is essentially the Oseen approximation.

to and far from the singularity of the perturbation ϵ. These expansions will be called the *outer* and *inner* expansions respectively. In the sphere problem, of the preceeding section, the outer expansion might be called the Oseen form and the inner expansion the Stokes form. The inner expansion is a straightforward expansion of Whitehead type in the perturbation parameter ϵ (Re for the sphere) for fixed values of the space variables made dimensionless by the *finite* length scale of the body. For the outer expansion the coordinate system is first *stretched* (recall the Carrier problem and the boundary layer equations) by some factor depending upon ϵ, in such a way that the length scale of variations in the sayptotic solution at a great distance from the body is finite in the new coordinates. In these new coordinates the length scale of the body is small and the *singularity* of the perturbation is moved to the inside of the body (origin of coordinates). The outer expansion is then an expansion in the parameter ϵ of fixed values of the new coordinates.

To be more specific we assume a particular form

$$L(h, \epsilon) = 0$$
$$B_i(h) = 0 \qquad (5.83)$$

where L is some differential operator, B_i, are the proper number of boundary conditions and ϵ is the parameter. With r, μ as the independent variables we assume that the inner (h) and outer (H) expansions have the form

$$h = \sum_{i=0}^{\infty} f_i(\epsilon) h_i(r, \mu) \qquad (5.84)$$

$$H = \sum_{i=0}^{\infty} F_i(\epsilon) H_i(\rho, \mu), \qquad \rho = \epsilon r \qquad (5.85)$$

where the functions $f_n(\epsilon)$ and $F_n(\epsilon)$ are not necessarily simple powers of ϵ, and are initially restricted only by the requirements that for all n

$$\lim_{\epsilon \to 0} \frac{f_{n+1}}{f_n} = 0, \qquad \lim_{\epsilon \to 0} \frac{F_{n+1}}{F_n} = 0. \qquad (5.86)$$

These solutions are so constructed that: (a) the inner expansion h satisfies the boundary conditions at the body surface; (b) the outer expansion H satisfies the boundary conditions at infinity; and (c) the two solutions match identically at some arbitrary distance from the

5.10 THE INNER AND OUTER EXPANSIONS

surface and both remain bounded as $\epsilon \to 0$. Of course we must follow the "rules," i.e., for example in obtaining the outer expansion the proper variable (measuring distance from the body) must be strectched and the resulting equation used.

The assumption that the inner expansion takes the form indicated is equivalent to the mild assumption that there is no singular dependence on ϵ in the finite part of the field. The requirement of boundedness often forces $f_0(\epsilon) = 1$ although this may not be necessary. Now the matching conditions must be explored. The inner expansion h is *invalid* far from the body so the boundary conditions at infinity (e.g., uniform stream conditions) must be replaced by the requirement that the expansion should be perfectly matched to the outer expansion which is valid in the outer region. The usual case is that both the inner and outer expansions hold in an *overlap* region. This overlap region can be used to match the inner and outer expansions. The matching conditions follow from the fact that the inner and outer expansions are both expansions of the same function for small values of ϵ and therefore are related. To obtain this relation and therefore the matching conditions we invoke the Lagerstrom-Kaplun *matching condition* which states that the two expansions are related to each other by the stretching transformation (of course one expansion may not determine the other uniquely). The common features of the inner expansion of the infinitely many functions that all have the same outer expansion should be possessed by *the* inner expansion of the outer expansion itself. That is to say if the outer expansion (5.85) is formally expanded about $\epsilon = 0$, for fixed values of r (by expanding $H_n(\rho, \mu)$ for small values of ρ and rearranging terms) the resulting expansion $\sum_{n=0}^{\infty} g_n(\epsilon) \phi_n(r, \mu)$ must be closely related to the inner expansion itself. The only differences might arise from the fact that the h_n of the inner expansion may have terms of the form e^{-r} which are important in the inner region but contribute little in the outer region. It therefore appears that $g_n(\epsilon) = f_n(\epsilon)$ and that $h_n(r, \mu)$ and $\phi_n(r, \mu)$ must have the same asymptotic expansion for large r. Summarizing we see that the expansions of the outer solution coefficients (for small ρ) determine *uniquely* the expansions of the inner solution coefficients for large values of r and conversely. Mathematically we can state this as

$$h(r \to \infty, \mu) = H(\rho \to 0, \mu). \tag{5.87}$$

We have only attempted to make the matching principle plausible in the above argument. The Lagerstrom-Kaplun *"theory of intermediate limits"* goes a long way towards further insight in the process. But at the present stage in development this theory does not yet prove the matching principle nor delineate its region of validity. Nevertheless the technique has been successfully used in a number of important problems.

Before giving a detailed example we briefly discuss the advantages of this method over that of a uniformly valid expansion (such as Whitehead's). (a) The mathematical structure is simpler. The inner-outer expansions are usually power series, or simple extensions, in ϵ. Uniformly valid approximations usually depend on ϵ in more complicated ways since functions of both stretched and unstretched coordinate systems are involved. (b) Except for the first approximation, uniformly valid approximation are not usually of much physical interest. (c) Lagerstrom and Cole [24] point out that when dealing with asymptotic expansions for small Re (extension to small ϵ is obvious) one should restrict attention to those expansions that can in principle be derived from the exact solution by the application of formal limit processes which may be defined a priori. The discussion of error and the domain of applicability is then more easily accomplished. The inner and outer expansions are of this type since for ϵ the parameter and x the dimensionless position vector the limiting process $\epsilon \to 0$ for either fixed x or fixed ϵx defines the inner or outer expansions. On the other hand a uniformly valid expansion cannot usually be derived from the exact solution in this manner. It can be derived from the exact solution when that solution is known, or it can be defined as an iterative process (the usual case) based upon the governing equation. Both of these concepts are cumbersome and since uniformly valid approximations may be constructed from the inner-outer expansions it is more satisfactory to proceed via this route.

5.11 EXAMPLES

The utility of the inner-outer expansion for the solution of singular perturbation problems has received great interest in several areas since the fundamental papers of Lagerstrom and Cole, Kaplun, and Proudman and Pearson (*op. cit.*). Erdelyi [31] gives a summary of the procedures as applicable to ordinary differential equations and lists some fundamental papers in this area. He also summarizes, elsewhere [32, 33], considerable work in partial differential equations. Wasow [34] considers the singular perturbation problem for second order nonlinear *ordinary* differential equations. A group of elasticity problems which fall into the singular perturbation area (boundary layer category) are summarized by by Friedrichs [35]. Several examples are detailed by Bromberg and Stoker [36], Friedrichs and Stoker [37], and Reissner [38]. None of the latter articles use the inner-outer expansion procedure although the procedure certainly applies. Further, most of these problems concern nonlinear ordinary differential equations.

The subject procedure has been recently applied to problems involving partial differential equations (in addition to Lagerstrom [24] and Kaplun [26]) by a handful of researchers. Among these are Yakura [39] who considered the hypersonic flow associated with power law shocks. The method was used to obtain uniformly valid solutions far downstream from the blunt nose of slender bodies. The inner expansion describes the flow in the entropy layer and the outer expansion the flow external to the entropy layer. Martin [40] uses the singular perturbation method to investigate the vibrations of a circular plate under uniform tension. Acrivos [41] settles some of the existing controversy about heat and mass transfer from a solid sphere into low Reynolds number flow by by using the subject expansion procedure. Van Dyke in two fundamental papers [42, 43] obtained higher approximations in boundary layer theory and Brenner and Cox [44] studied the resistance to a particle of arbitrary shape in small Reynolds number flows. The *Journal of Fluid Mechanics* is currently the richest source of inner-outer expansion examples.[‡]

The example that we detail is based on that one considered by Proudman and Pearson [25] for flow past a sphere. There seems no doubt that this article is already a classic—certainly it has been a strong stimulus to numerous researchers.

Flow past a sphere is governed by the equation for the dimensionless stream function ψ

$$\frac{1}{r^2} \frac{\partial(\psi, \nabla_r^2 \psi)}{\partial(r, \mu)} + \frac{2}{r^2} \nabla_r^2 \psi L_r \psi = \frac{1}{\epsilon} \nabla_r^4 \psi \quad \dagger \tag{5.88}$$

where $\mu = \cos\theta$,

$$\nabla_r^2 = \frac{\partial^2}{\partial r^2} + \frac{1 - \mu^2}{r^2} \frac{\partial^2}{\partial \mu^2}, \qquad \epsilon = \text{Re}, \qquad L_r = \frac{\mu}{1 - \mu^2} \frac{\partial}{\partial r} + \frac{1}{r} \frac{\partial}{\partial \mu}$$

and the symbol $\partial(\,,\,)/\partial(\,,\,)$ is the Jacobian of the approriate quantities. We note that Eq. (5.88) is independent of the latitude angle usually designated by ϕ.

[†] See for example Pai, "Viscous Flow Theory" [30].
[‡] See also Van Dyke [89].

5. APPROXIMATE METHODS

In the Stokes region of the flow where $r = O(1)$ we assume an inner expansion of the form

and
$$\psi = f_0(\epsilon)\psi_0(r, \mu) + f_1(\epsilon)\psi_1(r, \mu) + \cdots$$
$$f_{n+1}/f_n \to 0 \quad \text{as} \quad \epsilon \to 0 \tag{5.89}$$

in accordance with the general expansions (5.84)–(5.86). Since the magnitude of the velocities are everywhere bounded we write $f_0(\epsilon) = 1$ and thus allow the possbility that $\psi_0 = 0$ although it seems obvious that ψ_0 is probably the Stokes solution (5.71). The inner expansion is required to satisfy Eq. (5.88) and the no slip conditions on the sphere. In addition a matching condition of the form (5.87) with the outer solution must hold.

As r increases the inner expansion becomes invalid because the inertia and viscous forces become *comparable*. This suggests that, for the outer solution, we should hunt for a transformation which removes the Reynolds number ϵ from the equation. Such a *stretching* transformation (completely analogous to that used in Section 5.7) thereby suits the new coordinate system to the fact that all terms in the equation for the outer expansion are of the same order of magnitude. There are undoubtedly many such transformations but we choose the simplest which stretches the distance to the sphere and scales the stream function. That is we introduce two functions $f(\epsilon)$ and $g(\epsilon)$ so that

$$\rho = f(\epsilon)r, \quad H = g(\epsilon)\psi \tag{5.90}$$

and these are to be determined so that ϵ should not appear in the governing equation and so that the dimensionless velocity is $O(1)$ in the region of validity of the outer expansion—that is where ρ and H are $O(1)$. The condition that ϵ should not appear is obtained from Eq. (5.88) as

$$\epsilon f(\epsilon) = g(\epsilon). \tag{5.91}$$

Since

$$u_r = \frac{1}{r^2 \sin\theta} \psi_\theta = \frac{f^2(\epsilon)}{g(\epsilon)} \frac{1}{\rho^2 \sin\theta} H_\theta \tag{5.92}$$

and for it to be $O(1)$ in the region of the outer expansion

$$f^2(\epsilon) = g(\epsilon). \tag{5.93}$$

Equations (5.91) and (5.93) thus imply that $\epsilon = f(\epsilon)$ so that the proper outer variables are

$$\rho = \epsilon r, \quad H = \epsilon^2 \psi. \tag{5.94}$$

5.11 EXAMPLES

The governing equation (5.88) takes the form

$$\frac{1}{\rho^2}\frac{\partial(H, \nabla_\rho^2 H)}{\partial(\rho, \mu)} + \frac{2}{\rho^2}(\nabla_\rho^2 H)L_\rho H = \nabla_\rho^4 H \tag{5.95}$$

where ∇_ρ^2 and L_ρ are the same operators as in Eq. (5.88) with ρ replacing r.

The outer expansion (in the Oseen region of the flow) is assumed to have the form

$$H = H_0(\rho, \mu) + F_1(\epsilon)H_1(\rho, \mu) + F_2(\epsilon)H_2(\rho, \mu) + \cdots$$

and
$$F_{n+1}/F_n \to 0 \quad \text{as} \quad \epsilon \to 0. \tag{5.96}$$

That the leading term is independent of ϵ follows from the choice of the outer variables (5.94). The outer expansion (5.96) is required to satisfy Eq. (5.95), the uniform stream condition at infinity and the matching conditions with the inner expansion.[†]

A word here about the actual calculation procedure is useful. It is necessary to solve for one term at a time and the method for an inner term would be as follows: The general solution for this term is obtained from the pertinent differential equation such that the no slip condition on the sphere is satisfied. The outer expansion of this term is then developed and the individual terms in this expansion are compared with the corresponding terms that have previously been calculated in the outer expansion of the full solution. According to

$$\lim_{r\to\infty} \psi(r, \mu) = \lim_{\rho\to 0} H(\rho, \mu)$$

a comparison of coefficients uniquely determines the arbitrary constants in the general solution for the inner term.

Carrying out the computation we substitute (5.89) into (5.88) and retain terms in $O(1)$. The equation for ψ_0 is

$$\nabla_r^4 \psi_0 = 0 \tag{5.97}$$

[†] Note that the sphere in the new coordinate system has shrunk to a sphere of radius ϵ, thereby *introducing* ϵ into the *boundary* conditions instead of in the differential equations.

a biharmonic equation, whose general solution, leading to $u_r = u_\theta = 0$ at $r = 1$ and vanishing at the stagnation point $u = \pm 1$, is

$$\psi_0 = \sum_{n=1}^{\infty} [A_n\{(2n-1)r^{n+3} - (2n+1)r^{n+1} + 2r^{-n+2}\}$$

$$+ B_n\{2r^{n+1} - (2n+1)r^{-n+2} + (2n-1)r^{-n}\}]Q_n(\mu) \qquad (5.98)$$

where A_n and B_n are constants, $P_n(\mu)$ is the Legendre polynomial of degree n defined by the relations

$$P_0(\mu) = 1, \quad P_1(\mu) = \mu, \quad (n+1)P_{n+1}(\mu) = (2n+1)\mu P_n(\mu) - nP_{n-1}(\mu) \qquad (5.99)$$

$$n \geqslant 1, \quad \text{and} \quad Q_n(\mu) = \int_{-1}^{\mu} P_n(x)\, dx.$$

In the outer solution the situation regarding H_0 appears to require the solution of a nonlinear partial differential equation. However, motivated by the Oseen example of setting the uniform stream $\mathbf{U}$ into $(\mathbf{u} \cdot \nabla)\mathbf{u}$ so that the inertia terms read $(\mathbf{U} \cdot \nabla)\mathbf{u}$ we can easily show that

$$H_0 = \tfrac{1}{2}\rho^2(1-\mu^2) \qquad (5.100)$$

i.e., the uniform stream, is the exact solution of the nonlinear equation for H_0. There is a simple physical argument for H_0. The outer radial variable ρ is defined by

$$\rho = \epsilon r = \operatorname{Re} r = \frac{Ua}{\nu} \frac{\bar{r}}{a} = \frac{U\bar{r}}{\nu} \qquad (5.101)$$

where $\bar{r}$ is the dimensional radius. This quantity is independent of a. Therefore for fixed U and ν a fixed value of ρ corresponds to a fixed position on some circle in space. Hence, if the limiting process $\epsilon \to 0$ is interpreted as the limiting process $a \to 0$ it follows that the flow at the fixed point ultimately becomes that of the free stream.

We are now in a position to apply the matching criterion. In Eq. (5.98) we obtain the outer expansion by replacing r by $\rho\epsilon^{-1}$ and obtain $\psi_0(r \to \infty, \mu)$ as

$$\epsilon^{-2} \sum_{n=1}^{\infty} [A_n\{(2n-1)\epsilon^{-n-1}\rho^{n+3} - (2n+1)\epsilon^{-n+1}\rho^{n+1} + 2\epsilon^n\rho^{-n+2}\}$$

$$+ B_n\{2\epsilon^{-n+1}\rho^{n+1} - (2n+1)\epsilon^n\rho^{-n+2} + (2n-1)\epsilon^{n+2}\rho^{-n}\}]Q_n(\mu). \qquad (5.102)$$

5.12 FLOW PAST A SPHERE

Since $H = \epsilon^2 \psi$, $\psi = \epsilon^{-2} H$ and the requirement that this contribution to H should not contain terms of *greater* order than unity† (as $\epsilon \to 0$) forces us to choose

$$A_n = 0 \quad \text{for all} \quad n = 1, 2, \ldots$$
$$B_n = 0 \quad \text{for all} \quad n = 2, \ldots \quad (5.103)$$

Thus since $Q_1(\mu) = \int_{-1}^{\mu} P_1(\mu)\, d\mu = -\tfrac{1}{2}(1 - \mu^2)$ the contribution of Eq. (5.102) to H is

$$-\tfrac{1}{2} B_1 \{2\rho^2 - 3\epsilon\rho + \epsilon^3 \rho^{-1}\}(1 - \mu^2). \quad (5.104)$$

The requirement that the term of order $O(1)$ represents the uniform form stream Eq. (5.100) gives

$$B_1 = -\tfrac{1}{2}$$

so that the Stokes solution

$$\psi_0 = \tfrac{1}{4}(2r^2 - 3r + r^{-1})(1 - \mu^2) \quad (5.105)$$

is recovered as expected.

5.12 HIGHER APPROXIMATIONS FOR FLOW PAST A SPHERE

In order to obtain the equation for H_1 we shall assume $F_1(\epsilon) = \epsilon$, although by this assumption we shall *not* preclude the possibility that the constants in the solution may themselves depend upon ϵ and thus act to modify the form. Using this form of the outer expansion (5.96) we substitute into Eq. (5.95) and collect terms in $O(\epsilon)$ obtaining the equation

$$\left(\frac{1 - \mu^2}{\rho}\right) \frac{\partial \nabla_\rho^2 H_1}{\partial \mu} + \mu \frac{\partial \nabla_\rho^2 H_1}{\partial \rho} = \nabla_\rho^4 H_1 \quad (5.106)$$

which really is Oseen's equation (5.73). Now this equation may be written as

$$-\frac{\sin \theta}{\rho} \frac{\partial T}{\partial \theta} + \cos \theta \frac{\partial T}{\partial \rho} = \nabla_\rho^2 T \quad (5.107)$$

---
† In this case terms of $O(\epsilon^n)$ $n \geq 0$ are acceptable but $n < 0$ are not.

with $T = \nabla_\rho^2 H_1$. Equation (5.107) is of the "energy" form

$$-\nabla\phi \cdot \nabla T = \nabla^2 T \qquad (5.108)$$

where $\phi = -\rho \cos \theta = -\rho\mu$ is the potential. This is a special case of a general theory discussed by de la Cuesta and Ames [45, 46][†] and used by Goldstein [28] in this special case. From the results of the footnote we see that ϕ is harmonic (it is $-x$) and therefore the transformation

$$T = \exp(-\phi/2)S = \exp(\tfrac{1}{2}\rho\mu)S \qquad (5.109)$$

reduces Eq. (5.108) to

$$\nabla_\rho^2 S - \tfrac{1}{4}S = 0 \qquad (5.110)$$

since $(\nabla\phi)^2 = 1$ and $\nabla^2\phi = 0$. The general solution for Eq. (5.110) that vanishes as $\rho \to \infty$ and when $\mu = \pm 1$ (the latter condition follows from the form of ψ) is obtained by separation of variables to be (recall $\nabla_\rho^2 H_1 = \exp(\tfrac{1}{2}\rho\mu)S$)

$$\nabla_\rho^2 H_1 = \exp[\tfrac{1}{2}\rho\mu] \sum_{n=1}^{\infty} A_n(\rho/2)^{\frac{1}{2}} K_{n+\frac{1}{2}}(\rho/2) Q_n(\mu) \qquad (5.111)$$

where $K_{n+\frac{1}{2}}(\rho/2)$ is a modified Bessel function whose half integral order allows a closed form expression

$$(\rho/2)^{\frac{1}{2}} K_{n+\frac{1}{2}}(\rho/2) = (\pi/2)^{\frac{1}{2}} e^{-\rho/2} \sum_{m=0}^{n} \frac{(n+m)!}{(n-m)!m!} \rho^{-m}. \qquad (5.112)$$

As $\rho \to 0$ the asymptotic form from Eq. (5.112) is

$$(\rho/2)^{\frac{1}{2}} K_{n+\frac{1}{2}}(\rho/2) \approx (\pi/2)^{\frac{1}{2}} \frac{(2n)!}{n!} \rho^{-n}. \qquad (5.113)$$

[†] The pertinent theorems are:
(1) The substitution $T = h \exp(-\phi/2k)$ transforms $\partial T/\partial t - \nabla\phi \cdot \nabla T = k\nabla^2 T$ into the Schroedinger equation

$$h_t + k \left\{ \left(\frac{\nabla\phi}{2k}\right)^2 + \frac{\nabla^2\phi}{2k} \right\} h = k\nabla^2 h.$$

(2) $\nabla \cdot (K \nabla T) = 0$ reduces to a Schroedinger equation by $T = K^{-1/2}h$.
(3) $\nabla \cdot (K \nabla T) = 0$ has the solution $T = K^{-1/2}h$ where h is harmonic if $K^{1/2}$ is harmonic.
(4) Those functions ϕ leading to separable equation are completely determined in two dimensions and partially classified in three.

5.12 FLOW PAST A SPHERE

Equation (5.111) is an inhomogeneous differential equation for H_1 whose determination is troublesome (see Goldstein [28]). As previously mentioned the matching requirement has not been rigorously mathematically justified—and neither is the next step taken here. But at this point we apply the matching condition in the partially integrated form (5.111). Application of the matching conditions at this point is reasonable because of our ultimate goal of a uniformly valid approximation (combining the inner and outer solutions) as discussed in Section 5.9. At this level the matching criterion takes the form

$$\nabla_r^2 \psi_0(r \to \infty, \mu) = \nabla_\rho^2 H_0(\rho \to 0, \mu) + \epsilon \nabla_\rho^2 H_1(\rho \to 0, \mu)$$

where again we recall $H = \epsilon^2 \psi$.

To obtain the inner expansion of $\nabla_\rho^2 H_1$ we set $\rho = \epsilon r$ and get for $\epsilon \nabla_\rho^2 H_1$

$$\epsilon \left[1 + \frac{\epsilon r \mu}{2} + \cdots \right] \sum_{n=1}^{\infty} A_n \left(\frac{\epsilon r}{2} \right)^{\frac{1}{2}} K_{n+\frac{1}{2}}\left(\frac{\epsilon r}{2} \right) Q_n(\mu), \tag{5.114}$$

but as $\rho \to 0$ we see from Eq. (5.113) that Eq. (5.114) is asymptotic to

$$\epsilon \left[1 + \frac{\epsilon r \mu}{2} + \cdots \right] \sum_{n=1}^{\infty} A_n \left(\frac{\pi}{2} \right)^{\frac{1}{2}} \frac{(2n)!}{n!} \left(\frac{2}{\epsilon r} \right)^n Q_n(\mu). \tag{5.115}$$

From the matching condition the requirement that this contribution to $\nabla_r^2 \psi$ be $O(1)$ is that

$$A_n = 0 \quad \text{for all} \quad n \geqslant 2$$

so that the contribution to $\nabla_r^2 \psi$ is

$$-A_1 \left(\frac{\pi}{2} \right)^{\frac{1}{2}} r^{-1}(1 - \mu^2). \tag{5.116}$$

Now to obtain A_1 we find from ψ_0 that

$$\nabla_r^2 \psi_0 = \frac{3}{2r}(1 - \mu^2) + \frac{1 - \mu^2}{r^2}(1 - r^2)$$

and as $r \to \infty$ this is[†]

$$\nabla_r^2 \psi_0 = \frac{3}{2r}(1 - \mu^2) + o(1) \tag{5.117}$$

[†] The notation $o(1)$ means "something that is bounded as $r \to \infty$"; in this case $(1 - \mu^2)(1 - r^2)/r^2 \to -(1 - \mu^2)$ as $r \to \infty$.

Thus

$$A_1 \left(\frac{\pi}{2}\right)^{\frac{1}{2}} = -\frac{3}{2} \tag{5.118}$$

and therefore the equation for $\nabla_\rho^2 H_1$ becomes

$$\nabla_\rho^2 H_1 = \frac{3}{4}\left(1 + \frac{2}{\rho}\right) \exp[-\rho(1-\mu)/2](1-\mu^2). \tag{5.119}$$

The particular integral is

$$H_1 = -\tfrac{3}{2}(1+\mu)(1 - \exp[-\rho(1-\mu)/2]), \tag{5.120}$$

which is the second term of the Oseen approximation, Eq. (5.74). Our choice of $F_1(\epsilon) = \epsilon$ is thus justified.

The second term of the inner expansion is calculated in much the same fashion as discussed above. To obtain the equation for ψ_1 we first approximate the left hand side of Eq. (5.88) (the nonlinear terms) by ψ_0 and noting that $Q_2(\mu) = \int_{-1}^{\mu} P_2(\mu)\, d\mu = -\mu(1-\mu^2)/2$ we have

$$\frac{f_1(\epsilon)}{\epsilon} \nabla_r^4 \psi_1 = \frac{9}{2}\left[\frac{2}{r^2} - \frac{3}{r^3} + \frac{1}{r^5}\right] Q_2(\mu). \tag{5.121}$$

Again we assume $f_1(\epsilon) = \epsilon$ allowing for the possibility that the arbitrary constants in the integration of Eq. (5.121) may be functions of ϵ. In order for this solution not to contribute to the leading term of the inner solution these functions must be of smaller order than $1/\epsilon$. There is no point in considering functions of smaller order than 1 because these will be considered in later terms. Thus it appears that the coefficients will be functions of *order* lying between $O(1/\epsilon)$ *and* $O(1)$ (if at all) such as $O(\ln \epsilon)$, etc.

A particular integral of Eq. (5.121) satisfying the boundary conditions at $r = 1$ and $\mu = \pm 1$ is

$$\frac{3}{16}\left\{2r^2 - 3r + 1 - \frac{1}{r} + \frac{1}{r^2}\right\} Q_2(\mu) \tag{5.122}$$

and in view of Eq. (5.98) the general solution is

$$\psi_1 = (\text{Eq. 5.98}) + (\text{Eq. 5.122}). \tag{5.123}$$

The matching condition in this case reads

$$\psi_0(r \to \infty, \mu) + \epsilon \psi_1(r \to \infty, \mu) = H_0(\rho \to 0, \mu) + \epsilon H_1(\rho \to 0, \mu).$$

5.12 FLOW PAST A SPHERE

We are therefore interested in the contribution that ψ_1 makes to H_1 (the leading terms of the inner expansion and outer expansion have already been matched.) The contribution of the general solution (5.123) to H must be such that no term is of order greater than ϵ so that $A_n = 0$, $B_n = 0$ $n \geqslant 2$, and $B_1 = O(1)$. This leads to

$$\psi_1 = 3/32\,[2r^2 - 3r + r^{-1}](1 - \mu^2) - 3/32\,[2r^2 - 3r + 1 - r^{-1} + r^{-2}]\mu(1 - \mu^2). \quad (5.124)$$

Higher order terms in the inner and outer expansions are *not* proportional to simple powers of ϵ. To illustrate this more complicated dependence we examine the character of ψ_2. If we set $f_2(\epsilon) = \epsilon^2$ and allow (if necessary) the constants obtained in the integration for ψ_2 to be functions of ϵ then one obtains the general solution for ψ_2 as

$$\psi_2 = -\frac{3}{40}\Big[C_1 r^2 + C_2 r + C_3 r^{-1} - r^3 + 3r^2 \log r - \frac{3}{4}$$
$$- \frac{3 \log r}{5r} - \frac{7}{24r^2} + \frac{1}{40 r^3}\Big] Q_1(\mu)$$
$$+ \frac{27}{32}\Big[C_4 r^3 + C_5 + C_6 r^{-2} + \frac{r^2}{3} - \frac{r}{2} - \frac{1}{6r}\Big] Q_2(\mu)$$
$$+ \frac{9}{20}\Big[C_7 r^{-1} + C_8 r^{-3} + \frac{1}{9} r^3 - \frac{43 r^2}{120} + \frac{11 r}{24} - \frac{1}{3} + \frac{4 \log r}{35 r} + \frac{\log r}{42 r^3}\Big] Q_3(\mu)$$
$$(5.125)$$

with the C_n as constants. Evaluation of the C's proceeds from the matching criterion. Setting $r = \rho/\epsilon$ we examine the $Q_1(\mu)$ term and find that contributions to H of the term in $Q_1(\mu)$ is

$$-3/40[C_1 \epsilon^2 \rho^2 - \epsilon \rho^3 + 3\epsilon^2 \rho^2 \log \rho - 3(\epsilon^2 \log \epsilon)\rho^2] Q_1(\mu) + o(\epsilon^2). \quad (5.126)$$

This term contains $\epsilon^2 \log \epsilon$. It is easy to show that no such term exists in the outer expansion H_2 so that C_1 must be such that the term is also not present in Eq. (5.126). Thus

$$C_1 = 3 \log \epsilon + 0(1). \quad (5.127)$$

Similarly C_2 and C_3 must also be multiples of $\log \epsilon$. We are finally led to the conclusion that $f_2(\epsilon) \neq \epsilon^2$ but should be

$$f_2(\epsilon) = \epsilon^2 \log \epsilon. \quad (5.128)$$

236   5. APPROXIMATE METHODS

The remaining details are of a computational nature and will be left for the references. Let us summarize by detailing the inner and outer expansions:

$$\psi = \psi_0 + \epsilon\psi_1 + \epsilon^2 \log \epsilon \psi_2 + \cdots$$

$$H = H_0 + \epsilon H_1 + \cdots.$$

Similar expansions have been carried out by Acrivos and Taylor [41] for heat and mass transfer about spheres. Their equation, the energy equation, is

$$\nabla_r^2 h = \epsilon[U_r h_r + (U_\theta/r)h_\theta] \tag{5.129}$$

where

$$\nabla_r^2 = \frac{1}{r^2}\frac{\partial}{\partial r}\left(r^2 \frac{\partial}{\partial r}\right) + \frac{1}{r^2}\frac{\partial}{\partial \mu}\left[(1-\mu^2)\frac{\partial}{\partial \mu}\right]$$

$$\mu = \cos\theta, \qquad \epsilon = \tfrac{1}{2}\text{Pe} \quad \text{(Peclet number)}$$

where

$$\text{Pe} = 2aU_\infty \bar{\rho}C_p/k \text{ for heat transfer}$$

$$= 2aU_\infty/D \text{ for mass transfer}$$

and the Stokes velocity components are

$$U_r = \left(1 - \frac{3}{2r} + \frac{1}{2r^3}\right)\mu, \qquad U_\theta = -\left(1 - \frac{3}{4r} - \frac{1}{4r^3}\right)(1-\mu^2)^{\frac{1}{2}}.$$

They use the stretching transformation $\rho = \epsilon r$, $h = H$ and obtain the inner solution

$$h = \frac{1}{r} + (\epsilon + \tfrac{1}{2}\epsilon^3 \ln \epsilon)h_1 + \tfrac{1}{2}(\epsilon^2 \ln \epsilon)\left(\frac{1}{r} - 1\right) + \epsilon^2 h_2 + \cdots$$

and outer solution

$$H = \frac{1}{\rho}\exp\left[\frac{\rho}{2}(\mu - 1)\right] + \cdots$$

which again points out the dependence of the f_i on functions of ϵ other than powers of ϵ.

5.13 ASYMPTOTIC APPROXIMATIONS

Numerous articles have expounded the idea that equations while often insolvable analytically could yield information by examining the solution to those equations obtained as some parameter in the system goes to a limit, usually zero or infinity. In fact the inner-outer expansion falls into this general category although the idea is certainly more general than that. Often these asymptotic equations can be exactly solved and the results *sometimes* show enough of the trend of the full solution so that they may be pieced together. The result of this piecing is an approximate and sometimes accurate picture of what is happening as in the work of Acrivos [47] in convection. The use of asymptotic solutions near zero and infinite values of a parameter or variable has been considerably exploited in theoretical physics as discussed in Morse and Feshbach [15]. Plasma physics is especially rich in such solutions (see the review by Gottschalk, Feeny, Lutz, and Ames [48].)The theory of asymptotic expansions about small or large values of a variable is a rather well developed theory for ordinary differential equations as discussed by Erdelyi [49] and Jeffreys [50] for example. These concepts may be applied to linear problems governed by partial differential equations whose solutions are obtainable by means of reduction to ordinary differential equations and they should be applicable to NLPDE that are reduced to ordinary form by, say, similarity. This area is little developed—indeed the the methods are not well understood for nonlinear ordinary differential equations.

In this section we consider asymptotic equations developed for a large or small parameter. The major tool in this development is the "boundary layer stretching" previously discussed. The three basic papers of interest here are Morgan and Warner [51], Morgan, Pipkin, and Warner [52] and Acrivos [47], where we have chosen to discuss the energy equation

$$\psi_y t_x - \psi_x t_y = (\text{Re } \sigma)^{-1}[t_{xx} + t_{yy}] + K_1 \text{Re}^{-1}[4(\psi_{xy})^2 + (\psi_{yy} - \psi_{xx})^2] \tag{5.130}$$

for forced convection. Here ψ is the stream function obtained from the momentum equations, $\text{Re} = U_0 L/\nu$ is the Reynolds number, $\sigma = c_p \mu/k$ is the Prandtl number, $K_1 = U_0^2/[c_p(t_0^* - t_\infty)]$ is another dimensionless number and t is a dimensionless temperature $= (t_1 - t_\infty)/(t_0^* - t_\infty)$. We assume ψ is known and is such that

$$\psi(x, 0) = \frac{\partial \psi(x, 0)}{\partial y} = 0, \quad \frac{\partial \psi(x, \infty)}{\partial y} = U_\infty(x), \quad t(x, \infty) = 0$$

and
$$t(x, 0) = \theta(x).$$

From Eq. (5.130) we desire to determine the behavior of the Nusselt number as a function of the Prandtl number and Reynolds number for large values of the Prandtl number. The Nusselt number N is defined by

$$N = Q/kb(t_0{}^* - t_\infty) = -\int_0^1 \frac{\partial t}{\partial y}\bigg|_{y=0} dx \qquad (5.131)$$

in terms of the dimensionless coordinates and temperature. To determine N, we first change the equations into the boundary layer equations. The normal coordinate y is changed by "*boundary layer stretching*" (for forced convection set $\xi = \mathrm{Re}^{\frac{1}{2}}y$, for free convection $\xi = G^{\frac{1}{4}}y$, G the Grashof number) and the boundary layer form of the energy equation (5.130) is obtained.

Specifically, for forced convection, we set $\xi = \mathrm{Re}^{\frac{1}{2}}y$, $\psi = \mathrm{Re}^{-\frac{1}{2}}w(x, \xi)$, $t(x, y) = T(x, \xi)$ and the thermal boundary layer equation is

$$w_\xi T_x - w_x T_\xi = \frac{1}{\sigma} T_{\xi\xi} + K_1(w_{\xi\xi})^2 \qquad (5.132)$$

and the Nusselt number becomes

$$N = -\mathrm{Re}^{\frac{1}{2}} \int_0^1 \frac{\partial T}{\partial \xi}\bigg|_{\xi=0} dx. \qquad (5.133)$$

For large Prandtl number, σ, the thermal boundary layer equation has the same basic form in all cases. In each case the effect of the dissipation term $K_1(w_{\xi\xi})^2$ must be considered. However, in any case since the highest order derivative is multiplied by the small parameter σ^{-1} (when σ is large) the problem is a singular perturbation problem in the parameter σ^{-1}.

To treat this problem we must consider a further transformation of the normal coordinate ξ of the form $\eta = \sigma^m \xi$, $m > 0$. This provides a further stretching of the *thermal* boundary layer based upon the Prandtl Number in a manner directly analogous to the original stretching in $\mathrm{Re}^{\frac{1}{2}}$ of the velocity and thermal layers. We also observe that it must be assumed that any functions $f(x, \xi)$ dependent upon σ and appearing in the boundary layer equation can be transformed by $f(x, \xi) = \sigma^n F(x, \eta)$ in such a way that F and its first two η derivatives are independent of

5.13 ASYMPTOTIC APPROXIMATIONS

σ to a first approximation. Lastly we note that for large σ the thermal boundary layer is much thinner than the velocity layer. Thus in the thermal layer we may replace the stream function $w(x, \xi)$ by the first term in its ξ power series development in the energy equation. The error is $O(\xi^3)$.

We proceed by making the stretching transformations in the energy equation, balance terms with respect to their σ dependence—thereby obtaining m and n and the asymptotic equation.

As an example assume forced convective flow over a body with arbitrary free stream velocity distribution $U_\infty(x)$, arbitrary body surface temperature $T(x, 0) = \theta(x)$ and of negligible dissipation term so $K_1 = 0$. Equation (5.132) is thus

$$w_\xi T_x - w_x T_\xi = \sigma^{-1} T_{\xi\xi}. \tag{5.134}$$

For σ large, $w(x, \xi) \cong A(x)\xi^2$ and the free stream velocity affects the form of $A(x)$ but not the ξ dependence of w.

Applying the Prandtl number stretch to the thermal boundary layer let

$$\eta = \sigma^m \xi, \qquad T(x, \xi) = \sigma^n S(x, \eta). \tag{5.135}$$

Now $T(x, 0) = \sigma^n S(x, 0) = \theta(x)$ does not depend on σ so $n = 0$. Substituting Eq. (5.135) into Eq. (5.134) with $w(x, \xi) = A(x)\xi^2$ we have

$$2A(x)\sigma^{-m}\eta(\partial S/\partial x) - A'(x)\sigma^{-m}\eta^2(\partial S/\partial \eta) = \sigma^{-1+2m}\, \partial^2 S/\partial \eta^2 \tag{5.136}$$

and for the convective and conductive terms to be of the same order of magnitude (for large σ) $-m = 2m - 1$, i.e., $m = \tfrac{1}{3}$. Thus Eq. (5.136) becomes

$$2A(x)\eta\,(\partial S/\partial x) - A'(x)\eta^2\,(\partial S/\partial \eta) = \frac{\partial^2 S}{\partial \eta^2} \tag{5.137}$$

and

$$N = -\mathrm{Re}^{\frac{1}{2}}\sigma^{\frac{1}{3}} \int_0^1 \frac{\partial S}{\partial \eta}\bigg|_{\eta=0} dx. \tag{5.138}$$

We note that

(a) $A(x) = \dfrac{\partial u}{\partial \xi}\bigg|_{\xi=0}$;

and

(b) the dependence of N on σ is obtained without actually solving the energy equation!

By a similar analysis we can also show that as $\sigma \to 0$ Eq. (5.132) takes the form

$$U_\infty(x) \frac{\partial S}{\partial x} - \frac{dU_\infty}{dx} \eta \frac{\partial S}{\partial \eta} = \frac{\partial^2 S}{\partial \eta^2}. \tag{5.139}$$

The solution of the two equations gives the behavior of S for large σ and small σ.

5.14 ASYMPTOTIC SOLUTIONS IN DIFFUSION WITH REACTION

The use of asymptotic methods can be further illustrated with a system consisting of one substance of concentration designated by $a(x, t)$ diffusing into a medium containing a second substance $b(x, t)$ which may also diffuse within the medium. Substances A and B react according to the *second*-order mechanism

$$A + B \xrightarrow{k} C.$$

The diffusion, assumed to occur in the semifinite medium $x \geqslant 0$, is governed, for the species a and b, by the coupled equations

$$a_t = D_a a_{xx} - kab \tag{5.140}$$

$$b_t = D_b b_{xx} - kab \tag{5.141}$$

with the auxiliary conditions as

$$\begin{aligned}
&\text{Initial} \quad t = 0: \quad a = 0 \quad \text{for all} \quad x > 0; \\
&\phantom{\text{Initial} \quad t = 0:} \quad b = C_0 \quad \text{for all} \quad x \geqslant 0 \\
&\text{Boundary} \quad t > 0: \quad a = C^*, \quad \partial b/\partial x = 0, \quad x = 0; \\
&\phantom{\text{Boundary} \quad t > 0:} \quad a \to 0, \quad b \to C_0 \quad \text{as} \quad x \to \infty.
\end{aligned} \tag{5.142}$$

When the dimensionless variables

$$\alpha = a/C^*, \quad \beta = b/C_0, \quad \theta = kC^*t, \quad \xi = (kC^*/D_a)^{\frac{1}{2}}x$$

and parameters

$$\Delta = D_b/D_a, \quad \Gamma = C_0/C^*$$

5.14 ASYMPTOTIC SOLUTIONS IN DIFFUSION WITH REACTION

are introduced, Eqs. (5.140) and (5.141) become

$$\frac{\partial \alpha}{\partial \theta} = \frac{\partial^2 \alpha}{\partial \xi^2} - \Gamma \alpha \beta \tag{5.143}$$

$$\frac{\partial \beta}{\partial \theta} = \Delta \frac{\partial^2 \beta}{\partial \xi^2} - \alpha \beta \tag{5.144}$$

with auxiliary conditions

$$\theta = 0: \alpha = 0, \quad \text{for all} \quad \xi > 0; \quad \beta = 1, \quad \xi \geqslant 0 \tag{5.145}$$
$$\theta > 0: \alpha = 1, \quad \partial \beta / \partial \xi = 0, \quad \xi = 0; \quad \alpha = 0, \beta \to 1 \quad \text{as} \quad \xi \to \infty.$$

Before treatment of the system (5.143)–(5.145) we first note the special case $\Delta = 1, \Gamma = 1$ which is equivalent to $D_a = D_b$, $C_0 = C^*$. When these relations occur we may subtract Eq. (5.143) from Eq. (5.144) and obtain

$$\Omega_\theta = \Omega_{\xi\xi}$$

with $\Omega = \beta - \alpha$. The initial condition from Eq. (5.145) carries over directly to $\Omega(0, \xi) = 1$ but the boundary condition at $\xi = 0$ must be modified to specify either $\partial \alpha / \partial \xi$ or β. Adopting the former we obtain the boundary conditions $\Omega_\xi(\xi,0) = \beta_\xi(\xi,0) - \alpha_\xi(\xi,0)$, and $\Omega(\xi,\infty) = 1$. The new problem in Ω is solvable as a standard diffusion problem for the difference $\beta - \alpha$. This solution is useful in the evaluation of the rate constant k and the ratio kC^*/D_a but it does not appear that we can proceed to obtain either α or β from it.

The development of the asymptotic solutions discussed here rests upon the research of Pearson [53], Ferron [54], and Brian, Hurley, and Hasseltine [55]. The latter authors use a predominately numerical procedure. In examining the problem described by Eqs. (5.143)–(5.145) we can distinguish essentially three limiting cases.

Case 1. For sufficiently small time θ there is little α concentration, the reaction is essentially absent and β is nearly equal to 1. Equation (5.143) is essentially one of *pure A diffusion*. Similar remarks apply if Γ is sufficiently small. To treat this case we apply the similarity transformation

$$\xi = \eta \theta^{\frac{1}{2}}, \tag{5.146}$$

which has a form similar to the stretching transformations used in the boundary layer work, so that the α Eq. (5.143) becomes

$$\alpha_{\eta\eta} + \tfrac{1}{2}\eta\alpha_\eta = (\theta\Gamma)\alpha\beta. \tag{5.147}$$

For small $\Gamma\theta$ a solution for α of the form

$$\alpha = \alpha_0(\eta) + (\theta\Gamma)\bar{\alpha}(\eta, \theta) \tag{5.148}$$

is suggested by the form of Eq. (5.147). Upon substitution we find that α_0 satisfies

$$(\alpha_0)_{\eta\eta} + \tfrac{1}{2}\eta(\alpha_0)_\eta = 0$$

whose solution, as an ordinary differential equation, consistent with conditions (5.145) is

$$\alpha_0(\eta) = \mathrm{erfc}(\eta/2)$$

which represents simple diffusion of A into the semi-infinite medium $x \geqslant 0$.

From Eq. (5.148) it follows that, at least for $\Gamma \approx 1$,

$$\alpha(\xi, \theta) = \alpha_0(\xi/\theta^{\frac{1}{2}}) + O(\Gamma\theta), \qquad \Gamma\theta \ll 1. \tag{5.149}$$

Case 2. The case in which $D_b \to \infty$ or equivalently $\varDelta \to \infty$ means essentially the same thing as $\Gamma \gg 1$ that is physically the proportional depletion of B due to the inward diffusion of A is negligible in the early stages. This can either occur because of a much more rapid diffusion of B ($D_b \gg D_a$) or an initially high concentration of B ($\Gamma \gg 1$). We might therefore call this the the *first order reaction* case.

We consider the case $\Gamma \gg 1$. To make the third term of Eq. (5.143) have the same order as the first and second terms we introduce the new (stretched) variables

$$\tau = \Gamma^n\theta, \qquad \eta = \Gamma^m\xi \qquad \text{for the } \alpha \text{ equation}$$

and θ, $\xi^1 = \xi/\varDelta^p$ for the β equation. After determining $n = 1$, $m = \tfrac{1}{2}$, $p = \tfrac{1}{2}$ we find the Eqs. (5.143) and (5.144) become

$$\frac{\partial \alpha}{\partial \tau} = \frac{\partial^2 \alpha}{\partial \eta^2} - \alpha\beta(\tau\Gamma^{-1}, \eta\Gamma^{-\frac{1}{2}}\varDelta^{-\frac{1}{2}}) \tag{5.150}$$

$$\frac{\partial \beta}{\partial \theta} = \frac{\partial^2 \beta}{\partial (\xi^1)^2} - \alpha(\theta\Gamma, \xi^1\Gamma^{\frac{1}{2}}\varDelta^{\frac{1}{2}})\beta \tag{5.151}$$

with the initial and boundary conditions as in the original formulation. When $\tau \ll \Gamma$ and $\eta \ll (\Gamma\Delta)^{\frac{1}{2}}$ we can assert (see (5.149)) that

$$\beta(\tau\Gamma^{-1}, \eta(\Gamma\Delta)^{-\frac{1}{2}}) \simeq \beta(0, 0) = 1$$

so that our asymptotic approximation is

$$\frac{\partial \alpha_0}{\partial \tau} = \frac{\partial^2 \alpha_0}{\partial \eta^2} - \alpha_0 \tag{5.152}$$

which is of first order reaction form, hence the name. The solution, subject to the boundary condtions (5.145) is

$$\alpha_0 = \tfrac{1}{2} e^{-\eta} \operatorname{erfc}[\eta/(2\tau)^{\frac{1}{2}} - \tau^{\frac{1}{2}}] + \tfrac{1}{2} e^{\eta} \operatorname{erfc}[\eta/(2\tau)^{\frac{1}{2}} + \tau^{\frac{1}{2}}]. \tag{5.153}$$

Further considerations to Case 3, that of instantaneous reaction ($k \to \infty$) are discussed by Pearson. He also tabulates the various asymptotic forms as to their range of validity in terms of the physical parameters of the system.

5.15 WEIGHTED RESIDUAL METHODS: GENERAL DISCUSSION

A desire for direct methods in the calculus of variations led Rayleigh (for eigenvalue problems) and Ritz (for equilibrium problems) [56] to developed powerful methods which have been widely used in linear problems. Later Galerkin [57] in 1915 developed the first true weighted residual method. The background for these methods and examples of their applications to linear problems are available in many sources—here we content ourselves with listing three, Kantorovich and Krylov [58], Crandall [59] and Collatz [60].

With x the vector of independent variables we suppose the problem is formulated in a domain D as

$$\begin{aligned} L[u] &= f(x) \\ B_i[u] &= g_i(x), \quad = 1, 2, ..., p \end{aligned} \tag{5.154}$$

where L is a nonlinear differential operator (of course it could be linear), B_i represent the appropriate number of boundary conditions and f, g_i are functions of the coordinates employed.

We seek an approximate solution to problem (5.154) in the linear form

$$\bar{u}(x) = \sum_{j=1}^{n} C_j \varphi_j + \varphi_0 \tag{5.155}$$

where the φ_j, $j = 1, 2, ..., n$ is a set of "trial" functions chosen beforehand. The φ_j are *often* chosen to satisfy the boundary conditions, indeed the original Galerkin method was of this type. This requirement may be modified and its modification will be discussed subsequently. We can always consider the chosen functions to be linearly independent and to represent the first n functions of some set of functions $\{\varphi_i\}$, $i = 1, 2, ...$ which is *complete*[†] in the given region of integration. The φ_j may be functions of all the independent variables in which case the C_j are *undetermined parameters*. One or more of the variables may not be included in the choice of the φ_j in which case the C_j are *undetermined functions*. There are two basic types of criteria for fixing the C_j. In the weighted residual method the C_j are so chosen as to make a weighted average of the equation residual vanish. In the other (which in reality is not a weighted residual method but has so many common features that we treat it here), the C_j (as undetermined parameters) are chosen to give a stationary value to a functional related to the problem (5.154). This functional is usually obtained via the calculus of variation. In both cases, for undetermined parameters, one obtains a set of n simultaneous (nonlinear) algebraic equations for the C_j. In the undetermined function case one obtains a set of n simultaneous differential equations for the C_j.

In the weighted residual methods the trail solution (5.155) is *usually*[‡] chosen to satisfy all the boundary conditions in both equilibrium and initial value problems. This can be accomplished in many ways. The suggested method is to choose the φ_j to satisfy

$$\begin{aligned} B_i[\varphi_0] &= g_i, & i &= 1, ..., p \\ B_i[\varphi_j] &= 0, & i &= 1, ..., p; \quad j \neq 0. \end{aligned} \tag{5.156}$$

It is then clear that $\bar{u}$ satisfies all the boundary conditions. However, in the case of *initial value problems* the initial conditions often cannot also be satisfied and a separate initial residual is established.

[†] In a function space (e.g., the set of all continuous functions in $0 \leqslant x \leqslant 1$) a complete set $\{\varphi_i\}$ is defined as a set such that no function $F(x)$ exists in the space that cannot be expanded in terms of the $\{\varphi_i\}$.

[‡] A discussion of possible generalizations is found in Section 5.17.

5.15 WEIGHTED RESIDUAL METHODS

For the stationary functional method it is only necessary that $\bar{u}$ satisfy the *essential*[†] boundary conditions.

When the trial solution (5.155) is substituted into Eq. (5.154) the equation R becomes

$$R[C, \varphi] = f - L[\bar{u}] = f - L\left[\varphi_0 + \sum_{j=1}^{n} C_j \varphi_j(x)\right]. \quad (5.157)$$

where the notation $R[C, \varphi]$ indicates the dependence of R on the vector $C = (C_1, ..., C_n)$ and φ. When $\bar{u}$ is the exact solution R is identically zero. Within a restricted trial family a *good* approximation may be described as one in which R is small *in some sense*.

a. Stationary Functional Criterion (for Equilibrium Problems)

Let Ω be a functional (e.g., an integral) derived from the original problem. This (Ritz) method consists in inserting the trial family $\bar{u}$ (Eq. 5.155)) directly into Ω and asking for

$$\frac{\partial \Omega}{\partial C_j} = 0, \quad j = 1, 2, ..., n. \quad (5.158)$$

These n equations are solved for the C_j and the corresponding function $\bar{u}$ represents an approximate solution to the extremum problem. The n algebraic equations are generally nonlinear for a nonlinear equilibrium problem.

b. Weighted Residual Criteria (Stated for Equilibrium Problems)

Modifications for initial value problems are discussed later.

1. UNDETERMINED PARAMETERS (C_j)

The sense in which the residual R is small is that each of the weighted averages,

$$\int_D W_k R \, dD \quad k = 1, 2, ..., n, \quad (5.159)$$

[†] Essential boundary conditions can be described by the Collatz [see 50] condition: If the differential equation is of order $2m$, then the essential boundary conditions are those that can be expressed in terms of u and its first $m - 1$ derivatives. In elasticity the essential conditions are those of geometric compatibility and the natural ones are those of force balance.

of R with respect to the weighting function W_k, $k = 1, 2, ..., n$ should vanish. This requirement provides n algebraic, usually nonlinear equations for the nC_j if the trial solution (5.155) is selected in the linear form. However, in some cases nonlinear (in the C_j) trial solutions may be chosen that will lead to linear equations. Various cases can be distinguished and *certainly this list is not complete*.

(i) **Method of Moments.** In this case $W_k = P_k(\bar{x})$, so Eq. (5.159) becomes $\iint_D P_k(\bar{x}) R \, dD = 0$ which is equivalent to asking for the vanishing of the first n "moments" of R to vanish. Here $P_k(\bar{x})$ are orthogonal polynomials in the vector $\bar{x}$ over the domain D. This procedure is probably most useful in one space dimension where the theory of orthogonal polynomials is well understood. In one dimension (x) some authors use $x^k = W_k$ but these are not orthogonal on the interval $0 \leqslant x \leqslant 1$ and better results would be obtained if they were orthogonalized before use. The use of x^k gives the method its name "method of moments."

Sometimes we may use a product of separate one dimensional orthogonal polynomials for the $P_k(\bar{x})$. For example, in three dimensions,

$$P_k(x, y, z) = R_k(x) S_k(y) T_k(z)$$

where R_k, S_k, T_k represent orthogonal polynomials over (say) $0 \leqslant x, y, z \leqslant 1$.

(ii) **Collocation.** In this case we choose n points in the domain D, say p_i, $i = 1, 2, ..., n$, and let

$$W_k = \delta(p - p_k)$$

where δ represents the unit impulse or Dirac delta which vanishes everywhere except at $p = p_k$ ($p = p(x_1, x_2, ..., x_l)$, $l =$ space dimension) and has the property that $\int_D \delta(p - p_k) R \, dD = R(p_k)$. This criterion is thus equivalent to setting R equal to zero at n points in the domain D. The point location is arbitrary but is usually such that D is covered by a simple pattern. Special circumstances of the particular problem may dictate other patterns.

(iii) **Subdomain.** Here the domain D is subdivided into n subdomains, not necessarily disjoint, say $D_1, D_2, ..., D_n$. The weighting functions W_k are chosen as

$$W_k(D_k) = 1, \qquad W_k(D_j) = 0 \qquad j \neq k,$$

so that the closeness criterion (5.159) becomes

$$\int_{D_k} R \, dD = 0 \qquad k = 1, 2, ..., n.$$

(iv) **Least Squares.** The integral of the square of the residual is minimized with respect to the undetermined parameters to provide the n simultaneous equations for the C_j. Then

$$\frac{\partial}{\partial C_k} \int_D R^2 \, dD = 2 \int_D \frac{\partial R}{\partial C_k} R \, dD = 0, \qquad k = 1, 2, ..., n$$

so that we can infer that $W_k = \partial R / \partial C_k$.

(v) **Galerkin's Method.** Here we choose $W_k = \varphi_k$, where φ_k are the portion of the complete (and hopefully orthogonal) set used to construct the trial solution $\bar{u}$. Thus Galerkin's method asks for

$$\int_D \varphi_k R \, dD = 0 \qquad k = 1, 2, ..., n.$$

When the problem is an initial value (propagation) problem the trial function should be so selected that the initial conditions are satisfied. Since the range of the time variable is infinite an estimate of the steady state (if any) and when it is approximately achieved will be helpful in establishing what time interval need be considered. If the steady solution is obtainable it may be used as the asymptotic solution for $t \to \infty$ and and approximate solution can be calculated for small t by one of these weighted residual methods.

2. UNDETERMINED FUNCTIONS

The foregoing procedures transformed a continuous problem into an approximately equivalent lumped parameter system whose form is obtained by the solution of a set of (nonlinear) algebraic equations. We will now describe analogous procedures which may be applied to replace multidimensional continuous problems with an approximately equivalent system consisting of a finite number of lower dimensional (most often one or two) problems. Kantorovich [61, 62] is usually credited with the introduction of this idea and its first application.

For the case of undetermined functions a trial family analogous to

Eq. (5.155) is attempted except that in this case the C_j are undetermined functions of (usually one or more independent variables. Thus

$$\bar{u} = \varphi_0 + \sum_{j=1}^{n} C_j(x_i)\varphi_j \qquad (5.160)$$

and the φ_j are predetermined functions of the remaining independent variables or more generally of all the independent variables. When the φ_j are chosen to satisfy the boundary condition, suitable orthogonality and completeness conditions, and the proper boundary conditions are imposed on the C_j, then Eq. (5.160) will satisfy the boundary conditions of the problem. Application of the foregoing weighted residual methods or the appropriate calculus of variations algorithm will determine a set of n differential equations for the C_j.

When the undetermined functions approach is applied to an initial value problem the initial conditions may be difficult to satisfy identically. In such cases an *initial residual* R_0 is introduced for the purpose of obtaining initial values for the C_j. Any of the above methods may be applied to this residual.

It is clear from the preceding discussion that the most important and most difficult step, in all these methods, is the selection of the trial solution (5.155) or (5.160). Application of the criteria to this trial solution has the effect of selecting the "best" approximation from the given family. One must insure that good approximations are included within the trial family. In selecting the φ_j one should carefully insure that the functions are (a) linearly independent; (b) members of a complete set; and (c) incorporating special characteristics of the problem such as symmetry, monotonicity, etc., which may be known.

When the system (5.154) is *linear* the algebraic and differential equations for the C_j's obtained by any of the weighted residual methods will be linear. Moreover, in the case of undetermined parameters, the matrix of the coefficients of the C_j will always be symmetric in the least squares method but not generally for the other methods. However if the problem (5.154) is self-adjoint[†] then Galerkin's criterion will also generate symmetric equations. When the functional Ω is quadratic, the Ritz method generates symmetric linear equations for the C_j.

[†] When the system (5.154) is linear in both the equation and boundary conditions we say it is self-adjoint if for any two functions u and v satisfying the homogeneous boundary condition $B_i[u] = B_i[v] = 0$ for all i, we always have $\int_D uL[v]\, dD = \int_D vL[u]\, dD$.

When these methods are applied to a particular trial family satisfying all the boundary conditions they generally produce different approximations. In the case of a *linear* equilibrium problem having an equivalent extremum formulation the Ritz method and Galerkin's method (applied to the same trial family) give identical results (see Galerkin [57]). Thus in this case Galerkin's criteria provides the optimum of the true weighted residual methods in the sense that the approximation so obtained also renders Ω stationary.

Numerous investigations have been carried out comparing these methods on the same problem. Comparisons on linear problems in partial differential equations have been made (on system properties) in equilibrium problems by Crandall [59], p. 234] and point by point with the exact solution in Hetenyi [63]. Eigenvalue comparisons have been detailed by Crandall [59, p. 318] and by Frazer, Jones, and Skan [64]. Bickley [65] considered an initial value problem of heat transfer. His approximations using two undetermined functions compared collocation on both the equation and initial residual, collocation on the initial residual with moments on the equation residual, moments of both residuals and Galerkin on both residuals. All of these *linear* comparisons demonstrate the superiority of Galerkin's method (equivalent to Ritz in some cases). A similar comparison to that of Bickley has been made by Collings [66] on a nonlinear heat transfer problem where the resulting ordinary differential equations were solved on an analog computer. Except for possible accuracy loss at the initial condition the Galerkin method again proved superior. *All of this experience does not constitute proof and the reader is cautioned that there may be examples where other methods are superior.*

After discussing some examples we shall return to the question of convergence and error associated with these methods.

5.16 EXAMPLES OF THE USE OF WEIGHTED RESIDUAL METHODS

a. The Method of Moments

Yamada [67] and Fujita [68] have applied the elementary method of moments to the problems of diffusion. In illustration we consider the problem of diffusion into a plane sheet of width $2L$ in the general case of a concentration-dependent diffusion coefficient $D(C)$. The physical problem is governed by the equation

$$\frac{\partial C}{\partial t} = \frac{\partial}{\partial x}\left(D(C)\frac{\partial C}{\partial x}\right) \tag{5.161}$$

and we shall suppose that the initial conditions is

$$C(x, 0) = 0 \qquad -L \leqslant x \leqslant L \tag{5.162a}$$

and the boundary conditions are

$$C(-L, t) = C(L, t) = C_0 \quad \text{for all} \quad t > 0. \tag{5.162b}$$

If the dimensionless variables

$$\psi = C/C_0, \qquad \eta = x/L, \qquad \theta = D_0 t/L^2, \qquad F(\psi) = D(C/C_0)$$

where $D = D_0$ when $C = C_0$, are introduced into Eqs. (5.161) and (5.162) the problem is defined by

$$\frac{\partial \psi}{\partial \theta} = \frac{\partial}{\partial \eta}\left(F(\psi)\frac{\partial \psi}{\partial \eta}\right) \tag{5.163}$$

and

$$\psi(\eta, 0) = 0, \qquad \psi(1, \theta) = 1, \qquad \frac{\partial \psi}{\partial \eta}(0, \theta) = 0. \tag{5.164}$$

Note that the symmetry condition has been invoked so that $0 \leqslant \eta \leqslant 1$ instead of $-1 \leqslant \eta \leqslant 1$.

In the early stages of diffusion we proceed by noting that the concentration distance curve may be represented approximately by a curved portion near the boundary, followed by a horizontal portion coinciding with the η-axis as shown in Fig. 5-2. We recognize that, strictly speaking, the concentration may be finite, though it may be very small, everywhere in the sheet at the instant diffusion commences. It is therefore clear that the region over which the concentration may be assumed zero depends on the accuracy of working. *For a prescribed accuracy* we let $\eta_0(\theta)$ be the point at which the concentration becomes zero. Of course $\eta_0(\theta)$ depends upon "time" θ so that in particular $\eta_0(0) = 1$ and $\eta_0(\theta_1) = 0$. The quantity θ_1, not initially known, is that at which the concentration first becomes nonzero at the center of the sheet, to the accuracy of working. One of the results of the following analysis will be the value of θ_1.

5.16 EXAMPLES

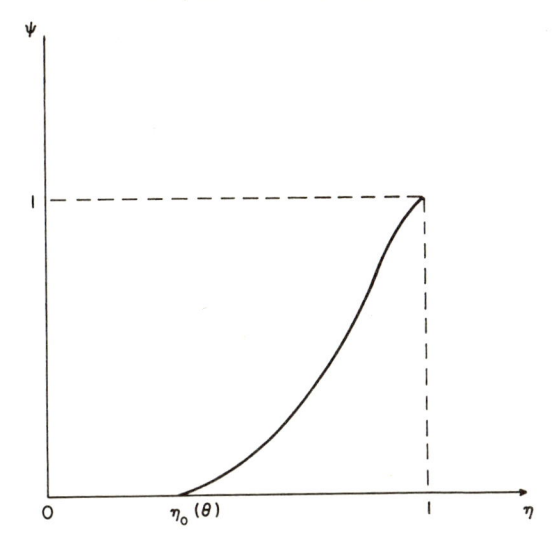

FIG. 5-2. A schematic of the short time solution of nonlinear diffusion.

We now consider the interval $\eta_0(\theta) \leqslant \eta \leqslant 1$ and select a trial family so that all boundary conditions and initial conditions are satisfied. Such a family as

$$\begin{aligned}\tilde{\psi}(\eta, \theta) &= 0, \quad 0 \leqslant \eta \leqslant \eta_0(\theta) \\ \tilde{\psi}(\eta, \theta) &= B(\theta)[\eta - \eta_0]^2 + E(\theta)[\eta - \eta_0]^3, \quad \eta_0 \leqslant \eta \leqslant 1\end{aligned} \quad (5.165)$$

where $B(\theta)$, $E(\theta)$ and $\eta_0(\theta)$ are to be determined, has the following desireable properties: (a) independence of the prescribed members $(\eta - \eta_0)^2$ and $(\eta - \eta_0)^3$; (b) if ψ is infinitely η differentiable it will have a Taylor series expansion of which Eq. (5.165) is a part—this gives a heuristic idea of possible completeness; (c) $\tilde{\psi}(\eta_0, \theta)$ as well as $\partial \tilde{\psi}/\partial \eta(\eta_0, \theta)$ are both zero insuring a tangential approach to the axis; and (d) The split of the range in θ will allow the possibility of more accuracy (and an easier task) than a single trial solution over the whole range of θ[†].

[†] By this point it is clear that one should construct the trial solution so that a maximum of information can be extracted with a minimum of computation. The more we know about the expected behavior of a solution the better a trial solution can be made. In propagation problems a preliminary study of the expected behavior is very important. It is often unwise to attempt to approximate the solution by a single approximate solution throughout. The solution domain should be broken up and the separate zones individually treated.

With Eq. (5.165) as the trial solution what must be done to insure satisfaction of the auxiliary conditions (5.164)? The condition $\partial \bar{\psi}/\partial \eta (0, \theta) = 0$ is satisfied while the condition $\bar{\psi}(\eta, 0) = 0$ is equivalent to

$$\eta_0(\theta) = 1 \quad \text{when} \quad \theta = 0. \tag{5.166}$$

Satisfaction of the condition $\psi(1, \theta) = 1$ can be insured by substituting into Eq. (5.165) obtaining

$$1 = B(\theta)[1 - \eta_0]^2 + E(\theta)[1 - \eta_0]^3.$$

For simplicity we set

$$B(\theta)[1 - \eta_0(\theta)]^2 = U(\theta), \quad E(\theta)[1 - \eta_0]^3 = V(\theta) \tag{5.167}$$

and because of the boundary condition $\bar{\psi}(1, \theta) = 1$, U and V are related by

$$U + V = 1. \tag{5.168}$$

At this point we pause to note that $B(\theta)$, $E(\theta)$ and $\eta_0(\theta)$ are to be determined and we might be tempted to specify that the first three moments vanish. However, Eq. (5.168) constitutes one relation among the three functions so that only the first two simple moments[†]

$$\int_0^1 R \, d\eta = 0, \quad \int_0^1 R\eta \, d\eta = 0 \tag{5.169}$$

of the residual $R = (\bar{\psi})_\theta - [F(\bar{\psi})\bar{\psi}_\eta]_\eta$ are required. The zero moment of the residual upon integration becomes

$$\frac{d}{d\theta}[(1 - \eta_0)(U/3 + V/4)] = \frac{2U + 3V}{1 - \eta_0} F(1) \tag{5.170}$$

which reduces to

$$\frac{d}{d\theta}[(1 - \eta_0)(1 + U/3)] = \frac{12}{1 - \eta_0}(1 - U/3)F(1) \tag{5.171}$$

since $U + V = 1$. The notation $F(1)$ denotes the value of F when $\psi = 1$.

[†] Choice of orthogonal polynomials on $0 \leqslant \eta \leqslant 1$ would improve the approximation.

A similar computation for the first moment gives the differential equation

$$\frac{d}{d\theta}[(1-\eta_0)^2(1+2U/3)] = 2 0 G(1) \tag{5.172}$$

where
$$G(\bar{\psi}) = \int_0^{\bar\psi} F(\psi)\, d\psi. \tag{5.173}$$

Since $G(1)$ is constant we may integrate and get

$$(1-\eta_0)^2(1+2U/3) = 2 0 G(1)\theta + \text{constant}$$

and since $\eta_0(0) = 1$ the constant $\equiv 0$. Thus we have the U, $\xi = 1 - \eta_0$ relation

$$U(\theta) = 3 0 G(1)\theta\xi^{-2} - \tfrac{3}{2}. \tag{5.174}$$

Substituting Eq. (5.174) into Eq. (5.171) gives the equation for $\xi = 1 - \eta_0$

$$\frac{d}{d\theta}\left[\frac{\xi}{2} + \frac{3 0 G(1)\theta}{3\xi}\right] = \frac{12}{\xi}\left(\frac{3}{2} - \frac{3 0 G(1)\theta}{3\xi^2}\right)F(1). \tag{5.175}$$

Since $\eta_0(0) = 1$, $\xi(0) = 0$ and therefore the solution of Eq. (5.175) subject to $\xi(0) = 0$ is

$$\xi = (\theta/\beta)^{\frac{1}{2}} \tag{5.176}$$

where β is an integration constant whose value is determined from the quadratic equation

$$[720 G(1)F(1)]\beta^2 + [30 G(1) - 108 F(1)]\beta + \tfrac{3}{2} = 0. \tag{5.177}$$

For Eq. (5.176) to have physical meaning $\beta > 0$. If one root is positive, or if both are equal there is no problem. If both roots are positive a question arises as to which one to select.[†] We defer this question until later.

The quantity θ_1 was the time at which $\eta_0 = 0$, that is $\xi(\theta_1) = 1 - \eta_0(\theta_1) = 1$. From Eq. (5.176) it is seen that

$$\theta_1 = \beta \tag{5.178}$$

and an alternate method of choosing the correct root would be to deter-

---

[†] This uniqueness question is apt to be troublesome in other applications of weighted residual techniques to nonlinear problems.

mine the time for the advancing front to reach $\eta = 0$ (to the accuracy desired).

To determine $B(\theta)$ and $E(\theta)$ we substitute Eq. (5.176) into Eq. (5.174) and obtain

$$U = 30G(1)\beta - \tfrac{3}{2}$$

and

$$V = -30G(1)\beta + \tfrac{5}{2}. \tag{5.179}$$

Since U and V are both constant, we have from Eq. (5.167)

$$B(\theta) = \frac{30G(1)\beta - \tfrac{3}{2}}{\theta/\beta}$$

$$E(\theta) = \frac{\tfrac{5}{2} - 30G(1)\beta}{(\theta/\beta)^{\frac{3}{2}}}$$

and consequently the approximate solution is

$$\bar{\psi}(\eta, \theta) = [30G(1)\beta - \tfrac{3}{2}]\frac{\beta}{\theta}[\eta - 1 + (\theta/\beta)^{\frac{1}{2}}]^2$$

$$-[30G(1)\beta - \tfrac{5}{2}]\left(\frac{\beta}{\theta}\right)^{\frac{3}{2}}[\eta - 1 + (\theta/\beta)^{\frac{1}{2}}]^3$$

$$\quad\text{for}\quad 1 - (\theta/\beta)^{\frac{1}{2}} \leqslant \eta \leqslant 1 \tag{5.180}$$

$$= 0 \quad\text{for}\quad 0 \leqslant \eta \leqslant 1 - (\theta/\beta)^{\frac{1}{2}}.$$

When the concentration just ceases to be zero at the center of the sheet, $\theta = \beta = \theta_1$, $\eta_0 = 0$, we have

$$\bar{\psi}(\eta, \theta_1) = [30G(1)\beta - \tfrac{3}{2}]\eta^2 - [30G(1)\beta - \tfrac{5}{2}]\eta^3, \quad 0 \leqslant \eta \leqslant 1. \tag{5.181}$$

Our discussion of the choice of β revolves around a suggestion of Fujita [68]. He proposes that one should consider the special case of a constant diffusion coefficient and determine which root gives best agreement (say r_1) with the exact solution. When the diffusion coefficient is concentration dependent he suggest that we select that root which tends to r_1 as $D \to D_0$. This is a reasonable argument especially in the case of slowly varying diffusion coefficients, but it is far from rigorous. For our special case $D(C) = C_0, F = 1, G = 1$ so that Eq. (5.177) becomes $1440\beta^2 - 156\beta + 3 = 0$ with roots $\beta = 1/12$ and $\beta = 1/40$. It is found that Eq. (5.180) is the best approximation when $\beta = 1/12$ so that if Fujita's method is adopted $r_1 = 1/12$.

The next step concerns the time $\theta \geqslant \beta = \theta_1$ i.e., to obtain an approximate solution which holds for later times when the concentration at the sheet's center has become appreciable to the accuracy of working. We shall not give the details but indicate the approach. Such a solution must satisfy the boundary conditions $\partial \psi/\partial \eta(0, \theta) = 0$, $\psi(1, \theta) = 1$ and the "initial condition" (for $\theta = \theta_1$) given by Eq. (5.181). A suitable trial solution might again be of the form

$$\psi_1 = A_1(\theta) + B_1(\theta)\eta^2 + E_1(\theta)\eta^3, \qquad 0 \leqslant \eta \leqslant 1$$

where A_1, B_1 and E_1 are to be determined. The first boundary condition is satisfied automatically and the second forces the relation

$$A_1 + B_1 + E_1 = 1.$$

Then the moment equations are used to determine the remaining two degrees of freedom.

It is instructive to rewrite Eq. (5.180) in terms of the original variables—thus the approximate solution is

$$\frac{C(x, t)}{C_0} = [30G(1)\beta - \tfrac{3}{2}]\left\{x\sqrt{\frac{\beta}{D_0 t}}\right\}^2 - [30G(1)\beta - \tfrac{5}{2}]\left\{x\sqrt{\frac{\beta}{D_0 t}}\right\}^3.$$

The total amount of diffusing substance M_t taken up by the sheet per unit area at time t is given by

$$M_t = 2\int_0^L C(x, t)\, dx = 2LC_0 \int_0^1 \psi(\eta, \theta)\, d\eta$$

and for early times $\theta < \theta_1$ we have

$$M_t = 2LC_0 \left[\frac{1}{8\beta^{\frac{1}{2}}} + \frac{30G(1)\beta^{\frac{1}{2}}}{12}\right]\theta^{\frac{1}{2}} \qquad 0 < \theta < \theta_1$$

and for later times

$$M_t = 2LC_0[1 - \tfrac{2}{3}B_1(\theta) - \tfrac{3}{4}E_1(\theta)].$$

We note that for early times, M_t is proportional to the square root of time, *irrespective of how the diffusion coefficient depends upon concentration*. This is a well known characteristic feature of concentration-dependent diffusion.

The results of this approximate solution have been compared with solutions obtained by finite difference methods for the case $D = D_0[1 + \alpha C]$ and are found to agree to an accuracy within 5% for the concentration-distance curves and within 1% for the total uptake. The greater accuracy of a system property (e.g., total uptake) than a local property is a characteristic feature of these methods. The accuracy of the solution is quite sensitive to the form of the diffusion coefficient.

b. Collocation

Various applications of collocation to linear problems are discussed by Crandall [59]. Little use has been made of this procedure for equations probably because of the *assumed* inferiority of the method to the more sophisticated procedures. Recently Jain [69] has utilized a modification of this method which is called "extremal point collocation" but his applications are restricted to the nonlinear ordinary differential equations obtained from the fluid equations by similarity transformations. However, the method has many useful features so we present it and the associated error analysis here.

Consider the boundary value problem in the form

$$L(u) = 0$$
$$B_i(u) = 0 \qquad (5.182)$$

where L is a function of the coordinates x_j, u and its derivatives and B_i are the boundary conditions. We shall find an approximation of the form Eq. (5.155), which we write as

$$u \approx \bar{u}(x_1, x_2, ..., x_p, C_1, ..., C_n),$$

in the p dimensional domain D, which depends upon the parameters C_j, $j = 1, 2, ..., n$ and is such that for arbitrary values of the C_j (a) the differential equation $L(\bar{u}) = 0$ is exactly satisfied; or (b) the boundary conditions are exactly satisfied; or (c) neither the differential equation nor the boundary conditions are satisfied by $\bar{u}$. Conditions (a) and (b) are obvious extensions of our previous discussion. According to our general discussion we then try to determine the parameters C_j so that $\bar{u}$ satisfies in case (a) the boundary conditions, in case (b) the differential equations as accurately as possible. The determination of the parameters and *estimation of the error* in the solution is made by some suitably chosen criterion. In the general discussion we asked for the residual R to vanish at n points but this tells us nothing about the intermediate errors.

What we need to estimate the intermediate error is some measure of the "distance" between the true solution and the approximate solution $\bar{u}$. Such distance function or *norms*, as they are usually called, are supplied in large numbers by the abstract *function spaces* as discussed for example in Collatz [60] from the approximation position, theoretically in Dunford [70], Banach [71], Friedman [72], Halmos [73] and many others. *We shall replace the "vanishing residual" criterion for determination of the free parameters by a norm criterion which will also allow us to estimate the error of the approximation.*

What norm shall we use? In the L^p function spaces (Hilbert spaces) the norm is

$$\|f - g\| = \left[\int_D |f - g|^p \, dD \right]^{1/p}. \tag{5.183}$$

While theoretically very important the Hilbert norm is cumbersome to calculate and it has the further disadvantage that $\|f - g\|$ may be very small yet $f - g$ may in fact be far apart over small intervals. Thus Eq. (5.183) is discarded. Any other candidate must be carefully analyzed. One good possibility is the maximum norm (see Banach [71])

$$\|f - g\| = \text{maximum} |f - g| \quad \text{in } D \tag{5.184}$$

used for the Banach space of continuous functions. Collatz [74] has found this useful in error estimates.

The *extremal point collocation* method requires that the residual $R(\mathbf{x}, \mathbf{C})$ (or $R_i(\mathbf{x}, \mathbf{C})$ for the boundary conditions) should possess *alternate* maximum and minimum values. We choose $(n + 1)$ distinct points $P_0, P_1, ..., P_n$ ($P = P(x_1, x_2, ..., x_p)$) in the hypersurface D and ask that

$$R(P_k, \mathbf{C}] - (-1)^k R[P_0, \mathbf{C}] = 0 \tag{5.185}$$

where R is the residual error either in the differential equation or in the boundary conditions. This is a system of n equations in the n unknowns $C_j (j = 1, 2, ..., n)$ whose solution is obtained by some iteration method such as the generalized Newton method [60]. The general ideas of convergence and error are discussed in Section 5.18.

c. Galerkin's Method (see also Protusevich (75))

The often observed (but unproved) superiority of Galerkin's method has prompted many researchers to utilize it in approximating solutions to equations. Applications have been especially frequent in elasticity. In the recent NASA Technical Note TN D-1510 "Collected Papers on Instability of Shell Structures" (1962) appeal to this method is made no less than five times. One also notes that several authors (e.g., Sylvester [76] and Budiansky and Roth [77]) suggest that other, more natural, less *ad hoc*, methods may be better. Nowinski [78, 79, 80] has made extensive use of Galerkin's method in papers concerned with large amplitude oscillations of elastic structures. Mushtari and Galimov [81] make extensive use of the approach in nonlinear thin elastic shells.

None of the weighted residual methods have been very popular in fluid mechanics although several authors have utilized Galerkin's method. We mention (in addition to Jain's work [69]) Frederiksen's [82] study of the resonance behavior of nonlinear one dimensionel gas vibrations and the calculation of fluid velocity profiles by Snyder, Spriggs, and Stewart [83]. Schetz [84] applied collocation techniques to viscous flow problems in nozzles.

Budiansky and Roth [77] have made a typical application of this method in examining the axisymmetric dynamic buckling of clamped spherical shells. The basic equations for equilibrium in the vertical direction and compatibility of membrane strains are respectively [85]

$$(\xi w_{\xi\xi})_{\xi\xi} - (\xi^{-1} w_\xi)_\xi - \lambda^2 (\xi \phi)_\xi + \lambda^4 (\xi w_{\tau\tau}) = (\phi w_\xi)_\xi + 4\lambda^4 p \xi \quad (5.186)$$

$$(\xi \phi_\xi)_\xi - \xi^{-1} \phi + \lambda^2 \xi w_\xi = -\tfrac{1}{2}(w_\xi)^2. \quad (5.187)$$

The variables here are $\xi = r/a$, τ nondimensional time, nondimensional displacement w, ϕ is a dimensionless stress function and p is a pressure parameter. λ is a geometrical parameter involving several geometrical constants. The boundary conditions for the clamped edge at $\xi = 1$ are

$$w = w_\xi = 0$$

and

$$\phi_\xi - \nu\phi = 0$$

(5.188)

where ν is Poisson's ratio.

In the radial variable r we know of the *completeness* of the Bessel functions so we select a trial solution of the form (which are also proportional to the *natural* modes of axisymmetric vibration of a circular clamped flat plate of radius a)

$$w(\xi, \tau) = \sum_{n=1}^{N} a_n(\tau) w_n(\xi) \tag{5.189}$$

where

$$w_n(\xi) = \frac{I_0(\Omega_n \xi)}{I_0(\Omega n)} - \frac{J_0(\Omega_n \xi)}{J_0(\Omega n)} \tag{5.190}$$

(see McLachlan [86]) (clearly $w_n(1) = (w_n)_\xi(1) = 0$) where J_0 is the Bessel function of zero order, first kind, I_0 is the modified Bessel function of zero order, first kind and Ω_n are the roots of

$$J_0(\Omega_n) I_1(\Omega_n) + I_0(\Omega_n) J_1(\Omega_n) = 0.$$

Upon substituting Eq. (5.189) into Eq. (5.186) and following Galerkin's procedure we have

$$\Omega_m^4 a_m(\tau) + \lambda^4 \frac{d^2 a_m}{d\tau^2} = \lambda^2 \int_0^1 (\xi \phi)_\xi w_m \, d\xi + \int_0^1 (\phi w_\xi)_\xi w_m \, d\xi + 4\lambda^4 p N_m$$

$$m = 1, 2, ..., N \quad (5.191)$$

where

$$N_m = \int_0^1 \xi w_m \, d\xi = -\frac{2 J_1(\Omega_m)}{\Omega_m J_0(\Omega_m)}$$

and use has been made of the orthonormality conditions for the flat plate vibration modes

$$\int_0^1 \xi w_i w_j \, d\xi = \begin{cases} 1 & i = j \\ 0 & i \neq j. \end{cases}$$

To handle the integrals on the right-hand side of Eq. (5.191) we set $\phi = \phi_0 + \phi_1$ such that both ϕ_0 and ϕ_1 satisfy the boundary conditions of Eq. (5.188) and

$$[\xi(\phi_0)_\xi]_\xi - \xi^{-1}\phi_0 = -\lambda^2(\xi w_\xi) \tag{5.192a}$$

$$[\xi(\phi_1)_\xi]_\xi - \xi^{-1}\phi_1 = -\tfrac{1}{2}(w_\xi)^2. \tag{5.192b}$$

The solution of Eq. (5.192a), regular at $\xi = 0$, is

$$\phi_0 = -\frac{\lambda^2}{\xi} \int_0^\xi \eta w(\tau, \eta) \, d\eta - \lambda^2 \left[\frac{1+\nu}{1-\nu}\right] \xi \int_0^1 \eta w(\tau, \eta) \, d\eta \tag{5.193}$$

so that

$$\int_0^1 (\xi\phi_0)_\xi w_m \, d\xi = -2\lambda^2 \left[\frac{1+\nu}{1-\nu}\right] N_m \sum_{n=1}^N N_n a_n(\tau) - \lambda^2 a_m. \quad (5.194)$$

Equation (5.194) may be used directly in Eq. (5.191).
Next we write, in the spirit of Galerkin,

$$\phi_0 = A_0 \xi + \sum_n b_n^{(0)}(\tau) J_1(\lambda_n \xi) \quad (5.195a)$$

$$\phi_1 = A_1 \xi + \sum_n b_n^{(1)}(\tau) J_1(\lambda_n \xi) \quad (5.195b)$$

where the λ_n are the roots of $J_1(x) = 0$. Upon setting Eqs. (5.194) and (5.195) into Eq. (5.191) we find

$$\frac{d^2 a_m}{d\tau^2} + \left[1 + \left(\frac{\Omega m}{\lambda}\right)^4 + 2\left(\frac{1+\nu}{1-\nu}\right) N_m^2\right] a_m$$

$$+ 2\left(\frac{1+\nu}{1-\nu}\right) N_m \sum_n N_n a_n (1 - \delta_{nm})$$

$$= \frac{1}{\lambda^2}\left[2 A_1 N_m - \sum_n C_{mn} b_n^{(1)}\right] - \frac{1}{\lambda^4}\left[(A_0 + A_1) \sum_n D_{mn} a_n \right.$$

$$\left. + \sum_n \sum_p E_{mnp} a_n (b_p^{(0)} + b_p^{(1)})\right] + 4p N_m \quad m = 1, 2, \ldots \quad (5.196)$$

where

$$C_{mn} = \int_0^1 \xi J_1(\lambda_n \xi)(w_m)_\xi \, d\xi = \frac{\Omega_m^4 N_m \lambda_n J_0(\lambda_n)}{\lambda_n^4 - \Omega_m^4}$$

$$D_{mn} = \int_0^1 \xi (w_m)_\xi (w_n)_\xi \, d\xi = -(\Omega_n \Omega_m)^2 \left[\frac{N_m + N_n}{\Omega_m^2 + \Omega_n^2} + \frac{N_m - N_n}{\Omega_m^2 - \Omega_n^2}\right]$$

$$\hspace{10em} m \neq n$$

$$= \frac{\Omega_m^2 N_m^2}{4} + \frac{\Omega_m^2 N_m}{2} \quad m = n.$$

$E_{mnp} = \int_0^1 (w_m)_\xi (w_n)_\xi J_1(\lambda_p \xi) \, d\xi$ must be evaluated numerically.
Finally A_0, A_1, $b_n^{(0)}$ and $b_n^{(1)}$ must be expressed in terms of the a_n.

Setting Eq. (5.195a) into Eq. (5.192a), and Eq. (5.195b) into Eq. (5.196) gives respectively

$$b_n^{(0)} = \frac{2\lambda^2}{\lambda_n^2 J_0^2(\lambda_n)} \sum_m C_{mn} a_m \tag{5.197}$$

and

$$b_n^{(1)} = \frac{1}{\lambda_n^2 J_0^2(\lambda_n)} \sum_r \sum_s E_{rsn} a_r a_s. \tag{5.198}$$

From the boundary conditions $\phi_\xi = \nu\phi$ at $\xi = 1$ we find

$$A_1 = -\frac{1}{1-\nu} \sum_n b_n^{(1)} \lambda_n J_0(\lambda_n) \tag{5.199}$$

and from Eq. (5.193)

$$A_0 = \phi_0(1) = -\lambda^2 \left[1 + \frac{1+\nu}{1-\nu}\right] \int_0^1 \eta w(\eta, \tau)\, d\eta]$$

$$= -\frac{2\lambda^2}{1-\nu} \sum_n N_n a_n(\tau). \tag{5.200}$$

The constants $\Omega_n, \lambda_n, N_n, C_{mn}, D_{mn}$, and E_{mnp} are tabulated in Radowski [87] up to $m = n = p = 5$. The Eqs. (5.196) constitute a set of nonlinear second order equations for the a_n. Usually the initial conditions are $a_n(0) = \dot{a}_n(0) = 0$. These equations almost always require numerical (digital or analog) solution.

5.17 COMMENTS ON THE METHODS OF WEIGHTED RESIDUALS

The determination of the undetermined parameters or functions in all of these methods can be thought of as utilizing certain *error distribution principles*. These principles distribute the error in the approximate solution over the domain of the problem according to a prescribed criterion. The method of moments and the Galerkin method are special cases of "orthogonality" methods which specify that the error should be orthogonal to a chosen set of linearly independent weighting functions. The error distribution methods, as opposed to the Ritz and other stationary functional methods have the advantage that they work directly with the differential equations rather than with an equivalent variational problem.

We have pointed out that one should take the trial function members φ_j in order from a complete set. While any complete set may be used it is convenient to choose an expansion which identically satisfies either the boundary conditions (as we stated previously) or the partial differential equations. One can usually obtain an accurate approximation with few terms in such an event.

Basically there are three variations of the methods;

(1) Interior Method. The trial functions are chosen so that the boundary conditions are identically satisfied. Since only the interior errors R_E are not zero it is distributed according to one of the error distributions of Section 5.15.

(2) Boundary Method. The trial functions are chosen so that the differential equations are identically satisfied. The only sources of error are those of the boundary errors R_B and the initial errors R_I. These errors are distributed according to one of the error distributions of Section 5.15.

(3) Mixed Method. In some situations it may not be feasible to choose a set of trial functions which is of type (1) or (2). For such systems all three types of error must be distributed.

5.18 MATHEMATICAL PROBLEMS OF APPROXIMATE METHODS

Associated with any approximate procedure there is always a question regarding the accuracy of approximation. In many cases there are theorems to the effect that if an iteration is continued without limit, or if an increment size is decreased without limit, or if the number of parameters is increased indefinitely etc, then the process converges to the exact solution. Such results have value in that they encourage the analyst to adopt a procedure *but* it is seldom possible to obtain more than one or five or 200 steps (with a computer) of an infinite process. *Therefore a realistic error bound applicable at any stage of the calculation is of considerable more value than convergence theorems.* Convergence theorems are relatively easy to prove (we shall see several subsequently) while *realistic* error bounds are relatively rare because they are difficult to obtain, even in linear systems. If no error analysis is available a common procedure

5.18 MATHEMATICAL PROBLEMS OF APPROXIMATE METHODS

for examining the power of an approximate method is to apply it to a problem whose exact solution is already known. The error can then be exactly determined and the presumption is made that in similar problems the method will produce errors of the same order of magnitude. Such arguments are far from ideal but often the only practical way out.

For the weighted residual methods and the Ritz method, as applied to equations, convergence theorems and error bounds are rare. However some are available which will act as guides for further research.

First, consider the stationary functional (Ritz) method where the functional $\Omega(u)$ has the classical) form

$$\Omega(u) = \iint_D F(x, y, u, u_x, u_y)\, dD \qquad (5.201\text{a})$$

with the boundary condition

$$u = \varphi(s) \quad \text{on} \quad \Gamma \qquad (5.201\text{b})$$

where Γ is the boundary curve of D. We are interested in the minimum of Ω. Let $u^*(x, y)$ be the exact solution of this problem and $\Omega(u^*) = m$ the minimum. If we can succeed in constructing an approximate solution $\bar{u}(x, y)$ satisfying (5.201b) and such that $\Omega(\bar{u})$ is very close to m, then one would expect $\bar{u}$ to be a good approximation to the true solution u^*. Moreover, if we could find a sequence $\bar{u}_n$ satisfying Eq. (5.201b) and such that $\Omega(\bar{u}_n) \to m$ as $n \to \infty$ we would expect convergence of the $\bar{u}_n$ to u^*.

We then address the question of convergence for the Ritz method where the determination of the parameters in the trial solution $\phi(x, y, a_1, \ldots, a_n)$ is determined by solving the system of algebraic equations

$$\frac{\partial \Omega(\phi)}{\partial a_k} = 0, \quad k = 1, 2, \ldots, n. \qquad (5.202)$$

By means of criterion (5.202) a sequence of successively more exact approximations can be obtained. For this purpose we consider a number of families $\phi_n(x, y, a_1, \ldots, a_n)$ $n = 1, 2, \ldots$ each of which includes *all* the functions of the preceding plus an additional parameter. Let $\bar{u}_n = \phi_n(x, y, \bar{a}_1, \ldots, \bar{a}_n)$ be the nth approximation, i.e., the function giving the least value to Ω from the trial family $\phi_n(x, y, a_1, \ldots, a_n)$, where the $\bar{a}_j$ are determined by means of Eq. (5.202). Since each successive family contains *all* the functions of the preceding, the successive minimums are nonincreasing, i.e.,

$$\Omega(\bar{u}_1) \geqslant \Omega(\bar{u}_2) \geqslant \cdots .$$

The following results are known [58, 60, 64]:

(1) A sufficient condition for $\lim_{n \to \infty} \Omega(\bar{u}_n) = \Omega(u^*) = m$ is that the system of families $\phi_n(x, y, a_1, a_2, ..., a_n)$ are *relatively complete*.

A sequence is said to be *relatively complete* if for any function u such that u, u_x, u_y are continuous, $u = \varphi(s)$ on Γ, and for any $\epsilon > 0$ there exists an n and a function of the nth family

$$u_n^* = \phi_n(x, y, a_1^*, ..., a_n^*)$$

such that

$$|u_n^* - u| < \epsilon, \quad \left|\frac{\partial u_n^*}{\partial x} - \frac{\partial u}{\partial x}\right| < \epsilon, \quad \left|\frac{\partial u_n^*}{\partial y} - \frac{\partial u}{\partial y}\right| < \epsilon \quad (5.203)$$

everywhere in D. Stated otherwise the sequence is relatively complete if any admissible function together with its partial derivatives may be approximated as closely as one pleases by means of functions of the given families.

One cannot overemphasize the fundamental importance of completeness —violation of these requirements can lead to gross error.

(2) The last two conditions can be somewhat weakened for the linear Poisson equation, $u_{xx} + u_{yy} = f$, to

$$\iint_D \left(u_x - \sum_1^n a_k \frac{\partial \varphi_k}{\partial x}\right)^2 dD < \epsilon$$

$$\iint_D \left(u_y - \sum_1^n a_k \frac{\partial \varphi_k}{\partial y}\right)^2 dD < \epsilon. \quad (5.203a)$$

(3) As the functions $\varphi_k(x, y)$, for problems with two independent variables, we usually choose different combinations of trigonometric functions[†], or polynomials satisfying the boundary conditions. The completeness of the system can usually be verified by recourse to the generalized famous *Weierstrass Approximation Theorem*.

---
† In infinite regions negative exponentials, etc., are often used.

5.18 MATHEMATICAL PROBLEMS OF APPROXIMATE METHODS

Theorem. If the function $u(x, y)$ and $\partial u/\partial x$, $\partial u/\partial y$ are continuous in a closed and bounded region D, then for any $\epsilon > 0$ there exists a polynomial $p(x, y)$ such that in D

$$|u - p(x, y)| < \epsilon, \quad \left|\frac{\partial u}{\partial x} - \frac{\partial p}{\partial x}\right| < \epsilon, \quad \left|\frac{\partial u}{\partial y} - \frac{\partial p}{\partial y}\right| < \epsilon.$$

(A proof may be found in de la Vallee-Poussin [88].)

In particular this approximation theorem may be used to prove the completeness of various systems. In our first example we suppose that we have performed a transformation to change $u = \varphi(s)$ on Γ to the Dirichlet condition $u = 0$ on Γ. If $\omega(x, y)$ is a continuous function having continuous and bounded derivatives in D, and such that $\omega(x, y) > 0$ within D, $\omega(x, y) = 0$ on Γ then the set

$$\varphi_0 = \omega, \quad \varphi_1 = \omega x, \quad \varphi_2 = \omega y, \quad \varphi_3 = \omega x^2, \quad \varphi_4 = \omega xy, \ldots \quad (5.204)$$

is complete in D. (For proof see Kantorovich [58, p. 276].)

With the aid of changes of variable (e.g., $x = r\cos\theta$, $y = r\sin\theta$) one can extend these results to other systems.

How is ω constructed? Often we construct a function describing the boundary of the region D. For example if D is the rectangle $-a \leqslant x \leqslant a$, $-b \leqslant y \leqslant b$ one can adopt $\omega(x, y) = (x^2 - a^2)(y^2 - b^2)$. For a circle of radius r center at a, b one can set $\omega(x, y) = r^2 - (x - a)^2 - (y - b)^2$. Some general rules are possible for the Dirichlet case $u = 0$ on Γ.

Case a. If the bounding contour Γ is describable by an equation of the form $B(x, y) = 0$ where B and its partial derivatives are continuous one can adopt $\omega(x, y) = \pm B$.

Case b. For convex polygons whose sides are describable by $e_i x + h_i y + k_i = 0$ one can choose $\omega(x, y) = \prod_i^n (e_i x + h_i y + k_i)$. Regions bounded by curved lines can be treated in the same manner.

(4) The estimation of the maximum error that is obtained at the nth step in the approximate solution by Ritz's method and the order, relative to $1/n$, with which this error $\to 0$ as $n \to \infty$ has been extensively investigated by Krylov (see [58], p. 336) for ordinary differential equations.

(5) Convergence theorems for equations of (linear) elliptic type are available. Typical are the following theorems (p. 342—357 in [58]).

Theorem a. In the region D bounded by a rectifiable contour Γ let u be the solution of the elliptic equation

$$\frac{\partial}{\partial x}\left(a\frac{\partial u}{\partial x}\right) + \frac{\partial}{\partial y}\left(b\frac{\partial u}{\partial y}\right) - cu = f \qquad (a, b > 0 \quad c \geqslant 0)$$

and $u = 0$ on Γ. We assume that

$$\Omega(u) = \iint_D [a(u_x)^2 + b(u_y)^2 + cu^2]\, dx\, dy < \infty$$

and

$$\int_\alpha^\beta (u_y)^2\, dy \leqslant K_0,$$

α, β being such that the segment joining (x, α) and (x, β) is in D. Let $u_n(x, y)$ be a sequence of functions which vanish on Γ, are continuous in the closed region $\Gamma + D$ and have continuous derivatives within D. Furthermore suppose

$$\Omega(u_n) - \Omega(u)$$
$$= \iint_D \left[a\left(\frac{\partial u_n}{\partial x} - \frac{\partial u}{\partial x}\right)^2 + b\left(\frac{\partial u_n}{\partial y} - \frac{\partial u}{\partial y}\right)^2 + c(u_n - u)^2\right] dx\, dy = \epsilon_n$$

$$\int_\alpha^\beta \left(\frac{\partial u_n}{\partial y}\right)^2 dy \leqslant K_n$$

$$\lim_{n\to\infty} \epsilon_n = 0, \qquad \lim_{n\to\infty} [\epsilon_n \ln K_n] = 0.$$

Then the sequence of functions u_n converges uniformly to $u(x, y)$ in the region D and the order of convergence is

$$|u_n - u| = O\left[\epsilon_n^{\frac{1}{2}} + \left|\epsilon_n \ln \frac{K_n + K_0}{\epsilon_n}\right|^{\frac{1}{2}}\right].$$

Theorem b. Make the same assumptions on D and u as in Theorem a. Further assume that the contour Γ has a finite curvature at the finite number of extremal abscissa points. Let $u_n = 0$ on Γ and have the form

$$u_n = a_0(x) + a_1(x)y + \cdots + a_n(x)y^n$$

where a_i and a_i' are continuous and

$$\lim_{n\to\infty} \epsilon_n \ln n = 0.$$

Then the sequence u_n converges uniformly to u with

$$|u_n - u| = O[(\epsilon_n \ln \frac{n}{\epsilon_n})^{\frac{1}{2}}].$$

Theorem c. Let D be the region bounded by $x = 0$, $x = l$, $y = g(x)$, $y = h(x)$, $h(x) > g(x)$ where g, h and their first and second derivatives are defined on $0 \leqslant x \leqslant l$ and the latter bounded on $0 \leqslant x < l$. Let u be defined in D, 0 on Γ and such that u_{yy} and u_{xy} are square integrable in D. Let a, b, c be bounded continuous functions. Then there may be constructed functions $\bar{u}_n$, $\tilde{u}_n$ of the form

$$\bar{u}_n(x, y) = \sum_{k=1}^{n} f_k(x) \sin \frac{k\pi[y - g(x)]}{h(x) - g(x)} \qquad (5.205)$$

$$\tilde{u}_n(x, y) = \sum_{k=1}^{n} f_k(x) y^{k-1} [y - g(x)][y - h(x)] \qquad (5.206)$$

where $f_k(0) = 0$, $f_k(l) = 0$ and such that

$$\bar{\epsilon}_n^{\text{or}\sim} = \iint_D [a(u_x - \bar{u}_{nx})^2 + b(u_y - \bar{u}_{ny})^2 + c(u - \bar{u}_n)^2] \, dx \, dy$$

are of $O(1/n^2)$.

(i). Let D be as in Theorem c. Let u satisfy the conditions of Theorem b and c. Then the sequence of approximations $\bar{u}_n$ or $\tilde{u}_n$ of Eqs. (5.205) or (5.206) is uniformly convergent to u and the order of the error of the nth approximation is $O(\sqrt{\ln n}/n)$.

The proofs of these results depend upon conversion to a variational problem. Similar convergence theorems and error bounds do not appear to be available for other linear systems and certainly not for nonlinear partial differential equations. However, the above results are useful in orienting ones thinking and in suggesting possible trial functions.

References

1. Poincaré, H., "New Methods in Celestial Mechanics," Vol. 1, Chapter 2, 3, and 4. Gauthier-Villars, Paris, 1892.
2. Stoker, J. J., "Nonlinear Vibrations." Wiley (Interscience), New York, 1950.
3. Minorsky, N., "Introduction to Nonlinear Mechanics." Edwards, Ann Arbor, Michigan, 1947.

4. Nowinski, J. L., and Ismail, I. A., Application of a multi-parameter perturbation method to elastostatics, *Univ. Delaware (Newark, Del.) Dept. Mech. Eng., Tech. Rept.* No. 24 (1963).
5. Bellman, R. E., "Perturbation Techniques in Mathematics, Physics and Engineering" (Athena Ser.). Holt, New York, 1964.
6. Benney, D. J., *in* "Nonlinear Problems" (R. E. Langer, ed.), Univ. of Wisconsin Press, Madison, Wisconsin, 1963; see also *J. Math. Phys.* **41**, 254 (1962).
7. Dolph, C. L., *in* "Nonlinear Problems" (R. E. Langer, ed.), pp. 13-46. Univ. of Wisconsin Press, Madison, Wisconsin, 1963.
8. Chandrasekhar, S., "Plasma Physics," p. 151. Univ. of Chicago Press, Chicago, Illinois, 1960.
9. Bohm, D., and Gross, E. P., *Phys. Rev.* **75**, 1851 (1949).
10. Gross, E. P., *Phys. Rev.* **82**, 232 (1951).
11. Chien, W. Z., *Chinese J. Phys.* **7**, 102 (1947).
12. Nash, W. A., and Cooley, I. D., *J. Appl. Mech.* **26**, 291 (1959).
13. Moon, P., and Spencer, D. E., "Field Theory for Engineers." Van Nostrand, Princeton, New Jersey, 1961.
14. Moon, P., and Spencer, D. E., "Field Theory Handbook." Springer, Berlin, 1961.
15. Morse, P. M., and Feshbach, H., "Methods of Theoretical Physics," Part II. McGraw-Hill, New York, 1953.
16. Ames, W. F., and Laganelli, T. L., The generalized Hopf transformation in fluid mechanics (Lecture), *A.I.Ch.E. Meeting, Buffalo, May 1963*.
17. Ames, W. F., Nonlinear partial differential equations with exact solutions (1963, unpublished).
18. Carrier, G. F., *Advan. Appl. Mech.* **3**, 1-18 (1953).
19. von Kármán, T., and Millikan, C. B., On the theory of laminar boundary layers involving separation, *Natl. Advisory Comm. Aeron. Tech. Rept.* No. NACA-504 (1934).
20. Munk, W. H., *J. Meterol.* **7**, 79 (1950).
21. Munk, W. H. and Carrier, G. F., *Tellus* **2**, 158 (1950).
22. Lighthill, M. J., *Phil. Mag.* [7] **40**, 1179 (1949).
23. Kuo, Y. H., *J. Math. Phys.* **32**, 83 (1953).
24. Lagerstrom, P. A., and Cole, J. D., *J. Rational Mech. Anal.* **4**, 817 (1955).
25. Proudman, I., and Pearson, J. R. A., *J. Fluid Mech.* **2**, 237 (1957).
26. Kaplun, S., *J. Math. Mech.* **6**, 585 (1957).
27. Oseen, C. W., *Arkiv Mat. Astron. Fys.* **6**, No. 29 (1910).
28. Goldstein, S., *Proc. Roy. Soc.* **A123**, 225 (1929).
29. Whitehead, A. N., *Quart. J. Math.* **23**, 143 (1889).
30. Pai, S. I., "Viscous Flow Theory," Vol. I, p. 239. Van Nostrand, Princeton, New Jersey (1956).
31. Erdelyi, A., *J. Soc. Ind. Appl. Math.* **11**, No. 1, 105 (1963).
32. Erdelyi, A., *Atti. Accad. Sci. Torino, Cl. Sci. Fis. Mat. Nat.* **95**, 651 (1960-1961).
33. Erdelyi, A., *Bull. Am. Math. Soc.* **68**, 420 (1962).
34. Wasow, W., *Commun. Pure Appl. Math.* **9**, 93 (1956).
35. Friedrichs, K. O., The edge effect in bending and buckling with large deflec-

tions, *Proc. 1st Symp. Appl. Math., Providence, Rhode Island, 1947* p. 188. Am. Math. Soc., Providence, Rhode Island, 1949.
36. Bromberg, E., and Stóker, J. J., *Quart. Appl. Math.* **3**, 246 (1945).
37. Friedrichs, K. O., and Stoker, J. J., *J. Appl. Mech.* **9**, 7 (1942); *Am. J. Math.* **63**, 839 (1941).
38. Reissner, E. (problem formulation only), *Proc. 1st Symp. Appl. Math., Providence, Rhode Island, 1947*, p. 213. Am. Math. Soc. Providence, Rhode Island, 1949.
39. Yakura, J. K., *Am. Rocket Soc.* Preprint No. 1983-61 (1961).
40. Martin, C. J., *Proc. 4th U.S. Congr. Appl. Mech., Berkeley, California, 1962* p. 277. ASME, New York.
41. Acrivos, A., and Taylor, T. D., *Phys. Fluids* **5**, No. 4, 387 (1962).
42. Van Dyke, M., *J. Fluid Mech.* **14**, 161 (1962).
43. Van Dyke, M., *J. Fluid Mech.* **14**, 481 (1962).
44. Brenner, H., and Cox, R. G., *J. Fluid Mech.* **17**, 561 (1963).
45. de la Cuesta, H., and Ames, W. F., *Ind. Eng. Chem. Fund.* **2**, 21 (1963).
46. Ames, W. F., and de la Cuesta, H., *J. Math. Phys.* **42**, 301 (1963).
47. Acrivos, A., *Chem. Eng. Sci.* **17**, 457 (1962).
48. Gottschalk, W. M., Feeny, H. F., Lutz, B. C., and Ames, W. F., Oscillations and wave propagation in plasma, Tech. Repts. 1, 2, and 3; AEC Contract AT(30-1)-2440, University of Delaware (1961).
49. Erdelyi, A., "Asymptotic Expansions." Dover, New York, 1956.
50. Jeffreys, H., "Asymptotic Approximations." Oxford Univ. Press, London and New York, 1962.
51. Morgan, G. W., and Warner, W. H., *J. Aeron. Sci.* **23**, 937 (1956).
52. Morgan, G. W., Pipkin, A. C., and Warner, W. H., *J. Aeron. Sci.* **25**, 173 (1958).
53. Pearson, J. R. A., *Appl. Sci. Res.* **A11**, 321 (1963).
54. Ferron, J. R., Approximations for mass transfer with chemical reaction, *A.I.Ch.E. Journal* in press (1965).
55. Brian, P. L. T., Hurley, J. F., and Hasseltine, E. H., *A.I.Ch.E. Journal* **7**, 226 (1961).
56. Ritz, W., *J. Reine Angew. Math.* **135**, 1 (1908).
57. Galerkin, B. G., *Vestn. Inzhenerov i Tekhnikov* p. 879 (1915).
58. Kantorovich, L. V., and Krylov, V. I., "Approximate Methods of Higher Analysis." Wiley (Interscience), New York, 1958.
59. Crandall, S. H., "Engineering Analysis." McGraw-Hill, New York, 1956.
60. Collatz, L., "The Numerical Treatment of Differential Equations." Springer, Berlin, 1960 (English edition).
61. Kantorovich, L. V., *Bull. Acad. Sci. USSR* **7**, 647 (1933).
62. Kantorovich, L. V., *Prikl. Mat. i Mekhan* **6**, 31 (1942).
63. Hetenyi, M., "Beams on Elastic Foundations." p. 60. Univ. of Michigan Press, Ann Arbor, Michigan, 1946).
64. Frazer, R. A., Jones, W. P., and Skan, S. W., Approximations to functions and to the solutions of differential equations, *Aeron. Res. Comm. Rept. and Mem.* No. 1799 (1937).
65. Bickley, W. G., *Phil. Mag.* [7] **32**, 50 (1941).
66. Collings, W. Z., The method of undetermined functions as applied to non-

linear diffusion problems, M.M.E Thesis, University of Delaware (1962).
67. Yamada, H., *Rept. Res. Inst. Fluid Eng. Kyushu Univ.* **3**, No. 3, 29 (1947).
68. Fujita, H., *Mem. Coll. Agr., Kyoto Univ.* **59**, 31 (1951).
69. Jain, M. K., *Appl. Sci. Res.* **A**11, 177 (1962).
70. Dunford, N., and Schwartz, J. T., "Linear Operators," 2 vols. Wiley (Interscience), New York, 1958.
71. Banach, S., "Théorie des opérations linéaires," Monogr. Matem. Warsaw, Poland, 1932.
72. Friedman, B., "Principles and Techniques of Applied Mathematics." Wiley, New York, 1956.
73. Halmos, P. R., "Introduction to Hilbert Space." Chelsea, New York, 1951.
74. Collatz, L., *Z. Angew. Math. Mech.* **33**, 116 (1953).
75. Protusevich, Ya. A., "Variational Methods in Structural Mechanics" (in Russian). GTTI, Moscow, 1948.
76. Sylvester, R. J., Stability problems in missile structures, *Natl. Aeron. Space Admin. Rept.* No. NASA-TN-D-1510, p. 11 (1962).
77. Budiansky, B., and Roth, R. S., Axisymmetric dynamic buckling of clamped shallow spherical shells, *Natl. Aeron. Space Admin. Rept.* No. NASA-TN-D-1510, p. 597 (1962).
78. Nowinski, J., *Rozpr. Inzh.* **9**, 331 (1957).
79. Nowinski, J., *Z. Angew. Math. Phys.* **14**, 112 (1963).
80. Nowinski, J., Snap-through and large amplitude oscillations of oblique panels with an initial curvature, *Univ. Delaware (Newark, Del.) Dept. Mech. Eng. Tech. Rept.* No. 25 (1963).
81. Mushtari, Kh. M., and Galimov, K. Z., "Nonlinear Theory of Thin Elastic Shells." U. S. Dept. Commerce, Washington, D.C., 1957 (English edition).
82. Frederiksen, E., *Ing.-Arch.* **25**, 100 (1957).
83. Snyder, L. J., Spriggs, T. W., and Stewart, W. E., Solution of the equations of change by Galerkin's method (Lecture), *A.I.Ch.E. Meeting, Buffalo April 1963*.
84. Schetz, J. A., On the approximate solution of viscous-flow problems, *ASME Paper* No. 63-APM-3 (1963).
85. Budiansky, B., *Proc. International Union of Theoretical and Applied Mechanics Symp. Theory of Thin Shells, Delft, 1959* p. 64. North-Holland, Amsterdam, 1960.
86. McLachlan, N. W., "Bessel Functions for Engineers." Oxford Univ. Press, London and New York, 1948.
87. Radowski, P. P., Humphreys, J. S., Bodner, S. R., Payton, R. G., and Budiansky, B., Studies on the dynamic response of shell structures to a pressure pulse, *Air Force Spec. Weapons Ctr. Rept.* No. AFSWC-TR-61-31 (II), AVCO RAD (July 1961).
88. de la Vallée-Poussin, Ch. J., "Cours d'analyse infinitésimale," 2nd ed., Vol. II, p. 120. Gauthier-Villars, Paris, 1912.
89. Van Dyke, M., "Perturbation Methods in Fluid Mechanics." Academic Press, New York, 1964.

CHAPTER **6**

Further Approximate Methods

6.0 INTRODUCTION

The approximate methods discussed in Chapter 5 fell into the two categories *asymptotic* and *weighted residual*. While these procedures could involve successive improvement by iteration their special character warrants separation from the methods in this chapter, some of which we classify as *iterative*. Admittedly these classifications are not mutually exclusive nor indeed are they exhaustive. We continue our discussion of approximate methods and description by example.

6.1 INTEGRAL METHODS IN FLUID MECHANICS

In Section 4.12 we introduced the concept of conversion of partial differential equations to integral equations and using an iterative procedure to obtain the exact solution. Unfortunately the exact solution is usually unattainable but the method is nevertheless an important approximate procedure. To illustrate the principle of this method, where iteration is not used, consider the two dimensional incompressible boundary layer equations, as treated by von Kármán [1],

$$u_t + uu_x + vu_y = -\frac{1}{\rho}p_x + \frac{\mu}{\rho}u_{yy} \qquad (6.1\text{a})$$

$$p_y = 0 \quad \text{(i.e., } p = F(x)) \tag{6.1b}$$

$$u_x + v_y = 0. \tag{6.1c}$$

We now *assume* the boundary layer has a definite thickness $\delta = \delta(x)$ and that u satisfies the same boundary conditions at $y = \delta(x)$ as it does at $y = \infty$. In particular we assume $u = u_\infty(x, t)$ at $y = \delta(x)$. Upon integrating Eq. (6.1a) with respect to y, from $y = 0$ to $y = \delta(x)$ we get

$$\int_0^\delta u_t \, dy + \int_0^\delta (uu_x + vu_y) \, dy = -\delta \rho^{-1} \frac{dp}{dx} - \frac{\mu}{\rho} \frac{\partial u}{\partial y}\bigg|_{y=0}. \tag{6.2}$$

From the second term we find that

$$\int_0^\delta vu_y \, dy = uv \bigg|_0^\delta - \int_0^\delta uv_y \, dy \tag{6.3}$$

by integration by parts. From continuity $v_y = -u_x$, so that upon using the no-slip condition at the wall, $v)_{y=\delta} = -\int_0^\delta u_x \, dy$. Equation (6.3) becomes

$$\int_0^\delta vu_y \, dy = -u_\infty \int_0^\delta u_x \, dy + \int_0^\delta uu_x \, dy. \tag{6.3a}$$

Now $-\rho^{-1} p_x = (u_\infty)_t + u_\infty (u_\infty)_x$ so that

$$-\delta \rho^{-1} p_x = \int_0^\delta \rho^{-1} p_x \, dy = \int_0^\delta (u_\infty)_t \, dy + (u_\infty)_x \int_0^\delta u_\infty \, dy, \tag{6.4}$$

since $u_\infty = u_\infty(x, t)$. Substitution of Eqs. (6.3a) and (6.4) into Eq. (6.2) gives

$$\frac{\mu}{\rho} \left(\frac{\partial u}{\partial y}\right)_{y=0} = \int_0^\delta (u - u_\infty)_t \, dy + u_\infty \frac{\partial}{\partial x} \int_0^\delta u \, dy - \frac{\partial}{\partial x} \int_0^\delta u^2 \, dy$$

$$+ (u_\infty)_x \int_0^\delta u_\infty \, dy. \tag{6.5}$$

The interchange of $\partial/\partial x$ and $\int_0^{\delta(x)}$ in the second and third terms on the right-hand side of Eq. (6.5) generates two extra terms which cancel, i.e.,

$$-\frac{\partial}{\partial x} \int_0^\delta u^2 \, dy + u_\infty \frac{\partial}{\partial x} \int_0^\delta u \, dy = \int_0^\delta (u^2)_x \, dy + u_\infty^2 \delta_x - u_\infty \int_0^\delta u_x \, dy - u_\infty^2 \delta_x.$$

Equation (6.5) may be a convenient stopping point for some investigations. Generally, however, we usually wish to calculate the "displacement thickness" δ^* defined by

$$\delta^* = \int_0^\delta \left(1 - \frac{u}{u_\infty}\right) dy, \qquad (6.6)$$

the "momentum thickness" λ defined by

$$\lambda = \int_0^\delta \frac{u}{u_\infty} \left(1 - \frac{u}{u_\infty}\right) dy \qquad (6.7)$$

and the shear stress at the wall

$$\tau_w = \mu \left(\frac{\partial u}{\partial y}\right)_{y=0}. \qquad (6.8)$$

To do this it is simply an algebraic matter to rewrite Eq. (6.5) as

$$\frac{\tau_w}{\rho u_\infty^2} = \frac{1}{u_\infty^2} \frac{\partial}{\partial t}(u_\infty \delta^*) + \frac{\partial \lambda}{\partial x} + \frac{1}{u_\infty}[2\lambda + \delta^*]\frac{\partial u_\infty}{\partial x}. \qquad (6.9)$$

A number of studies have been made using the integral equation (6.5) or its equivalent Eq. (6.9). Some of these are discussed in Rosenhead [2], Pai [3] and by others. Plausible assumptions are now made about the velocity distribution in the boundary layer and Eq. (6.9) then gives values of δ and how they depend upon x. This method abandons the attempt to satisfy the boundary layer equations for every particle of the fluid but chooses a reasonable velocity which satisfies the integrated momentum equation.

In the special case of incompressible flow over a flat plate *without* a pressure gradient von Kármán [1] showed that the agreement of this method with the Blasius similarity solution was excellent with an assumed linear velocity profile and the agreement was improved by selecting a cubic profile which could satisfy both the wall boundary conditions and the boundary conditions at the outer edge of the boundary layer.

Polhausen [4] adapted this method to treat the two dimensional steady flow of an incompressible fluid with a pressure gradient. From Eq. (6.9) the general momentum equation is

$$\rho^{-1}\tau_w = u_\infty^2 \frac{d\lambda}{dx} + u_\infty[2\lambda + \delta^*]\frac{du_\infty}{dx}. \qquad (6.10)$$

6. FURTHER APPROXIMATE METHODS

The choice of a suitable expression for the velocity distribution across the boundary layer depends upon the boundary conditions which are

$$u = 0 \text{ at } y = 0; \quad u = u_\infty(x), \quad \frac{\partial u}{\partial y} = 0 \text{ at } y = \delta(x). \tag{6.11}$$

We shall consider the boundary layer thickness $\delta(x)$ to be an undetermined parameter. Since a pressure gradient is present the chosen velocity form must allow for a point of inflection in retarded flow (adverse gradient) and no point of inflection in accelerated flow (favorable gradient). Finally, it should be possible for the velocity profile to have a separation point, i.e., an x for which $(\partial u/\partial y)_{y=0} = 0$.

If one adopts a velocity profile in the general form

$$u = u_\infty(x) f(y/\delta(x)) = u_\infty f(\eta) \tag{6.12}$$

the choice of $f(\eta)$ is made with the suggestions of the previous paragraph in mind. These are

$$\begin{aligned}
&\eta = 0, \quad \text{i.e.,} \quad y = 0: \quad u = 0, \quad \nu u_{yy} = \frac{1}{\rho}\frac{dp}{dx} = -u_\infty \frac{du_\infty}{dx} \\
&\eta = 1, \quad \text{i.e.,} \quad y = \delta: \quad u = u_\infty, \quad u_y = 0, \quad u_{yy} = 0
\end{aligned} \tag{6.13}$$

which lead to a polynomial of degree four in η, without a constant term,

$$f(\eta) = a\eta + b\eta^2 + c\eta^3 + d\eta^4 \tag{6.14}$$

$0 \leq \eta = y/\delta(x) \leq 1$. Clearly $f(0) = 0$ and the constants a, b, c, d are determined to be

$$a = 2 + \frac{\alpha}{6}, \quad b = -\alpha/2 \tag{6.15}$$
$$c = -2 + \frac{\alpha}{2}, \quad d = 1 - \alpha/6$$

from the boundary conditions (6.13). The dimensionless velocity profile form parameter $\alpha = \alpha(x)$ is

$$\alpha = \frac{\delta^2(x)}{\nu}\frac{du_\infty}{dx} = -\frac{dp}{dx}\frac{\delta}{\mu u_\infty/\delta}. \tag{6.16}$$

In terms of α we can write

$$\frac{u}{u_\infty} = f(\eta) = F(\eta) + \alpha G(\eta) \tag{6.17}$$

where
$$F(\eta) = 2\eta - 2\eta^3 + \eta^4 \tag{6.18a}$$
$$G(\eta) = (\eta/6)(1-\eta)^3. \tag{6.18b}$$

If $\alpha(x) = $ constant, a similar solution of the boundary layer equation exists. To find what value of α corresponds to the separation profile we obtain $\partial u/\partial y)_{y=0}$ as

$$\left.\frac{\partial u}{\partial y}\right|_{y=0} = \delta^{-1}\left.\frac{\partial u}{\partial \eta}\right|_{\eta=0} = a/\delta(x) = \frac{12+\alpha}{6\delta}$$

so that $\alpha = -12$ corresponds to the separation profile.

When $\alpha > 12$, $u/u_\infty > 1$ which is incorrect physically so we must restrict $-12 \leqslant \alpha \leqslant 12$. With Eqs. (6.17) and (6.18) we can now calculate δ^*/δ and λ/δ from Eqs. (6.6) and (6.7) as

$$\delta^* = \int_0^\delta \left(1 - \frac{u}{u_\infty}\right) dy = \delta \int_0^1 \left(1 - \frac{u}{u_\infty}\right) d\eta$$

so

$$\delta^*/\delta = \int_0^1 (1 - F - \alpha G)\, d\eta = \frac{3}{10} - \frac{\alpha}{120} \tag{6.19}$$

and

$$\lambda/\delta = \int_0^1 (F + \alpha G)(1 - F - \alpha G)\, d\eta = \frac{37}{315} - \frac{\alpha}{945} - \frac{\alpha^2}{9072}. \tag{6.20}$$

Similarly

$$\frac{\tau_w \delta}{\mu u_\infty} = \left.\frac{\partial u}{\partial y}\right)_{y=0} = \frac{12+\alpha}{6}. \tag{6.21}$$

Thus λ, δ^* and τ_w are all expressible in terms of α, i.e., δ^* and τ_w can be eliminated from Eq. (6.10) and λ calculated, etc. We shall not dwell on the details except to note that universal functions can be calculated as given in Pai [3] and other references.

The Kármán-Pohlhausen method has been extensively used and extended to the case of compressible flow by Kalikhman [5] and Cope and Hartree [6]. Three-dimensional problems were probably first discussed by Tetervin [7]. Extensive use of the method is still being made. Barua [8] considers secondary flow in stationary curved pipes by the Pohlhausen approach.

276   6. FURTHER APPROXIMATE METHODS

The aforementioned method is limited by the lack of a built in correction mechanism. An iteration method, based upon the general idea of the classical Picard method (see e.g. Courant and Hilbert [9]), converts the differential equation into integral form which provides for a method of successive approximations. We must add that this approach is not that of the classical integral equation method for linear systems. In that case a Green's function (known also as an influence function etc) is utilized. The lack of linearity negates such procedures.

To illustrate the general ideas we consider several examples. First, let us examine the steady form of Eq. (6.1) for incompressible flow with $1/\rho \, dp/dx = \nu f(x)$. Our concern will be with the general problem.

Upon integrating once with respect to y, Eq. (6.1a) becomes

$$\nu u_y = \nu f(x) y + \int_0^y (u u_x + v u_s) \, ds + H(x).$$

A second integration gives

$$\nu u = \nu f(x)(y^2/2) + \int_0^y (y - s)(v u_s - u v_s) \, ds + H(x) y + G(x) \qquad (6.22)$$

where we have applied the relation (see Hildebrand [10])

$$\underbrace{\int_a^x \cdots \int_a^x}_{n \text{ times}} f(x) \underbrace{dx \cdots dx}_{n \text{ times}} = \frac{1}{(n-1)!} \int_a^x (x - y)^{n-1} f(y) \, dy. \qquad (6.23)$$

In Eq. (6.22) we have also used the continuity equation to replace u_x by $-v_y$. Coupled with Eq. (6.23) is the integrated form of the continuity equation

$$v = -\int_0^y u_x(x, s) \, ds + J(x). \qquad (6.24)$$

The no slip condition at the wall $y = 0$ requires that $J(x) \equiv 0$, $G(x) \equiv 0$. By definition [equation (6.8)] $\tau_w = \mu(u_y)_{y=0}$ which implies that $H(x) = \rho^{-1} \tau_w(x)$ i.e., $H(x)$ is essentially the shear stress at the wall.

If we assume $\tau_w(x)$ and $f(x)$ known we may now describe an iterative method, based upon Eqs. (6.22) and (6.24), to find u and v. Select an initial approximation $u^{(0)}(x, y)$ and using this calculate

$$v^{(0)}(x, y) = -\int_0^y u_x^{(0)}(x, s) \, ds$$

6.1 INTEGRAL METHODS IN FLUID MECHANICS

and then successively calculate $u^{(n+1)}$ from

$$u^{(n+1)}(x, y) = f(x)(y^2/2) + \frac{1}{\nu}\left\{\rho^{-1}\tau_w(x)y + \int_0^y (y-s)[v^{(n)}u_s^{(n)} - u^{(n)}v_s^{(n)}]\,ds\right\} \quad (6.25)$$

and

$$v^{(n+1)}(x, y) = -\int_0^y u_x^{(n+1)}(x, s)\,ds.$$

Whether this method can be adapted to the general boundary layer problem is not known in view of the problem of satisfying the boundary conditions at ∞ (or $\delta(x)$). Beginnings of applications of these ideas are observable in Curle [11 see pp. 74–79].

An alternate point of view, concerning the further development of boundary layer profiles, has been taken by Goldstein [12], Prandtl [13] and Goertler [14, 15]. Their procedure is to calculate the profile at $x = x_0 + \Delta x$, from the initial profile, by the equation

$$u[x_0 + \Delta x, y] = u[x_0, y] + \frac{\partial u}{\partial y}(x_0, y)\,\Delta x \quad (6.26)$$

and in general

$$u[x_0 + (n+1)\Delta x, y] = u[x_0 + n\Delta x, y] + \frac{\partial u}{\partial y}[x_0 + n\Delta x, y]\,\Delta x \quad (6.27)$$

i.e., a step by step procedure beginning with the initial profile. Beginning with Eqs. (6.1) in the steady state, we have from application of the continuity equation to the momentum equation that

$$uu_x + vu_y = vu_y - uv_y = -u^2\frac{\partial}{\partial y}\left(\frac{v}{u}\right). \quad (6.28)$$

When Eq. (6.28) is inserted into Eq. (6.1a) and an integration, with respect to y, is performed we see that

$$v = vu\int_0^y \frac{1}{u^2}[f(x) - u_{yy}]\,dy. \quad (6.29)$$

Finally, from the equation of continuity and Eq. (6.29), the expression for $\partial u/\partial x$ is obtained as

$$\frac{\partial u}{\partial x} = \nu\frac{\partial}{\partial y}\left\{u\int_0^y \frac{1}{u^2}[u_{yy} - f(x)]\,dy\right\}. \quad (6.30)$$

If $\partial u/\partial x$ is bounded, Eq. (6.30) may be used in the iteration process (6.27). However, if u passes through zero, both Eqs. (6.29) and (6.30) show that the process fails unless $u_{yy}-f(x)$ quadratically vanishes, thus removing the singularity. These problems with the singularities reduce the effectiveness of this method. The four papers, referenced above, study the effect of these singularities.

Some authors prefer to transform the boundary layer equations into von Mises form, with x and ψ as new independent variables and use these coordinates for their iterative method.

6.2 NONLINEAR BOUNDARY CONDITIONS

Problems wherein the nonlinearities occur in the boundary conditions can often be transformed into a (true) nonlinear integral equation and methods such as that of successive approximations applied to obtain an approximate solution and/or solution bounds. Such problems occur in radiation as discussed, for example, by Chandrasekhar [16] and Mann and Wolf [17]. Lin [18] finds nonlinear integral equations an important tool in superfluidity.

The details can proceed from the formulation for heat conduction in a semi-infinite medium,

$$\frac{\partial u}{\partial t} = \frac{\partial^2 u}{\partial x^2}, \qquad x > 0, \qquad t > 0$$

$$u(x, 0) = 0$$

$$\frac{\partial u}{\partial x}(0, t) = \phi[u(0, t) - f(t)], \qquad t > 0. \qquad (6.31)$$

The function ϕ with argument $u(0, t) - f(t)$, has the form $k[u(0, t) - c \sin t]^3$ in Lin's problem [18] and is also a power function in radiation problems. In most physical situations ϕ and f are continuous and often $f(t)$ is periodic, representing a pulsating energy source.

From Carslaw and Jaeger [19] the problem described by $u_t = u_{xx}$, $u(x, 0) = 0$, $u_x(0, t) = \psi(t)$ in a semi-infinite solid has the solution

$$u(x, t) = -\frac{1}{\sqrt{\pi}} \int_0^t \frac{\psi(\eta)}{(t-\eta)^{\frac{1}{2}}} \exp\left[\frac{-x^2}{4(t-\eta)}\right] d\eta$$

6.2 NONLINEAR BOUNDARY CONDITIONS

and the generalization to our case is trivially

$$u(x, t) = -\frac{1}{\sqrt{\pi}} \int_0^t \frac{\phi[u(0, \eta)] - f(\eta)]}{(t - \eta)^{\frac{1}{2}}} \exp\left[-\frac{x^2}{4(t - \eta)}\right] d\eta \quad (6.32)$$

with the integral equation for $u(0, t)$ as

$$u(0, t) = -\frac{1}{\sqrt{\pi}} \int_0^t \frac{\phi[u(0, \eta)] - f(\eta)]}{\sqrt{t - \eta}} d\eta. \quad (6.33)$$

In order to reduce Eq. (6.33) to a standard form we define

$$F(t) = u(0, t) - f(t) \quad (6.34)$$

so that Eq. (6.33) becomes

$$F(t) = -f(t) - \pi^{-\frac{1}{2}} \int_0^t (t - \eta)^{-\frac{1}{2}} \phi[F(\eta)] d\eta. \quad (6.35)$$

Equation (6.35) is a singular integral equation of Volterra type.

The unknown function $F(t)$ of Eq. (6.35) may be obtained approximately by the method of successive approximations as described by say Tricomi [20]. In essence this procedure is similar to the iteration procedure of Eq. (6.25). To find $F(t)$ we select an initial approximation $F_0(t)$ (often taken as a constant) and calculate successively

$$F_{n+1}(t) + f(t) = -\pi^{-\frac{1}{2}} \int_0^t (t - \eta)^{-\frac{1}{2}} \phi[F_n(\eta)] d\eta. \quad (6.36)$$

An undo amount of labor may be involved here so approximate or numerical methods may be in order. However, to follow this trend leaves the scope of this work. We shall be content with references to Tricomi [20], Bückner [21], and Fox [22]. In addition, since Eq. (6.36) has a convolution kernel $(t - \eta)^{-\frac{1}{2}}$, we mention a convergence theorem of Aziz [23]. Under very mild assumptions Azziz shows that the successive approximations

$$\phi_0(t) = \psi(t) \quad (\psi(t) \text{ an arbitrary continuous function on } 0 \leqslant t \leqslant a)$$

$$\phi_{n+1}(t) = \int_0^t k(t - \eta) G[\eta, \phi_n(\eta)] d\eta + g(t)$$

converge to a *unique continuous* solution of

$$\phi(t) = g(t) + \int_0^t k(t - \eta) G[\eta, \phi(\eta)] d\eta.$$

Laminar boundary layer flows with surface reactions have been considered by Chambre and Acrivos [24] and Acrivos and Chambre [25] using the vehicle of nonlinear integral equations. Their problem is related to Eq. (6.31) in that the nonlinearity occurs in the boundary condition. The formulation begins with the diffusion equation in the boundary layer approximation as

$$uC_x + vC_y = DC_{yy} \tag{6.37}$$

with boundary conditions

$$\begin{gathered} C = a \quad \text{for} \quad x < 0, \quad C = a \quad \text{for} \quad y = \infty \\ D\frac{\partial C}{\partial y} = R(C) \quad \text{at} \quad y = 0. \end{gathered} \tag{6.38}$$

The problem is simplified by the fact that complete solution of Eq. (6.37) is not necessary; only the surface concentration and the local rate of the surface reaction is required. Their integral equation is the Volterra equation for the surface concentration C_0

$$R(C_0) = -A[\tau(x)]^{\frac{1}{2}} \int_0^x [\Gamma(x) - \Gamma(x_1)]^{-\frac{1}{3}} \frac{dC_0}{dx_1} dx_1 \tag{6.39}$$

and $C_0 = a$ when $x = 0$. The quantity $\Gamma(x)$ is related to $\tau(x)$ (shear stress) through

$$\Gamma(x) = \int_0^x [\tau(x)]^{\frac{1}{2}} dx.$$

In Chambre [24] the integral equation (6.39) is solved by a series approach which is equivalent to the method of successive approximation. In Chambre [25] the authors use a numerical procedure analogous to Wagner's method [26].

6.3 INTEGRAL EQUATIONS AND BOUNDARY LAYER THEORY

The conversion of equations to nonlinear ordinary differential equations by means of similarity transformations has been extensively discussed in Chapter 5. The resulting problem is usually a boundary value problem, with the range of the new independent variable from 0 to ∞.

6.3 INTEGRAL EQUATIONS AND BOUNDARY LAYER THEORY

Typical of these is the reduction of the boundary layer equations (Blasius' Problem)

$$uu_x + vu_y = \nu u_{yy} \tag{6.40a}$$
$$u_x + v_y = 0 \tag{6.40b}$$

with boundary conditions

$$y = 0: \quad u = v = 0; \quad y = \infty: \quad u = U, \tag{6.40c}$$

by means of the similarity transformations

$$\eta = y(U/\nu x)^{\frac{1}{2}}, \quad \psi = 2(\nu x U)^{\frac{1}{2}} f(\eta)$$

to

$$f''' + f f'' = 0 \tag{6.41a}$$

with boundary conditions

$$\eta = 0: \quad f = f' = 0; \quad \eta = \infty: \quad f' = 2. \tag{6.41b}$$

Initial value problems are especially suited for iteration methods and computer solution since all we must do is "march" the solution out from the specified initial conditions. We would therefore like to transform the boundary value problem described by Eqs. (6.41a) and (6.41b) into an initial value problem. This new problem can then be easily solved on a computer or transformed to an integral equation. We first discuss the Blasius approach [see 27] and then the generalization of Klamkin [28].

If $f''(0) = \lambda$ the solution of Eq. (6.41) may be expanded into the series

$$f(\eta) = \frac{\lambda \eta^2}{2!} - \frac{\lambda^2 \eta^5}{5!} + \frac{11\lambda^3 \eta^8}{8!} - \frac{375\lambda^4 \eta^{11}}{11!} + \cdots \tag{6.42}$$

in which the conditions at the origin are satisfied. If $F(\eta)$ represents the solution corresponding to $\lambda = 1$ then Eq. (6.42) may be written as

$$f(\eta) = \lambda^{\frac{1}{3}} F(\lambda^{\frac{1}{3}} \eta) \tag{6.43}$$

consequently,

$$2 = \lim_{\eta \to \infty} f'(\eta) = \lambda^{\frac{2}{3}} \lim_{\eta \to \infty} F'(\lambda^{\frac{1}{3}} \eta) = \lambda^{\frac{2}{3}} \lim_{\eta \to \infty} F'(\eta)$$

so that

$$\lambda = \left\{ \frac{2}{F'(\infty)} \right\}^{\frac{3}{2}}.$$

Examining what has been accomplished we see that the original boundary value problem (6.41) has been reduced to a pair of initial value problems described by

$$F''' + FF'' = 0$$
$$F(0) = 0, \quad F'(0) = 0, \quad F''(0) = 1 \tag{6.44}$$

and

$$f''' + ff'' = 0$$
$$f(0) = 0, \quad f'(0) = 0, \quad f''(0) = \left\{\frac{2}{F'(\infty)}\right\}^{\frac{3}{2}}. \tag{6.45}$$

Klamkin [28] investigates several system of differential equations and shows how to reduce the boundary value problem to an initial value problem. As an example of his procedure we discuss the second order differential equation

$$\sum_{m,n,r,s} A_{mnrs}(y'')^m (y')^n y^r \eta^s = 0 \tag{6.46}$$

with boundary conditions $y(0) = 0$, $y^{(d+1)}(\infty) = k$. Here, m, n, r, s are arbitrary indices, d is an integer and A_{mnrs} are constants. Let $y'(0) = \lambda$ and assume that y is expressible in the form

$$y(\eta) = \lambda^{1-\alpha} F(\lambda^\alpha \eta) \tag{6.47}$$

where $F(\eta)$ satisfies Eq. (6.46) but subject to the initial conditions $F(0) = 0$, $F'(0) = 1$. For both y and F to satisfy Eq. (6.46) we must have a certain dimensional homogeneity. This conditions on m, n, r, and s is obtained by substituting Eq. (6.47) into Eq. (6.46) to get

$$\sum_{m,n,r,s} A_{mnrs} [F''(\lambda^\alpha \eta)]^m [F'(\lambda^\alpha \eta)]^n [F(\lambda^\alpha \eta)]^r (\lambda^\alpha \eta)^s \lambda^C = 0$$

where

$$C = (1+\alpha)m + n + (1-\alpha)r - \alpha s. \tag{6.48}$$

In order that F satisfy this equation, C must be constant for all m, n, r, s so that Eq. (6.46) reduces to

$$\sum_{m,r,s} A_{mrs}(y'')^m (y')^n y^r \eta^s = 0$$

where n is given by Eq. (6.48).

6.3 INTEGRAL EQUATIONS AND BOUNDARY LAYER THEORY

Then
$$y^{(d+1)}(\infty) = k = \lambda^{1+d\alpha} F^{(1+d)}(\infty)$$

or
$$\lambda = \left\{ \frac{k}{F^{(d+1)}(\infty)} \right\}^{1/(1+d\alpha)}$$

and the conversion is completed.

Third-order problems and simultaneous equations are also treated by Klamkin.

Problems of conversion of the equations of boundary layer theory to integral equation form have been investigated by other researchers for example Weyl [29] (see also Rosenhead [2]) whose approach is that of transformation of the differential equation into an integral equation. Siekmann [30, 31] has adapted Weyl's idea to several problems. In both of Siekmann's examples he arrives at an integral equation of the form

$$g(\eta) = T\{g(\eta)\} \tag{6.49}$$

where T is an integral operator.

In illustration of the Weyl-Siekmann method we consider the example of steady laminar incompressible flow over the upper surface of a semi-nfinite system shown in Fig. 6-1. The portion of the plate $0 \leqslant x \leqslant \lambda$

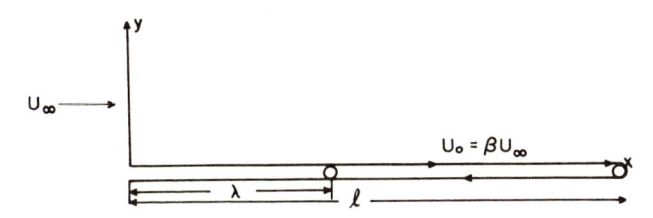

FIG. 6-1. Boundary layer flow over a moving surface.

is at rest; the surface of the plate between $\lambda \leqslant x \leqslant l$ is moving to the right with the constant velocity $U_0 = \beta U_\infty$. Under similarity transformation the boundary layer equations become

$$f'''(\eta) + f(\eta) f''(\eta) = 0$$
and for this problem the *initial* conditions are \qquad (6.50)
$$f(0) = 0, \quad f'(0) = \alpha \quad (\alpha \geqslant 0), \quad f''(0) = 1.$$

To solve this initial value problem we define the auxiliary function

$$g(\eta) = f''(\eta)$$

so that the nonlinear differential equation becomes

$$g' + g\left[\int_0^\eta \int_0^\eta g(\tau)\,d\tau\,d\tau + \gamma\eta + \delta\right] = 0, \tag{6.51}$$

since

$$f(\eta) = \int_0^\eta \int_0^\eta g(\tau)(d\tau)^2 + \gamma\eta + \delta.$$

By virtue of the initial conditions of Eq. (6.50), $\delta \equiv 0$ and $\gamma = \alpha$. Substituting these values into Eq. (6.51) and integrating gives

$$\ln g = -\int_0^\eta g(d\tau)^3 + \tfrac{1}{2}\alpha\eta^2$$
$$\text{(3 times)}$$

or by virtue of Eq. (6.23)

$$g(\eta) = \exp\left[-\tfrac{1}{2}\left\{\int_0^\eta (\eta-\tau)^2 g(\tau)\,d\tau - \alpha\eta^2\right\}\right] \tag{6.52}$$

which has the general form (6.49) with T the exponential of the integral. Beginning with $g_0 = 1$ one obtains

$$g_{n+1} = \exp\left[-\tfrac{1}{2}\int_0^\eta (\eta-\tau)^2 g_n(\tau)\,d\tau + \tfrac{1}{2}\alpha\eta^2\right] \tag{6.53}$$

as the iteration scheme. Integral equation methods are particularly useful since the auxiliary conditions are automatically incoporated and in addition are readily adapted to computers. Actual computation of Eq. (6.53) was done on a high speed digital computer by Siekmann [30].

6.4 ITERATIVE SOLUTIONS FOR $\nabla^2 u = bu^2$

The Dirichlet problem for systems of equation of the form

$$\nabla^2 u_i = b_i \prod_{k=1}^j u_k^{n_k} \qquad i = 1, 2, ..., j \tag{6.54}$$

arises in diffusion controlled chemical reaction problems. What is required in this constant temperature, nth order reaction, is the concentration u_i of of the ith species. Here $\sum_1^j n_k = n$ and the b_i are constant.

6.4 ITERATIVE SOLUTIONS

The equation of the section title

$$\nabla^2 u = bu^2 \tag{6.55}$$

and its "limited" modification

$$\nabla^2 u = bu(a - u) \tag{6.56}$$

occur frequently. The problem we shall discuss here is the construction of the solution of Eq. (6.55) for a dimensionless u ($0 \leqslant u \leqslant 1$) in a given bounded domain R and taking the value f on the boundary of R which we call S. Since the underlying problem is chemical the boundary values and b as well as the desired solution are nonnegative.

Three iterative methods are discussed by Ablow and Perry [32] for this problem. Each of these methods is generalizable to the more complicated problem as described by Eq. (6.54) and they also apply to the development of numerical techniques as we shall discuss in Chapter 7. Each of the Ablow-Perry methods are special cases of the general iteration[†]

$$\nabla^2 u_{k+1} = bu_k[cu_{k+1} + (1 - c)u_k]. \tag{6.57}$$

We shall let s and t be vectors of elements in R, ds and dt differential elements and $G(s, t)$ the Green's functions for R. Further we suppose that $h(s)$ is the harmonic function in R assuming the values $f(s)$ on the boundary S of R.

Case 1. *Newton's Method:* $c = 2$

If b is so *small* that

$$\frac{2b^2 G_1^2}{(1 - 2bG_1)^2} \leqslant \frac{1}{2}$$

where G_1 is an upper bound for $\int_R G(s, t)\, dt$ then the sequence

$$\begin{aligned} u_0(s) &= h(s) \\ \nabla^2 u_{k+1} &= bu_k[2u_{k+1} - u_k] \end{aligned} \tag{6.58}$$

or the equivalent integral form

$$u_0(s) = h(s)$$
$$u_{k+1} + 2b \int_R G(s, t) u_k(t) u_{k+1}(t)\, dt = h(s) + b \int_R G(s, t) u_k^2\, dt \tag{6.59}$$

[†] These relations are sometimes called *overrelaxation* iterations and c is a relaxation parameter.

converges to the solution u of Eq. (6.55), the solution being unique in a certain interval. Moreover, Kantorovich [33] shows the convergence of the sequence $\{u_k\}$ is almost quadratic, i.e., for large k,

$$\| u - u_{k+1} \|/\| u - u_k \|^p < \infty$$

for any nonnegative $p < 2$.

We note here that an upper bound for the $\int_R G(s, t)\, dt$ over a bounded domain R is obtainable by the methods of Bergman and Schiffer [34] for they show that this integral increases if the domain increases. Thus the easiest way is to take R' as a sufficiently symmetric domain containing R, the Green's functions G' for R' will be simple, the integration of G' can be performed explicitly and the inequality

$$\int_R G\, dt \leqslant \int_{R'} G'\, dt$$

supplies the upper bound which we have called G_1.

Case 2. *The Natural Iteration*: $c = 0$ (Picard)

The successive iterates in this case are

$$u_0 = h(s) \qquad \nabla^2 u_{k+1} = b u_k^2 \qquad k = 0, 1, 2, \ldots \tag{6.60}$$

or equivalently

$$u_{k+1}(s) = h(s) - b \int_R G(s, t) u_k^2(t)\, dt.$$

By appeal to the *maximum principle* ([9], Chapter 4), i.e., if $\nabla^2 w \geqslant 0$ in R then w takes its maximum value on the boundary of R; one may show that the successive iterates are alternately greater than and less than the solution. The convergence of the sequence is assured if $2bG_1 < 1$ and this proceeds as follows

$$0 \leqslant u_1 \leqslant u_3 \cdots \leqslant u_{2k+1} \leqslant \cdots \leqslant u \leqslant \cdots \leqslant u_{2k} \leqslant \cdots \leqslant u_2 \leqslant u_0. \tag{6.61}$$

Case 3. $c = 1$ (*Useful for arbitrary values of* $b \geqslant 0$)

The successive approximations in this case are

$$u_0 = h$$
$$\nabla^2 u_{k+1} = b u_k u_{k+1} \tag{6.62}$$

or equivalently
$$u_{k+1}(s) = h(s) - b \int_R G(s, t) u_k(t) u_{k+1}(t)\, dt.$$

This iterative scheme is shown to converge for arbitrary $b > 0$. It is therefore preferable for practical application to the previous cases.

6.5 THE MAXIMUM OPERATION

The maximum operation has been extensively used in the postwar (W. W. II) period thanks to the efforts of the New York University school and in particular Bellman and Lax. The techniques we discuss were first used by Bellman [36] who obtained a representation for the solution of the initial value problem of the Riccati (nonlinear ordinary) differential equation
$$u' = u^2 + b(x)u + c(x). \tag{6.63}$$

Kalaba [37] has developed this method, in some detail, for both nonlinear ordinary and partial differential equations. The method can be applied to obtain approximate solutions or (more easily) numerical solutions. We shall discuss its application to numerical solution of equations in Chapter 7.

Let x be the vector of independent variables and L be a linear partial differential operator. Consider equations of the form
$$L[u] = f(u, x) \tag{6.64}$$

where f is a *strictly convex*[†] function of u for all x in a domain D and for all u under consideration. Actually, for our purposes, we may think of strict convexity as meaning
$$\frac{\partial^2 f}{\partial u^2} > 0 \tag{6.65}$$

(which is implied by strict convexity).

[†] Let E^k denote a k dimensional real Euclidean space. A set C in E^k is convex if $\alpha x + (1 - \alpha)y$ is in C whenever x, y are in C and $0 \leqslant \alpha \leqslant 1$. That is if x, y are any two points of C the line segment joining them is also in C. If C is a convex set in E^k and f is a function defined on C taking real values or $+\infty$ then we say f is *convex* if for any x, y in C, $f[\alpha x + (1 - \alpha)y] \leqslant \alpha f(x) + (1 - \alpha) f(y)$ where $0 \leqslant \alpha \leqslant 1$. If only the inequality is allowed f is *strictly* convex.

6. FURTHER APPROXIMATE METHODS

The paramount property of strictly convex functions, important to us here, is:

If $f(u, x)$ is a strictly convex function of u and twice differentiable in any finite u interval then $f(u, x)$ may be represented in the form

$$f(u, x) = \max_{v} \left[f(v, x) + (u - v) \frac{\partial f}{\partial u} (v, x) \right] \tag{6.66}$$

where the maximization is over all v. The maximum is attained when $v = u$.

A geometric realization of this result in two dimensions is that for a strictly convex curve $(f''(u) > 0)$ the point P at which an arbitrary tangent cuts a given parallel to the ordinate axis is *not* above the point Q where the curve cuts the given parallel. (See Fig. 6-2.)

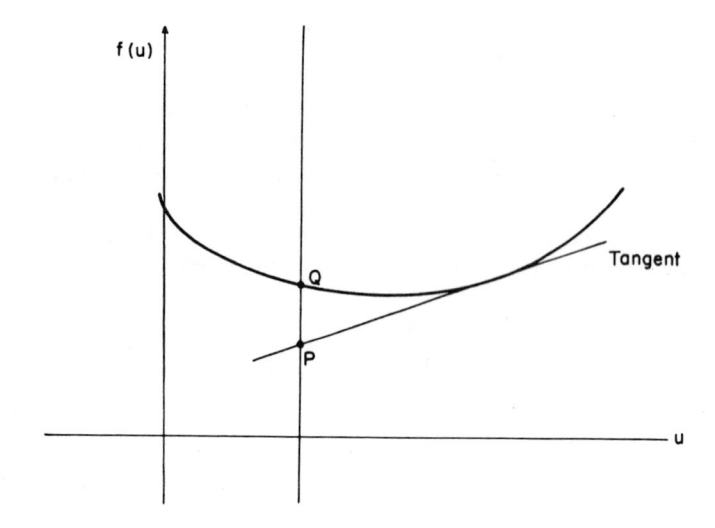

FIG. 6-2. Geometric meaning of the maximum principle.

The derivation is a simple application of the mean value theorem so we write $f(u)$ (note: we shall not specifically insert the x dependency—although it is not discarded)

$$f(u) = f(v) + (u - v) \frac{\partial f}{\partial u} (v) + (u - v)^2 \frac{\partial^2 f}{\partial u^2} (v_1) \tag{6.67}$$

6.6 EQUATIONS OF ELLIPTIC TYPE

where v_1 lies between v and u. The strict convexity implies $\partial^2 f/\partial u^2 > 0$ so that from Eq. (6.67)

$$f(u) - \left[f(v) + (u - v) \frac{\partial f}{\partial u}(v) \right] \geq 0 \qquad (6.68)$$

with equality for $u = v$.

For strictly *concave* functions $\partial^2 f/\partial u^2 < 0$) the corresponding representation is

$$f(u) = \min_v \left[f(v) + (u - v) \frac{\partial f}{\partial u}(v) \right]$$

and for strictly convex twice differentiable functions of n variables $f(u_1, u_2, ..., u_n)$

$$f(u_1, u_2, ..., u_n) = \max_{\mathbf{v}} \left[f(v_1, v_2, ..., v_n) + \sum_{i=1}^{n} (u_i - v_i) \frac{\partial f}{\partial v_i} \right].$$

Our applications to equations will be to the strictly convex case although an analogous theory is possible for the concave case. Kalaba [37] treats ordinary differential equations as well but this is not our concern here.

6.6 EQUATIONS OF ELLIPTIC TYPE AND THE MAXIMUM OPERATION

We now examine the equation of elliptic type

$$\begin{aligned} u_{xx} + u_{yy} &= f(u, x, y) \quad \text{in} \quad D \\ u &= \bar{u} \quad \text{on the boundary } B \text{ of } D. \end{aligned} \qquad (6.69)$$

In particular we desire to obtain a representation for the solution of Eq. (6.69) in terms of the maximum operation applied to the solution of an associated linear equation. We shall then use this linear problem to develop a *monotone sequence* of functions which converges to the actual solution u.

The argument requires that $f(u, x, y)$ is a strictly convex function of u for all finite u and for all x, y in D and such that f, $\partial^2 f/\partial u^2$ is continuous in all three arguments. The boundary B is assumed to have a continuously turning tangent. Now for $\bar{u}$ continuous on B we can find u_1 such that

$\nabla^2 u_1 = 0$ and $u_1 = \bar{u}$ on B. Then the function $u_2 = u - u_1$ satisfies the differential equation

$$\nabla^2 u_2 = \nabla^2 u = f(u) = f(u_1 + u_2) = g(u_2)$$
$$u_2 = 0 \quad \text{on } B$$

again not indicating the explicit dependence of g on x and y. We shall therefore consider only the problem

$$\nabla^2 u = f(u)\dagger \quad \text{in } D, \quad u = 0 \quad \text{on } B. \tag{6.70}$$

By Eq. (6.66) we may write Eq. (6.70) as

$$\nabla^2 u = \max_v [f(v) + (u - v)f'(v)] \quad u = 0 \quad \text{on } B \tag{6.71}$$

and $f'(v) = \partial f/\partial u(v)$. The maximization is over all functions v defined on D and occurs for $v = u$. From the non-negativity, shown in Eq. (6.68), we may write

$$\nabla^2 u = f(v) + (u - v)f'(v) + p \tag{6.72}$$

where p is nonnegative, whatever v. Let $w = w(x, y; v)$ be the solution of the associated linear equation

$$\nabla^2 w = f(v) + (w - v)f'(v) \quad w = 0 \quad \text{on } B. \tag{6.73}$$

Then the difference function

$$z = u - w$$

satisfies

$$\nabla^2 z - f'(v)z = p \quad z = 0 \quad \text{on } B.$$

The mechanism above allows the proof of the basic theorem:

The solution of Eq. (6.70) may be represented as

$$u = \min_{m_1 \leqslant v \leqslant m_2} w[x, y; v]$$

if $m_1 \leqslant u \leqslant m_2$; $f''(v) > 0$ *and D is small enough so that* $f'(v) \geqslant d - \lambda_1$, $d > 0$ *(for* $m_1 \leqslant v \leqslant m_2$*) where λ_1 is the smallest eigenvalue of* $\nabla^2 u + \lambda u = 0$, $u = 0$ *on B.*

---

† Special cases of Eqs. (6.69) and (6.70) occur in kinetics with $f = u^2$ and other powers of u (see Section 6.4). Also of interest is $f(u) = e^u$ as occurs in kinetics, nebular theory, vorticity (Section 4.16) and electrohydrodynamics (see Keller [38, 39]).

6.6 EQUATIONS OF ELLIPTIC TYPE

Actual calculation of u from Eq. (6.73) proceeds from Newtonlike recurrence relations which can be shown to be monotone decreasing and quadratically convergent. Based on Eq. (6.73), we select an initial function $v_0(x, y)$ and determine $u_0(x, y)$, as the solution of

$$\nabla^2 u_0 = f(v_0) + (u_0 - v_0)f'(v_0).$$

Then successively calculate

$$\nabla^2 u_{n+1} = f(u_n) + (u_{n+1} - u_n)f'(u_n), \quad u_k = 0 \quad \text{on } B. \quad (6.74)$$

This formulation is equivalent to the integral equation

$$u_{n+1} = \iint_D G(x, y; \eta, \theta)[f(u_n) + (u_{n+1} - u_n)f'(u_n)] \, d\eta \, d\theta \quad (6.75)$$

where $G(x, y; \eta, \theta)$ is the Green's function for the Laplace operator in D. Kalaba [37] proves, in terms of the max norm,

$$\| u \| = \max_{(x,y) \in D+B} | u |$$

for continuous u in the *closed* region $B + D$, that
(a) $\{u_n\}$ is uniformly bounded, i.e.,

$$\| u_{n+1} \| \leq \frac{2km}{1 - km} \leq 1, \quad km < \tfrac{1}{3}$$

on a sufficiently small domain D where

$$m \geq \max_{|v| \leq 1} \{|f(v)|, \ |f'(v)|, \ \tfrac{1}{2}|f''(v)|\},$$

and

$$k = \max_{(x,y) \in D} \iint_D | G | \, d\eta \, d\theta.$$

(b) $\{u_n\}$ is monotone decreasing and bounded from below.
(c) $\{u_n\}$ is quadratically convergent, i.e.,

$$\| u - u_{n+1} \| \leq \frac{km}{1 - km} \| u - u_n \|^2$$

and for $km < \tfrac{1}{2}$ is uniformly convergent on D.

6.7 OTHER APPLICATIONS OF THE MAXIMUM OPERATION

a. The Equation $u_t - u_{xx} = f(u, x, t)$

Equations of this nonlinear parabolic form occur in turbulent flow, diffusion with chemical reaction and in neutron multiplication problems (cf. Chapter 1). This equation will be discussed under the usual assumption of strict convexity of f as a function of u, continuity of f in all arguments and a bounded $\partial^2 f/\partial u^2$ for all values of u, x, t under consideration. For simplicity we do not indicate the specific dependence of f on x and t and denote partial derivatives with respect to u by primes.

The problem considered is

$$u_t - u_{xx} = f(u, x, t) \tag{6.76}$$

with the domain of integration the rectangle R having vertices at $(0, 0)$, $(1, 0)$, $(1, T)$, and $(0, T)$. The assumed boundary conditions are

$$\begin{array}{lll} x = 0, & 0 \leqslant t \leqslant T: & u = 0 \\ x = 1, & 0 \leqslant t \leqslant T: & u = 0 \\ t = 0, & 0 \leqslant x \leqslant 1: & u = 0. \end{array} \tag{6.77}$$

The representation of Eq. (6.76) follows the same line of reasoning as the sequence (6.70)–(6.73). Thus the associated linear equation for (6.76) in terms of $w[x, t; v]$ is

$$w_t - w_{xx} = f(v) + (w - v)f'(v) \tag{6.78}$$

and $w = 0$ on the boundary B described in Eq. (6.77).

Actual calculation proceeds from the Newton iteration sequence

$$\begin{array}{l}(u_0)_t - (u_0)_{xx} - f'(v_0)u_0 = f(v_0) - v_0 f'(v_0); \quad u_0 = 0 \text{ on } B \\ (u_{n+1})_t - (u_{n+1})_{xx} - f'(u_n)u_{n+1} = f(u_n) - u_n f'(u_n); \quad u_{n+1} = 0 \text{ on } B.\end{array} \tag{6.79}$$

The development of the equivalent integral equation follows from the integral solution

$$u = \int_0^1 dy \int_0^t K(x, y; t - s) f(y, s) \, ds \tag{6.80}$$

to the problem

$$u_t - u_{xx} = f(x, t), \quad u = 0 \quad \text{on } B$$

6.7 OTHER APPLICATIONS OF THE MAXIMUM OPERATION

(cf. Carslaw and Jaeger [19] or Doetsch [40]). The Kernel function of Eq. (6.80) is

$$K(x, y; t) = \frac{1}{2}\left[\theta_3\left(\frac{x-y}{2}, t\right) - \theta_3\left(\frac{x+y}{2}, t\right)\right] \quad (6.81)$$

where the theta function is

$$\theta_3(\eta, t) = 1 + 2\sum_{k=1}^{\infty} e^{-k^2\pi^2 t} \cos 2k\pi\eta.$$

From this expression we can easily show that K can be written in the form

$$K(x, y; t) = \frac{1}{\sqrt{4\pi t}} \sum_{n=-\infty}^{\infty} \{\exp[-(x-y+2n)^2/4t] - \exp[-(x+y+2n)^2/4t]\}.$$

From Eqs. (6.80) and (6.78) it follows that the integral equation is

$$u_{n+1} = \int_0^1 dy \int_0^t K(x, y; t-s)[f(u_n) + f'(u_n)(u_{n+1} - u_n)] \, ds. \quad (6.82)$$

Once again the sequence $\{u_n\}$ can be shown to be *uniformly bounded* on a certain domain D i.e.,

$$\|u_{n+1}\| = \max |u_{n+1}(x, t)| \leqslant \frac{2km}{1-km} \leqslant 1$$

provided $km \leqslant \frac{1}{3}$. Here $\max\{|f(u_n)|, |f'(u_n)|\} \leqslant m < \infty$ and $k = \max_{0 \leqslant x, y \leqslant 1} \int_0^T K(x, y; T-s) \, ds$. For fixed m, the relation $km \leqslant \frac{1}{3}$ can be achieved by choosing T sufficiently small. Further, the sequence $\{u_n\}$ converges *monotonically, quadratically* and *uniformly* to u in D.

b. The Conservation Law Equation $u_t + [f(u)]_x = 0$

Let $f(u)$ be strictly convex and have a finite second derivative for finite u. Consider the initial value problem defined by

$$u_t + [f(u)]_x = 0 \quad u(x, 0) = g(x), \quad 0 \leqslant t \leqslant T. \quad (6.83)$$

The equation can be rewritten as

$$u_t + f'(u)u_x = 0 \quad (6.84)$$

whose "soft" solution (see Section 4.2), satisfying the prescribed conditions, is

$$u(x, t) = g[x - tf'(u)]. \tag{6.85}$$

For representation by the maximum operation it is convenient to transform Eq. (6.84) by setting

$$V_x = u, \quad V = V(x, t) \tag{6.86}$$

so that Eq. (6.48) becomes

$$V_{xt} + f'(V_x)V_{xx} = 0$$

or upon integration with respect to x

$$V_t + f(V_x) = h(t). \tag{6.87}$$

From Eq. (6.86)

$$V = \int_0^x u \, ds + C(t) \quad \text{or at } t = 0$$

$$V(x, 0) = \int_0^x u(s, 0) \, ds + C(0) = \int_0^x g(s) \, ds + C(0).$$

We can now modify V by

$$V = U + H(t) \tag{6.88}$$

so that selecting $H'(t) = h(t)$, $H(0) = C(0)$ leaves the problem for U as

$$U_t + f(U_x) = 0 \quad U(x, 0) = \int_0^x g(s) \, ds = G(x). \tag{6.89}$$

In the usual way the associated linear equation is

$$w_t + f(v) + (w_x - v)f'(v) = 0 \quad w(x, 0) = G(x). \tag{6.90}$$

The difference function $z = U - w$ satisfies

$$z_t + z_x f'(v) = -p \quad z(x, 0) = 0$$

where $p \geqslant 0$ for each choice of v. Hence

$$z = U - w \leqslant 0, \quad t \geqslant 0$$

so that

$$U = \min_v w[x, t; v]$$

and the *minimizing* function is $v = u$.

The Newtonian approximating sequence is

$$(U_0)_t + f(v_0) + ((U_0)_x - v_0)f'(v_0) = 0$$
$$(U_{n+1})_t + [(U_{n+1})_x - (U_n)_x]f'[(U_n)_x] + f[(U_n)_x] = 0 \qquad (6.91)$$
$$U_k(x, 0) = G(x) \qquad k = 0, 1, 2, \ldots.$$

Application to nonlinear integral equations and systems of differential equations is also possible.

6.8 SERIES EXPANSIONS

The solution of linear partial differential equations by series expansions in orthogonal functions is a familiar procedure. In carrying out this method one proceeds by obtaining fundamental solutions and then applies superposition. For nonlinear equations this procedure can no longer be followed. However, numerous successful studies of these equations have been made using various series expansions or rather truncations of them. Usually these take the form, observed before in the weighted residual methods,

$$\sum_{n=0}^{\infty} \phi_n(y) f_n(x) \qquad (6.92)$$

where the $\{f_n\}$ are preselected and the $\{\phi_n\}$ are determined so that the partial differential equation is satisfied. Usually the ordinary differential equations for the ϕ_n are nonlinear (although not all need be). The selection of the f_n may be dictated by the boundary conditions, e.g., the form of the velocity at infinity in boundary layer theory; or by other physical considerations such as symmetry.

Numerous studies in fluid mechanics have utilized series of the form (6.92). In all of these one must be careful to note the progressive loss of accuracy (true of any truncated series), as one moves away from the expansion center, due to the neglected terms. In all of these expansions one strives to employ universal functions, that is functions which are independent of boundary conditions, parameters, etc., and which may therefore be tabulated once and for all.

In boundary layer theory Goldstein [27] and Rosenhead [2] record several applications of direct series solution of equations. Hassan [41] uses a form of Eq. (6.92) with $f_n = x^n$ to study unsteady flow. Because of transformations used to eliminate the time dependence the Hassan approach is limited (in unsteady flow) to the case where the free stream velocity u_∞ has the form $x^m t^n$.

Tipei [42] makes some use of series methods in obtaining solutions of the Reynolds equation in Lubrication. Large deflection of plates have been considered by several investigators using double Fourier Series. Specifically we mention the efforts of Levy [43, 44, 45] and Coan [46].

In Chapter 2 and 4 we saw that the class of similar solutions of the boundary layer equations is rather narrow. Blasius [47], Hiemenz [48] and Howarth [49] have treated the general case of the boundary layer on a cylindrical body placed in a stream which is perpendicular to its axis. Goertler [50] developed a generalization of all previously known expansions and his series is generally considered (1964) as the latest word in this area.

Because of its historical interest and motivation for Goertler's series we discuss the Blasius series for the steady boundary layer flow about a symmetric cylinder. That is, the cylinder is symmetric with respect to the direction of the undisturbed flow. The velocity of the outer (potential) flow is assumed to have the form of a power series in x where x denotes *the distance from the stagnation point measured along the contour.* Thus in general

$$u_\infty(x) = a_1 x + a_2 x^2 + a_3 x^3 + \cdots, \tag{6.93}$$

but for this symmetric case all the coefficients of the even power of x are zero because the pressure p on the body is an even function of x, then dp/dx and u_∞ must be odd.

The velocity profile in the boundary layer is also represented as a similar power series in x, where the coefficients are assumed to be functions of the coordinate y, measured at right angles to the wall. A substitution is then obtained (contribution of Howarth) which permits the universal validity of the y dependent coefficients, i.e., by a suitable transformation the coefficients are made independent of the particulars of the cylindrical body. The resulting functions can then be evaluated and presented in the form of tables. We now give the details and discuss the limitations later.

6.8 SERIES EXPANSIONS

With
$$u_\infty = a_1 x + a_3 x^3 + a_5 x^5 + \cdots \tag{6.94}$$

the coefficients a_{2n+1} depend only on the body shape and are assumed to be known. Thus the pressure term in the boundary layer equations (6.1) becomes

$$-\frac{1}{\rho}\frac{dp}{dx} = u_\infty \frac{du_\infty}{dx} = a_1{}^2 x + 4a_1 a_3 x^3 + (6a_1 a_5 + 3a_3{}^2)x^5$$
$$+ (8a_1 a_7 + 8a_3 a_5)x^7 + (10a_1 a_9 + 10a_3 a_7 + 5a_5{}^2)x^9 + \cdots. \tag{6.95}$$

The continuity equation is integrated by means of a stream function $\psi(x, y)$ in the usual way so that the momentum equation now reads $(u = \psi_y, \ v = -\psi_x)$

$$\psi_y \psi_{xy} - \psi_x \psi_{yy} = u_\infty \frac{du_\infty}{dx} + \nu \psi_{yyy} \tag{6.96}$$

as we have previously noted. It now remains to make a suitable assumption about the form of the stream function (and hence of the velocity components). In analogy with Eq. (6.95) it appears reasonable to assume a series of that form but with coefficients that depend upon y, determined so that Eq. (6.96) is satisfied. If the series for u_∞ is terminated with x^n, the pressure term ends in x^{2n-1} so that the series for ψ must continue as far as x^{2n-1}. If earlier termination is carried out for ψ an error ensues which grows as the distance from the stagnation point increases.

With
$$\eta = y \left[\frac{a_1}{\nu}\right]^{\frac{1}{2}} \tag{6.97}$$

the distance from the wall is dimensionless. A series for ψ is then assumed having the form

$$\psi = \sum_{j=0}^{\infty} b_{2j+1} x^{2j+1} f_{2j+1}(\eta). \tag{6.98}$$

In x, η coordinates, Eq. (6.96) becomes

$$\psi_\eta \psi_{x\eta} - \psi_x \psi_{\eta\eta} = \left(\frac{\nu}{a_1}\right) u_\infty u_{\infty x} + \nu \left(\frac{a_1}{\nu}\right)^{\frac{1}{2}} \psi_{\eta\eta\eta}. \tag{6.99}$$

Combining the knowledge of the forms (6.99) and (6.95) makes it plausible to adopt the series for ψ

$$\psi = \left(\frac{\nu}{a_1}\right)^{\frac{1}{2}} \left[a_1 x f_1(\eta) + \sum_{j=1}^{\infty} (2j+2) a_{2j+1} x^{2j+1} f_{2j+1}(\eta) \right] \quad (6.100)$$

i.e.,

$$b_1 = \left(\frac{\nu}{a_1}\right)^{\frac{1}{2}} a_1, \quad b_{2j+1} = \left(\frac{\nu}{a_1}\right)^{\frac{1}{2}} (2j+2) a_{2j+1}, \quad j > 0.$$

When Eq. (6.100) is substituted into Eq. (6.99) we see that the coefficient functions f_{2j+1} can be made independent of the particular problem, that is independent of the a_n's, if we decompose the f's as follows: the decomposition begins with f_5:

$$f_5 = g_5 + \frac{a_3^2}{a_1 a_5} h_5$$

$$f_7 = g_7 + \frac{a_3 a_5}{a_1 a_7} h_7 + \frac{a_3^2}{a_1^2 a_7} k_7 \quad (6.101)$$

$$f_9 = g_9 + \frac{a_3 a_7}{a_1 a_9} h_9 + \frac{a_5^2}{a_1 a_9} k_9 + \frac{a_3^2 a_5}{a_1^2 a_9} j_9 + \frac{a_3^4}{a_1^3 a_9} q_9$$

$$\vdots$$

where the *universal* functions are

$$f_1, f_3, g_5, h_5, g_7, h_7, k_7, \ldots .$$

Inserting Eqs. (6.95), (6.100), and (6.101) into Eq. (6.99), extracting the coefficients of terms having like powers of x, yields the system of ordinary differential equations for the universal functions

$$\begin{aligned}
(f_1')^2 - f_1 f_1'' &= 1 + f_1''' \\
4 f_1' f_3' - 3 f_1'' f_3 - f_1 f_3'' &= 1 + f_3''' \\
6 f_1' g_5' - 5 f_1'' g_5 - f_1 g_5'' &= 1 + g_5''' \\
6 f_1' h_5' - 5 f_1'' h_5 - f_1 h_5'' &= \tfrac{1}{2} + h_5''' - 8(f_3'^2 - f_3 f_3''),
\end{aligned} \quad (6.102)$$

with the boundary conditions

$\eta = 0$: $\quad f_1 = f_1' = 0; \quad f_3 = f_3' = 0; \quad g_5 = g_5' = 0; \ldots$

$\eta = \infty$: $\quad f_1' = 1; \quad f_3' = \tfrac{1}{4}; \quad g_5' = \tfrac{1}{6}; \quad h_5' = 0; \ldots .$

These Eqs. (6.102) are all third order *and only* the first is nonlinear.

The solution proceeds $f_1 \to f_3 \to g_5$ or h_5 etc. Tables and references for the calculation of u and v are given in Schlichting [51, pp. 150–151] and are attributed to Tifford [52] who made extensive corrections and improvements to already existing tables. Applications to specific problems are available in Rosenhead [2], Pai [3], and Schlichting [51].

The utility of the Blasius Series (like any series method) is restricted by the fact that a large number of terms may be required to achieve global accuracy. This is precisely the problem in the important case of very slender body shapes where a prohibitively large number of terms are required. In spite of this limitation the Blasius series method is of great importance because, in cases when its convergence is insufficient to reach the point of separation, it can be used to obtain with great accuracy the initial portion of the boundary layer near the stagnation point. The calculation can then be continued by means of numerical methods. We shall continue this case in Chapter 7.

A modification of the Blasius Series was carried out by Howarth [53] and Tani [54] for the case $u_\infty(x) = u_0 - ax^n$, $n = 1, 2, 3, \ldots,$. Other authors have carried out special expansions but it remained for Goertler [50] to obtain a generalization of all hitherto (1957) known series expansions. Further Goertler's work also succeeds (in most cases) in improving the convergence of the new series as compared with its predecessors.

6.9 GOERTLER'S SERIES

The success of the series of this section derives in part from the introduction of new variables in such a manner that the first term of the new series satisfies the boundary conditions, giving the exact value of the velocity just outside the boundary layer, for all values of x. The first term of the new series therefore constitutes a good approximation for a considerable distance downstream. One therefore hopes that improved convergence will result although this is by no means certain without proof (which is still lacking in 1964).

The new variables are

$$\xi = \frac{1}{\nu} \int_0^x u_\infty(x)\, dx \qquad (6.103)$$

$$\eta = \frac{y u_\infty(x)}{\left[2\nu \int_0^x u_\infty(x)\, dx \right]^{\frac{1}{2}}} \qquad 6.104$$

and upon introducing these and a stream function

$$\psi(x, y) = \nu(2\xi)^{\frac{1}{2}} F(\xi, \eta) \tag{6.105}$$

into Eq. (6.99) we find, for F,

$$F_{\eta\eta\eta} + F F_{\eta\eta} + \beta(\xi)(1 - F_\eta^2) = 2\xi(F_\eta F_{\eta\xi} - F_\xi F_{\eta\eta}) \tag{6.106}$$

where

$$\beta(\xi) = 2 \frac{u_\infty'(\xi)}{u_\infty^{2}(\xi)} \int_0^\xi u_\infty(x)\, dx. \tag{6.107}$$

The boundary conditions on F are

$$\begin{aligned}\eta = 0: \quad & F = F_\eta = 0 \\ \eta = \infty: \quad & F_\eta = 1.\end{aligned} \tag{6.108}$$

In Eq. (6.106) the specific data for a particular problem occur only in the form of a contracted coefficient function $\beta(\xi)$ which Goertler calls the "principal function". If δ^* is the displacement thickness of Eq. (6.6), it is clear that

$$\beta(\xi) = \frac{1}{[\eta_0(\xi)]^2} \frac{[\delta^*]^2 u_\infty'(\xi)}{\nu}$$

where

$$\eta_0(\xi) = \lim_{\eta \to \infty} \{\eta - F(\xi, \eta)\}.$$

In the case $\beta(\xi) = \beta_0 = $ constant, simple wedge solutions are obtained, with β_0 the wedge angle [cf. 50].

If the free stream velocity has the general form

$$u_\infty(x) = x^m \sum_{j=0}^\infty a_{j/2} x^{j(m+1)/2} \tag{6.109}$$

($m \neq -1$, $a_0 \neq 0$) then the principal functions $\beta(\xi)$ has the form

$$\beta(\xi) = \beta_0 + \beta_{\frac{1}{2}} \xi^{\frac{1}{2}} + \beta_1 \xi + \beta_{\frac{3}{2}} \xi^{\frac{3}{2}} + \cdots \tag{6.110}$$

where, for example,

$$\beta_0 = \frac{2m}{m+1}$$

and succeeding coefficients can be calculated from the coefficients of $u_\infty(x)$. The stream function is assumed to have the form

$$F(\xi, \eta) = F_0(\eta) + F_{\frac{1}{2}}(\eta)\xi^{\frac{1}{2}} + F_1(\eta)\xi + F_{\frac{3}{2}}(\eta)\xi^{\frac{3}{2}} + \cdots. \qquad (6.111)$$

To arrive at the universal function coefficients it is necessary that we split up the terms as

$$F_{\frac{1}{2}} = \beta_{\frac{1}{2}} f_{\frac{1}{2}}, \qquad F_1 = \beta_{\frac{1}{2}}^2 f_{\frac{1}{2}\frac{1}{2}} + \beta_1 f_1$$

$$F_{\frac{3}{2}} = \beta_{\frac{1}{2}}^3 f_{\frac{1}{2}\frac{1}{2}\frac{1}{2}} + \beta_{\frac{1}{2}}\beta_1 f_{\frac{1}{2}1} + \beta_{\frac{3}{2}} f_{\frac{3}{2}}$$

and the universal functions $f_{\frac{1}{2}}$, $f_{\frac{1}{2}\frac{1}{2}}$ ... can be evaluated once for all. Some tables are to be found in Goertler [55].

This new series includes the similar solutions, the Blasius series and the Howarth-Tani problem. Considerable further work remains.

6.10 SERIES SOLUTIONS IN ELASTICITY

As previously mentioned in Section 6.8 series have been a useful tool in nonlinear elasticity. Levy [43] (see also [44]) has led the way in obtaining solutions of the von Kármán equations for the lateral deflection w and the Airy stress function F for thin flat plates. The von Kármán equations (see Eqs. (1.42) and (1.43)) are

$$\nabla^4 F = \frac{\partial^4 F}{\partial x^4} + 2\frac{\partial^4 F}{\partial x^2 \partial y^2} + \frac{\partial^4 F}{\partial y^4} = E\left[\left(\frac{\partial^2 w}{\partial x \partial y}\right)^2 - \frac{\partial^2 w}{\partial x^2}\frac{\partial^2 w}{\partial y^2}\right] \qquad (6.112)$$

$$\nabla^4 w = \frac{p}{D} + \frac{h}{D}\left[\frac{\partial^2 F}{\partial y^2}\frac{\partial^2 w}{\partial x^2} + \frac{\partial^2 F}{\partial x^2}\frac{\partial^2 w}{\partial y^2} - 2\frac{\partial^2 F}{\partial x \partial y}\frac{\partial^2 w}{\partial x \partial y}\right] \qquad (6.113)$$

and accompanied by a variety of boundary conditions, four sets being discussed by Levy and associates. We restrict our attention to the boundary conditions describing the physical case of *simply supported edges*; i.e., both the deflections and the bending moments along the edges are zero, or

$$x = 0, \quad x = a: \quad w = 0, \quad \frac{\partial^2 w}{\partial x^2} = 0$$

$$y = 0, \quad y = b: \quad w = 0, \quad \frac{\partial^2 w}{\partial y^2} = 0. \qquad (6.114)$$

If the tensile load applied on the sides $x = 0, a$ is P_x in the x-direction and P_y in the y-direction then

$$h[(F_y)_{y=b} - (F_y)_{y=0}] = P_x, \qquad h[(F_x)_{x=a} - (F_x)_{x=0}] = P_y. \quad (6.115)$$

In accordance with conditions in airplane structures the plate is considered rigidly framed, all edges remaining straight after deformation. This implies that the elongation of the plate δ_x in the x-direction is independent of y and conversely. By definition

$$\delta_x = \int_0^a u_x \, dx = \int_0^a \left[\frac{1}{E} (F_{yy} - \nu F_{xx}) - \tfrac{1}{2}(w_x)^2 \right] dx \quad (6.116)$$

$$\delta_y = \int_0^b v_y \, dy = \int_0^b \left[\frac{1}{E} (F_{xx} - \nu F_{yy}) - \tfrac{1}{2}(w_y)^2 \right] dy \quad (6.117)$$

(see Timoshenko and Woinowsky-Krieger [56]) the first being independent of y and the second of x.

Probably motivated by the linear theory the lateral deflection is chosen in the double Fourier Series form[†]

$$w = \sum_{m=1}^{\infty} \sum_{n=1}^{\infty} w_{mn} \sin \frac{m\pi x}{a} \sin \frac{n\pi y}{b}. \quad (6.118)$$

Equation (6.118) has the advantage that it satisfies the boundary conditions with regard to the deflections and bending moments for any, yet unknown, values of the coefficients w_{mn}.

The lateral pressure p distributed over the plate's surface consists of two parts p' and p''. The part p'' corresponds with pressure near the edges of the plate whose purpose is to apply edge moments in those cases where the edges are built in ($p'' = 0$ in our case). The part p' corresponds to the given pressure distribution on all parts of the plate except the boundary. The pressure is expanded in a double Fourier series of the same general form as Eq. (6.118)

$$p = \sum_{m=1}^{\infty} \sum_{n=1}^{\infty} p_{mn} \sin \frac{m\pi x}{a} \sin \frac{n\pi y}{b}, \quad (6.119)$$

[†] This assumed form is also applicable to other cases if suitable relations among the coefficients are derived (see Levy [45]).

6.10 SERIES SOLUTIONS IN ELASTICITY

(we ignore the p'' in all that follows) with

$$p_{mn} = \frac{4}{ab} \int_0^a \int_0^b p \sin \frac{m\pi x}{a} \sin \frac{n\pi y}{b} \, dx \, dy.$$

In order to satisfy Eq. (6.112) and the boundary conditions (6.115) the Airy stress function is chosen in the form

$$F(x, y) = \frac{P_x y^2}{2bh} + \frac{P_y x^2}{2ah} + \sum_{m=0}^{\infty} \sum_{n=0}^{\infty} f_{mn} \cos \frac{m\pi x}{a} \cos \frac{n\pi y}{b}. \quad (6.120)$$

Substituting Eqs. (6.120) and (6.118) simultaneously into Eq. (6.112) and equating coefficients of like trigonometric terms yields the following relation between the coefficients

$$f_{mn} = \frac{E}{4\left[\frac{m^2 b}{a} + \frac{n^2 a}{b}\right]^2} \sum b_{rsjk} w_{rs} w_{jk} \quad (6.121)$$

where the summation includes all products for which $r + j = m$ or $r - j = m$ and for which n is either $s + k$ or $s - k$. The coefficients b_{rsjk} are given by the expression

$$b_{rsjk} = 2rsjk \pm (r^2 k^2 + s^2 j^2) \quad (6.122)$$

where the sign before the parenthesis is positive if $m = r + j$, $n = s - k$ or $m = r - j$, $n = s + k$. It is negative otherwise. As an example consider a square plate ($a = b$) so that

$$f_{2,4} = \frac{E}{1600} [-4 w_{1,1} w_{1,3} + 36 w_{3,3} w_{1,1} + 36 w_{1,1} w_{1,5} + \cdots].$$

It still remains to establish a relation between the deflections, the stress functions and the lateral loading. The necessary relation between the coefficients w_{mn}, Eq. (6.118), p_{mn}, Eq. (6.119) and f_{mn}, Eq. (6.120) is obtained when we substitute these relations simultaneously into the lateral equilibrium equation (6.113). Equating coefficients of like trigonometric terms on the two sides of Eq. (6.113) gives the relation

$$p_{mn} = D w_{mn} \pi^4 \left[\frac{m^2}{a^2} + \frac{n^2}{b^2}\right]^2 + P_x w_{mn} \frac{m^2 \pi^2}{a^2 b}$$

$$+ P_y w_{mn} \frac{n^2 \pi^2}{ab^2} + \frac{h \pi^4}{4 a^2 b^2} \sum d_{rsjk} f_{rs} w_{jk}. \quad (6.123)$$

This time the summation includes all products for which $r \pm j = m$ and $s \pm k = m$ and the coefficients d_{rsjk} are given by

$$d_{rsjk} = \pm(rk \pm sj)^2 \quad \text{if} \quad r \neq 0, \quad s \neq 0 \tag{6.124}$$

and are twice value if either r or s is zero. The signs in Eq. (6.124) are determined by the following rules: the first sign is $+$ if *either* $r - j = m$ or $s - k = n$ but *not* if both conditions are true. It is negative in all other cases. The second sign is positive if $r + j = m$ and $s - k = n$ or $r - j = m$ and $s + k = n$ and it is negative otherwise.

For the case $m = 1, n = 3$ the quantity $p_{1,3}$ is given by

$$p_{1,3} = Dw_{1,3}\pi^4(a^{-2} + 9b^{-2})^2 + P_x w_{1,3} \frac{\pi^2}{a^2 b} + P_y w_{1,3} \frac{9\pi^2}{ab^2}$$
$$+ \frac{h\pi^4}{4a^2 b^2}(-8f_{0,2}w_{1,1} - 8f_{0,2}w_{1,5} + 100f_{2,4}w_{3,1} - 64f_{2,2}w_{3,1} + \ldots). \tag{6.125}$$

Using the series for w (Eq. 6.118) and F (Eq. 6.120) we find from Eqs. (6.116) and (6.117) the expressions for the elongations. δ_x and δ_y as

$$\delta_x = \frac{P_x a}{bhE} - \frac{\nu P_y}{hE} - \frac{\pi^2}{8a} \sum_{m=1}^{\infty} \sum_{n=1}^{\infty} m^2 w_{mn}^2$$

$$\delta_y = \frac{P_y b}{ahE} - \frac{\nu P_x}{hE} - \frac{\pi^2}{8b} \sum_{m=1}^{\infty} \sum_{n=1}^{\infty} n^2 w_{mn}^2$$

and both of these are independent of y and x. Thus the boundary conditions for straight sides are satisfied.

A discussion of the computation with the above solution is given in the literature. Levy [43] found that in one case a single term gave answers accurate to three significant figures. The accuracy was judged by observing the change in the answer as the number of w_{mn} coefficients is increased.[†] In any event the knowledge of the exact solution, in the form given above, is already in the form for digital computer solution. The details for a number of cases are depicted for example in Levy [44].

The von Kármán equations (6.112) and (6.113) have been solved approximately by the Ritz (energy) method (see Chapter 5) by several investigators. We content ourselves with four references; Foppl [57], Way [58], Chien and Yeh [59] and Yusuff [60] consider various cases.

[†] In another case 36 coefficients were needed [45].

6.11 "TRAVELING WAVE" SOLUTIONS BY SERIES

In Section 5.3 the elementary idea of a traveling wave [see Eq. (5.26)] was used to extract a dispersion relation in plasma oscillation. Traveling wave solutions have been widely exploited in this area as discussed in the review article by Dolph [61]. Here we wish to introduce the Stuart-Watson [62, 63] formal *"traveling wave"* series (generalized Fourier Series) procedure for solving appropriate problems in equations. Their immediate concern was the investigation of stability problems in parallel fluid flow. The method combines a harmonic spatial analysis with an expansion in powers of time-dependent amplitudes hence reducing the equation to relatively simple nonlinear ordinary differential equations.

To be specific we consider Poiseuille flow under a pressure gradient p between two parallel planes which are at a distance $2h$ apart. We let x be the dimensionless coordinate parallel to the planes and z be normal to them where h is the reference length. In laminar undisturbed flow, a uniform pressure gradient produces a velocity distribution, independent of x, with maximum value U_0 at the channel center (see Figure 6.3).

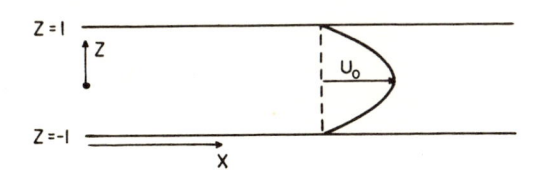

FIG. 6-3. Schematic of Poiseuille flow.

The corresponding dimensionless velocity components we designate by u, w, ψ is the stream function and t is time.

The governing equation of the two dimensional incompressible flow are

$$u_t + uu_x + wu_z = -p_x + \frac{1}{R}(u_{xx} + u_{zz}) \qquad (6.126)$$

$$w_t + uw_x + ww_z = -p_z + \frac{1}{R}(w_{xx} + w_{zz}) \qquad (6.127)$$

$$u_x + w_z = 0 \qquad (6.128)$$

where $R = hU_0/\nu$ is the Reynolds number. In the nondimensional process we have used U_0 as the reference velocity, h/U_0 as the reference time and ρU_0^2 is the reference pressure.

For flow under a pressure gradient between fixed planes (plane Poiseuille flow) the fundamental solutions are

$$u = 1 - z^2, \quad w = 0, \quad -p_x = 2/R, \quad -p_z = 0. \quad (6.129)$$

Our object is to examine the stability of the velocity profile Eq. (6.129) with respect to an infinitesimal disturbance which is travelling in the direction of the flow (the positive x-direction), that is the *traveling wave*, with a stream function of the form

$$\psi = C\psi_1(z) \exp[i\alpha(x - ct)] + \tilde{C}\tilde{\psi}_1(z) \exp[-i\alpha(x - \tilde{c}t)] \quad (6.130)$$

where $c = c_r + ic_i$ ($c_r \geqslant 0$) is a "wave velocity," $\alpha > 0$ is the wave number, C is an arbitrary constant and the symbol $\sim$ denotes a complex conjugate.

The linear equation[†] for ψ_1, together with the no slip boundary on the planes, constitute an eigenvalue problem to determine c as a function of α and R. In the general case of the linear theory there will be a sequence of values of c, and of corresponding eigenfunctions. For Poiseuille flow with given α, R in the supercritical region[‡] it appears that there is only one eigenvalue c with $c_i > 0$; furthermore this corresponds to an eigenfunction which is an even function of z. In the *nonlinear* theory the convergence of the solution obtained is expected to be most rapid when that c is chosen where c_i has the smallest magnitude. Corresponding to this eigenvalue is an eigenfunction and the associated infinitesimal disturbance is used as the basis for construction of the series. In *subcritical* Poiseuille flow the desired eigenvalue c might be expected to correspond to an eigenfunction which is even in z (also).

The stream functions representing infinitesimal disturbances in Poiseuille flow involve the sum of terms of the form $f(z, t)e^{i\alpha x}$, $F(z, t)e^{-i\alpha x}$. These stream functions satisfy the linear equations exactly but when the nonlinearities are not neglected each disturbance reacts with itself and with the main flow generating higher harmonics of the form

$$f_n(z, t)e^{ni\alpha x} \quad (n = \pm 2, \pm 3, ...).$$

[†] The linearized theory of instability is based on the neglect of all terms which are quadratic or involve products. The full details of the linear theory are given in the works of Lin [64] and Chandrasekhar [65].

[‡] The term *supercritical* is used to denote a region wherein the disturbance is amplified for small amplitudes; *subcritical* denotes conditions where a small disturbance does not amplify but is damped.

6.11 "TRAVELING WAVE" SOLUTIONS BY SERIES

It therefore appears reasonable to expand the stream function for the flow, when nonlinearity is included, as a general Fourier series in x as

$$\psi(x, z, t) = \bar{\phi} + \phi^1 = \bar{\phi}(z, t) + \sum_{n=1}^{\infty} \{\phi_n(z, t)e^{ni\alpha x} + \tilde{\phi}_n(z, t)e^{-ni\alpha x}\} \quad (6.131)$$

so that

$$u = \bar{u} + u^1 = \bar{u}(z, t) + \sum_{n=1}^{\infty} \{u_n^1(z, t)e^{ni\alpha x} + \tilde{u}_n^1(z, t)e^{-ni\alpha x}\} \quad (6.132)$$

$$w = w^1 = \sum_{n=1}^{\infty} \{w_n^1(z, t)e^{ni\alpha x} + \tilde{w}_n^1(z, t)e^{-ni\alpha x}\} \quad (6.133)$$

where the notation is

$$\bar{u}(z, t) = \bar{\phi}_z, \quad u_n^1(z, t) = (\phi_n)_z, \quad w_n^1(z, t) = -ni\alpha\phi_n. \quad (6.134)$$

The sum on the right-hand side of Eq. (6.131) represents the disturbance while $\bar{\phi}$ is the mean stream function where the mean is taken with respect to x over the wavelength of the disturbance $2\pi/\alpha$. The quantity $\bar{u} = \bar{\phi}_z$ is the mean velocity and it is not the same as the u of Eq. (6.129) since the mean flow and the disturbance interact. To determine the form of coefficients in Eq. (6.132) and (6.133) we substitute these into Eqs. (6.126) and (6.127) and equate like Fourier components.

For the pressure gradients to balance the remaining terms we readily find that the pressure must have a similar series form to that of Eq. (6.132) namely

$$p = xp^*(t) + p^{**}(z, t) + p^1(x, z, t) =$$

$$xp^*(t) + p^{**}(z, t) + \sum_{n=1}^{\infty} \{p_n(z, t)e^{ni\alpha x} + \tilde{p}_n(z, t)e^{-ni\alpha x}\}. \quad (6.135)$$

The boundary conditions are

(i) the mean velocity $\bar{u}$ assumes the same wall values as the undisturbed velocity u;

(ii) the disturbance velocities u^1, w^1 vanish at the walls; and

(iii) a suitable condition is imposed on the mean pressure gradient in the x-direction, or equivalently on the mean velocity. In our case of Poiseuille flow these become

$$\bar{u} = u_n^1 = w_n^1 = 0 \quad \text{at} \quad z = \pm 1 \quad \text{for} \quad n = 1, 2, \ldots$$
or
$$\bar{u} = \phi_{nz} = \phi_n = 0 \quad \text{at} \quad z = \pm 1. \quad (6.136)$$

We now substitute Eqs. (6.132), (6.133), and (6.135) into Eqs. (6.126) and (6.127) and equate like Fourier components. Alternatively, Stuart [66, 67] found the equations arising from equating the terms independent of x, by taking the means of Eqs. (6.126) and (6.127). These are

$$\overline{u_t} + \overline{u^1 u_x^1} + \overline{w^1 u_z^1} = -p^* + \frac{1}{R} \bar{u}_{zz} \tag{6.137}$$

$$\overline{u^1 w_x^1} + \overline{w^1 w_z^1} = -p_z^{**} \tag{6.138}$$

which, using the definitions (6.132) and (6.133), may be written as

$$\bar{u}_t + (\overline{u^1 w^1})_z = -p^* + \frac{1}{R} \bar{u}_{zz} \tag{6.139}$$

$$(\overline{(w^1)^2})_z = -p_z^{**}. \tag{6.140}$$

The nonlinear terms of Eqs. (6.139) and (6.140) are the Reynolds stress terms and they represent the effect of the disturbance on the mean motion; in the linear theory they are neglected. The disturbance equations are found upon subtracting Eq. (6.137) from Eq. (6.126) and Eq. (6.138) from Eq. (6.127). These are

$$u_t^1 + \bar{u} u_x^1 + w^1 \bar{u}_z + [u^1 u_x^1 + w^1 u_z^1 - \overline{u^1 u_x^1} - \overline{w^1 u_z^1}] = -p_x^1 + \frac{1}{R}(u_{xx}^1 + u_{zz}^1) \tag{6.141}$$

$$w_t^1 + \bar{u} w_x^1 + [u^1 w_x^1 + w^1 w_z^1 - \overline{u^1 w_x^1} - \overline{w^1 w_z^1}] = -p_z^1 + \frac{1}{R}(w_{xx}^1 + w_{zz}^1). \tag{6.142}$$

The bracketed expressions in Eqs. (6.141) and (6.142) are nonlinear parts of the disturbance equations, neglected in the linear theory.

To satisfy continuity the stream function (6.131) is utilized and Eq. (6.139) becomes

$$\bar{u}_t + i\alpha \sum_{n=1}^{\infty} n\{\phi_{nz}\tilde{\phi}_n - \phi_n\tilde{\phi}_{nz}\}_z = -p^* + \frac{1}{R}\bar{u}_{zz}. \tag{6.143}$$

Finally, we eliminate p^1 between Eqs. (6.141) and (6.142) by differentiating the first with respect to z, the second with respect to x and subtracting one from the other. The stream function Eq. (6.131) is

6.11 "TRAVELING WAVE" SOLUTIONS BY SERIES

inserted into the resulting equation. The nth $(n \geq 1)$ component has the form

$$L(n\alpha)\phi_n = \frac{1}{n}\Bigg[\sum_{m=n+1}^{\infty} (m-n)\phi_m'\{\tilde{\phi}_{m-n}'' - (m-n)^2\alpha^2\tilde{\phi}_{m-n}\}$$

$$- \sum_{m=1}^{n-1} (n-m)\phi_m'\{\phi_{n-m}'' - (n-m)^2\alpha^2\phi_{n-m}\}$$

$$- \sum_{m=1}^{\infty} (n+m)\tilde{\phi}_m'\{\phi_{n+m}'' - (n+m)^2\alpha^2\phi_{n+m}\}$$

$$- \sum_{m=1}^{\infty} m\tilde{\phi}_m\{\phi_{n+m}''' - (n+m)^2\alpha^2\phi_{n+m}'\}$$

$$+ \sum_{m=1}^{n-1} m\phi_m\{\phi_{n-m}''' - (n-m)^2\alpha^2\phi_{n-m}'\}$$

$$+ \sum_{m=n+1}^{\infty} m\phi_m\{\tilde{\phi}_{m-n}''' - (m-n)^2\alpha^2\tilde{\phi}_{m-n}'\}\Bigg], \qquad (6.144)$$

where the operator

$$L(n\alpha) = \left[\bar{u} - \frac{i}{n\alpha}\frac{\partial}{\partial t}\right]\left[\frac{\partial^2}{\partial z^2} - n^2\alpha^2\right] - \bar{u}'' + \frac{i}{n\alpha R}\left[\frac{\partial^2}{\partial z^2} - n^2\alpha^2\right]^2 \qquad (6.145)$$

and the (') in Eqs. (6.144) and (6.145) means differentiations with respect to z.

The next step is to determine ϕ_n and $\bar{u}$ from the infinite set of equations (6.143) and (6.144) in two variables. We shall indicate the direction that the analysis is taking although at this writing the work is still in progress. We shall concern ourselves only with the Poiseuille flow case. Couette flow is also considered in Watson [63] and Segel [68] reports an investigation of stability for the case of cellular thermal convection utilizing this method.

We seek a solution which represents a small finite disturbance with time-dependent amplitude and with the property that, as the amplitude $\to 0$, the disturbance tends through the infinitesimal disturbance (6.130) to zero (as $t \to \pm \infty$, according to whether the flow is subcritical or supercritical). As the amplitude tends to zero, the disturbance stream function ϕ^1, defined in Eq. (6.131), must approach the value given by

Eq. (6.130). Thus we find, on comparing components, $\phi_1 \sim C\psi_1(z)e^{-i\alpha ct}$, while for $n > 1$, $\phi_n \to 0$ more rapidly. This observation suggests that a solution might be obtained in which ϕ_n ($n \geq 1$) is of separable form. We therefore expect that the highest-order term in ϕ_1, that which approaches $C\psi_1(z)e^{-i\alpha ct}$ as the amplitude goes to zero, will be of the form $A(t)\psi_1(z)$, where $A(t)$ behaves like $Ce^{-i\alpha ct}$ as the amplitude goes to zero. If there exist finite disturbances which are in equilibrium, we expect to find a set of such disturbances such that as the neutral curve is approached the amplitude tends to zero.

Therefore, in any case we look for a solution in which $|A|$ is small. As $A \to 0$, $\phi_2 \to 0$ more rapidly than ϕ_1 so ϕ_2 must be of smaller order than $|A|$. Examining Eq. (6.144), for $n = 2$, we see that the terms depending on ϕ_1 only occur in the form of the product of A^2 times a function of z (the derivatives are with respect to z on the right-hand side of Eq. (6.144). If we assume that the remaining terms are of lower order in A, then the main term in ϕ_2 is expected to be of the form $A^2\psi_2(z)$. From the proportionality of $A(t)$ to $e^{-i\alpha ct}$ as the amplitude $\to 0$ it follows that[†]

$$\frac{1}{A}\frac{dA}{dt} \to -i\alpha c \quad \text{as} \quad A \to 0 \tag{6.146}$$

and hence we look for a solution for which $1/A \, dA/dt$ is a function of A and $\tilde{A}$ only in the form

$$\frac{dA}{dt} = -i\alpha c A + \text{(smaller order terms)}. \tag{6.147}$$

Let us turn our attention to the departure of the mean velocity $\bar{u}$ from the undisturbed laminar value $\bar{u}_l = 1 - z^2$. To the highest order this departure $\bar{u} - \bar{u}_l$ is found by retaining only the highest order term in ϕ_1 so that the approximation reads

$$\bar{u}_t + i\alpha[\phi_{1z}\tilde{\phi}_1 - \phi_1\tilde{\phi}_{1z}] = -p^* + \frac{1}{R}\bar{u}_{zz}.$$

We readily observe that for the difference $\bar{u} - \bar{u}_l$ to be separable it must have the form

$$\bar{u} - \bar{u}_l = A\tilde{A}f_1(z) + \text{smaller order terms}.$$

[†] We shall now use the sloppier notation, as $A \to 0$ instead of the more precise statement "as the amplitude of the disturbance goes to zero."

6.11 "TRAVELING WAVE" SOLUTIONS BY SERIES

Returning now to Eq. (6.144) for $n = 1$ we see that the highest order terms on the right hand side are of the form of a product $A^2\tilde{A}$ times a function of z. These arise, for example, from terms like $\phi_2'\phi_1''$, etc., where again the assumption is made that the remaining terms are of smaller order. With $n = 1$ we divide Eq. (6.144) by A and let $A \to 0$ (the left-hand side is of order A) obtaining the Orr-Sommerfeld equation

$$(\bar{u}_l - c)(\psi_1'' - \alpha^2\psi_1) - \bar{u}_l''\psi_1 + \frac{i}{\alpha R}(\psi_1^{(4)} - 2\alpha^2\psi_1'' + \alpha^4\psi_1) = 0 \quad (6.148)$$

for the eigenfunction ψ_1, which we would expect to be the coefficient of A. Hence Eq. (6.144) serves to determine the second highest order term in ϕ_1 which is of the form $A^2\tilde{A}\psi_{11}(z)$, provided that dA/dt is of the form

$$\frac{dA}{dt} = -i\alpha c A + a_1 A^2 \tilde{A} + \text{(smaller order terms)} \quad (6.149)$$

where a_1 is some constant.

In turn we can easily verify that the solutions for ϕ_1, ϕ_2 and $\bar{u}$ have the form†

$$\begin{aligned}\phi_1 &= A\psi_1 + A^2\tilde{A}\psi_{11} + \text{(lot)} \\ \phi_2 &= A^2\psi_2 + \text{(lot)} \\ \bar{u} &= \bar{u}_l + A\tilde{A}f_1 + \text{(lot)} \\ \frac{dA}{dt} &= -i\alpha c A + a_1 A^2 \tilde{A} + \text{(lot)}.\end{aligned} \quad (6.150)$$

Further investigation of (6.143) and (6.144) suggests that we look for a solution of the form

$$\phi_n = A^n \left\{\psi_n + \sum_{m=1}^{\infty} |A|^{2m}\psi_{mm}(z)\right\} \quad n \geqslant 1 \quad (6.151)$$

$$\bar{u} = \bar{u}_l + \sum_{m=1}^{\infty} |A|^{2m} f_m \quad (6.152)$$

with

$$\frac{dA}{dt} = A \sum_{m=0}^{\infty} a_m |A|^{2m} \quad (a_0 = -i\alpha c) \quad (6.153)$$

where we recall that $A\tilde{A} = |A|^2$.

---

† (lot) = lower order terms.

The differential equation for $|A|^2$ is easily obtained from (6.153) whose conjugate is

$$\frac{d\tilde{A}}{dt} = \tilde{A} \sum_{m=0}^{\infty} \tilde{a}_m |A|^{2m}. \tag{6.154}$$

Upon multiplying Eq. (6.154) by A and Eq. (6.153) by $\tilde{A}$ and adding we find that

$$\frac{d|A|^2}{dt} = |A|^2 \sum_{m=0}^{\infty} (a_m + \tilde{a}_m) |A|^{2m}$$

$$= 2|A|^2 \sum_{m=0}^{\infty} \mathrm{Re}(a_m) |A|^{2m} \tag{6.155}$$

where Re stand for real part.

Before solutions can be obtained the equations must be analyzed to obtain the equations for the ψ_{ij}. The details are lengthy and may be found in the cited papers.

The relationship of this type of solution to the general Fourier Series seems obvious. The details, like those of the elasticity problem of Section 6.10 are very tedious and the analyst, before undertaking the task, should carefully weigh whether the procedure is justified. One easily observes the basic idea of expanding each unknown in similar series, whose form is suggested by simple related systems and physical information. Once the model is established the work is relatively straightforward.

REFERENCES

1. von Kármán, T., *Z. Angew. Math. Mech.* **1**, 244 (1921).
2. Rosenhead, L. (ed.), "Laminar Boundary Layers." Oxford Univ. Press, London and New York, 1963.
3. Pai, S. I., "Viscous Flow Theory," Vol. I. Van Nostrand, Princeton, New Jersey, 1956.
4. Pohlhausen, K , *Z. Angew. Math. Mech.* **1**, 235 (1921).
5. Kalikhman, B. L., *Natl. Advisory Comm. Aeron. Rept.* No. NACA-TM-1229 (1949).
6. Cope, W. F., and Hartree, D. R., *Phil. Trans. Roy. Soc.* **A241**, 1 (1948).
7. Tetervin, N., *Natl. Advisory Comm. Aeron. Rept.* No. NACA-TN-1479 (1947).
8. Barua, S. N., *Quart. J. Mech. Appl. Math.* **16**, 61 (1963).

9. Courant, R., and Hilbert, D., "Methods of Mathematical Physics," Vol. II. Wiley (Interscience), New York, 1963.
10. Hildebrand, F. B., "Methods of Applied Mathematics." Prentice-Hall, Englewood Cliffs, New Jersey, 1952.
11. Curle, N., "The Laminar Boundary Layer Equations." Oxford Univ. Press, London and New York, 1962.
12. Goldstein, S., *Proc. Cambridge Phil. Soc.* **26**, 1 (1930).
13. Prandtl, L., *Natl. Advisory Comm. Aeron. Rept.* No. NACA-TM-959 (1940).
14. Goertler, H., *Z. Angew. Math. Mech.* **19**, 129 (1939).
15. Goertler, H., *J. Roy. Aeron. Soc.* **45**, 35 (1941).
16. Chandrasekhar, S., "Radiative Transfer." Dover, New York, 1960.
17. Mann, W. R., and Wolf, F., *Quart. Appl. Math.* **9**, 163 (1951).
18. Lin, C. C., *Phys. Rev. Letters* **2**, 245 (1959).
19. Carslaw, H., and Jaeger, J., "Conduction of Heat in Solids," 2nd ed., p. 76. Oxford Univ. Press, London and New York, 1959.
20. Tricomi, F. G., "Integral Equations." Academic Press, New York, 1957.
21. Bückner, H. F., *in* "A Survey of Numerical Analysis" (J. Todd, ed.), Chapter 12. McGraw-Hill, New York, 1962.
22. Fox, L. (ed.), "Numerical Solution of Ordinary and Partial Differential Equations," Chapter 2. Addison-Wesley, Reading, Massachusetts, 1962.
23. Azziz, A. K., On a nonlinear integral equation, *Not. AMS* **11**, Abstr. 64T-114 (1964).
24. Chambre, P. L., and Acrivos, A., *J. Appl. Phys.* **27**, 1322 (1956).
25. Acrivos, A., and Chambre, P. L., *Ind. Eng. Chem.* **49**, 1025 (1957).
26. Wagner, C., *J. Math. Phys.* **32**, 289 (1954).
27. Goldstein, S., "Modern Developments in Fluid Dynamics," Vol. I, p. 135. Oxford Univ. Press, London and New York, 1938.
28. Klamkin, M. S., *SIAM Rev.* **4**, 43 (1962).
29. Weyl, H., *Ann. Math.* **43**, 381 (1942).
30. Siekmann, J., The laminar boundary layer along a flat plate, *in* "Partial Differential Equations and Continuum Mechanics" (R. E. Langer, ed.), Abstr., p. 380. Univ. of Wisconsin Press, Madison, Wisconsin, 1961.
31. Siekmann, J., *in* "Nonlinear Problems" (R. E. Langer, ed.), Abstr., p. 298. Univ. of Wis. Press, Madison, Wisconsin, 1963.
32. Ablow, C. M., and Perry, C. L., *J. Soc. Ind. Appl. Math.* **7**, 459 (1959).
33. Kantorovich, L. V., Functional analysis and applied mathematics, *Usp. Mat. Nauk* **3**, 89-185 (1948); translated by C. D. Benster, Natl. Bur. Std., 1952.
34. Bergman, S., and Schiffer, M., *Duke Math. J.* **15**, 535 (1948).
35. Schuh, H., *in* "50 Jahre Grenzschichtforschung" (H. Görtler and W. Tollmien, eds.), p. 149. Vieweg, Braunschweig, 1955.
36. Bellman, R., *Proc. Natl. Acad. Sci. U.S.* **41**, 482–485 and 743–746 (1955).
37. Kalaba, R., *J. Math. Mech.* **8**, 519 (1959).
38. Keller, J., On solutions of $\nabla^2 u = f(u)$, *NYU Inst. Math. Sci. Res. Rept.* No. MME-1 (1957).
39. Keller, J., *J. Rational Mech. Anal.* **5**, 715 (1956).
40. Doetsch, G., "Theorie und Anwendung der Laplace Transformation," p. 358. Dover, New York, 1943.
41. Hassan, H. A., *J. Fluid Mech.* **9**, 300 (1960).

42. Tipei, N., "Theory of Lubrication." Stanford Univ. Press, Stanford, California, 1962.
43. Levy, S., *Natl. Advisory Comm. Aeron. Rept.* No. NACA-TN-846 (1942).
44. Levy, S., *Proc. Symp. Appl. Math.* 1, 197 (1949).
45. Levy, S., *Natl. Advisory Comm. Aeron. Rept.* Nos. NACA-TN-847 and 852 (1942).
46. Coan, J. M., *J. Appl. Mech.* 18, 143 (1951).
47. Blasius, H., *Z. Math. Phys.* 56, 1 (1908); see also *Natl. Advisory Comm. Aeron. Rept.* No. NACA-TM-1256 (1947).
48. Hiemenz, K., *Dinglers Polytech. J.* 326, 32 (1911).
49. Howarth, L., *Gt. Brit. Aeron. Res. Council Rept.* No. ARC-R&M-1632 (1935).
50. Goertler, H., *J. Math. Mech.* 6, 1 (1957).
51. Schlichting, H., "Boundary Layer Theory" (transl. by J. Kestin), 4th ed., p. 150. McGraw-Hill, New York, 1960.
52. Tifford, A. N., *Wright Air Develop. Ctr. (WADC) Tech. Rept.* No. 53-288, Part 4 (August 1954).
53. Howarth, L., *Proc. Roy. Soc.* **A164**, 547 (1938).
54. Tani, I., *J. Phys. Soc. (Japan)* 4, 149 (1949).
55. Goertler, H., Zahlentafeln universeller Funktionen, zur neuen Reihe für die Berechnung laminarer Grenzschichten. *Ber. Deut. Ver. Luftfahrt* 34 (1957).
56. Timoshenko, S. P., and Woinowsky-Krieger, S., "Theory of Plates and Shells," 2nd ed., pp. 415-420. McGraw-Hill, New York, 1959.
57. Foppl, A. L., *Drang u. Zwang* 1, 226 (1924).
58. Way, S., *Proc. 5th Intern. Congr. Appl. Mech., Cambridge, Mass., 1938*, p. 123. Wiley, New York (1939).
59. Chien, W. Z. and Yeh, K. Y., *Proc. 9th Intern. Congr. Appl. Mech., 1957* Vol. 6, p. 403.
60. Yusuff, S., *J. Appl. Mech.* 19, 446 (1952).
61. Dolph, C. L., *in* "Nonlinear Problems" (R. E. Langer, ed.), pp. 13-46. Univ. of Wisconsin Press, Madison, Wisconsin, 1963.
62. Stuart, J. T., *J. Fluid Mech.* 9, 353 (1960).
63. Watson, J., *J. Fluid Mech.* 9, 371 (1960).
64. Lin, C. C., "The Theory of Hydrodynamic Stability." Cambridge Univ. Press, London and New York, 1955.
65. Chandrasekhar, S., "Hydrodynamic and Hydromagnetic Stability." Cambridge Univ. Press, London and New York, 1963.
66. Stuart, J. T., *J. Aeron. Sci.* 23, 86 (1956).
67. Stuart, J. T., *Z. Angew. Math. Mech.* Sonderheft, p. S32 (1956).
68. Segel, L. A., *in* "Nonlinear Problems" (R. E. Langer, ed.), Abstr., p. 295. Univ. of Wisconsin Press, Madison, Wisconsin, 1963.

CHAPTER 7

Numerical Methods

7.0 INTRODUCTION

The post World War II development of high speed digital and analog computing devices has spurred a substantial growth of the mathematical science of numerical analysis. We are herein interested in numerical methods for solving these equations. Many of these methods are suggested from procedures originally developed for linear equations. Extension of these ideas to these equations is very difficult if a thorough analysis of stability, convergence and error is carried through. However, several cases have been completely detailed, and these will be discussed. In addition some hitherto unpublished experiments concerned with direct application of linear methods to these equations will be given.

There are a number of numerical methods for solving partial differential equations. Of these our concern will be with those of finite differences and also (for hyperbolic systems) the method of characteristics. Of the methods only that of finite differences stands out as being universally applicable to both linear and nonlinear systems.

In recent years (1957–1964) we have witnessed an upsurge of monographs devoted specifically to numerical methods for differential equations. For our purposes those of Forsythe and Wasow [1], Fox [2], Young and Frank [3], Richtmyer [4] and Todd [5] give excellent reviews of much of the current material although they are (by necessity) predominantly linearly biased.

The first few sections summarize basic concepts, followed by a summary of methods developed and analyzed for linear systems. Examples of equations using these methods are then given and the limitations discussed. These are followed by methods specifically designed for these equations and the chapter is concluded by a discussion of the method of characteristics.

7.1 TERMINOLOGY AND COMPUTATIONAL MOLECULES[†]

Whenever a continuous operator, such as $u(\partial u/\partial x)$, is replaced by a finite difference approximation an error, called the *truncation error*, is introduced into the resulting equation. The partial derivative $\partial u/\partial x$ can be approximated in a number of ways. Suppose $u = u(x, y)$ and let the points $P_{i,j}$ form a grid as in Fig. (7–1). If the network is square of

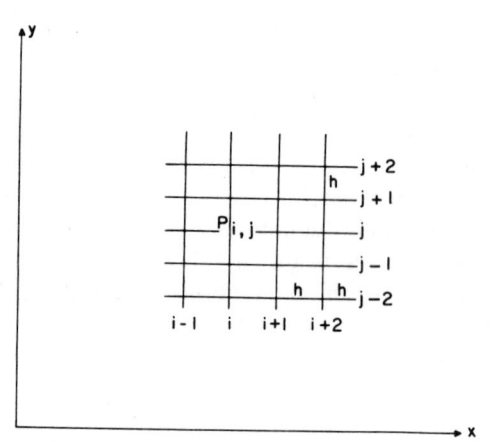

FIG. 7-1. Notation for the rectangular mesh.

spacing h then we can approximate $\partial u/\partial x \vert_{i,j}$ in the following simple ways, with the indicated truncation errors:

$$\text{(a)} \quad \frac{\partial u}{\partial x}\bigg|_{i,j} = \frac{1}{h}[u_{i+1,j} - u_{i,j}] + O(h) \tag{7.1}$$

[†] This term is due to Bickley [6] and is particularly useful for visualizing the course of the computation. Other authors use the term computational stencil or lozenge.

7.1 TERMINOLOGY AND COMPUTATIONAL MOLECULES

$$\text{(b)} \quad \left.\frac{\partial u}{\partial x}\right|_{i,j} = \frac{1}{h}[u_{i,j} - u_{i-1,j}] + O(h) \tag{7.2}$$

$$\text{(c)} \quad \left.\frac{\partial u}{\partial x}\right|_{i,j} = \frac{1}{2h}[u_{i+1,j} - u_{i-1,j}] + O(h^2) \tag{7.3}$$

where $u_{i,j}$ represents the value of u at the point $P_{i,j}$. An elementary application of the Taylor's series generates Eqs. (7.1)–(7.3). Equation (7.3) may be pictorially represented by a computational molecule as shown in Fig. 7-2. The most commonly used molecules for square

$$\left.\frac{\partial u}{\partial x}\right|_{i,j} = \frac{1}{2h}\{(-1) \quad (0)_{i,j} \quad (1)\} + O(h^2)$$

$$\left.\frac{\partial u}{\partial y}\right|_{i,j} = \frac{1}{2h}\left\{\begin{matrix}(1)\\(0)_{i,j}\\(-1)\end{matrix}\right\} + O(h^2)$$

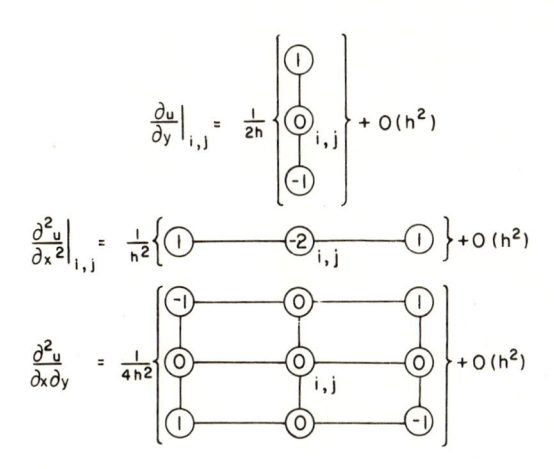

$$\left.\frac{\partial^2 u}{\partial x^2}\right|_{i,j} = \frac{1}{h^2}\{(1) \quad (-2)_{i,j} \quad (1)\} + O(h^2)$$

$$\left.\frac{\partial^2 u}{\partial x \partial y}\right|_{i,j} = \frac{1}{4h^2}\left\{\begin{matrix}(-1) & (0) & (1)\\(0) & (0)_{i,j} & (0)\\(1) & (0) & (-1)\end{matrix}\right\} + O(h^2)$$

FIG. 7-2. Computational molecules for the partial derivatives (Eq. (7.3)).

networks, of spacing h, are shown in Fig. 7-3. If the spacing is different say h and l the ideas are easily extended. For example the Laplacian $\nabla^2 u = 0$ with uneven spacing has the form

$$\nabla^2 u \big|_{i,j} = \frac{1}{h^2}\{u_{i+1,j} - 2u_{i,j} + u_{i-1,j}\}$$
$$+ \frac{1}{l^2}\{u_{i,j+1} - 2u_{i,j} + u_{i,j-1}\} + O(h^2 + l^2) \tag{7.4}$$

with a corresponding computational molecule.

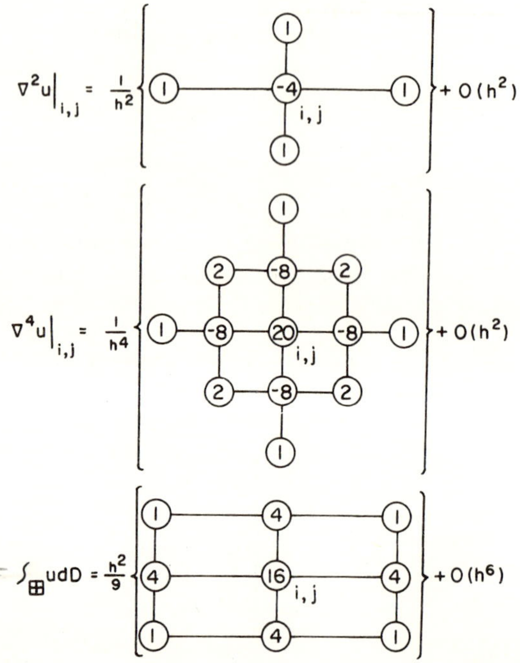

$$\nabla^2 u \Big|_{i,j} = \frac{1}{h^2} \left\{ \begin{array}{c} \text{(1)} \\ \text{(1)}--\text{(-4)}_{i,j}--\text{(1)} \\ \text{(1)} \end{array} \right\} + O(h^2)$$

$$\nabla^4 u \Big|_{i,j} = \frac{1}{h^4} \left\{ \begin{array}{c} \text{(1)} \\ \text{(2)--(-8)--(2)} \\ \text{(1)--(-8)--(20)}_{i,j}\text{--(-8)--(1)} \\ \text{(2)--(-8)--(2)} \\ \text{(1)} \end{array} \right\} + O(h^2)$$

$$\text{AREA} = \int_{y-h}^{y+h}\int_{x-h}^{x+h} u\, dx\, dy = \int_{\boxplus} u\, dD = \frac{h^2}{9} \left\{ \begin{array}{c} \text{(1)--(4)--(1)} \\ \text{(4)--(16)}_{i,j}\text{--(4)} \\ \text{(1)--(4)--(1)} \end{array} \right\} + O(h^6)$$

FIG. 7-3. Computational molecules for some common differential operators (two-dimensional).

Strictly speaking the error, discussed above, is the truncation error of the equation and not of the solution. We make this point here because as in the analytical approach the boundary conditions are vital to the solution. These too, (if they are not of the Dirichlet form, $u = f$, on the boundary), must be approximated by finite differences thereby introducing a truncation error in the boundary conditions. The error resulting from the equation and boundary condition truncation has been called the *discretization error* by Wasow [7]. In addition to this error, resulting from replacing the continuous problem by a discrete model, there is an additional error whenever the discrete equations are not solved exactly. This error, called *round-off error*, is present in computer iterative solutions since the iteration is only continued until there is no change out to a certain number of decimal places. The interval size h affects the discretization error and round-off error in the opposite sense. The first decreases as h decreases, while the round-off error generally increases. It is for this reason that one cannot generally assert that

7.1 TERMINOLOGY AND COMPUTATIONAL MOLECULES

decreasing the mesh size always increases the accuracy. From this discussion we can infer that the growth of error cannot be tolerated in a numerical technique. Therefore *error analysis* is one of the prime considerations of a numerical method. As one might expect this subject is under extensive cultivation.

In addition to the error analysis there are two other fundamental concepts, those of *convergence* and *stability* which are often interrelated. To discuss these concepts we suppose that the partial differential equation is[†]

$$L(u) = 0 \quad \text{in a region } S, \quad u = g \quad \text{on} \quad B \quad (7.5)$$

where B is the boundary of the region S. Associated with Eq. (7.5) is a finite difference system, whose network, no matter the configuration, depends upon certain parameters, usually the interval sizes. We suppose there is only one, say h, and write the finite difference problem as

$$L_h(U) = 0 \quad \text{in } S, \quad U = g_h \quad \text{on} \quad B.$$

This is clearly not the most general situation for it is possible for S and B to depend upon h—for example in the case where a curvilinear boundary is replaced by straight line segments. We say the *finite difference scheme converges if $U(P)$ converges to the solution $u(P)$, with the same boundary values, as $h \to 0$.* Later we shall demonstrate a useful approach for proving convergence in a class of nonlinear partial differential equations.

The concept of stability of difference approximations has been often discussed in detail but precise definition appears difficult. The definition given by most authors will be adopted here. Let $U(x, t)$ be the solution of a given difference approximation, solvable step by step in the t-direction. The effect of a mistake or round-off error may replace the value $U(x_0, t_0)$ by $U(x_0, t_0) + \epsilon$ at the grid point (x_0, t_0). If the solution procedure is continued with the value $U(x_0, t_0) + \epsilon$, without new errors being introduced, and if at subsequent points the value $U^*(x, t)$ is obtained then we call $U^*(x, t) - U(x, t)$ the *departure of the solution resulting from the error ϵ at (x_0, t_0)*. When errors are committed or introduced at more than one point *cumulative departures* occur which are not additive except in linear problems. If we designate δ as the maximum absolute error, i.e., $|\epsilon(x, t)| \leqslant \delta$, and h the interval size, then most authors call a *procedure stable* if the cumulative departure

[†] This formulation is used only for the sake of simplicity. Often the boundary conditions are more complicated but the arguments used here can be extended to such cases with relative ease.

tends to zero as $\delta \to 0$ and does not increase faster than some power of h^{-1} as $h \to 0$.

A *qualitative* description of instability is often illustrative. To this end we consider a finite difference approximation for the simple ordinary differential equation

$$y' = y - x \tag{7.6}$$

whose general solution is

$$y = Ae^x + x + 1. \tag{7.7}$$

A is an arbitrary constant to be determined by an initial condition. The condition $y(0) = 1$ requires that $A = 0$, so that the exact solution is a simple linear term growing slowly when compared to the exponential term. In an approximate solution of Eq. (7.6) this exponential is likely to be introduced due to round-off errors. Even if the local multiple of e^x is very small the resulting error will soon obliterate the true solution. Any finite difference scheme which allows the growth of error, eventually "swamping" the true solution, is unstable. To avoid such occurrences we usually have to restrict the interval size or other parameter. This restriction avoids the undesirable exponential growth in $1/h$. Amplification of this discussion may be found in Forsythe and Wasow [1, 29–37 *et seq.*].

Finite difference approximations are often labeled with the titles *explicit* or *implicit*. An explicit formula provides for a noniterative "marching" process for obtaining the solution at each present point in terms of known preceding and boundary points. Since parabolic and hyperbolic equations characteristically have open integration domains, explicit methods are applicable to these problems. Questions of stability are most critical for explicit schemes. Implicit procedures are usually iterative simultaneous calculations of many present values in terms of known preceding values and boundary conditions.

A. PARABOLIC EQUATIONS

7.2 EXPLICIT METHODS FOR PARABOLIC SYSTEMS

Consider a region $R: 0 \leqslant x' \leqslant L, \; t' \geqslant 0$ in which we seek the solution of

$$a \frac{\partial U}{\partial t'} = \frac{\partial}{\partial x'}\left(K \frac{\partial U}{\partial x'}\right) + \phi(x', t', U) \tag{7.8}$$

7.2 EXPLICIT METHODS FOR PARABOLIC SYSTEMS

with appropriate boundary conditions. Herein we assume these to be

$$U(0, t') = U_0 f(t'), \qquad \frac{\partial U}{\partial x'}(L, t') = 0 \tag{7.9}$$

and the initial condition to be

$$U(x', 0) = U_0 g(x'), \qquad 0 \leqslant x' \leqslant L. \tag{7.10}$$

For the moment we shall assume a and K to be constants. This problem may be brought into dimensionless form by setting

$$x = x'/L, \qquad t = \frac{t'}{(a/K)L^2}, \qquad u = U/U_0. \tag{7.11}$$

Hence the equation for u is[†]

$$u_t = u_{xx} + \lambda \phi(x, t, u) \tag{7.12}$$

with auxiliary conditions

$$u(0, t) = f(t), \qquad \frac{\partial u}{\partial x}(1, t) = 0$$
$$u(x, 0) = g(x) \qquad 0 \leqslant x \leqslant 1. \tag{7.13}$$

Finite difference representation of the terms in Eq. (7.12) require that a set of mesh points be defined in the x, t plane. Let $\Delta x = h$, $\Delta t = k$ so that a finite difference approximation of $u(x, t)$ is to be obtained at the points (ih, jk) where $x = ih$, $t = jk$, $i = 1, 2, ..., H - 1$ ($H = 1/h$) and $j = 1, 2, ..., \infty$. The boundaries are specified by $i = 0$, $i = H$ and any "false" boundaries by $i = -1, -2, ...,$ and $i = H + 1$, $H + 2$. The initial line is designated $j = 0$, and the numerical approximation at $x = ih$, $y = jk$ is designated $u_{i,j}$.

One of the simplest methods for solving problems defined by Eqs. (7.12) and (7.13) is to approximate the time derivative $\partial u/\partial t$ by a forward difference in analogy with Eq. (7.1)

$$u_t \vert_{i,j} = \frac{u_{i,j+1} - u_{i,j}}{k} + O(k) \tag{7.14}$$

[†] When a specific form for ϕ is given, a parameter may multiply this term. For the general discussion we call this λ.

and u_{xx} by the relation

$$u_{xx}|_{i,j} = \frac{u_{i+1,j} - 2u_{i,j} + u_{i-1,j}}{h^2} + O(h^2). \tag{7.15}$$

Upon substituting these expressions into Eq. (7.12) and solving for $u_{i,j+1}$ results in

$$u_{i,j+1} = r[u_{i+1,j} + u_{i-1,j}] + (1 - 2r)u_{i,j} + \lambda r h^2 \phi(ih, jk, u_{ij}) \tag{7.16}$$

where

$$r = k/h^2. \tag{7.17}$$

The initial condition is expressed as

$$u_{i,0} = g(ih) \tag{7.18a}$$

and the boundary condition at $x = 0$ by

$$u_{0,j} = f(jk) \tag{7.18b}$$

while the condition at $x = 1$ needs special consideration.

Through the parameter r we observe that $k = rh^2$ so both Eqs. (7.14) and (7.15) have $O(h^2)$ truncation errors. In order to maintain a discretization error of the same order we use the approximation given in Eq. (7.3). This representation at $x = 1$, that is $i = H$, becomes

$$u_x|_{H,j} = \frac{1}{2h}[u_{H+1,j} - u_{H-1,j}] = 0 \tag{7.19}$$

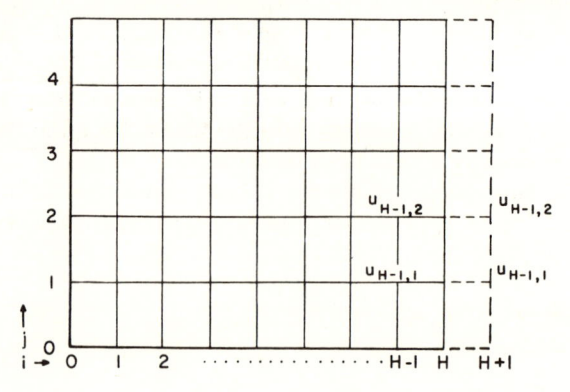

FIG. 7-4. Rectangular mesh with false boundary approximation of a gradient boundary condition.

7.2 EXPLICIT METHODS FOR PARABOLIC SYSTEMS

thereby requiring that auxiliary points (a false boundary) be introduced to the right of $x = 1$ (see Fig. 7-4) designated by $P_{H+1,j}$. From Eq. (7.19) we see that the boundary condition is satisfied to the order h^2 if

$$u_{H+1,j} = u_{H-1,j}. \tag{7.20}$$

Equations (7.16), (7.18a), (7.18b), and (7.20) can be used recursively to determine all $u_{i,j}$ for $1 \leqslant i \leqslant H-1$ and $j \geqslant 1$. The indicated use of false boundaries is the most common method of handling boundary conditions (or initial conditions) which involve partial derivatives.

Numerous researchers have investigated the stability of the aforediscussed explicit scheme. In the previous section stability was discussed. We recall that, roughly speaking, a numerical procedure for solving a parabolic equation is stable if it has the property that when an error is introduced in the numerical solution at some time t_0, and if no further errors are made, then the error will remain bounded for all $t > t_0$. The value of the ratio

$$r = k/h^2 = \frac{\Delta t}{(\Delta x)^2} = \frac{K}{a}\frac{(\Delta t')}{(\Delta x')^2} \tag{7.21}$$

is intimately related with the stability of finite difference methods for parabolic equations. We find several alternate proofs in the cited literature indicating that $r = \tfrac{1}{2}$ is the boundary between stability and instability. For the boundary conditions of this problem the system is stable for all values of $r \leqslant \tfrac{1}{2}$ while it is stable only for $r < \tfrac{1}{2}$ if the condition at $x = 0$ is replaced by

$$u(0\ t) - \frac{\partial u}{\partial x}(0\ t) = 0. \tag{7.13a}$$

Richtmyer [4] proves that the solution of the simple explicit difference equation (7.16) actually converges to the true solution of the initial value problem defined by Eqs. (7.12) and (7.13) if Δt and Δx go to zero in such a way that

$$r = \frac{\Delta t}{(\Delta x)^2} < \tfrac{1}{2}. \tag{7.22}$$

The truncation error associated with Eq. (7.16) is easily calculated by using Taylor's series. Expanding $u_{i,j+1}$ about $u_{i,j}$ we have

$$u_{i,j+1} = u_{i,j} + \frac{\partial u_{i,j}}{\partial t}\Delta t + \frac{1}{2}\frac{\partial^2 u_{i,j}}{\partial t^2}(\Delta t)^2 + \cdots$$

or
$$\frac{\partial u_{i,j}}{\partial t} = \frac{u_{i,j+1} - u_{i,j}}{\Delta t} - \frac{\partial^2 u_{i,j}}{\partial t^2} \frac{\Delta t}{2} + O[(\Delta t)^2]. \tag{7.23}$$

We next expand $u_{i+1,j}$ and $u_{i-1,j}$ about $u_{i,j}$, add the results and divide by $(\Delta x)^2$ to obtain

$$\frac{\partial^2 u_{i,j}}{\partial x^2} = \frac{u_{i+1,j} - 2u_{i,j} + u_{i-1,j}}{(\Delta x)^2} - \frac{\partial^4 u_{i,j}}{\partial x^4} \frac{(\Delta x)^2}{12} + O[(\Delta x)^4]. \tag{7.24}$$

Utilizing Eqs. (7.23) and (7.24) we find that the truncation error, in the terms of order Δt and $(\Delta x)^2$, is

$$\left\{ \frac{\Delta t}{2} \frac{\partial^2 u}{\partial t^2} - \frac{(\Delta x)^2}{12} \frac{\partial^4 u}{\partial x^4} \right\}\bigg|_{i,j}. \tag{7.25}$$

Now for the case where $\phi \equiv 0$, Milne [8] noted that u satisfies $u_t = u_{xx}$ so that

$$[u_t]_t = [u_t]_{xx} = [u_{xx}]_{xx} = u_{xxxx}.$$

Thus $u_{tt} = u_{xxxx}$ and we may rewrite Eq. (7.25) as

$$\frac{1}{2}\left\{ \Delta t - \frac{(\Delta x)^2}{6} \right\} \frac{\partial^4 u}{\partial x^4}\bigg|_{i,j} \tag{7.26}$$

which vanishes if $r = \Delta t/(\Delta x)^2 = \frac{1}{6}$. With this choice of r the truncation error becomes $O[h^4]$ but the discretization error remains $O[h^2]$ unless the boundary conditions are also approximated with truncation error $O[h^4]$. If the boundary conditions involve no approximation at all, e.g., for Dirichlet conditions, then the case $r = \frac{1}{6}$ gives $O[h^4]$ discretization error.

When $\phi \not\equiv 0$, the choice of $r = \frac{1}{6}$ does not yield the same results but is a good initial selection.

7.3 SOME NONLINEAR EXAMPLES

The simple explicit scheme based upon approximations (7.14) has been applied to several equations of the form

$$u_t = \frac{\partial}{\partial x}\left[K(u) \frac{\partial u}{\partial x} \right] \tag{7.27}$$

and the computations carried out for long time on a digital computer. The heuristic approach to the stability question is to note that $K(u)$ is the effective diffusion coefficient (called K/a in the linear case) which varies

7.3 SOME NONLINEAR EXAMPLES

with x and t. Instabilities, when they occur, manifest themselves by rapid oscillations of a local character. Hence one would expect the simple explicit method to be stable up to time $t' = t_0'$ if and only if

$$\frac{K(u)\,\Delta t'}{(\Delta x')^2} < \frac{1}{2} \tag{7.28}$$

—a result inferred from Eq. (7.21).

For nonlinear problems stability depends not only upon the structure of the finite difference procedure *but also on the solution* being obtained. For a given solution the system may be stable for some values of t' and not for others. In the practical solution of these equations it is highly desirable, if the difference equations are not unconditionally stable, for the machine program to keep a constant check on stability. One such check could be the testing of the inequality (7.28) either stopping when the condition is violated or altering Δt to restore stability.

The first case concerns the problem of conduction given by

$$\frac{\partial}{\partial x'}\left[k_0 \exp\left(-\frac{\alpha T}{T_0}\right)\frac{\partial T}{\partial x'}\right] = \beta \exp(-\gamma T/T_0)\frac{\partial T}{\partial t'} \tag{7.29}$$

with the auxiliary conditions

$$T(x', 0) = T_1, \qquad T(0, t') = T_0, \qquad T(L, t') = T_0.$$

Equation (7.29) is reduced to dimensionless form by setting

$$\theta = T/T_0, \qquad x = x'/L, \qquad t = \frac{k_0 t'}{\beta L^2}$$

so that

$$\theta_{xx} - \alpha(\theta_x)^2 = \theta_t \exp[(\alpha - \gamma)\theta] \tag{7.30}$$

with the auxiliary conditions

$$\theta(x, 0) = T_1/T_0, \qquad \theta(0, t) = \theta(1, t) = 1. \tag{7.30a}$$

Application of the simple explicit procedure of Section 7.2 gives the finite difference approximation

$$\theta_{i,j+1} = \theta_{i,j} + r\exp[(\gamma - \alpha)\theta_{i,j}]\,[\theta_{i+1,j} - 2\theta_{i,j} + \theta_{i-1,j}$$
$$- \frac{\alpha}{4}(\theta_{i+1,j} - \theta_{i-1,j})^2] \tag{7.31}$$

with $r = \Delta t/(\Delta x)^2$ as usual. Several cases were considered.

Case 1. $\gamma = 0$. In this situation Eq. (7.30) is exactly of the form Eq. (7.27) with $K(\theta) = e^{-\beta\theta}$. This case was calculated via Eq. (7.31) for the boundary conditions $\theta(0, t) = 0$, $\theta(1, t) = 0$ and the initial condition $\theta(x, 0) = 1$. Equation (7.28) can be rewritten in the form

$$r = \frac{\Delta t}{(\Delta x)^2} < \tfrac{1}{2} e^{\beta\theta} \tag{7.32}$$

as the bound for stability. In this problem $\theta \leqslant 1$ for all x and t and furthermore $\theta \to 0$ as $t \to \infty$. Near $t = 0$ we would expect to begin with $r < \tfrac{1}{2} e^{\beta}$ and as $t \to \infty$ the step size should be decreased so that, eventually, $r < \tfrac{1}{2}$. In practical calculations we found these arguments essentially correct although several times (for programming economy) it proved better to begin with lower values of r (with $\beta = 1$, near $t = 0$ we began with $r = \tfrac{3}{4}$ rather than $\tfrac{1}{2}e$). Richtmyer [4, p. 109] reports a computation with the equation $u_t = [5u^4 u_x]_x$ and asserts that in all cases the predictions of the heuristic stability argument (Eq. (7.28)) seemed to be verified.

— like Cole-Hopf potential

Case 2. $\gamma \neq 0$. Here we set $R = e^{-\gamma\theta}$ so that Eq. (7.30), in terms of R, becomes

$$\frac{\partial}{\partial x}\left[R^{(\alpha/\gamma)-1} \frac{\partial R}{\partial x} \right] = \frac{\partial R}{\partial t} \tag{7.33}$$

so that the previous arguments apply to R. θ is easily calculated from R by means of $\theta = -1/\gamma \ln R$.

If $\alpha = \gamma$: $\theta_{xx} - \gamma(\theta_x)^2 = \theta_t \to \frac{\partial^2 R}{\partial x^2} = \frac{\partial R}{\partial t}$

7.4 ALTERNATE EXPLICIT METHODS

A variety of explicit finite difference procedures have been investigated for linear parabolic equations. We discuss them here since they prove useful in stimulating new developments in nonlinear partial differential equations.

First we note that *extreme care* must be exercised when using a certain scheme for a particular problem. Suppose for the time derivative u_t, in the linear equation $u_t = u_{xx}$, we adopt the difference approximation

$$u_t \big|_{i,j} = \frac{u_{i,j+1} - u_{i,j-1}}{2\Delta t} + O((\Delta t)^2). \tag{7.34}$$

↳ central difference in time

7.4 ALTERNATE EXPLICIT METHODS

At first glance one might expect the resulting explicit finite difference scheme

$$u_{i,j+1} = u_{i,j-1} + 2r[u_{i+1,j} - 2u_{i,j} + u_{i-1,j}], \qquad r = \Delta t/(\Delta x)^2 \qquad (7.35)$$

to be better (in some sense) than the method of Section 7.2. However, this method of Richardson [9][†] is *unstable for all values of r*, thereby negating any gains obtained by using the approximation (7.34) with higher order truncation error. Instability, for all r, was first demonstrated by O'Brien, Hyman, and Kaplan [10]. An alternate proof of instability is presented by Young and Frank [3, p. 49]. A quick cursory check of stability can be based upon what we call the "*positive test.*" Positive coefficients assure stability while negative coefficients *may* lead to instability. However, the presence of negative coefficients *does not necessarily imply instability* as we shall presently see. The simple finite difference scheme, Eq. (7.16), is stable for $r < \frac{1}{2}$, i.e., all coefficients are positive. On the other hand the coefficient of $u_{i,j}$ in Eq. (7.35) is always negative.

If we replace $u_{i,j}$ in Richardson's method, Eq. (7.35), by the arithmetic

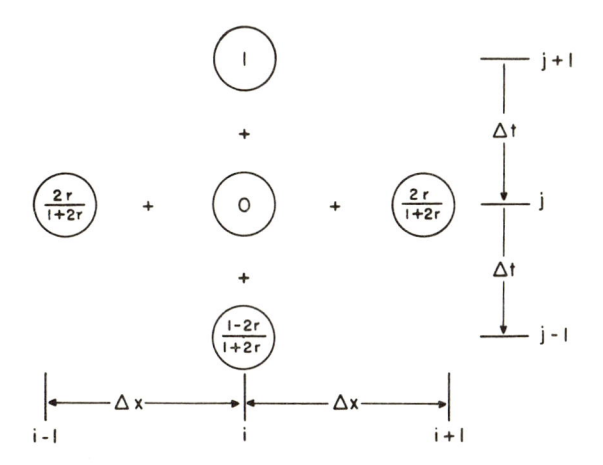

FIG. 7-5. The Dufort-Frankel explicit computation molecule for the diffusion equation.

[†] This pioneering paper of 1910 was a *major* contribution for it demonstrated successful methods for equilibrium and eigenvalue problems. The procedure (7.35) for the parabolic equation was shown to be unstable much later (1951).

average $\frac{1}{2}(u_{i,j+1} + u_{i,j-1})$ we obtain a scheme of Dufort and Frankel [11]

$$u_{i,j+1} = \frac{1}{1+2r}\{u_{i,j-1} + 2r[u_{i+1,j} - u_{i,j-1} + u_{i-1,j}]\} \tag{7.36}$$

whose computational molecule is shown in Fig. 7-5. Like the Richardson procedure this method utilizes the two preceding lines ($j-1$ and j) to march ahead to line $j+1$. And here also, since initial data are given on only one line, the first row ($j=1$) of values must be obtained by another method. Various proofs have been given that this method is *stable for all values of r*. Besides Dufort and Frankel [11], alternate proofs are found in Young and Frank [3, p. 51] and Richtmyer [4, p. 85]. For any sufficiently often differentiable function $u(x, t)$ the truncation error is easily obtained from Taylor series expansions and we find that

$$\frac{u_{i,j+1} - u_{i,j-1}}{2\Delta t} - \frac{u_{i+1,j} - u_{i,j+1} - u_{i,j-1} - u_{i-1,j}}{(\Delta x)^2} - \left[\frac{\partial u}{\partial t} - \frac{\partial^2 u}{\partial x^2}\right]_{i,j}$$
$$= \left(\frac{\Delta t}{\Delta x}\right)^2 \frac{\partial^2 u}{\partial t^2} + O[(\Delta t)^2] + O[(\Delta x)^2] + O\left[\frac{(\Delta t)^4}{(\Delta x)^2}\right] \tag{7.37}$$

i.e., of order $(\Delta t/\Delta x)^2$. Consistency of Eq. (7.36) with $u_t = u_{xx}$ requires that $\Delta t/\Delta x \to 0$ as $\Delta t \to 0$. That is Δt must go to zero *faster* than Δx. If this were not the case but $\Delta t/\Delta x$ is kept fixed, say $= \alpha$, then Eq. (7.36) is consistent not with the diffusion equation $u_t = u_{xx}$ but with the (hyperbolic) damped wave equation[†]

$$u_t - u_{xx} + \alpha u_{tt} = 0! \; . \tag{7.38}$$

Thus Δt cannot be chosen proportional to Δx. Consistency conditions are discussed in greater detail in the basic references (see [1–5]).

In many respects the more general equation

$$u_t = a_0(x, t)u_{xx} + a_1(x, t)u_x + a_2(x, t)u \tag{7.39}$$

involves very little extra complication than the discussion above. A simple explicit formula can be obtained by the approximation (7.14)

[†] Thereby suggesting an explicit method for that equation! This relatively simple example should serve as a *warning against blind* belief in the results of a numerical "solution" of an equation.

7.4 ALTERNATE EXPLICIT METHODS

and (7.15), by Eq. (7.15) and the Dufort-Frankel average or by other methods. Using (7.14) and (7.15) a simple explicit formula is

$$u_{i,j+1} = u_{i,j} + a_{0i,j}(k/h^2)[u_{i+1,j} - 2u_{i,j} + u_{i-1,j}]$$
$$+ a_{1i,j}(k/2h)[u_{i+1,j} - u_{i-1,j}]$$
$$+ ka_{2i,j}u_{i,j}. \tag{7.40}$$

John [12] has investigated such explicit schemes which he writes in the more general form

$$u_{i,j+1} = \sum_{s=-N}^{N} C^{(s)}(x, t, \Delta t)u_{i+s,j} \tag{7.41}$$

where the $C^{(s)}$ are known coefficient functions, N is the number of x levels used and $a_0(x, t) \geqslant$ constant > 0. He assumes $\Delta t/(\Delta x)^2 =$ constant as the mesh size is refined. The *consistency condition*[†] of the finite difference expression (7.41) with Eq. (7.39) is seen by Taylor's series methods to be

$$\lim_{\Delta t \to 0} \frac{1}{\Delta t} \left\{ \sum_{s=-N}^{N} [C^{(s)}(x, t, \Delta t) - \delta_{s0}] \right\} = a_2(x, t) \tag{7.42a}$$

$$\lim_{\Delta t \to 0} \frac{\Delta x}{\Delta t} \left\{ \sum_{-N}^{N} sC^{(s)}(x, t, \Delta t) \right\} = a_1(x, t) \tag{7.42b}$$

$$\lim_{\Delta t \to 0} \frac{(\Delta x)^2}{\Delta t} \left\{ \sum_{-N}^{N} \tfrac{1}{2}s^2 C^{(s)}(x, t, \Delta t) \right\} = a_0(x, t). \tag{7.42c}$$

We shall assume these conditions are satisfied. In terms of the maximum norm (see equation (5.184) for the definition) John proves that the difference approximation (7.41) is stable if there exists a constant $M > 0$ such that

$$\left| \sum_{-N}^{N} C^s(x, t, 0)e^{is\beta} \right| \leqslant e^{-M\beta^2} \quad \text{for} \quad |\beta| \leqslant \pi. \tag{7.43}$$

Further, he showed that if a sufficiently smooth solution $u(x, t)$ of Eq. (7.39) and an initial condition $u(x, 0) = f(x)$ exists, and if the stability condition (7.43) and consistency condition are satisfied then the solution of the difference Eq. (7.41) converges to $u(x, t)$ as $\Delta t \to 0$.

[†] John calls this, together with other requirements, a compatibility condition.

For the simple explicit form given in Eq. (7.40) the results of John leads to the condition

$$\text{least upper bound}_{(x,t)} \left\{ a_0(x, t) \frac{\Delta t}{(\Delta x)^2} \right\} < \tfrac{1}{2}. \tag{7.44}$$

Forsythe and Wasow [1, p. 108] also discuss the work of John and enlarge, somewhat, on his results.

7.5 THE QUASI-LINEAR PARABOLIC EQUATION

The results of John [12] apply also to the quasilinear parabolic equation

$$u_t = a_0(x, t)u_{xx} + a_1(x, t)u_x + d(x, t, u) \tag{7.45}$$

examples of which occur in diffusion with chemical reaction (of first and higher order). The explicit difference approximations are of the form

$$u_{i,j+1} = \sum_{s=-N}^{N} C^{(s)}(x, t, \Delta t)u_{i+s,j} + \Delta t d_{i,j} \tag{7.46}$$

where the functions $C^{(s)}$ involve the coefficient functions from Eq. (7.45). If we replace u_t and u_x by their respective forward finite difference quotients, analogous to Eq. (7.14), and u_{xx} by the second central difference quotient (7.15) then $N = 1$ and

$$C^{(-1)} = ra_0(x, t) \tag{7.47a}$$
$$C^{(0)} = 1 - 2ra_0(x, t) - r(\Delta x)a_1(x, t) + r(\Delta x)^2 a_2(x, t) \tag{7.47b}$$
$$C^{(1)} = ra_0(x, t) + r(\Delta x)a_1(x, t). \tag{7.47c}$$

The approximation, (7.46), to Eq. (7.45), is also convergent if the stability condition (7.43) and consistency conditions (7.42), formulated for the linear problem, are satisfied.

An example of the type Eq. (7.45) will be considered later when the simultaneous solution of two equations of this form will be given.

7.6 SINGULARITIES

In rectangular coordinates the most common form of singularity in the parabolic problem arises at the corners of the integration domain where the initial line meets the vertical boundary. For purposes of

7.6 SINGULARITIES

discussion let the initial line be $t = 0$, and the vertical boundaries be $x = 0$ and $x = 1$ as shown in Fig. 7-6. The easiest initial and boundary values have the form

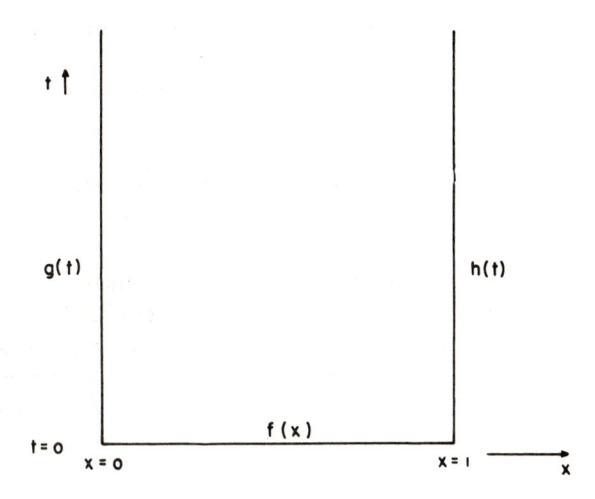

$$u(x, 0) = f(x), \quad u(0, t) = g(t), \quad u(1, t) = h(t). \tag{7.48}$$

FIG. 7-6. A boundary singularity in a parabolic problem.

We usually take these conditions to mean

$$\lim_{t \to 0} u(x, t) = f(x), \quad 0 < x < 1$$

$$\lim_{x \to 0} u(x, t) = g(t), \quad t > 0, x > 0, \tag{7.49}$$

$$\lim_{x \to 1} u(x, t) = h(t), \quad t > 0, x < 1.$$

The corners at $(0, 0)$ and $(1, 0)$ will often involve some form of discontinuity. One of the most "violent" of these is a difference in function value, i.e.,

$$\lim_{x \to 0} f(x) \neq \lim_{t \to 0} g(t)$$

or

$$\lim_{t \to 0} h(t) \neq \lim_{x \to 1} f(x). \tag{7.50}$$

Weaker types of discontinuities can also occur and will cause some trouble. As an example if $f(x) = 1$ and $g(t) = e^{-t}$ it is clear that

$f(0) = 1 = g(0)$. However, if the equation is $u_t = u_{xx}$ it is clear that $\partial g/\partial t \neq \partial^2 f/\partial x^2$ at $x = 0$, $t = 0$.

Finite difference methods have questionable value in the neighborhood of the points of discontinuity for a "region of infection" lies adjacent to such points. The effect of such discontinuities does not penetrate deeply into the field of integration *provided the methods we use are stable* in which case the errors introduced by the discontinuity decay. There are serious difficulties in obtaining accurate solutions near the points of discontinuity and usually we must abandon finite difference processes to obtain this accuracy.

When the discontinuity is of the form (7.50) the sensitivity of the procedure can be reduced by selecting the value at (0, 0) to be the mean value of $f(0)$ and $g(0)$. Crandall [13] has investigated the accuracy of approximation of the linear diffusion equation with crude networks and found it quite sensitive to the values assigned at points of discontinuity.

In some problems removal of a singularity is possible by the adoption of new independent variables. The transformation is to be so devised that the singular point is *expanded into a line or curve*. Thus a sudden change of the type (7.50) will become a smooth change along the line or curve in the new independent variables. The similarity transformations of Chapter 4 are often useful or suggestive of other transformations to achieve the stated goal.

As an example of the removal of a singularity consider the diffusion equation

$$u_t = u_{xx} \tag{7.51}$$

with a simple discontinuity at the origin of the form

$$\lim_{x \to 0} f(x) \neq \lim_{t \to 0} g(t) \tag{7.52}$$

and an integration domain as shown in Fig. 7-6. We attempt a transformation, modeled after the similarity form,

$$\theta = x^a t^b, \quad \eta = t^c \tag{7.53}$$

with the goal of expanding the singular point into a line and the further aim of having a relatively easy equation to work with along this line. The selection of a, b, and c is not unique. One useful set is $a = 1$, $b = -\frac{1}{2}$, and $c = \frac{1}{2}$ so that Eq. (7.53) becomes

$$\theta = xt^{-\frac{1}{2}}, \quad \eta = t^{\frac{1}{2}} \tag{7.54}$$

7.6 SINGULARITIES

and the differential equation (7.51) in the new coordinates is

$$u_{\theta\theta} + \tfrac{1}{2}\theta u_\theta = \tfrac{1}{2}\eta u_\eta. \tag{7.55}$$

From Eq. (7.53) it is clear that the boundary line $x = 0$, $t > 0$ in the original plane becomes the line $\theta = 0$ in the (θ, η) plane. The point of discontinuity $x = 0$, $t = 0$ becomes the line $\eta = 0$ and the line $t = 0$, $x > 0$ becomes the point at infinity on the line $\eta = 0$. (see Fig. 7-7).

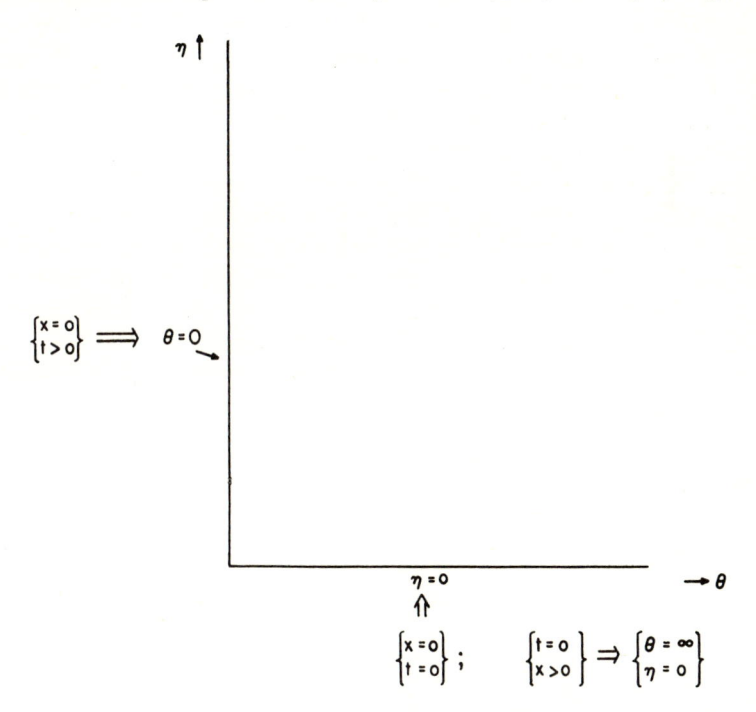

FIG. 7-7. The new integration domain when the singularity is eliminated by a "similarity" type transformation.

The effect of the mapping is to expand the origin into a line and the jump from $f(0)$ to $g(0)$ becomes a smooth change along this line.

Along $\eta = 0$ the initial values are given by the solution of the boundary value problem

$$\frac{d^2 u}{d\theta^2} + \frac{\theta}{2}\frac{du}{d\theta} = 0, \quad u(0) = g(0), \quad u(\infty) = f(0). \tag{7.56}$$

The solution of Eq. (7.56) for $u(\theta, \eta)$ with $\eta = 0$ is

$$u(\theta, 0) = f(0)\pi^{-\frac{1}{2}} \int_0^\theta e^{-p^2/4} \, dp + g(0). \tag{7.57}$$

Having determined the initial values (7.57) we then express the partial differential equation (7.55) in a finite difference approximation, say by an explicit technique, and march a few steps in the η-direction sufficient to take us away from the neighborhood of the discontinuity. Rather than proceed further in the semi-infinite (θ, η) domain it is usually desirable to return to the original plane. However, for equidistant grid points in the (x, t) plane, an interpolation is usually required (at least in one direction) to obtain the transformed values of u.

Alternative methods for small values of t are sometimes possible and perhaps more attractive. They generally involve some asymptotic approximation of the partial differential equation such as keeping constant, in a small region, coefficients which are variable. Sometimes the outright removal of nonlinearities in a small region will pay dividends. Solutions for small t, in both cases, are then found by standard methods. Exponential type solutions (of the diffusion equation) in t are generally unsatisfactory since they are usually much too large for small t. However, another solution of the equation $u_t = u_{xx}$ is of "source type"

$$u = t^{n/2} i^n \operatorname{erfc}[x/4t^{\frac{1}{2}}] \tag{7.58}$$

where

$$\operatorname{erfc} \eta = 1 - \operatorname{erf} \eta = \frac{2}{\sqrt{\pi}} \int_\eta^\infty e^{-p^2} \, dp$$

$$i^n \operatorname{erfc} \eta = \int_\eta^\infty i^{n-1} \operatorname{erfc} p \, dp. \tag{7.59}$$

A series of complementary error functions (erfc) is rapidly convergent for large $x/4t^{\frac{1}{2}}$, i.e., for small t. The details on such series may be found in Carslaw and Jaeger [14].

In the next section we consider a nonlinear parabolic equation from boundary layer theory and discuss how the singularity is handled in the solution of that problem.

7.7 A TREATMENT OF SINGULARITIES (EXAMPLE)

Consider a flat plate of length L held fixed in a steady uniform stream of viscous incompressible fluid so that the plate is edgewise to the flow. If the coordinates are chosen so that x is along the plate and y is normal

7.7 A TREATMENT OF SINGULARITIES

to the plate (see Figure 7-8) and the pressure is constant throughout the flow then the boundary layer theory furnishes us the equations

$$u\frac{\partial u}{\partial x} + v\frac{\partial u}{\partial y} = \nu\frac{\partial^2 u}{\partial y^2} \tag{7.60}$$

$$\frac{\partial u}{\partial x} + \frac{\partial v}{\partial y} = 0. \tag{7.61}$$

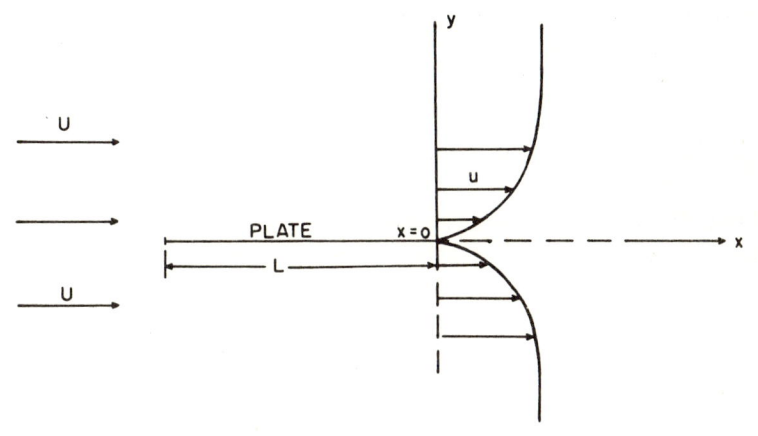

FIG. 7-8. Boundary layer wake of a flat plate.

The solution for the flow *along* the plate was given by Blasius (1908) [15] with the boundary conditions

$$\begin{aligned} u = 0, \quad & v = 0 \quad \text{at} \quad y = 0 \\ u = U, \quad & v = 0 \quad \text{at} \quad y = \infty. \end{aligned} \tag{7.62}$$

We here take the Blasius solution as known (at L) and consider the problem of continuing the solution into the wake region of the plate—the calcuation to be done by explicit finite difference methods. In this new problem $x = 0$ corresponds to the right hand end of the plate.

The wake problem is characterized by Eq. (7.60) and (7.61), subject to an "initial" condition at $x = 0$ furnished by the Blasius solution, and the boundary conditions

$$\begin{aligned} u = U, \quad & v = 0 \quad \text{at} \quad y = \infty \\ \frac{\partial u}{\partial y} = 0, \quad & v = 0 \quad \text{at} \quad y = 0. \end{aligned} \tag{7.63}$$

Several methods of handling this problem are possible. One of these is to introduce a stream function ψ, thereby satisfying (7.61) automatically and then to perform the von Mises transformation [Chapter 1, Equations (1.23)–(1.28)]. Taking x and ψ as independent variables we obtain the equation

$$uu_x = vu(uu_\psi)_\psi \tag{7.64}$$

for $u = u(x, \psi)$. Now $u = \psi_y$, $v = -\psi_x$ so that *along* $y = 0$ we must have $\psi = $ constant which we may select as zero if we ask for $\psi = 0$ at the origin. As y increases ψ increases without limit. The boundary conditions for Eq. (7.64) are therefore

$$\frac{\partial u}{\partial \psi} = 0 \quad \text{for} \quad \psi = 0; \quad u = U \quad \text{for} \quad \psi = \infty, \quad x > 0 \tag{7.65}$$

and the initial condition at $x = 0$ will be taken from the Blasius solution.

Lastly, we introduce the dimensionless variables

$$\eta = x/L, \quad \xi = \frac{\psi}{UL}\left(\frac{UL}{\nu}\right)^{\frac{1}{2}}, \quad \Gamma = 1 - (u/U)^2 \tag{7.66}$$

so that our problem becomes that of finding Γ so that

$$\frac{\partial \Gamma}{\partial \eta} = (1 - \Gamma)^{\frac{1}{2}} \frac{\partial^2 \Gamma}{\partial \xi^2} \tag{7.67}$$

subject to

$$\left. \begin{array}{l} \Gamma(0, \xi) \quad \text{given by Blasius Solution} \\ \dfrac{\partial \Gamma}{\partial \xi} = 0 \quad \text{at} \quad \xi = 0 \\ \Gamma = 0 \quad \text{at} \quad \xi = \infty \end{array} \right\} \tag{7.67a}$$

The integration domain is shown in Fig. 7-9.

Our interest herein lies in the calculation of Γ in the neighborhood of the plate's trailing edge. Not only is Eq. (7.67) nonlinear but there is a singularity at the origin which physically corresponds to that point where the flow reunites at the rear of the plate. With these problems in mind we use finite difference approximations analogous to Eqs. (7.14) and (7.15) for Γ_η and $\Gamma_{\xi\xi}$ respectively, to obtain the explicit formula

$$\Gamma_{i+1,j} = \Gamma_{i,j} + \frac{\Delta\eta}{(\Delta\xi)^2}(1 - \Gamma_{i,j})^{\frac{1}{2}}[\Gamma_{i,j+1} - 2\Gamma_{i,j} + \Gamma_{i,j-1}] \tag{7.68}$$

7.7 A TREATMENT OF SINGULARITIES

for marching to the right in Fig. 7-9. Our "positive" coefficient test and heuristic stability argument (Section 7.3) suggest that for stability the quantity

$$(1 - \Gamma_{i,j})^{\frac{1}{2}} \frac{\Delta\eta}{(\Delta\xi)^2} < \frac{1}{2}. \tag{7.69}$$

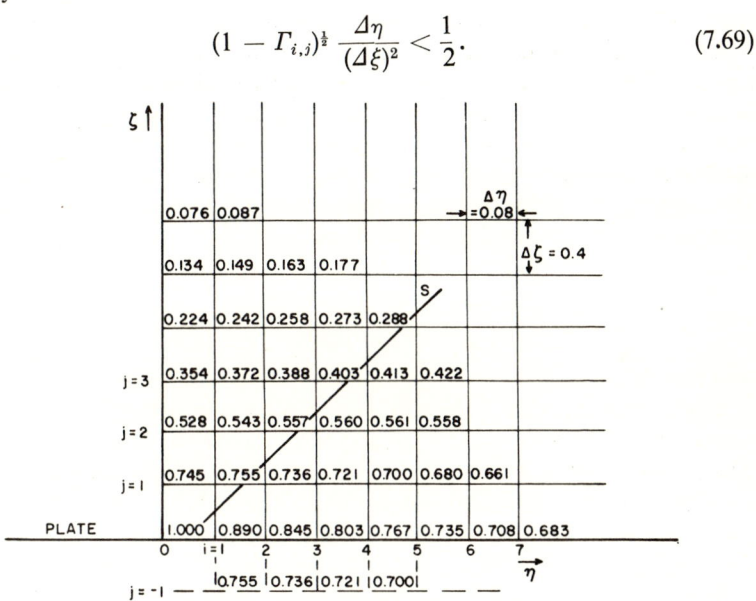

FIG. 7-9. Explicit numerical solution of the boundary layer wake problem illustrating the treatment of the singularity at the origin.

Now initially (and in fact everywhere, except at (0, 0), of interest in this calculation) $0 \leqslant \Gamma < 1$ so that a choice of $\Delta\eta/(\Delta\xi)^2 = \frac{1}{2}$ should be safe. In fact this is true so we choose $\Delta\eta = 0.08$ and $\Delta\xi = 0.4$. The initial values are known out to $\xi = \infty$ from the Blasius solution as tabulated by Luckert [16] for Γ. The first few of these are entered in the first column of Fig. 7-9 as the values of Γ at the grid points to the immediate left of the entry.

Using Eq. (7.68) there is no difficulty in obtaining that part of the solution *above* the solid line labeled S. The calculation of the solution *below* the line S must proceed by using values on the boundary $\xi = 0$. Normally, for conditions of the form $\partial \Gamma/\partial \xi = 0$ at $\xi = 0$, we introduce a false boundary and proceed to calculate $\psi_{i,0}$ via the same difference equation as before, namely Eq. (7.68). Thus for $\psi_{1,0}$ it follows that

$$\psi_{1,0} = \psi_{0,0} + \tfrac{1}{2}(1 - \psi_{0,0})^{\frac{1}{2}}[\psi_{0,1} - 2\psi_{0,0} + \psi_{0,-1}]$$

or since $\psi_{0,0} = 1$, we find $\psi_{1,0} = 1$. Continuing down the line we find that all the $\psi_{i,0}$ are unity. The corresponding solution represents a *continuation* of the boundary layer and *not* the wake. Clearly, the difference equation (7.68) is useless for this boundary calculation.

The nature of the difficulty is readily observable from the differential equation (7.67). When $\Gamma = 1$, $\partial \Gamma/\partial \eta$ must vanish *unless* $\partial^2 \Gamma/\partial \xi^2$ becomes infinite. This is precisely what happens at the origin. For when the flow first reunites there is a sharp discontinuity in $\partial \Gamma/\partial \xi$ so that $\partial^2 \Gamma/\partial \xi^2$ becomes arbitrarily large. We therefore abandon the finite difference approach and utilize an asymptotic solution, obtained by Goldstein [17] for small values of $\theta = \xi \eta^{-\frac{3}{2}}$. When $\xi \to 0$ the leading terms of Goldstein's series are

$$\Gamma(\eta, 0) = 1 - 0.596 \eta^{\frac{2}{3}} + O(\eta^{4/3}), \qquad (7.70)$$

clearly indicating that although Γ is continuous, its slope is very steep near $\eta = 0$. Equation (7.70) is used to calculate the value $\Gamma_{1,0} = 0.890$. The computation is continued by returning to the finite difference system with use of the false boundary procedure. When these results are compared with those of Goldstein and Luckert the error is seen to be less than 2% in a wake region of length $(\frac{3}{4})L$.

Alternate treatments of this problem are reported by Rosenhead and Simpson [18]. Later in this chapter we will consider the general problem of solving the boundary layer equations by finite difference methods. The state of development will be surveyed to 1964.

Thus far we have confined our attention to explicit methods for parabolic systems. Such procedures are often unsatisfactory because of the small step sizes required for ensuring stability and convergence of the finite difference scheme. Stability problems are usually eliminated by using implicit techniques.

7.8 IMPLICIT PROCEDURES

An implicit finite difference formula is one in which two or more unknown values in the $j + 1$ row (j represents the marching variable) are expressed in terms of *known* values in the preceding row or rows by a single application of the formula. If there are N unknown values in the $(j + 1)$ row we must apply the recurrence formula N times across the

7.8 IMPLICIT PROCEDURES

whole length of the row. The unknown values are then given implicitly by the set of N simultaneous equations. As previously mentioned there is considerable interest in these methods in which stability for all $r > 0$ is ensured at the price of greater complexity of computation. The difficulty with many nonlinear partial differential equations is that the resulting algebraic equations are nonlinear also. A few techniques avoid this complexity.

Implicit finite difference recurrence formulas for the quasilinear problem (see Eq. (7.45)) are easily illustrated for the problem

$$u_t = u_{xx} + f(x, t, u) \tag{7.71}$$

with initial and boundary conditions, which we take in the form,

$$u(x, 0) = 1, \quad 0 \leqslant x \leqslant 1$$
$$u_x(1, t) = 0 \tag{7.71a}$$
$$u(t, 0) - u_x(0, t) = 0.$$

These auxiliary conditions are used for illustrative purposes only.

O'Brien et al. [10] suggested approximating the derivative u_{xx} in the $(j+1)$st row instead of the jth row to achieve an implicit scheme. Earlier Crank and Nicolson [19] had suggested approximating u_{xx} by an average of approximations in the j and $(j+1)$st rows. More generally one can introduce a "relaxation factor λ" and use a weighted average to approximate u_{xx} as

$$u_{xx}|_{i,j} = \frac{\lambda}{(\Delta x)^2} [u_{i-1,j+1} - 2u_{i,j+1} + u_{i+1,j+1}]$$
$$+ \frac{(1-\lambda)}{(\Delta x)^2} [u_{i-1,j} - 2u_{i,j} + u_{i+1,j}] \tag{7.72}$$

with $0 \leqslant \lambda \leqslant 1$. If u_t is still approximated by the forward difference $(u_{i,j+1} - u_{i,j})/\Delta t$ and $\Delta t/(\Delta x)^2 = r$ the resulting finite difference equation is

$$-r\lambda u_{i-1,j+1} + (1 + 2r\lambda)u_{i,j+1} - r\lambda u_{i+1,j+1}$$
$$= r(1-\lambda)u_{i-1,j} + [1 - 2r(1-\lambda)]u_{i,j} + r(1-\lambda)u_{i+1,j}$$
$$+ \Delta t f(u_{i,j}, i\,\Delta x, j\,\Delta t). \tag{7.73}$$

A single application of Eq. (7.73) equates a linear combination of three unknown values in the $j + 1$ row to four known values in the j row.

340   7. NUMERICAL METHODS

Crandall [20] has investigated stability, oscillation and truncation error properties of Eq. (7.73). The limiting results, as the number of spatial subdivisions increases, are shown in Figure 7-10 where each point

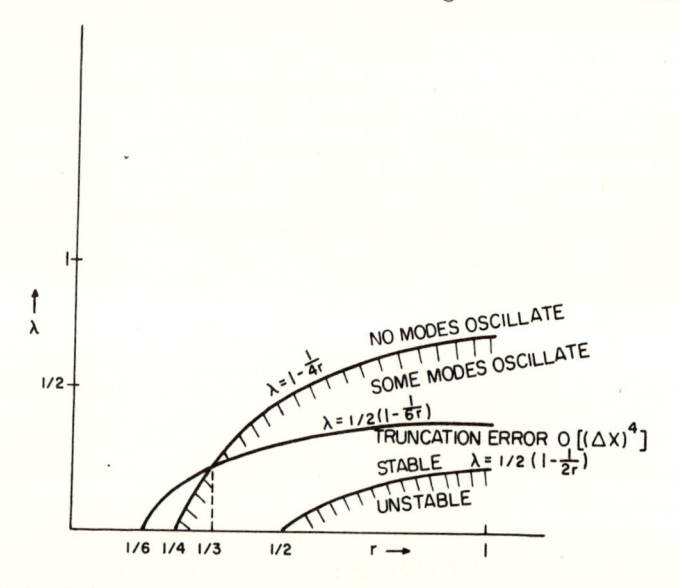

FIG. 7-10. Stability, oscillation, and truncation error properties of various finite difference approximations to the linear diffusion equation.

(r, λ) represents a different finite difference scheme. We note that $\lambda = 0$, represents the explicit schem, $\lambda = \frac{1}{2}$ the Crank-Nicolson method and $\lambda = 1$ the formula of O'Brien, Hyman, and Kaplan. With increasing λ the stability limit increases. For $\lambda \geqslant \frac{1}{2}$ the implicit formula (7.73) is stable for all r. An alternate proof for stability of the Crank-Nicolson procedure may be found in the Young-Frank article [3, p. 52].

The Crank-Nicolson method ($\lambda = \frac{1}{2}$) is stable for any positive value of r, hence for any positive values of Δx and Δt. Convergence of this method has been examined by Juncosa and Young [21] for the linear case

$$u_t = u_{xx}; \quad u(0, t) = 0, \quad u(1, t) = 0, \quad u(x, 0) = f(x). \quad (7.74)$$

The finite difference scheme, for this example, is convergent provided $f(x)$ is piecewise continuous and $\Delta t = O[\Delta x/|\ln \Delta x|]$ as $\Delta x \to 0$. This result implies convergence for all positive $r = \Delta t/(\Delta x)^2$. The truncation error, $O[(\Delta t)^2] + O[(\Delta x)^2]$ is the smallest for this implicit method.

7.8 IMPLICIT PROCEDURES

For the linear diffusion equation a variety of implicit methods are tabulated by Richtmyer [4, pp. 94 and 95]. Questions of stability and truncation error are also answered in that table. These results may be suggestive in the development of finite difference techniques for equations of parabolic form.

We should also remark that the solution of Eq. (7.73) is particularily easy in the quasilinear case.[†] This results because the equations have a "tridiagonal" form. We rewrite the equations as

$$a_k u_{i-1} + b_k u_i + c_k u_{i+1} = d_k \tag{7.75}$$

where the coefficients include those of Eq. (7.73) together with boundary condition modifications and d_k includes the right-hand side of Eq. (7.73). Notice that we have dropped the $j+1$ index on the unknowns. All known quantities have been lumped in d_k. When the computational molecule is centered such that u_{i-1} is known (either from boundary values or false boundary values) then $a_1 u_{i-1}$ is known. Similarity $c_n u_{n+1}$ is known. Thus our system has the tridiagonal appearance (for each mesh row)

$$\begin{aligned} b_1 u_1 + c_1 u_2 & = d_1 \\ a_2 u_1 + b_2 u_2 + c_2 u_3 & = d_2 \\ 0\, u_1 + a_3 u_2 + b_3 u_3 + c_3 u_4 &= d_3 \\ &\vdots \\ a_{n-1} u_{n-2} + b_{n-1} u_{n-1} + c_{n-1} u_n &= d_{n-1} \\ a_n u_{n-1} + b_n u_n &= d_n \end{aligned} \tag{7.76}$$

with zeros everywhere except on the main diagonal and on the two diagonals parallel to it on either side.

As we shall see this system can be solved explicitly for the unknowns, thereby eliminating any matrix operations. The method we describe was discovered independently by many people and has been called the Thomas [22] algorithm by Young. Its general description first appeared in widely distributed published form in an article by Bruce, Peaceman, Rachford, and Rice [23].

[†] The same approach applies to the more general quasilinear system Eq. (7.45) except the coefficients are now functions of x and t.

The Gaussian elimination process, i.e., successive subtraction of a suitable multiple of each equation from the following equation, transforms the system into a simpler one of *upper bidiagonal* form. The coefficients of this new system we designate by a_k', b_k', c_k', d_k' and in particular we note that

$$a_k' = 0 \quad k = 2, 3, ..., n$$
$$b_k' = 1 \quad k = 1, 2, ..., n. \tag{7.77}$$

The coefficients c_k', d_k' are calculated successively from the relations

$$c_1' = c_1/b_1, \qquad d_1' = d_1/b_1 \tag{7.78}$$

$$c_{k+1}' = \frac{c_{k+1}}{b_{k+1} - a_{k+1}c_k'}$$
$$d_{k+1}' = \frac{d_{k+1} - a_{k+1}d_k'}{b_{k+1} - a_{k+1}c_k'} \quad k = 1, 2, ..., n-1 \tag{7.79}$$

and of course $c_n = 0$.

Having performed the elimination we examine the new system and see that the nth equation is now

$$u_n = d_n' \tag{7.80}$$

and substituting this value into the $(n-1)$st equation

$$u_{n-1} + c_{n-1}'u_n = d_n'$$

we have

$$u_{n-1} = d_n' - c_{n-1}'u_n.$$

Thus starting with u_n (Eq. 7.80) we have successively the solution for u_k as

$$u_k = d_k' - c_k'u_{k+1}, \quad k = n-1, n-2, ..., 1. \tag{7.81}$$

When performing this calculation one must be sure that $b_1 \neq 0$ and that $b_{k+1} - a_{k+1}c_k' \neq 0$. If $b_1 = 0$ we can solve for u_2 and thus reduce the size of our system. If $b_{k+1} - a_{k+1}c_k' = 0$ we can again reduce the system size by solving for u_{k+2}. A serious source of round off error can arise if $|b_{k+1} - a_{k+1}c_k'|$ is small. Modifications of the Gauss elimination process, often called pivoting, can help to avoid this error.

An easy comparison of the implicit method and the explicit method can be made. On most machines the calculation of all grid point for fixed t should not take more than four times as long by the implicit method as by the explicit method. This gives us an estimate of the value of r to make the implicit method attractive i.e., to say the implicit method with $r > 2$ is preferable to the explicit method with $r = \frac{1}{2}$.

7.9 A SECOND-ORDER METHOD FOR $Lu = f(x, t, u)$

The success of the tridiagonal solution (7.80) and Eq. (7.81) of the implicit finite difference scheme (7.73) depends upon the use of the Picard type of approximation for f, i.e., $f(u_{i,j})^\dagger$. If this were changed to involve, also, the $j + 1$ level then the system of Eq. (7.76) would become nonlinear and the "linear" method of tridiagonal matrices would not apply. Of course this new system could be solved by iterative methods such as successive overrelaxation—a method we discuss later (Section 7.17).

Previously (Section 6.7) we discussed the application of the maximum operation in the development of iterative methods for $u_t - u_{xx} = f(u, x, t)$. More generally we consider the system

$$Lu = f(u) \qquad (7.82)$$

where L is a *linear parabolic* partial differential operator. One may linearize Eq. (7.82) by using Picard's method which introduces a sequence of functions $\{u^{(k)}\}$ which satisfy the same boundary conditions as required on u and the linear partial differential equation

$$Lu^{(k+1)} = f[u^{(k)}]. \qquad (7.83)$$

When the sequence $\{u^{(k)}\}$ converges the convergence is linear, i.e., as $k \to \infty$

$$u^{(k+1)} - u = O[u^{(k)} - u]. \qquad (7.84)$$

Alternatively, if f is differentiable, Bellman, Juncosa, and Kalaba [24] suggest a different linearization, used also by others for ordinary differential equations. If one replaces the right-hand side of Eq. (7.83) by the expansion about $u^{(k)}$

$$f[u^{(k)}] + [u^{(k+1)} - u^{(k)}]f'[u^{(k)}] \qquad (7.85)$$

---

† For simplicity we again neglect to write the explicit dependence of f on x and t.

then a new linear partial differential equation results, namely

$$Lu^{(k+1)} - f'[u^{(k)}]u^{(k+1)} = f[u^{(k)}] - u^{(k)}f'[u^{(k)}] \tag{7.86}$$

and this sequence, when convergent is usually quadratically convergent, i.e.,

$$u^{(k+1)} - u = O[(u^{(k)} - u)^2] \quad \text{as} \quad k \to \infty. \tag{7.87}$$

For parabolic equations the stable numerical methods have truncation errors of low order in the mesh size (eg. $(\Delta x)^2$). The numerical treatment of most multidimensional parabolic cases (and Elliptic problems) results in systems of algebraic equations which must be solved by various numerical means such as overrelaxation[†] or other methods. Consequently, there is considerable interaction between the truncation errors of the discrete analog, the rate of convergence (7.84) or (7.87), depending upon the choice of Eq. (7.83) or (7.86), and the rate of convergence of the numerical procedure. Analytic representation of total error bounds appear to be difficult to obtain in such cases. The paper of Bellman *et al.* [24] reports the results of some numerical experiments using Eq. (7.83) and (7.86) on a parabolic equation and an elliptic one. In both cases they remark the superiority of Eqs. (7.86) over (7.83) with respect to the error interaction discussed above.

Consider the parabolic equation

$$u_t - u_{xx} = (1 + u^2)(1 - 2u) \tag{7.88}$$

over the two triangles: $0 \leqslant t \leqslant 1 - x$, $0 \leqslant x \leqslant 1$, and $0 \leqslant t \leqslant 1.5 - x$, $0 \leqslant x \leqslant 1.5$. In each case boundary conditions were chosen to give the unique exact solution $u = \tan(x + t)$. To avoid stability questions, a discrete analog of Eq. (7.88) was constructed by the Crank-Nicolson form of Eq. (7.73) where $\lambda = \frac{1}{2}$. Setting $u_{i,j} = u(i\Delta x, j\Delta t)$ with $\Delta x = \Delta t = 0.01$ we have $r = 100$. The solution of the Picard and Newton forms of Eq. (7.88) will be obtained by iteration. We let k_j be the number of iterations required to obtain an acceptable approximation to $u_{i,j}$ at the grid points on the line $j\Delta t$. The criterion for acceptance of an approximation was that

$$\max_j \left| \frac{u_{i,j}^{(k)} - u_{i,j}^{(k-1)}}{u_{i,j}^{(k)}} \right| \leqslant 10^{-6} \tag{7.89}$$

[†] Suppose $u = F(u)$ is to be solved for a root $\bar{u}$. Selecting an initial value u_0 if we calculate successively $u_{n+1} = (1 - \omega)u_n + \omega F(u_n)$ then we say that ω is a relaxation factor. We shall discuss these ideas in more detail in our discussion of Elliptic problems.

so that $k_j = \min k$ such that Eq. (7.89) is true. This is a common criterion.

The function $f[u^{(k)}]$ on the right-hand side of Eqs. (7.83) and (7.86) was replaced by the average value $\frac{1}{2}\{f[u_{i,j+1}^{(k)}] + f[u_{i,j}^{(k_j)}]\}$ which gives superior accuracy. Finally the iteration schemes are for the Picard method,

$$Lu^{(k+1)} = \frac{u_{i,j+1}^{(k+1)} - u_{i,j}^{(k_j)}}{\Delta t} - \frac{1}{2(\Delta x)^2}[u_{i+1,j+1}^{(k+1)} - 2u_{i,j+1}^{(k+1)} + u_{i-1,j+1}^{(k+1)}$$
$$+ u_{i-1,j}^{(k_j)} - 2u_{i,j}^{(k_j)} + u_{i+1,j}^{(k_j)}]$$
$$= (1 + \tfrac{1}{2}\{[u_{i,j+1}^{(k)}]^2 + [u_{i,j}^{(k_j)}]^2\})(1 - u_{i,j+1}^{(k)} - u_{i,j}^{(k_j)})$$

and a corresponding expression for the Newton form.

The significant results were that while both methods required about the same amount of work in the triangle $0 \leqslant t \leqslant 1 - x$, $0 \leqslant x \leqslant 1$ where there are *no steep gradients* the situation is radically different in the larger triangle. In $0 \leqslant t \leqslant 1.5 - x$, $0 \leqslant x \leqslant 1.5$ the boundary condition on $t = 1.5 - x$ is $u = \tan 1.5$ which is quite large. In this triangle the Picard method required nine times as many iterations as the Newton method for the same accuracy. A similar result occurred for the nonlinear elliptic equation $u_{xx} + u_{yy} = e^u$.

Based on the above "experimental" evidence we conclude that if a nonlinear problem has no steep gradients there appears to be no advantage of the Newton method over the Picard approach. However, when steep gradients occur there is a decided advantage on the side of the Newton method.

7.10 PREDICTOR CORRECTOR METHODS

When the equation of parabolic type has the general character

$$u_{xx} = F\left(x, t, u, \frac{\partial u}{\partial x}, \frac{\partial u}{\partial t}\right) \quad (7.90)$$

the implicit finite difference formula analogous to Eq. (7.73) will give rise to *nonlinear* algebraic equations at each time step, whose solution may be difficult—multiplicity of roots is only one of the possible troubles. These nonlinearities can be avoided in several large classes of equations extracted from the general system (7.90) by using predictor-corrector modifications.

Predictor-Corrector methods have been successfully used by many in the numerical solution of *ordinary* differential equations. A discussion of some of these is to be found in Milne [8] and Fox [2, p. 28] variously labeled as the Adams-Bashforth method, methods of Milne and so forth. The general approach starts from known or previously computed results at previous pivotal points, up to and including the point x_n, by "predicting" results at x_{n+1} with formulae which needs no knowledge at x_{n+1}. These predicted results are relatively inaccurate. They are then improved by the use of more accurate "corrector" formulae which require information at x_{n+1}. This amounts to computing results at x_{n+1} from a nonlinear algebraic equation, to the solution of which the "predictor" gives a first approximation and the "corrector" is used repeatedly, if necessary, to obtain the final result.

For the equation

$$y' = f(x, y) \tag{7.91}$$

one such simple technique to obtain a finite difference approximation $y_n = y(n \Delta x)$ to y is to go through the two steps

$$y_{n+\frac{1}{2}} = y_n + \tfrac{1}{2} h f(x_n, y_n) \tag{7.92a}$$
$$y_{n+1} = y_n + h f(x_n + \tfrac{1}{2} h, y_{n+\frac{1}{2}}) \tag{7.92b}$$

where $x_n = n \Delta x$. The method (7.92) is of second order, i.e., $|y_n - y| = O(h^2)$ and further it is algebraically explicit.

Douglas and Jones [25] have considered Eq. (7.90) on $0 \leqslant x \leqslant 1$, $0 \leqslant t \leqslant T$ with $u(x, 0)$, $u(0, t)$, and $u(1, t)$ as specified boundary conditions. The standard Crank-Nicolson form for Eq. (7.90) is rewritten below so some notation can be introduced:

$$\tfrac{1}{2} \Delta_x^2 (u_{i,j+1} + u_{i,j}) = F[x_i, t_{j+\frac{1}{2}}, \tfrac{1}{2}(u_{i,j+1} + u_{i,j}), \tfrac{1}{2} \delta_x (u_{i,j+1} + u_{i,j}),$$
$$(u_{i,j+1} - u_{i,j})/k] \tag{7.93}$$

where the following notation is used,

$$\begin{aligned} \Delta_x^2 (u_{i,j}) &= h^{-2}(u_{i+1,j} - 2u_{i,j} + u_{i-1,j}) \\ \delta_x u_{i,j} &= (2h)^{-1}(u_{i+1,j} - u_{i-1,j}) \\ x_i &= ih, \quad t_j = jk, \quad u_{ij} = u(x_i, t_j) \\ h &= N^{-1}, \quad k = TM^{-1}, \quad N \text{ and } M \text{ positive integers.} \end{aligned} \tag{7.94}$$

7.10 PREDICTOR CORRECTOR METHODS

If either

$$F = f_1(x, t, u)\frac{\partial u}{\partial t} + f_2(x, t, u)\frac{\partial u}{\partial x} + f_3(x, t, u) \tag{7.95}$$

or

$$F = g_1\left(x, t, u, \frac{\partial u}{\partial x}\right)\frac{\partial u}{\partial t} + g_2\left(x, t, u, \frac{\partial u}{\partial x}\right) \tag{7.96}$$

a predictor-corrector modification of the Crank-Nicolson procedure is possible *so that the resulting algebraic problem is linear*. This is a significant advance since the class (7.95) includes the Burger's equation

$$u_{xx} = uu_x + u_t$$

of turbulence and suggests extension to higher order systems in fluid mechanics. The class (7.96) includes the equation

$$\frac{\partial}{\partial x}\left(K(u)\frac{\partial u}{\partial x}\right) = \alpha(u)\frac{\partial u}{\partial t}$$

of nonlinear diffusion.

If F is of the form (7.95) the following predictor-corrector analogue combined with the boundary date $u_{i,0}$, $u_{0,j}$ and $u_{N,j}$ leads to *linear algebraic equations*.

The *predictor*, in the notation of Eq. (7.94), is

$$\Delta_x^2 u_{i,j+\frac{1}{2}} = F[x_i, t_{j+\frac{1}{2}}, u_{i,j}, \delta_x u_{i,j}, (u_{i,j+\frac{1}{2}} - u_{i,j})/k/2] \tag{7.97a}$$

for $i = 1, 2, ..., N - 1$, followed by the *corrector*

$$\tfrac{1}{2}\Delta_x^2(u_{i,j+1} + u_{i,j}) = F[x_i, t_{j+\frac{1}{2}}, u_{i,j+\frac{1}{2}}, \tfrac{1}{2}\delta_x(u_{i,j+1} + u_{i,j}), (u_{i,j+1} - u_{i,j})/k]. \tag{7.97b}$$

Equation (7.97a) is a backward difference equation utilizing the intermediate time points $(j + \tfrac{1}{2})\Delta t$. Since Eq. (7.95) involves $\partial u/\partial t$ only linearly the calculation into the $(j + \tfrac{1}{2})$ time row is a linear algebraic problem. To move up to the $(j + 1)$st time row we use Eq. (7.97b) and by virtue of the linearity of Eq. (7.95) in $\partial u/\partial x$ this problem also is a linear algebraic problem.

Instead of Eq. (7.97a) one may use

$$\tfrac{1}{2}\Delta_x^2(u_{i,j+\frac{1}{2}} + u_{i,j}) = F[x_i, t_{j+\frac{1}{2}}, u_{i,j}, \delta_x u_{i,j}, (u_{i,j+\frac{1}{2}} - u_{i,j})/k/2].$$

Douglas and Jones prove that the predictor-corrector method defined by Eqs. (7.97a) and (7.97b) converges uniformly to the solution of Eq. (7.90), with F from Eq. (7.95), with an error of the order $O[h^2 + k^2]$. If Eq. (7.97b) is replaced by

$$\tfrac{1}{2}\Delta_x^2(u_{i,j+1} + u_{i,j}) = F[x_i, t_{j+\frac{1}{2}}, u_{i,j+\frac{1}{2}}, \delta_x u_{i,j+\frac{1}{2}}, (u_{i,j+1} - u_{i,j})/k] \quad (7.98)$$

then the predictor-corrector system (7.97a) and Eq. (7.98) does produce linear algebraic equations for the calculation of the finite difference approximation. Again convergence of the finite difference approximation $u_{i,j}$ to obtain the solution of Eq. (7.90), with F from Eq. (7.96) and the error is $O[h^2 + k^{3/2}]$.

Extension of these concepts to the higher dimensional problem

$$\nabla \cdot [a(x, y, u) \nabla u] = b(x, y, u)\frac{\partial u}{\partial t} \quad (7.99)$$

seems possible although basic theory is lacking at this writing.

Finally mention of several other papers pertinent to this idea will complete this section. Douglas [26] applies predictor-corrector methods to the mildly nonlinear parabolic equation

$$\frac{\partial u}{\partial t} = Lu + f(x, t, u) \quad (7.100)$$

where L is a linear elliptic operator. He also gives an excellent survey of numerical methods for parabolic differential equations in *Advances in Computers*, Vol. II [27].

7.11 TRAVELING WAVE SOLUTIONS

On occasion we can utilize the running or traveling wave solution to check special cases of our finite difference calculations. Solutions obtainable by this technique may not be physically reasonable but they do act as a check on the numerical accuracy of the approximate solution. Solutions of this type were discussed in Section 6.11 (see also Eq. (4.19)) as they applied to equations of stability theory in fluid mechanics.

Traveling wave solutions are attempted by looking for solutions of the equation in which u depends on x and t only through the combination $x - vt$ where v is a constant. For example let n be a positive integer and consider the case

$$u_t = (u^n)_{xx} \quad (7.101)$$

Let us assume a traveling wave solution of the form

$$u = F[x - vt] \tag{7.102}$$

with v constant. Upon substituting Eq. (7.102) into Eq. (7.101) we find that F must satisfy the nonlinear ordinary differential equation

$$(F^n)'' + vF' = 0 \tag{7.103}$$

where the prime refers to differentiation with respect to the grouping $(x - vt)$. One integration yields

$$(F^n)' + vF = A \tag{7.104}$$

where A is a constant of integration. The final integration is easily accomplished yielding the implicit solution

$$\sum_{j=0}^{n-2} \frac{(A/v)^j u^{n-1-j}}{n-1-j} + \left(\frac{A}{v}\right)^{n-1} \ln(u - A/v) = \frac{v}{n}(vt - x + B) \tag{7.105}$$

If one wishes to obtain the inverse function, i.e., $u = \psi[v(vt - x + B)]$ a graph may be plotted or a table constructed by some numerical procedure.

On occasion more complicated equations are transformable to the same type as Eq. (7.101). We encountered such an equation in (7.29) which under the transformation $R = \exp[-\gamma\theta]$ became

$$\frac{\partial R}{\partial t} = \frac{\partial}{\partial x}\left[R^{(\alpha/\gamma)-1}\frac{\partial R}{\partial x}\right]. \tag{7.33}$$

If $(\alpha/\gamma) - 1$ is an integer, or approximately an integer, then the above analysis can be used to generate a travelling wave solution.

7.12 FINITE DIFFERENCES APPLIED TO THE BOUNDARY LAYER EQUATIONS

A number of studies have been directed at obtaining solutions for the laminar boundary layer equations by finite difference methods. We shall give a partial list of some of the recent studies.

The solution of the boundary layer equations in the physical plane (x, y coordinates) is described in Flugge-Lotz and Eichelbrenner [28] first written in 1948. Soon thereafter, Flugge-Lotz [29] noted that the application of finite differences is more straight forward when the equations are first transformed to a new system having x *and* the x component of velocity, u, as the independent variables. This new system is obtained by means of the Crocco transformation [see 30] discussed below.

The steady two dimensional flow of a compressible fluid with variable density and viscosity in a laminar boundary layer is described by the equations of conservation of mass, momentum and energy which for a perfect gas with constant Prandtl number and specific heat take the form

$$(\rho u)_x + (\rho v)_y = 0 \qquad (7.106)$$

$$\rho[u u_x + v u_y] = -p' + (\mu u_y)_y \qquad (7.107)$$

$$\rho[u i_x + v i_y] = u p' + \mu u_y{}^2 + \frac{1}{Pr}(\mu i_y)_y . \qquad (7.108)$$

These are supplemented by the equation of state for a perfect gas

$$p = \rho R T = \frac{\gamma - 1}{\gamma} \rho i \qquad (7.109)$$

and the Sutherland viscosity law [31] from the kinetic theory of gases

$$\frac{\mu}{\mu_0} = \left(\frac{i}{i_0}\right)^{\frac{3}{2}} \frac{i_0 + c_p S}{i + c_p S} . \qquad (7.110)$$

The variables x, y are space coordinates parallel and perpendicular to the surface, u and v are the corresponding velocity components while ρ, μ and i are density, viscosity and enthalpy respectively. $p = p(x)$ is a known "outer edge" pressure and the prime refers to x differentiation. The quantity S is a known constant in the Sutherland viscosity law, Eq. (7.110).

Crocco introduced x and u as independent variables such that

$$y = y(\xi, u), \qquad \xi = x \qquad (7.111)$$

From Eq. (7.111) we have upon differentiating with respect to x and then with respect to u

$$\frac{\partial u}{\partial x} = -\frac{\partial y}{\partial \xi} \Big/ \frac{\partial y}{\partial u}, \qquad \frac{\partial u}{\partial y} = 1 \Big/ \frac{\partial y}{\partial u} . \qquad (7.112)$$

7.12 FINITE DIFFERENCES

By the chain rule we also have

$$\frac{\partial}{\partial x} = \frac{\partial}{\partial \xi} + \frac{\partial u}{\partial x}\frac{\partial}{\partial u} = \frac{\partial}{\partial \xi} - \frac{y_\xi}{y_u}\frac{\partial}{\partial u}$$

$$\frac{\partial}{\partial y} = \frac{1}{y_u}\frac{\partial}{\partial u}.$$

(7.113)

The corresponding relations for the inverse transformation are found in the same way to be

$$\frac{\partial}{\partial \xi} = \frac{\partial}{\partial x} - \frac{u_x}{u_y}\frac{\partial}{\partial y}$$

$$\frac{\partial}{\partial u} = \frac{1}{u_y}\frac{\partial}{\partial y}.$$

(7.114)

The transformation ceases to be one to one when either of the Jacobians

$$J = \frac{\partial(x, y)}{\partial(\xi, u)} = y_u = \frac{\mu}{\tau}$$

$$\frac{1}{J} = \frac{\partial(\xi, u)}{\partial(x, y)} = u_y = \frac{\tau}{\mu}$$

(7.115)

are zero. Here τ is the shear stress so that these vanish when τ is infinite or zero. The first singularity, i.e., infinite shear stress occurs at the apex of flat plates or wedges (included angle less than 180°). The shear stress is zero in several cases. The most obvious of these occurs at the outer edge of the boundary layer, where the flow achieves main stream characteristics, and on the wall at a point of separation of the boundary layer. It also vanishes at a stagnation point and at the point of maximum velocity in cases where overshoot[†] is present. In numerical solutions some care must be exercised near these singularities (see Sections 7.6 and 7.7).

When the transformation (7.113) is applied to Eqs. (7.106) and (7.107) we find

$$y_u(\rho u)_\xi - y_\xi(\rho u)_u + (\rho v)_u = 0$$

(7.116)

and

$$-\rho u y_\xi + \rho v = \tau_u - p' y_u.$$

(7.117)

[†] When velocities in the boundary layer exceed the outer edge value we say overshoot occurs. In this case the Crocco transformation leads to double valued functions of u. We therefore abandon the Crocco equations in favor of other procedures if "overshoot" occurs.

The quantity ρv can be eliminated by differentiating Eq. (7.117) with respect to u and substituting. Thus

$$u(\rho y_u)_\xi + \tau_{uu} - p'y_{uu} = 0 \qquad (7.118)$$

suggesting that the shear stress τ might be a better variable. Since $y_u = \mu/\tau$, Eq. (7.118) becomes

$$u\left(\frac{\rho\mu}{\tau}\right)_\xi + \tau_{uu} - p'\left(\frac{\mu}{\tau}\right)_u = 0$$

or

$$\tau^2 \tau_{uu} + \mu p' \tau_u + [u(\rho\mu)_x - \mu_u p']\tau = u\rho\mu\tau_x . \qquad (7.119)$$

In a similar manner we can write the energy equation (7.108) as

$$\frac{1-\Pr}{\Pr} i_u \tau \tau_u + \frac{i_{uu} + \Pr}{\Pr} \tau^2 + \mu p'(u + i_u)$$
$$= u\rho\mu i_x \qquad (7.120)$$

where we recall that the Prandtl number $\Pr = c_p \mu/k$ is assumed to be constant. When Eq. (7.109) and (7.110) are solved for density ρ and viscosity as functions of enthalpy i and the results are substituted into Eqs. (7.119) and (7.120) we obtain a pair of coupled, nonlinear parabolic equations for $\tau(x, u)$ and $i(x, u)$.

This system of equations has been the subject of an intensive study by Flugge-Lotz and Baxter [32] and Baxter [33] for a variety of physical problems. They utilize an *explicit* finite difference technique with the approximations (7.14) for single x derivatives, Eq. (7.34) for single u derivatives and Eq. (7.15) for second derivatives. The quantity $(\rho\mu)_x$ appearing in Eq. (7.119) is replaced by

$$(\rho\mu)_x = \rho\mu \left[\frac{\gamma p'}{(\gamma-1)\rho i} - \frac{1}{2}\left(\frac{i - c_p S}{i + c_p S}\right) \frac{i_x}{i} \right]$$

obtainable from Eqs. (7.109) and (7.110).

The resulting algebraic equations can be solved for $i_{m+1,n} = i[(m+1)\Delta x, n\Delta u]^\dagger$ and $\tau_{m+1,n}$ as

$$i_{m+1,n} = f[\tau_{m,n+1}; \tau_{m,n}; \tau_{m,n-1}; i_{m,n+1}; i_{m,n}; i_{m,n-1}; n; p'; \Pr, \Delta x, \Delta u]$$
$$\tau_{m+1,n} = g[\tau_{m,n+1}; \tau_{m,n}; \tau_{m,n-1}; i_{m,n+1}; i_{m,n}; i_{m,n-1}; i_{m+1,n}; n; p'; \Delta x; \Delta u].$$

---

† Marching takes place in the x-direction from an "initial" condition at $x = 0$.

7.12 FINITE DIFFERENCES

The quantity $i_{m+1,n}$ must be obtained first, at each step, since it is required in the τ calculation. In practice the computation is straightforward except at the wall ($n = 0$) and the boundary layer edge.

At the wall τ_x and i_x are *indeterminate* and the finite difference approximations are abandoned in favor of series expansions in a manner similar to that employed in Section 7.7. The series used for τ^2 is

$$\tau^2 = \tau_w^2 + 2\mu_w p' u + (\mu_u)_w p' u^2 + \cdots$$

and a similar expression for i. The details may be found in the references.

A stability criterion is difficult to obtain for this situation but the authors assert that the relations

$$\frac{2\tau^2}{\Pr u \rho \mu} \frac{\Delta x}{(\Delta u)^2} < 1, \qquad (7.121)$$

based on computational evidence, appears to be a sufficient condition that the explicit scheme be stable. The first mesh points from the wall are critical and there a sufficient condition for stability was found to be

$$\frac{\Delta x}{x} < \frac{2\Pr(\Delta u/u_e)^3}{(C_f \sqrt{\operatorname{Re}_x/C})^2} \qquad (7.122)$$

where u_e is the exterior flow velocity and the quantity $C_f\sqrt{\operatorname{Re}_x/C}$ is the skin friction parameter.

Realistic estimates with Eq. (7.122) show that the explicit method is not completely satisfactory because of the *small* step sizes needed to ensure stability and convergence. It was natural to turn to the use of implicit methods and indeed since 1958 a variety of successful efforts have been reported. Kramer and Lieberstein [34] solve the same Crocco equations, (7.119) and (7.120), by an implicit method but this procedure is still not valid when "overshoot" occurs. Overshoot occurs for certain cases of heated walls with favorable pressure gradient and helium injection.

In an attempt to avoid the overshoot difficulty Flugge-Lotz and Yu [35] studied an explicit finite difference scheme for solving the boundary layer equations in the physical (x, y) plane. The method proved unsatisfactory at high Mach numbers and with a heated wall since the step-size requirements became so severe that it was well nigh impossible to obtain stable solutions. Wu [36] has also used an explicit finite difference technique for solving these equations in the physical plane.

Wu [36], Pallone [37] and Blottner [38] apply various features of the Howarth-Dorodnitsyn transformation, combined some times with other devices, to avoid some of the aforementioned problems. This transformation has the occasionally desireable feature of stretching the coordinate normal to the wall by setting

$$\xi = x, \qquad \eta = \int_0^y \rho \, dy. \tag{7.123}$$

The relations between the differential operators in the old and new coordinates are

$$\frac{\partial}{\partial x} = \frac{\partial}{\partial \xi} + \eta_x \frac{\partial}{\partial \eta}, \qquad \frac{\partial}{\partial y} = \rho \frac{\partial}{\partial \eta}. \tag{7.124}$$

When the transformations are applied to Eqs. (7.106)–(7.108) a new dependent variable

$$V = \eta_x u + \rho v, \tag{7.125}$$

together with u, is convenient to use. Thus Eqs. (7.106)–(7.108) become

$$u_\xi + V_\eta = 0 \tag{7.126}$$

$$u u_\xi + V u_\eta = -p'/\rho + \bar{\rho}\bar{\imath}\left(\frac{\mu}{\bar{\imath}}\right)u_{\eta\eta} + \bar{\rho}\bar{\imath}\frac{d}{d\bar{\imath}}(\mu/\bar{\imath})i_\eta u_\eta \tag{7.127}$$

$$u i_\xi + V i_\eta = u p'/\rho + \bar{\rho}\bar{\imath}(\mu/\bar{\imath})[u_\eta{}^2 + i_{\eta\eta}/\mathrm{Pr}] + \frac{\bar{\rho}\bar{\imath}}{\mathrm{Pr}}\left(\frac{\mu}{\bar{\imath}}\right)i_\eta{}^2 \tag{7.128}$$

and $\bar{\rho}$, $\bar{\imath}$ refer to conditions at the main stream edge of the boundary layer.

Wu [36] uses the H-D transformation (7.123) to improve the stability of the explicit scheme. Blottner [38] develops two implicit methods analogous to the O'Brien-Kaplan ($\lambda = 1$) and the Crank-Nicolson form ($\lambda = \frac{1}{2}$) of Eq. (7.73). For hypersonic flows along insulated walls the velocity profiles vary almost linearly except near the outer edge of the boundary layer where there is a large change in the velocity gradient. For accurate results a particularily small grid size is required near the outer edge. Rather than having a varying grid size the coordinate normal to the wall is stretched with the H-D transformation. Thus smoother profiles are expected so that a uniform grid size can be used.

Pallone [37] uses the H-D equations (7.126)–(7.128) with a Polhausen integral method. He subdivides the boundary layer into a number of strips parallel to the wall and integrates the H-D equations from the wall to the various strips using a polynomial approximation for the velocity and enthalpy profiles. The result is a set of first order ordinary differential equations which must be integrated. No error or other analysis of the method is presented. Convergence needs to be analyzed as the number of strips increases indefinitely.

Blottner [38] linearizes his system of difference equations so that the tridiagonal scheme (see Section 7.8) can be used. He asserts that his method requires less computation time than all other methods (except Pallone's). De Santo and Keller [39] study the transition from laminar to turbulent flow over a flat plate by finite difference techniques.

7.13 OTHER NONLINEAR PARABOLIC EXAMPLES

Numerous successful calculations involving various parabolic equations have been carried out. Since our theoretical knowledge is very fragmentary it seems, at least for the present, that only numerical experimentation can decide what is the best finite difference procedure for a given parabolic problem. Consequently we examine some examples and discuss them briefly in the hope that they will provide useful guides for related problems.

a. Flow of Heat Generated by a Chemical Reaction

The flow of heat generated by a chemical reaction in a medium where the reaction rate depends upon the local temperature has been investigated by Crank and Nicolson [19]. If u is the temperature and w the concentration of the reactant species the equations to be solved on $0 \leqslant x \leqslant 1$, $t > 0$ are

$$u_t = u_{xx} - qw_t$$
$$w_t = -cw \exp[-A/u] \tag{7.129}$$

with the nonlinear boundary conditions

$$u(x, 0) = f(x), \quad w(x, 0) = g(x)$$
$$u_x(0, t) = h(u), \quad u_x(1, t) = k(u). \tag{7.130}$$

Here q, c, and A are known constants.

A number of methods are discussed in the paper which, as we previously remarked, was the origin of one of the first implicit schemes.

Their recommended method for the first equation of (7.129) is the implicit technique using the approximation (7.72), with $\lambda = \frac{1}{2}$ for u_{xx}. Thus

$$\frac{u_{i,j+1} - u_{i,j}}{\Delta t} = \frac{u_{i+1,j} - 2u_{i,j} + u_{i-1,j}}{2(\Delta x)^2}$$

$$+ \frac{u_{i+1,j+1} - 2u_{i,j+1} + u_{i-1,j+1}}{2(\Delta x)^2}$$

$$- q \frac{w_{i,j+1} - w_{i,j}}{\Delta t}. \tag{7.131}$$

We pause here to examine the formal error of this approximation. From Eq. (7.72) we know the truncation error is $O[(\Delta x)^2]$. The difference quotient $(\Delta t)^{-1}[u(x, t + \Delta t) - u(x, t)]$ approximates $u_t(x, t)$ with an error term of $O[\Delta t]$. However, if $u(x, t + \Delta t)$ is expanded about $(x, t + \frac{1}{2}\Delta t)$ instead of (x, t) we see that the difference quotient approximates $u_t(x, t + \frac{1}{2}\Delta t)$ *with an error of* $O[(\Delta t)^2]$. More generally, in choosing a finite difference approximation for a given equation of parabolic form some effort is usually devoted to achieving a symmetry about $(x, t + \frac{1}{2}\Delta t)$. Terms of the form $f(u)$ are often replaced by $f\{\frac{1}{2}[u(x, t + \Delta t) + u(x, t)]\}$. If f is a complicated function the numerical calculation required to find $u(x, t + \Delta t)$ may be made *very difficult* as compared with that when $f[u(x, t)]$ is used.

If the reason for choosing a more complicated difference equation is that the truncation error changes from $O[\Delta t + (\Delta x)^2]$ to $O[(\Delta t)^2 + (\Delta x)^2]$ it is implicitly assumed that $\lim_{\Delta x \to 0} (\Delta x)^2/\Delta t = 0$. Thus no improvement can be expected unless $(\Delta x)^2 \ll \Delta t$. A second hypothesis has also been made, i.e., that the discretization error is of the order of magnitude of the truncation error. This does not necessarily follow and therefore one should not always assume the computational superiority of a symmetric formula over a simpler unsymmetric one.

In the light of these last few remarks we see that Eq. (7.131) has a truncation error of the order $O[(\Delta x)^2 + (\Delta t)^2]$ at $(i, j + \frac{1}{2})$.

An equally accurate approximation for the second equation of (7.129) seems desirable. To obtain this we integrate that equation from t to $t + \Delta t$ and obtain

$$\ln \frac{w(x, t + \Delta t)}{w(x, t)} = - c \int_t^{t+\Delta t} \exp[-A/u(\eta)]d\eta. \tag{7.132}$$

7.13 OTHER NONLINEAR PARABOLIC EXAMPLES

The right-hand side of Eq. (7.132) can be approximated in a number of ways. Crank and Nicolson approximated it by aiming for the aforementioned symmetry so that

$$\ln \frac{w_{i,j+1}}{w_{i,j}} = -(\Delta t)c \exp[-2A/(u_{i,j+1} + u_{i,j})]$$

i.e.,

where

$$w_{i,j+1} = w_{i,j} e^R \quad (7.133)$$

$$R = -(\Delta t)c \exp[-2A/(u_{i,j+1} + u_{i,j})].$$

This approximation has a truncation error at the level $(i, j + \tfrac{1}{2})$ of $O[(\Delta t)^2]$.

The boundary conditions involving u_x are so discretized that their truncation error is of the same order as the preceding errors. Thus one introduces false boundaries at $x = 0$ (say $i = -1$) and at $x = 1$ (say $N + 1$) and the condition $u_x(0, t) = h(u)$, for example, becomes

$$\frac{u_{1,j} - u_{-1,j}}{2\Delta x} = h\left[\frac{u_{0,j+1} + u_{0,j}}{2}\right], \quad (7.134)$$

with a truncation error at $(i, j + \tfrac{1}{2})$ of $O[(\Delta x)^2 + (\Delta t)^2]$.

The determination of the values $u_{i,j+1}$ when values of u and w on the previous t level (j) are known requires the solution of the fairly complicated nonlinear equations (7.131) and (7.133). We must ask the question whether this complicated procedure is really warranted by sufficient gain in accuracy? The Crank-Nicolson scheme is stable for all mesh ratios, so Eq. (7.131) is *probably* also stable. If we assume the discretization error is of the order of the truncation errors then it is $O[(\Delta x)^2 + (\Delta t)^2]$. If $\Delta t = O[(\Delta x)^2]$ then the discretization error is $O[\Delta t]^\dagger$. In such a case the simpler form (for $w_t = -cw \exp[-A/u]$)

$$w_{i,j+1} - w_{i,j} = -c\,\Delta t\, w_{i,j} \exp[-A/u_{i,j}] \quad (7.135)$$

leads to a much simplified system. For in this case we can calculate $w_{i,j+1}$ directly from Eq. (7.135) in terms of known quantities from the preceding time row. However, the nonlinearity still remains because of the functions $h(u)$ and $k(u)$ on the boundaries. The final solution must still involve iteration. Of course if h and k are linear functions we again have a finite difference system which may be solved with the tridiagonal algorithm of Thomas.

---

† In this case $O[(\Delta t)^2 + \Delta t]$. But this is $O[\Delta t]$ when we recall the information suppressing role of the O notation.

b. Thermal Problems

A thermal-ignition problem is characterized by Hicks, Kelso, and Davis [40]. In such problems a combustible (or explosive) material is placed in contact with a hot gas. The temperature of the material rises rapidly. The objective of the analysis is to determine the time required to reach a certain temperature, known as the ignition temperature. The essentials of the problem involve the equation

$$u_t = u_{xx} + \exp[-A/u]$$

with the initial condition $u(x, 0) = u_0$ and the boundary conditions

$$u_x(0, t) = -\alpha[\beta - u(0, t)]$$
$$\lim_{x \to \infty} u_x(x, t) = 0$$

where A, α, β are known constants. In actual practice the second boundary condition was replaced by $u_x(B, t) = 0$ for finite, large B. This problem was solved by the Crank-Nicolson procedure and the tridiagonal algorithm.

Heat flow in shells and rockets is governed by nonlinear parabolic equations because the temperature ranges are such that the casing thermal properties vary appreciably. We meet similar equations in diffusion when the diffusion coefficient depends upon concentration. The general equation is of the form

$$[k(u)u_x]_x = \alpha(u)u_t \tag{7.136}$$

where $k(u) > 0$. Depending upon the initial and boundary conditions, it may be advantageous to introduce a new variable v, defined by

$$v = \int_{u_0}^{u} k(\eta)\, d\eta \tag{7.137}$$

whereupon Eq. (7.136) becomes

$$v_{xx} = \frac{\alpha(u)}{k(u)} v_t \tag{7.138}$$

where u is expressible in terms of v by Eq. (7.137). The reason for doing this, especially in heat transfer problems, is that α/k often changes slowly with u. For example, if $u_0 = 0$, $\alpha = u^s$, $k = u^r$ then $v = cu^{r+1}$

7.13 OTHER NONLINEAR PARABOLIC EXAMPLES

and $v_{xx} = dv^{(s-r)/(r+1)}v_t$. Boundary conditions involving the flux $k(u)u_x$ may be such that Eq. (7.137) cannot be applied. Such a problem has been treated by Eddy [41] where the boundary conditions are

$$u(x, 0) = u_0, \quad k(u)u_x = H(u, t) \quad \text{at} \quad x = 0, \quad k(u)u_x = 0$$

at $x = 1$.

Eddy's work illustrates several methods of reducing the difficulties of the nonlinearities. In order not to have to solve a complicated nonlinear system of equations for $u_{i,j+1}$ in terms of values at the j level the function u in $k(u)$ and $\alpha(u)$ is replaced by the extrapolated mean

$$\mu u = \tfrac{3}{4}[u_{i,j} + u_{i+1,j}] - \tfrac{1}{4}[u_{i,j-1} + u_{i+1,j-1}]$$

with truncation error $O[(\Delta x)^2 + (\Delta t)^2]$ at $(i + \tfrac{1}{2}, j + \tfrac{1}{2})$. The derivative u_x is approximated by

$$\delta u = \frac{1}{2\,\Delta x}[u_{i+1,j+1} + u_{i+1,j} - u_{i,j+1} - u_{i,j}]$$

so that when a *backward difference* of $k(\mu u)\,\delta u$ is used to approximate $(k(u)u_x)_x$ a discrete form of order $O[(\Delta x)^2 + (\Delta t)^2]$ at $(i, j + \tfrac{1}{2})$ is obtained.

A rocket gas burning problem is described by Forsythe and Wasow [1, p. 141] with the equation

$$u_t = cu_{xx} - \tfrac{1}{2}u_x^2 + d(x, t) \quad -\infty < x < \infty, \quad t > 0$$

and $u(x, 0) = f(x)$, $u(x + 2\pi, t) = u(x, t)$ where $f(x)$ and $d(x, t)$ are periodic functions of period 2π in x. They discuss possibilities for *approximating u_x^2* by *implicit methods*. As has been previously pointed out in Chapter 2 this problem can be *linearized* to

$$v_t = cv_{xx} - (2c)^{-1}d(x, t)v \tag{7.139}$$

with the auxiliary conditions

$$v(x, 0) = \exp[-(2c)^{-1}f(x)]$$
$$v(x + 2\pi, t) = v(x, t)$$

when we set $u = -2c \ln v$. We should therefore use the well understood linear form (7.139) instead of the nonlinear form.

c. Moving Boundary Problems

In Section 4.9 we introduced the concept of moving boundary problems in which some coefficients of the differential equation change discontinuously. This occurs, for example, where the medium suffers a change of state, say from solid to liquid. Usually the position of the surface of separation is not known in advance and has to be determined in the course of the computation. A number of authors have utilized the similarity transformation, or its modification (see Section 4.9) to immobilize the boundary by means of a transformation of the equations. The resulting ordinary differential equation is then solved. Unfortunately this method only applies in doubly or semi-infinite regions.

For purposes of discussion we consider the example of one dimensional heat conduction in a melting ice medium

$$\frac{\partial}{\partial x}\left(k \frac{\partial T}{dx}\right) = \rho c \frac{\partial T}{\partial t} \tag{7.140}$$

where k, ρ, c and T are respectively thermal conductivity, density, specific heat and temperature. At the solid-liquid interface we have

$$k_1 \frac{\partial T_1}{\partial x} - k_2 \frac{\partial T_2}{\partial x} = L\rho \frac{dX}{\partial t}, \qquad T_1 = T_2 \tag{7.141}$$

where the subscripts 1, 2 relate to the different regions, $X = X(t)$ is the equation for the interface and L is the latent heat of fusion of water.

Crank [42] has solved this problem by an explicit method. The novelty of his approach is that of the use of *unequal-interval* finite difference formulae, at points adjacent to the interface, for the computation of $\partial^2 T/\partial x^2$ and $\partial T/\partial x$. In this case we remark on the advantage of the explicit method for computational convenience. In fact small time steps are highly desirable and we therefore do not wish to take advantage of the extra stability of the implicit method.

Butler *et al.* [43] have treated similar problems, concerned with solidifying steel, on an analog computer. They introduce two transformations analogous to Eq. (7.137), i.e.,

$$v = \int_0^T k \, d\eta, \qquad w = \int_0^T \rho c \, d\eta \tag{7.142}$$

so that Eq. (7.140) becomes

$$\frac{\partial w}{\partial t} = \frac{\partial^2 v}{\partial x^2}. \tag{7.143}$$

The known thermal properties, i.e., $k = k(T)$ and $\rho c = g(T)$ give a graphical relation between v and w as functions of T. The effect of latent heat is represented by a known change in w with *no* corresponding change in T. Calculation of the solution proceeds by taking finite differences in the x-direction only for Eq. (7.143) and integrating the resulting simultaneous first order equations in t. Albasiny [44] applied a similar method but his integrations were carried out on a digital machine using the Runge-Kutta process (see for example Fox [2], p. 16). Other references include Ehrlich [45], and Douglas and Gallie [46]. A somewhat different approach is taken by Ting [47] which is discussed in example e below.

d. Diffusion-Second-Order Reaction

Asymptotic solutions, by Pearson, for the problem of diffusion—second-order reaction systems were discussed in Section 5.14. However, a large range of the parameters is not covered by these solutions. From Section 5.14 the governing equations are the simultaneous quasi-linear system

$$\frac{\partial \alpha}{\partial \theta} = \frac{\partial^2 \alpha}{\partial \xi^2} - \Gamma \alpha \beta \tag{7.144}$$

$$\frac{\partial \beta}{\partial \theta} = \Delta \frac{\partial^2 \beta}{\partial \xi^2} - \alpha \beta \tag{7.145}$$

subject to the initial and boundary conditions

$\theta = 0$: $\alpha = 0$ for all $\xi > 0$; $\beta = 1$, $\xi \geqslant 0$

$\theta > 0$: $\alpha = 1$, $\dfrac{\partial \beta}{\partial \xi} = 0$, $\xi = 0$; $\alpha = 0$, $\beta \to 1$ as $\xi \to \infty$.

The system (7.144)–(7.145) is treated by implicit finite difference methods using the Crank-Nicolson procedure, based upon Eq. (7.72) for $\partial^2 \alpha / \partial \xi^2$ and $\partial^2 \beta / \partial \xi^2$. Further we avoid the nonlinear difficulties by *not* choosing a formula centered about $\alpha_{i,j+\frac{1}{2}} = \alpha(i \Delta \xi, (j + \frac{1}{2}) \Delta \theta)$. Thus $\alpha \beta$ is replaced by $\alpha_{i,j} \beta_{i,j}$. Pearson reports no difficulties with the computation, obtaining good agreement with his known asymptotic solutions for limiting cases.

e. Higher Order Parabolic Equation (Ting [47])

The deformation of a cantilever beam, with strain rate sensitivity, subjected to impact loading at its base, is governed by the parabolic type of equation

$$[(v_{xx}/\alpha)^{1/p}]_{xx} + v_t = 0. \tag{7.146}$$

7. NUMERICAL METHODS

Equation (7.146) reflects the inclusion of the inertia forces in the plastic region. Here v, x, and t are dimensionless velocity, distance and time while p and α are constants in the expression of strain rate dependence of the yield stress—given by the moment-curvature relation

$$v_{xx} = K_t = \alpha(s-1)^p$$

or

$$s = 1 + (v_{xx}/\alpha)^{1/p} = 1 + A(x,t) \tag{7.147}$$

where K is the dimensionless curvature, s is the dimensionless bending moment and q the shear force of a cross section. The assumed initial conditions are

$$v(x,0) = v_0 = \text{constant}, \quad R(0) = 0 \tag{7.148}$$

where the length of the plastic segment $R(t)$ varies with t and hence gives a *floating boundary*. The boundary conditions are

$$v(0,t) = v_0[1 - f(t)] \tag{7.149a}$$

$$v_x(0,t) = v_{xx}[R(t),t] = 0 \tag{7.149b}$$

where $f(t)$ is a continuous function of t with the condition that $f(t) = 0$, $t < 0$, $f(t) = 1$ for $t > t_0$.

Two more conditions for the floating boundary are provided by the equations of conservation of angular and linear momentum of the whole beam. These are

$$(\tfrac{1}{2} + k)v_0 f(t) - \int_0^t s(0,\tau)\,d\tau = \int_0^R (1-x)\left[k + \tfrac{1}{6}(1-x)(2+x)\right]v_{xx}(x,t)\,dx \tag{7.150a}$$

$$(1 + k)v_0 f(t) - \int_0^t q(0,\tau)\,d\tau = \left[\cos \int_0^t v_x(R,\tau)\,d\tau\right]\phi(t) \tag{7.150b}$$

with

$$\phi(t) = \int_0^R (1-x)[k + \tfrac{1}{2}(1-x)]v_{xx}(x,t)\,dx.$$

From the definition of q we also have

$$q(x,t) = -s_x(x,t) = -[(v_{xx}/\alpha)^{1/p}]_x.$$

7.13 OTHER NONLINEAR PARABOLIC EXAMPLES

The numerical technique used to solve the problem characterized by Eqs. (7.146), (7.148)–(7.150) is analogous to the backward difference method of O'Brien, Hyman, and Kaplan [10] but it uses an *integral representation* instead of the difference expression.

Upon integrating Eq. (7.146) four times with respect to x we find that

$$v(x, t) = v(0, t) + \int_0^x \int_0^\sigma \alpha \left\{ A(0, t) - \int_0^\lambda [q(0, t) + \int_0^\eta v_t(\xi, t)\, d\xi]\, d\eta \right\}^p d\lambda\, d\sigma. \tag{7.151}$$

In Eq. (7.151) four values are necessary. These are $A(0, t)$, $q(0, t)$, $v(0, t)$ and $v_x(0, t) = 0$ which have already been introduced.

After approximating v_t in Eq. (7.151) by the backward difference

$$v_t(x, t) = [v(x, t) - v(x, t - \Delta t)]/\Delta t + O(\Delta t) \tag{7.152}$$

the numerical procedure consists of two stages resembling a predictor-corrector technique. First assume an approximate value for $q(0, t)$ and $K_t(0, t)$ (note: $K_t(x, t) = v_{xx}(x, t) = \alpha[A(x, t)]^p$) and solve Eq. (7.147). Second, correct the approximate values $q(0, t)$ and $K_t(0, t)$ by using the conditions for the floating boundary equations (7.150a) and (7.150b).

Stage one proceeds by successive approximations. Upon assuming the values of $q(0, t)$ and $A(0, t)$ we replace $v_t(x, t)$ in Eq. (7.151) by an initial approximation $v_t^{(0)}(x, t)$ and compute $v^{(1)}(x, t)$ and thence $v_t^{(1)}(x, t)$ by Eqs. (7.151) and (7.152). The iteration is continued until the desired accuracy is obtained. The computation of the integrals in Eq. (7.151) is carried out numerically in the space (x) direction for each value of time t. Convergence of the successive approximations is assured if

$$\frac{p\alpha(\Delta x)^4}{5!\, \Delta t} [A(x_j, t)]^{p-1} < 1. \tag{7.153}$$

Now all values from the first stage of the calculation can be checked by substituting into the Eqs. (7.150a, b). If the assumed values $q(0, t)$, $K_t(0, t)$ are incorrect, errors ΔI_1 and ΔI_2 will result where ΔI_1 arises from Eq. (7.150a) and ΔI_2 from Eq. (7.150b). Corrections δq and δK_t can be computed by an elementary process and step one repeated.

No comparison of this method with a finite difference procedure is given by the author. Difficulties here occur in the interaction of the truncation error of Eq. (7.151) with that of the numerical integration scheme.

f. Problems in Higher Dimensions

Equations of the type

$$a(\bar{x}, t, u) \frac{\partial u}{\partial t} + b(\bar{x}, t, u)u = \nabla \cdot [c(\bar{x}, t, u) \nabla u] + e(\bar{x}, t, u), \quad (7.154)$$

where $\bar{x}$ represents a vector of space coordinates, are quasilinear since a, b, c, and e do not contain partial derivatives of u. The right-hand side of Eq. (7.154) involves an elliptic operator, so at each time step an elliptic problem must be solved using the methods of the next portion of this chapter.

For the higher dimension problems the explicit methods suffer from the same drawback as in two dimensions, i.e., the value of Δt for stability is severely restricted. The implicit methods are most easily handled by the matrix techniques of point successive overrelaxation and alternating direction (see Section 7.16).

Calculation of non-steady incompressible viscous flow and dynamics and heat transfer in von Kármán wakes behind rectangular cylinders have been reported by Fromm [48] and Fromm and Harlow [49]. Their results closely resemble experimental data of Thom.

g. Percolation Problems

Soil moisture content changes and rates of water entry during rain infiltration into a semi-infinite soil column are governed by the parabolic type of equation

$$\frac{\partial w}{\partial t} = \frac{\partial}{\partial x} [D(w) \frac{\partial w}{dx} - K(w)]$$

subject to the initial and boundary conditions

$$t = 0, \quad x > 0; \quad w = w_i = \text{constant}$$

$$t > 0, \quad x = 0; \quad \frac{\partial w}{\partial x} = -\frac{R(t) - K(w)}{D(w)}.$$

The variables here are as follows: x is soil depth, $t =$ time, $w =$ soil moisture content, $K(w) =$ the soil hydraulic conductivity, $D(w)$ is the soil water diffusivity and $R(t)$ is the rain intensity. In addition the desired solution only applies to $w \leqslant$ moisture content of saturated soil.

Philip [50, 51, 52] bases his study of this problem on the Boltzman

7.14 FINITE DIFFERENCE FORMULA FOR ELLIPTIC EQUATIONS

similarity transformation which turns out not to be applicable for large values of t. An asymptotic solution gave a useable result in the case of $t \to \infty$. [see 52]. Rubin and Steinhardt [53] use a finite difference procedure to obtain approximate solutions. Their procedure is a slight modification of the von Neumann-Eddy [41] procedure previously discussed in thermal problems of this section.

B. ELLIPTIC EQUATIONS

7.14 FINITE DIFFERENCE FORMULA FOR ELLIPTIC EQUATIONS IN TWO DIMENSIONS

An equation of the form

$$au_{\xi\xi} + 2bu_{\xi\eta} + cu_{\eta\eta} = d(\xi, \eta, u, u_\xi, u_\eta), \quad (7.155)$$

where a, b, c are functions of ξ and η, is said to be

or
$$\begin{array}{ll} \text{hyperbolic if } b^2 - ac > 0; \\ \text{parabolic} \quad \text{if } b^2 - ac = 0; \\ \text{elliptic} \quad \text{if } b^2 - ac < 0 \end{array} \Bigg\} \quad (7.156)$$

The parabolic case has been considered in the preceding pages. In this and the ensuing sections we are interested in finite difference methods for nonlinear elliptic systems.

Elliptic equations of the form (7.155) can be transformed into

$$\alpha u_{xx} + \gamma u_{yy} = f(x, y, u, u_x, u_y) \quad (7.157)$$

by the introduction of new variables x and y obtained from the first order partial differential equations

$$\frac{\partial x}{\partial \xi} - (a/c)^{\frac{1}{2}} \frac{\partial x}{\partial \eta} = 0, \quad \frac{\partial y}{\partial \xi} + (a/c)^{\frac{1}{2}} \frac{\partial y}{\partial \eta} = 0. \quad (7.158)$$

The development of the canonical form (7.157) is discussed, for example in Chapter 3 of Garabedian's text [54]. In actual fact many of our problems naturally fall into the form (7.157) without requiring the construction of the transformation. The simplifying transformation requires the integration of equations of the form $d\eta/d\xi = (c/a)^{\frac{1}{2}}$ which may be difficult to perform. In what ensues we may assume without loss of generality that $\alpha > 0$, $\gamma > 0$ in the bounded connected plane region R in which we are to construct an approximate solution to Eq. (7.157) subject to appropriate boundary conditions.

We proceed to subdivide the plane domain of integration into a rectangular grid with the points of integration as the mesh points. The interval sizes will be denoted by $h = \Delta x$ and $k = \Delta y$ and the approximate values at the mesh points are denoted $u_{i,j} = u(ih, jk)$. There are three well-known methods for obtaining finite difference approximations to Eq. (7.157) together with specified boundary conditions. These methods are based upon variational formulations, Taylor series expansions and integral equations. Of these the Taylor's series is most commonly used and we shall restrict our attention to that process. The other methods have been most highly explored for *linear problems* by MacNeal [55], Forsythe and Wasow [1, p. 182] and Varga [56, 57].

As in Section 7.2 we shall often use the differences obtained by neglecting terms of order h^2 and k^2 in

$$[u_x]_{i,j} = \frac{1}{2h}[u_{i+1,j} - u_{i-1,j}] + O(h^2)$$

$$[u_y]_{i,j} = \frac{1}{2k}[u_{i,j+1} - u_{i,j-1}] + O(k^2) \qquad (7.159)$$

$$[u_{xx}]_{i,j} = \frac{1}{h^2}[u_{i+1,j} - 2u_{i,j} + u_{i-1,j}] + O(h^2).$$

Equations (7.159) can be used to develop the finite difference approximation

$$(\alpha/h^2)[u_{i+1,j} - 2u_{i,j} + u_{i-1,j}] + (\gamma/k^2)[u_{i,j+1} - 2u_{i,j} + u_{i,j-1}]$$
$$= f[ih, jk, u_{ij}, (u_{i+1,j} - u_{i-1,j})/2h, (u_{i,j+1} - u_{i,j-1})/2k] \qquad (7.160)$$

for Eq. (7.157).

Occasionally interior mesh points adjacent to the boundary of R are not exactly the mesh distances h and k away from the boundary. This situation is most apt to occur in irregular domains where the boundary is such that the mesh points do not always fall on the boundary. In such cases there are two procedures which are often used. The more *accurate* of these is to require the governing difference equation to apply at every internal point. At a point such as P in Fig. 7-11 we must modify the regular computational molecule by taking into account both the

7.14 FINITE DIFFERENCE FORMULA FOR ELLIPTIC EQUATIONS

boundary conditions and the nonstandard spacing $\lambda_R k$ and $\lambda_S h$. This method was used by Mikeladze [58] although it was also used earlier in manual relaxation procedures.

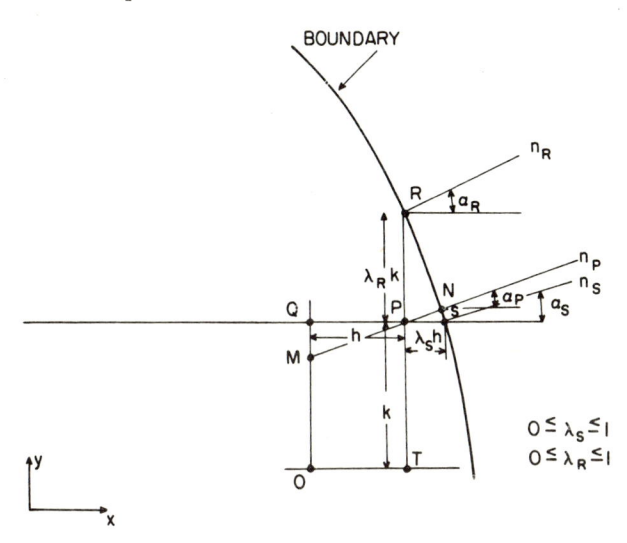

FIG. 7-11. Geometry of an irregular point near the boundary.

The second simpler, less accurate method, used by Gerschgorin [59] and Collatz [60] approximates the partial differential equation with the *standard* finite difference molecule only at those points where it can be fitted without changes. There remains an outer ring of mesh points, which may be interior or exterior, at which the u values are not required to satisfy the governing equation. Instead these values are determined by requiring them to satisfy finite difference approximations to the boundary conditions—the uneven spacings are included here.

To illustrate the construction of the irregular molecule we consider the situation in Fig. 7-11 and obtain approximations for u_x, u_y, u_{xx}, and u_{yy} simultaneously. Let us suppose that u is known on the boundary—so that u_R and u_S are given. If we expand $u(x, y)$ about P, which is assumed to be the origin for the time being, in a Taylor series we get

$$u(x, y) = u_P + x\left(\frac{\partial u}{\partial x}\right)_P + y\left(\frac{\partial u}{\partial y}\right)_P + \left(\frac{\partial^2 u}{\partial x^2}\right)_P \frac{x^2}{2}$$
$$+ \left(\frac{\partial^2 u}{\partial x\, \partial y}\right)_P xy + \left(\frac{\partial^2 u}{\partial y^2}\right)_P \frac{y^2}{2} + \cdots. \tag{7.161}$$

Thinking again of P as $(0, 0)$ the points R, S, T, and Q can be written as

$$(0, \lambda_R k), \quad (\lambda_S h, 0), \quad (0, -k), \quad \text{and} \quad (-h, 0) \qquad (7.162)$$

Substituting the four points of Eq. (7.162) into Eq. (7.161) we obtain the four equations [neglecting terms of $O(h^3)$ and $O(k^3)$]

$$\begin{bmatrix} 0 & \lambda_R k & 0 & \tfrac{1}{2}(\lambda_R k)^2 \\ \lambda_S h & 0 & \tfrac{1}{2}(\lambda_S h)^2 & 0 \\ 0 & -k & 0 & \tfrac{1}{2}k^2 \\ -h & 0 & \tfrac{1}{2}h^2 & 0 \end{bmatrix} \begin{bmatrix} (u_y)_P \\ (u_x)_P \\ (u_{yy})_P \\ (u_{xx})_P \end{bmatrix} = \begin{bmatrix} u_R - u_P \\ u_S - u_P \\ u_T - u_P \\ u_Q - u_P \end{bmatrix}$$

whose solution for the four unknowns is

$$(u_x)_P = h^{-1}\left[\frac{1}{\lambda_S(1 + \lambda_S)}u_S - \frac{\lambda_S}{(\lambda_S + 1)}u_Q - \frac{1 - \lambda_S}{\lambda_S}u_P\right] + O(h^2) \qquad (7.163)$$

$$(u_{xx})_P = 2h^{-2}\left[\frac{1}{\lambda_S(\lambda_S + 1)}u_S + \frac{1}{(\lambda_S + 1)}u_Q - \frac{1}{\lambda_S}u_P\right] + O(h) \qquad (7.164)$$

$$(u_y)_P = k^{-1}\left[\frac{1}{\lambda_R(1 + \lambda_R)}u_R - \frac{\lambda_R}{(\lambda_R + 1)}u_T - \frac{1 - \lambda_R}{\lambda_R}u_P\right] + O(k^2) \qquad (7.165)$$

$$(u_{yy})_P = 2k^{-2}\left[\frac{1}{\lambda_R(\lambda_R + 1)}u_R + \frac{1}{(\lambda_R + 1)}u_T - \frac{1}{\lambda_R}u_P\right] + O(k). \qquad (7.166)$$

Of course extensions can be obtained, by the same expansion, if points Q and T are also unevenly spaced.

The procedure for treating those boundary conditions involving the normal derivative $\partial u/\partial n$ hinge about the expression for the directional derivative

$$\left.\frac{\partial u}{\partial n}\right|_\alpha = \frac{\partial u}{\partial x}\cos\alpha + \frac{\partial u}{\partial y}\sin\alpha \qquad (7.167)$$

where the angle α (with the x-axis) specifies the direction. In this case we know $\partial u/\partial n$ along n_R, n_P, and n_S. Then by differentiation of Eq. (7.161) we have

$$u_x = (u_x)_P + x(u_{xx})_P + y(u_{xy})_P + \cdots \qquad (7.168\text{a})$$

$$u_y = (u_y)_P + x(u_{xy})_P + y(u_{yy})_P + \cdots. \qquad (7.168\text{b})$$

Using Eq. (7.168) we can express the normal derivatives at R, N, and

7.14 FINITE DIFFERENCE FORMULA FOR ELLIPTIC EQUATIONS

S in terms of the first and second derivatives of u at P. Thus for R we find

$$\left(\frac{\partial u}{\partial n}\right)_R = [(u_x)_P + \lambda_R k(u_{xy})_P] \cos \alpha_R + [(u_y)_P + \lambda_R k(u_{yy})_P] \sin \alpha_R \quad (7.169)$$

and related expressions at N and S. Adding to these three equations the two Taylors series for u_Q and u_T gives five equations for the derivatives u_x, u_y, u_{xy}, u_{xx}, and u_{yy} at P.

The second procedure of Gerschgorin will be discussed in terms of Fig. 7-11. The regular interior approximation is used at points Q and T. However, at P, a finite difference approximation to the boundary condition is employed. Suppose u is specified on the boundary so that u_S is known. By linear interpolation from S to Q we obtain

$$u_P = \frac{\lambda_S}{1 + \lambda_S} u_D + \frac{1}{1 + \lambda_S} u_S \quad (7.170)$$

as the expression for u at P. Of course we could equally well have interpolated in the other direction from R to T and obtain

$$u_P = \frac{\lambda_R}{1 + \lambda_R} u_T + \frac{1}{1 + \lambda_R} u_R.$$

Finally the average of these two interpolations can also be used. In some cases higher order interpolation may be justified to reduce the boundary error. Various interpolation schemes, are given, for example in Hildebrand [61].

In the case where $\partial u/\partial n$ is specified as the boundary condition we know $\partial u/\partial n]_N$ which may be approximated by $(u_P - u_M)/\overline{PM}$. The value u_M is obtained by (say) linear interpolation between Q and O, and the various lengths can be expressed in terms of h and α_P. Thus

$$u_P = u_O(1 - \tan \alpha_P) + u_0 \tan \alpha_P + h \left.\frac{\partial u}{\partial n}\right]_N \sec \alpha_P. \quad (7.171)$$

Other methods of dealing with normal boundary conditions are given by Shaw [62], Allen [63], Batschelet [64] and Viswanathan [65].

We now turn our attention to the algebraic equations of (7.160). In general these are nonlinear. Considerable effort has been expended to obtain rapidly convergent accurate iteration procedures for the *linear* system approximating the case when f is linear in u, u_x, and u_y or containing nonlinearities in u alone. Since these methods are suggestive and sometimes directly applicable we shall review them in the next sections as they apply to the linear elliptic equation

$$au_{xx} + cu_{yy} + du_x + eu_y + fu = g(x, y). \qquad (7.172)$$

7.15 LINEAR ELLIPTIC EQUATIONS

Consider the linear equation (7.172), in a rectangular region defined by $0 \leq x \leq \alpha$, $0 \leq y \leq \beta$, having Dirichlet type boundary conditions —i.e., $u =$ function of the coordinates (x, y) on each boundary. We suppose $a > 0$, $c > 0$ and $f \leq 0$ in the rectangular region and on its boundary. Using relations (7.159) with $h = k$ the finite difference equation for (7.172) becomes

$$\alpha_1 u_{i+1,j} + \alpha_2 u_{i-1,j} + \alpha_3 u_{i,j+1} + \alpha_4 u_{i,j-1} - \alpha_0 u_{i,j} = g_{i,j} \qquad (7.173)$$

where the α_i are functions of $x_i = ih$, $y_j = jh$ and are given by

$$\alpha_1 = a_{ij} + \tfrac{1}{2} h d_{ij}$$
$$\alpha_2 = a_{ij} - \tfrac{1}{2} h d_{ij}$$
$$\alpha_3 = c_{ij} + \tfrac{1}{2} h e_{ij}$$
$$\alpha_4 = c_{ij} - \tfrac{1}{2} h e_{ij}$$
$$\alpha_0 = 2(a_{ij} + c_{ij} - \tfrac{1}{2} h^2 f_{ij}) \qquad (7.174)$$

and the notation a_{ij} refers to the function evaluated at the point (ih, jh) where the five point computational molecule, shown in Fig. 7-12, is centered. h can be chosen so small that

$$0 < h < \min\left\{\frac{2a_{ij}}{|d_{ij}|}, \frac{2c_{ij}}{|e_{ij}|}\right\} \qquad (7.175)$$

where the minimum is taken over all points of the region and its boundary. Since $a > 0$, $c > 0$, $f \leq 0$ and they are bounded it is clear that a *positive minimum* exists and that for that h

$$\alpha_0 \geq \sum_{m=1}^{4} \alpha_m. \qquad (7.176)$$

This relation will be important in our discussion of iterative methods.

7.15 LINEAR ELLIPTIC EQUATIONS

Let us suppose the number of interior mesh points is N. The approximation (7.173) is used to represent $u_{i,j}$ at each mesh point resulting in a system of N linear equations whose matrix form is

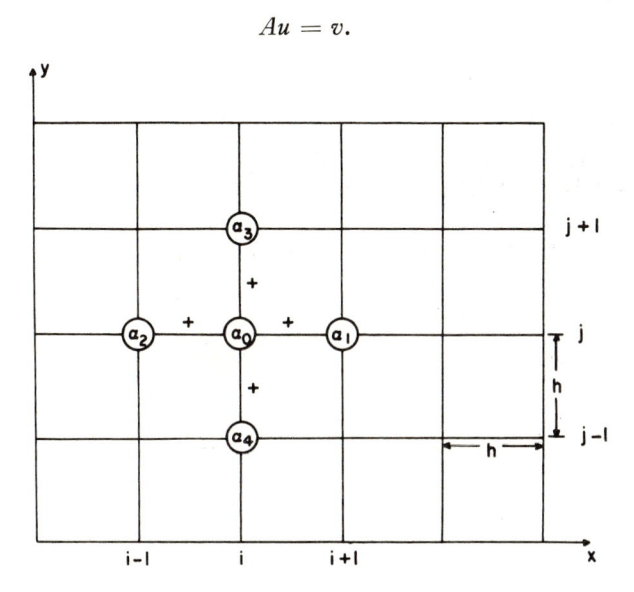

$$Au = v. \tag{7.177}$$

FIG. 7-12. A five point computational molecule for the general linear elliptic partial differential equations (7.172).

The vectors u and v consist of the N unknowns and the quantities $-g_{ij}$ together with boundary values respectively. The matrix A has real coefficients whose main diagonal elements are the α_0 of Eq. (7.174) and the off diagonal elements are the negatives of the α_i of (7.174) which do not correspond to boundary points.

In the ensuing work[†] the following properties of A will be important: $A = (a_{ij})$

(1) $a_{ii} > 0$, $\quad a_{ij} \leqslant 0 \quad i \neq j, \quad$ for all $\quad i, j$

(2) $a_{ii} \geqslant \sum_{\substack{j=1 \\ i \neq j}}^{N} |a_{ij}| \quad$ with strict inequality for some $i \quad$ (7.178)

(3) A is irreducible [see Section (7.16].

[†] In later work we shall often be concerned with this case. Thus our discussion of methods of solution for $Au = v$ will be restricted to matrices A with the properties (7.178) unless otherwise specified.

The first two conditions follow from Eq. (7.175) and the argument preceding it. The fact that inequality holds for some i follows by examining the molecule at an interior mesh point adjacent to a boundary.

As a simple illustration of the formulation we consider the equation

$$(x + 1)u_{xx} + (y + 1)^2 u_{yy} - u = 1$$

in the interior of the region $0 \leqslant x \leqslant 1$, $0 \leqslant y \leqslant 1$ with the boundary values $u(0, y) = y$, $u(1, y) = y^2$, $u(x, 0) = 0$, $u(x, 1) = 1$. With $h = \frac{1}{3}$ we then have four interior points as indicated in Fig. 7-13.

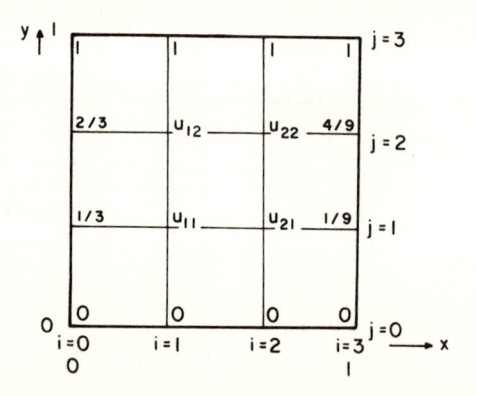

FIG. 7-13. The rectangular mesh for a linear elliptic problem.

The α_i of Eq. (7.174) in this special case are

$$\alpha_1 = (i + 3)/3, \qquad \alpha_2 = (i + 3)/3, \qquad \alpha_3 = [(j + 3)/3]^2$$
$$\alpha_4 = [(j + 3)/3]^2, \qquad \alpha_0 = 2\left\{(i + 3)/3 + [(j + 3)/3]^2 + \frac{1}{18}\right\}.$$

The four linear equations in the unknowns u_{11}, u_{12}, u_{21}, u_{22} are, in matrix form

$$\begin{bmatrix} \frac{57}{9} & -\frac{16}{9} & -4/3 & 0 \\ -\frac{25}{9} & \frac{25}{3} & 0 & -4/3 \\ -5/3 & 0 & 7 & -16/9 \\ 0 & -5/3 & -\frac{25}{9} & 9 \end{bmatrix} \begin{bmatrix} u_{11} \\ u_{12} \\ u_{21} \\ u_{22} \end{bmatrix} = \begin{bmatrix} -5/9 \\ 24/9 \\ -22/27 \\ 68/27 \end{bmatrix} \qquad (\alpha).$$

Several remarks about these equations are pertinent. First the main diagonal terms are dominant, i.e., in each case Eq. (7.176) holds *with strict inequality*. Second the matrix is not symmetric, i.e., $a_{ij} \neq a_{ji}$. Symmetric matrices have practical advantages from a computational viewpoint. If $a_{ij} = a_{ji}$ it is also true that the inverse matrix will be symmetric—thereby reducing the amount of computation.

The matrix A of Eq. (7.177) is symmetric[†] for Laplace's equation $u_{xx} + u_{yy} = 0$ and for Poisson's equation $u_{xx} + u_{yy} = g(x, y)$. Suppose the elliptic equation is *self-adjoint*, i.e., can be written as

$$(au_x)_x + (cu_y)_y + fu = g, \qquad (7.179)$$

or can be made self-adjoint[‡] by multiplying by a function $\phi(x, y)$ so that the form (7.179) is assumed. Then if the region has only regular mesh points the symmetry of A can be assured provided difference expressions of the form

$$[(au_x)_x]_{i,j} = a_{i+\frac{1}{2},j} h^{-2}[u_{i+1,j} - u_{i,j}] - a_{i-\frac{1}{2},j} h^{-2}[u_{i,j} - u_{i-1,j}] \qquad (7.180)$$

are used.

7.16 METHODS OF SOLUTION OF $Au = v$

The convergence of the finite difference scheme has been examined and proven under various situations. Little can be said about the accuracy for a given mesh size h although a number of analyses are recorded (see [1]–[5]). For this reason, and especially in nonlinear problems, one is inclined to use a small interval size to achieve a desired accuracy. This leads to a large number of *sparse* algebraic equations. Methods of solution of these systems fall into two categories—the *direct* and *iterative*

[†] We are assuming $h = k$.
[‡] An equation of the form (7.172) is said to be *essentially self-adjoint* when $\partial/\partial y[d - a_x/a] = \partial/\partial x[e - c_y/c]$ for in this case multiplication of both sides of Eq. (7.172) by $\phi(x, y)$, determined from $(\ln \phi)_x = d - a_x/a$, $(\ln \phi)_y = e - c_y/c$, makes the equation self-adjoint. A simple example is provided by the equation $u_{xx} + u_{yy} + x^2 u_x + y^2 u_y + u = 0$. The function ϕ is found to be $\phi(x, y) = \exp(x^3 + y^3/3)$ and the self-adjoint form is

$$\left[\exp\left(\frac{x^3 + y^3}{3}\right) u_x\right]_x + \left[\exp\left(\frac{x^3 + y^3}{3}\right) u_y\right]_y + \exp\left(\frac{x^3 + y^3}{3}\right) u = 0.$$

methods. In the linear case the direct methods are those which would give the exact answer in a finite number of steps, if there were no round off error. The algorithm for such a procedure is often complicated and nonrepetitive. Furthermore the *general idea* applies *only* to linear equations. Many such direct methods, for linear systems, are available in the literature, for example in Householder [66] and Bodewig [67], and research continues in this area. Some of the more useful of these methods are variants of the Gauss elimination procedure and the modified square root procedure. All these methods require excess computer storage and for this reason are usually not used.

On the other hand, iterative methods for sparse systems like Eq. (7.177) consist of repeated application of a simple algorithm taking advantage of the numerous zeros in Eq. (7.177) both in computer storage and operation. On the other hand they yield the exact answer *only* as a limit of a sequence, even without consideration of round off error. Iterative methods tend to be self correcting and their very structure allows modifications such as under and overrelaxation.

Our discussion in the following section will be restricted to iterative methods which fall into two classes: *point iterative* and *block iterative*.

In any iteration one begins with an arbitrary initial approximation to the solution and then successively modifies the approximation according to some rule. Convergence of the iteration is, of course, required. However, the method is not considered effective unless the convergence is quite rapid. Following a review of some of the current methods we shall compare their convergence characters on a particular problem.

For any iterative method to solve the nonsingular equation $Au = v$ (solution is $u = A^{-1}v$) we need a sequence $u^{(k)}$ defined with the hope that $u^{(k)} \to A^{-1}v$ as $k \to \infty$. In such a sequence the $u^{(k)}$ is a functions of A, v, $u^{(k-1)}$, $u^{(k-2)}$, ..., $u^{(k-r)}$ where r is said to be the degree of the iteration. To keep computer storage requirements low r is usually required to be small (often $r = 1$). With $r = 1$ we can write $u^{(k)} = F_k(A, v, u^{(k-1)})$. If F_k is independent of k the iteration is termed stationary and if F_k is *linear* in $u^{(k-1)}$ the iteration is linear.

The most general linear iteration is

$$u^{(k)} = G_k u^{(k-1)} + r_k \qquad (7.181)$$

where G_k is a matrix depending upon A and v and r_k is a column vector. For this to be useful the actual solution $u = A^{-1}v$ should be left invariant. Thus

$$A^{-1}v = G_k A^{-1}v + r_k$$

so that $r_k = (I - G_k)A^{-1}v$.[†] If we introduce $M_k = (I - G_k)A^{-1}$ our general linear iteration can be written

$$u^{(k)} = G_k u^{(k-1)} + M_k v$$

with $M_k A + G_k = I$. Convergence is studied by examining an error vector E_k defined in the oxvious way

$$\begin{aligned} E_k &= u^{(k)} - A^{-1}v = G_k u^{(k-1)} + M_k v - A^{-1}v \\ &= G_k u^{(k-1)} + M_k v - G_k A^{-1}v - M_k v \\ &= G_k E_{k-1}\,. \end{aligned} \qquad (7.182)$$

Equation (7.182) implies that E_k satisfies the basic iteration with $v = 0$. Thus $E_1 = G_1 E_0$, ..., so $E_k = G_k G_{k-1} \cdots G_1 E_0$. We shall not give convergence proofs but leave these for the literature references [1,2].

7.17 POINT ITERATIVE METHODS

In what follows the matrices A of $Au = v$ will be assumed to have the properties of Section 7.15 or when required more restrictive properties will be stated. Several of the best known iterative methods are built around a partition of A as

$$A = L + D + U \qquad (7.183)$$

where L contains the below diagonal elements, D contains the diagonal elements and U contains the above diagonal elements. For our purposes we lump L and U and write

$$A = D + C$$

i.e., D is diagonal and C has *zero* diagonal elements. Thus Eq. (7.177) becomes

$$(D + C)u = v. \qquad (7.184)$$

a. Jacobi Method (Simultaneous Displacements)

The diagonal elements $a_{ii} > 0$ for D so D is nonsingular, therefore $D^{-1} = [a_{ii}^{-1}]$ exists and upon premultiplying Eq. (7.184) by D^{-1} we find

$$u = -D^{-1}Cu + D^{-1}v. \qquad (7.185)$$

[†] This condition is often called a "consistency" condition.

Comparing this with the general linear iterative scheme (7.182) we see that the choice of

$$G_k = -D^{-1}C, \quad M_k = D^{-1} \tag{7.186}$$

gives the Jacobi method (simultaneous displacements), which is *stationary*

$$u^{(k)} = Gu^{(k-1)} + Mv. \tag{7.187}$$

We have discarded the subscripts in Eq. (7.187) to emphasize the stationary nature i.e., G_k is independent of k.

The concept of reducibility of a matrix is important here. It was introduced in Section 7.15. Setting $A = [a_{ij}]$ we say that the matrix A of the system $Au = v$ is *reducible* if some of the $N_1(N_1 < N)$ components of u are uniquely determined by some N_1 components of v. That is to say that *there exists* a permutation matrix P, having elements which are either 0 or 1, with exactly one element 1 in each row and column, such that

$$PAP^T = PAP^{-1} = \begin{bmatrix} A_1 & 0 \\ A_2 & A_3 \end{bmatrix}$$

i.e., a partitioned representation. In a boundary value problem this means that N_1 values of u are independent of part of the conditions on the boundary. Well posed linear elliptic problems almost always lead to irreducible matrices for their finite difference approximations.

Collatz [68] proves the following convergence theorem: *If A is diagonal dominant and irreducible* [properties (2) and (3) of (7.178)] *then the method of Jacobi converges.*

The actual execution of the Jacobi method, also called "iteration by total steps" is simple. From Eq. (7.187) set $G = [g_{ij}]$ and note that $G = -D^{-1}C$ so that $g_{ii} = 0$ for all i. A trial solution $u_{ij}^{(0)}$ is chosen at each mesh point. The basic single step of the iteration consists in replacing the current value $u_{ij}^{(k-1)}$ by the improved value obtained from the matrix operations

$$u^{(k)} = -D^{-1}Cu^{(k-1)} + D^{-1}v. \tag{7.188}$$

For Eq. (7.173) the value of the new solution $u_{ij}^{(k)}$ is obtained by an augmented weighted average of the old solution at the four neighbors of i, j:

$$u_{ij}^{(k)} = \frac{1}{\alpha_0}\left[\alpha_1 u_{i+1,j}^{(k-1)} + \alpha_3 u_{i-1,j}^{(k-1)} + \alpha_2 u_{i,j+1}^{(k-1)} + \alpha_4 u_{i,j-1}^{(k-1)} - g_{ij}\right]. \tag{7.189}$$

The *order* in which one solves for the components $u_{ij}^{(k)}$ is of no *consequence*. On a computer the procedure requires the saving of $u^{(k-1)}$ until the kth iteration is completed. If the values of the $(k-1)$st iteration are kept on cards for example the values of $u^{(k)}$ can be generated *independently*. It is not necessary to have them available until the kth iteration is completed.

b. Gauss-Seidel Method (Successive Displacements or Iteration by Single Steps)

A modification of the Jacobi method is based upon using the improved values *immediately* in computing the improvements for the next mesh point value. To systematize such a computation we must choose an *ordering* of the points—that is to say the order in which one solves for the components must be established beforehand. For an arbitrary but fixed ordering, which we designate by u_n,[†] ($n = 1, 2, ..., N =$ number of mesh points), the Gauss-Seidel is derivable from Eq. (7.188), by using the improved values when available, as

$$u_n^{(k)} = \sum_{j=1}^{n-1} g_{nj} u_j^{(k)} + \sum_{j=n+1}^{N} g_{nj} u_j^{(k-1)} + c_n \quad (7.190)$$

where $c = [c_n] = D^{-1} v$.

Suppose that our ordering has started with $u_{11}, u_{12}, ..., u_{2,1}, ...,$ i.e., from bottom to top of the mesh moving from left to right. Then the Gauss-Seidel form of Eq. (7.189) becomes

$$u_{ij}^{(k)} = \frac{1}{\alpha_0} \left[\alpha_1 u_{i+1,j}^{(k-1)} + \alpha_3 u_{i-1,j}^{(k)} + \alpha_2 u_{i,j+1}^{(k-1)} + \alpha_4 u_{i,j-1}^{(k)} - g_{ij} \right].$$

Thus we see that the latest estimates are used immediately upon becoming available.

In matrix form Eq. (7.190) becomes

$$u^{(k)} = Ru^{(k)} + Su^{(k-1)} + D^{-1} v \quad (7.191)$$

or

$$u^{(k)} = (I - R)^{-1} Su^{(k-1)} + (I - R)^{-1} D^{-1} v \quad (7.191a)$$

where

$$G = -D^{-1} C = R + S \quad (7.192)$$

with R a *lower triangular* matrix and S an upper triangular matrix.

[†] We order the u_{ij} in some fashion —eg $u_{11}, u_{12}, u_{13}, ..., u_{21}, u_{22}, ...$; and relabel this ordering as $u_1, u_2, ..., u_N$.

Convergence Theorem. *If A has diagonal dominance and is irreducible then the Gauss-Seidel method converges.*[†] Thus for the class of problems discussed herein the Jacobi method and the Gauss-Seidel method are either both convergent or both divergent and, as will be seen later, if they both converge the Gauss-Seidel method converges *faster*.

c. Successive Overrelaxation (The Young-Frankel Theory [70, 71, 72])

This method usually called SOR, first originated in the work of Frankel [70] and Young [71]. Young's work has been expounded and enlarged by Friedman [73], Keller [74], Arms, Gates, and Zondek [75] and others. The method grew out of the well known overrelaxation of "relaxation" methods used for paper-pencil computation in the pre-computer days. Often, the number of iterations required to reduce the error of an initial estimate of the solution of a system of equations by a predetermined factor can be substantially reduced by a process of extrapolation from previous iterants of the successive displacements method (Gauss-Seidel). The SOR method as given by Young accomplishes this extrapolation as follows. Let the u_{ij} be ordered as discussed in the Gauss-Seidel method, i.e., as u_n. Let $\bar{u}_n^{(k)}$ be the components of a new Gauss-Seidel iterant. The SOR technique is defined by

$$u_n^{(k)} = u_n^{(k-1)} + \omega[\bar{u}_n^{(k)} - u_n^{(k-1)}] = (1 - \omega)u_n^{(k-1)} + \omega\bar{u}_n^{(k)} \quad (7.193)$$

i.e., the *accepted value* for step k is extrapolated from the Gauss-Seidel value and the previous accepted value. If $\omega = 1$ the method reduces to Gauss-Seidel procedure. The parameter ω is called a *relaxation factor*, the choice of which determines the rapidity of convergence.

Substitution of Eq. (7.190) into Eq. (7.193) gives the computational form

$$u_n^{(k)} = (1 - \omega)u_n^{(k-1)} + \omega\left\{\sum_{j=1}^{n-1} b_{nj}u_j^{(k)} + \sum_{j=n+1}^{N} b_{nj}u_j^{(k-1)} + c_n\right\} \quad (7.194)$$

or in matrix notation

$$u^{(k)} = (1 - \omega)u^{(k-1)} + \omega[Ru^{(k)} + Su^{(k-1)} + D^{-1}v].$$

[†] Reich [69] proves the following theorem, which is *not* true for the Jacobi method of total steps; If A is symmetric, nonsingular and all $a_{ii} > 0$ then the method of Gauss-Seidel converges for all initial vectors $u^{(0)}$ if and only if A is positive definite.

An alternate matrix form, analogous to Eq. (7.191a), is easily found to be

$$u^{(k)} = (I - \omega R)^{-1}[(1 - \omega)I + \omega S]u^{(k-1)} + (I - \omega R)^{-1}\omega c. \quad (7.195)$$

Young's answer to the question of the worth of overrexlaxation is that *it does pay well* for matrices having property (A)—a property associated with many boundary value problems having elliptic difference equations. The matrix A is said to have property (A) if there is a suitable permutation matrix π such that $\pi A \pi^T$ has the form (not unique)

$$\pi A \pi^T = \begin{bmatrix} D_1 & E \\ F & D_2 \end{bmatrix} \quad (7.196)$$

where D_1 and D_2 are square diagonal matrices and E and F are rectangular matrices. If A has property (A) it can also be put into the form

$$\begin{bmatrix} D_1 & F_1 & 0 & \cdots & 0 & 0 \\ G_1 & D_2 & F_2 & \cdots & 0 & 0 \\ 0 & G_2 & D_3 & \cdots & 0 & 0 \\ \cdots & \cdots & \cdots & \cdots & 0 & 0 \\ 0 & \cdots & 0 \cdots & G_{p-2} & D_{p-1} & F_{p-1} \\ 0 & \cdots & 0 \cdots & 0 & G_{p-1} & D_p \end{bmatrix} \quad (7.197)$$

where the D_i are square matrices and the F_i and G_i are rectangular. This is accomplished by a permutation of the rows and corresponding columns. [1, 2].

Recent analysis and computation suggests that overrelaxation is also profitable for some matrices lacking property (A). Discussions of such examples are given by Forsythe and Wasow [1, p. 260] who discuss the work of Kahan [76] and Garabedian [77].

As we have seen each of the three methods Jacobi, Gauss-Seidel and SOR are linear, i.e., have the general form

$$u^{(k)} = Gu^{(k-1)} + r. \quad (7.181)$$

From Varga [57] it is well known that such a linear iterative method converges if and only if the spectral radius $\lambda(G)$, i.e.,

$$\text{Spectral radius of } G = \lambda(G) = \text{maximum of the absolute value of the eigenvalues of } G; \quad (7.198)$$

is less than one.

Convergence Theorem. If A is symmetric and $a_{ii} > 0$ for all $i = 1, 2, ..., N$, then $\lambda(L_\omega) < 1$ if and only if A is positive definite and $0 < \omega < 2$. [Here $L_\omega = (I - \omega R)^{-1}[(1 - \omega)I + \omega S]$, see Eq. (7.195)].

The fundamental question in Young's theory concerns the choice of ω so that convergence will be optimized. Some of the known results are as follows:

(a) If the eigenvalues λ_i of L_ω are all real with $|\lambda_i| < 1$ then the best over-relaxation factor ω_{opt} is[†]

$$\omega_{opt} = 1 + \left[\frac{\lambda_1}{1 + \sqrt{1 - \lambda_1^2}}\right]^2 = \frac{2}{1 + \sqrt{1 - \lambda_1^2}} \qquad (7.199)$$

where λ_1 is the spectral radius of the Gauss-Seidel method, i.e., $\lambda_1 = \lambda(G)$, G defined by Eq. (7.186).

(b) If A has property (A) the convergence of the Gauss-Seidel method is just *twice* that of the Jacobi method.

(c) For the case of the Dirichlet problem, for Laplace's equation, with A having property (A), and for a rectangle with sides $a = Rh$, $b = Sh$, R and S integers λ_1 is given by

$$\lambda_1 = \frac{1}{2}\left(\cos\frac{\pi}{R} + \cos\frac{\pi}{S}\right) \qquad (7.200)$$

(see Young [78]).

Young [78] also considered problems involving Laplace's equation for regions other than rectangles. If ω_{opt} is calculated as that for a circumscribing rectangle then the ω so obtained will be at least as large as the correct value. An *overestimation* of the optimum value of ω does not result in a serious reduction in the rate of convergence [78], but underestimation may. An alternate approach is to estimate the optimum ω by using a rectangle of approximately the same size and proportions.

For other equations the difficulty of obtaining λ_1 usually means some approximation is necessary. It may be possible to show that the proper value of ω will be close to that for Laplace's equation. If a large number of cases are to be considered one can experiment with different values of ω for different cases—thereby determining an estimate of the optimum ω. This is the approach used on a nonlinear problem to be described subsequently.

[†] ω_{opt} lies on the range $1 \leq \omega \leq 2$.

The great contribution of Young's method lies in its acceleration of the rate of convergence of an already convergent process. It does not create a convergent algorithm when the Gauss-Seidel method diverges.

A large class of related methods called *semi-iterative* have recently been intensively studied. Suppose the linear point iterative method is consistent [see material following Eq. (7.181)] and has the form

$$u^{(k)} = Gu^{(k-1)} + c. \tag{7.201}$$

Given a set of coefficients β_{kp}, $p = 0, 1, 2, ..., k$

$$\beta_{0,0}$$
$$\beta_{1,0} \quad \beta_{1,1}$$
$$\beta_{2,0}, \quad \beta_{2,1}, \quad \beta_{2,2}$$
$$...$$

such that

$$\sum_{p=0}^{k} \beta_{kp} = 1 \tag{7.202}$$

then

$$v^{(k)} = \sum_{p=0}^{k} \beta_{kp} u^{(p)} \tag{7.203}$$

defines a semi-iterative method with respect to the iterative method (7.201). If $\beta_{kp} = 0$ for $p < k$ and $\beta_{kk} = 1$ the semi-iterative method reduces to the original iterative method, i.e., $v^{(k)} = u^{(k)}$. Consequently, an optimum choice of the β_{kp} will give rise to a semi-iterative method which has a convergence behavior at least as good as the original method. In many cases the convergence rates are substantially improved by the use of a related semi-iterative method. Use of a semi-iterative method is similar to SOR with a variable relaxation factor ω although computer storage requirements differ.

d. General Semi-Iterative Method

Suppose G has *real eigenvalues* and that the iterative method (7.201) is convergent. Then by the convergence theorem following Eq. (7.198) the eigenvalues μ_i of G satisfy

$$-1 < a \leqslant \mu_i \leqslant b < 1$$

for some numbers a and b and for all i. By examining the error vector $f^{(k)} = v^{(k)} - u$ one can prove the following results (see Young and Frank [3], pp. 24–27).

Theorem a. Let G have real eigenvalues and suppose $b = \lambda(G) = -a$ (for example when the Jacobi method is used and the matrix A is symmetric having property (A)). Then the semi-iterative method is

$$v^{(0)} = u^{(0)}, \qquad v^{(1)} = u^{(1)}$$
$$v^{(k+2)} = \alpha_{k+2}\{Gv^{(k+1)} + c - v^{(k)}\} + v^{(k)} \qquad k \geqslant 0 \qquad (7.204)$$
$$\alpha_{k+2} = \frac{2P_{k+1}(1/\lambda)}{\lambda P_{k+2}(1/\lambda)}, \qquad \lambda = \lambda(G) = b.$$

The quantities P_k of Eq. (7.204) are the Chebyshev polynomials of degree k defined by

$$P_k(x) = \begin{cases} \cos(k \cos^{-1} x), & -1 \leqslant x \leqslant 1, \quad k \geqslant 0 \\ \cosh(k \cosh^{-1} x), & |x| > 1, \quad k \geqslant 0 \end{cases} \qquad (7.205)$$

satisfying the relation

$$P_{k+1}(x) = 2xP_k(x) - P_{k-1}(x), \qquad k \geqslant 1$$
$$P_0(x) = 1, \qquad P_1(x) = x. \qquad (7.206)$$

The recursion relation (7.204) allows one to calculate successive iterants of this semi-iterative process *without* computing the iterants $u^{(k)}$ of the underlying original method.

In general $b = \lambda(G) > -a$ [recall the definition of $\lambda(G)$, equation (7.198)]. In this case the following result holds:

Theorem b. Let G have real eigenvalues and $b = \lambda(G) > -a$, but a (smallest eigenvalue) can be estimated. Then the semi-iterative method is

$$v^{(0)} = u^{(0)},$$
$$v^{(1)} = \frac{2}{2 - (b + a)} [Gu^{(0)} + c] - \frac{b + a}{2 - (b + a)} u^{(0)}$$
$$v^{(k+2)} = v^{(k)} + \frac{4}{b - a} \frac{P_{k+1}(r)}{P_{k+2}(r)} [Gv^{(k+1)} + c - v^{(k)}] \qquad (7.207)$$
$$+ 2 \frac{b + a}{b - a} \frac{P_{k+1}(r)}{P_{k+2}(r)} [v^{(k)} - v^{(k+1)}],$$

$k \geqslant 0$, and $r = (2 - b + a)/(b - a)$.

e. Cyclic Chebyshev Method

Golub and Varga [79] examine the situation $Au = v$ when A has the properties, symmetry, positive definite, property (A), is consistently ordered with the form

$$A = \begin{bmatrix} D_1 & F \\ G & D_2 \end{bmatrix} \qquad (7.208)$$

with D_1, D_2 rectangular diagonal matrices. They obtain a semi-iterative scheme called the Cyclic Chebyshev method, derived in a similar manner to Eq. (7.207). They assert that there is considerable computational advantage using this method compared with any of the three iterative methods.

f. Other Procedures and Comments

(1) When the matrix A of $Au = v$ is symmetric and definite the gradient method (see Forsythe and Wasaw [1], p. 224) is an extension of the Jacobi method coinciding with that procedure when A has a scalar matrix (cI) as its diagonal.

(2) Richardson's method also provides that the relaxation factor ω may depend upon the iteration number, $\omega = \omega_k$. It is closely related to and in fact is the precursor of Eqs. (7.204) and (7.207).

(3) If the matrix A satisfies the conditions for the SOR to apply then the *optimum semi-iterative method relative to the SOR method is the SOR method itself* (see Varga [80]).

(4) A modification of SOR by Sheldon [81] is subject to acceleration of convergence by a semi-iterative method. Sheldon's method, called symmetric-successive over-relaxation (SSOR), is such that a complete iteration consists of two half iterations. The first of these is the SOR method. The second half iteration consists of the SOR applied to the equations in the reverse order, i.e., an interchange of the matrices R and S of Eq. (7.195). The net result is that the complete iteration has a symmetric associated matrix. Therefore the eigenvalues of the SSOR method are real and the semi-iterative methods may be applied using the above developments.

g. Comparison of the Point Iterative Methods

The convergence of an iterative method does not automatically qualify it as useful. To be useful convergence must be relatively fast.

Effectiveness of methods can be compared on two important bases—
the work per iteration and *the number of steps required for convergence*.
The comparisons illustrated have been gathered from the cited literature
plus the papers of Sheldon [82]. The number of steps required for
convergence, i.e., the number of iterations needed to reduce the error to
a specified factor ϵ of its original value, are valid for the solution of
Poisson's equation, with Dirichlet boundary conditions, on a square with
square mesh size h. The five point formula (7.173) is used.

Method	Number of steps	Work per iteration[a]
Jacobi	$2/h^2$	~ 1
Gauss-Seidel	$1/h^2$	~ 1
SOR (optimum)	$1/2h$	1
Sem-Iterative		
(a) General	$1/h$	~ 1.2
(b) Cyclic Chebyshev	$1/2h$	~ 1.2
(c) SSOR	$\sim 1/\sqrt{h}$	~ 2.5
Richardson	$1/0.6h$ to $1/h$	~ 1.2
(Depends on period k)		

[a] The work per iteration is normalized about the SOR method. This term is
modified to include the effect of extra computer storage requirements.

The number of steps required are only the reciprocals of the first terms
of a power series in h.

7.18 BLOCK ITERATIVE METHODS

A re-examination of the point iterative methods of Section 7.17
discloses that at each step the approximate solution is modified at a
single point of the domain. Furthermore the value of each component
in the kth iteration $u^{(k)}$ is determined by an *explicit* linear relation.
Such explicit methods have natural extensions to block iterative
methods in which groups of components of $u^{(k)}$ are modified simultaneously in such a way that we must solve a linear system for the whole
subset of modified components at once. Therefore individual components
are defined implicitly in terms of other components of the same group
so it is natural to call the method *implicit*.

The redefinition of an explicit method so that it becomes implicit
often leads to an appreciable increase in the convergence rate at the cost

of some complication in the computational algorithm. The blocks may be single rows (or columns) of the matrix, two rows etc. We discuss several examples.

a. Single Line Methods (Generalizations of Point Iterative Methods)

Consider a two dimensional rectangular domain R in which the Eq. (7.172) is discretized with a five-point[†] difference scheme like Eq. (7.173). In such a formula a point (i, j) is coupled only with *two* other points on a given line (row or column). If the domain R has m lines, with N/m unknowns on each line, then the matrix A of the system (7.173) can be written in block tridiagonal form as

$$\begin{bmatrix} D_1 & E_1 & 0 & 0 & \cdots & \cdots & 0 \\ C_1 & D_2 & E_2 & 0 & \cdots & \cdots & 0 \\ 0 & \cdots & \cdots & \cdots & \cdots & \cdots & 0 \\ 0 & \cdots & \cdots & C_{m-2} & D_{m-1} & E_{m-1} \\ 0 & \cdots & \cdots & 0 & C_{m-1} & D_m \end{bmatrix} \begin{bmatrix} U_1 \\ U_2 \\ \vdots \\ U_{m-1} \\ U_m \end{bmatrix} = \begin{bmatrix} K_1 \\ K_2 \\ \vdots \\ K_{m-1} \\ K_m \end{bmatrix} \quad (7.209)$$

where the square matrices D_i are tridiagonal, U_i are subvectors of the unknowns having the column dimension of the D_i.[‡] The subsystems obtained from Eq. (7.209) have the matrix form

$$C_{i-1} U_{i-1} + D_i U_i + E_i U_{i+1} = K_i \quad i = 1, 2, \cdots, m \quad (7.210)$$

which are tridiagonal. These may be solved directly by the Thomas Algorithm described in Section 7.8.

The solution of Eq. (7.209) is now considered by block iterative methods. The whole set of point iterative methods can be generalized to block methods. These are done for the Jacobi, Gauss-Seidel and SOR by Arms *et al.* [75], Friedman [73] and Keller [74].

The number of steps for convergence are listed below, again for the Dirichlet problem of Poisson's equation in a square.

[†] Other difference approximations are available. A nine point approximation is given in Forsythe and Warsaw [1, p. 194]. Garabedian [77] finds the optimum ω (for the nine point approximation of Laplace's equation) of SOR to be $2/[1 + 26/25\,\pi h]$.

[‡] An example of these procedures will be given for the Peaceman-Rachford method in two dimensions.

Method	Number of steps	Reference
Single line Jacobi	$1/h^2$	[73]
Single line Gauss-Seidel	$1/2h^2$	[73]
Single line SOR	$1/2\sqrt{2}h$	[3]
Two line SOR	$1/4h$	[81], [82]

Several remarks are pertinent.

(1) The optimum value of the relaxation factor ω_{opt} for the single line SOR method is the same as for the SOR method—see equation (7.199).

(2) Cuthill and Varga [83] and Varga [84] show that in many cases it is possible to perform both single line and two line SOR in approximately the same number of arithmetic operations per mesh point as the point SOR method requires. Therefore since the point SOR method requires $1/2h$ steps for convergence versus either $1/4h$ or $1/2\sqrt{2}h$ for the block methods it is clear that the block methods offer decreases in total computational effort.

b. Alternating Direction Implicit Methods (ADI)

These methods are somewhat similar to single line iterative methods but are complicated by alternating directions. Two variants are in vogue. These are similar and are known as the Peaceman-Rachford method [85] and the Douglas-Rachford [86] method. We also note that these methods are applicable to parabolic systems. The Peaceman-Rachford (PR) method is not directly generalizable to three dimensions while the DR method is.

Each iteration of the PR method consists of *two* half iterations—each of which is a single line (block) iteration. In the first half iteration all the values on all the rows are modified and in the second half all the values on all the columns are modified. To illustrate the procedure, and incidentally show what a line iteration is like, we consider an example. Let Laplace's equation $\nabla^2 u = 0$ be approximated by a five point difference equation in a rectangular domain with equal mesh sizes h. From Eq. (7.173) this formula when centered at (i,j) is

$$u_{i+1,j} + u_{i-1,j} + u_{i,j+1} + u_{i,j-1} - 4u_{i,j} = 0. \qquad (7.211)$$

As before we use an iteration. We select an arbitrary initial set $u_{i,j}^{(0)}$. In general having determined $u_{i,j}^{(n)}$ we then determine $u_{i,j}^{(n+\frac{1}{2})}$ by a single row (line) iteration and then $u_{i,j}^{(n+1)}$ by a single column iteration according to the equations

$$u_{i,j}^{(n+\frac{1}{2})} = u_{i,j}^{(n)} + \rho_n [u_{i+1,j}^{(n+\frac{1}{2})} + u_{i-1,j}^{(n+\frac{1}{2})} - 2u_{i,j}^{(n+\frac{1}{2})}]$$
$$+ \rho_n [u_{i,j+1}^{(n)} + u_{i,j-1}^{(n)} - 2u_{i,j}^{(n)}] \qquad (7.212a)$$

$$u_{i,j}^{(n+1)} = u_{i,j}^{(n+\frac{1}{2})} + \rho_n [u_{i+1,j}^{(n+\frac{1}{2})} + u_{i-1,j}^{(n+\frac{1}{2})} - 2u_{i,j}^{(n+\frac{1}{2})}]$$
$$+ \rho_n [u_{i,j+1}^{(n+1)} + u_{i,j-1}^{(n+1)} - 2u_{i,j}^{(n+1)}] \qquad (7.212b)$$

The quantities ρ_n are positive constants, which may depend upon n, called iteration parameters. They must be the same for both halves of any iterative step. After the calculation of $u^{(n+1)}$ the auxiliary vector $u^{(n+\frac{1}{2})}$ is discarded. To determine $u_{i,j}^{(n+\frac{1}{2})}$ from the first of these equations we must solve, for each value of j, a set of M simultaneous linear equations, where M = number of interior meshpoints on the row. The resulting matrix is tridiagonal thereby allowing the application of the Thomas Algorithm.[†] Similar statements apply to the single column implicit part (7.212b).

Equations (7.212) constitute the heart of the PR alternating direction implicit iterative method. Birkhoff, Varga, and Young [87] and Young and Wheeler [88] discuss the extension of this method to the self adjoint problem

$$[A(x, y)u_x]_x + [C(x, y)u_y]_y + fu = g \qquad (7.213)$$

and show that with the choice (7.180) for $(Au_x)_x$ the PR method again leads to tridiagonal matrices.

At least two methods have been given for choosing the iteration parameters ρ_n. Neither of these methods yield optimum values of the parameters, however application of them does make the PR method more effective. The first set, given by the originators Peaceman and Rachford [85], for the self-adjoint equation (7.213) are

$$\rho_n^{(P)} = \bar{b}(\bar{a}/\bar{b})^{(2n-1)/2m} \qquad n = 1, 2, \cdots, m \qquad (7.214)$$

[†] When the Thomas Algorithm is used the number of operations is $O(M)$ as opposed to $O(M^3)$ for straightforward matrix methods.

where
$$\bar{a} = \min(a, \alpha)$$
$$\bar{b} = \max(b, \beta)$$
$$a = A(x + \tfrac{h}{2}, y), \quad \alpha = C(x, y + h/2), \quad c = A(x - \tfrac{h}{2}, y),$$
$$b = \tfrac{1}{2}(a + c), \quad \gamma = C(x, y - h/2), \quad \beta = \tfrac{1}{2}(\alpha + \gamma).$$

Of course these functions are evaluated for $x = ih$, $y = jh$. In practice the quantity m, appearing in Eq. (7.214) is chosen as the smallest positive integer so that

$$(\sqrt{2} - 1)^{2m} \leqslant (\bar{a}/\bar{b}). \tag{7.215}$$

The set of number $\rho_n^{(P)}$ determined by Eqs. (7.214) and (7.215) is used in the cyclic order $\rho_1^{(P)}, \rho_2^{(P)}, \ldots \rho_m^{(P)}; \rho_1^{(P)}, \ldots$.

A second analysis by Wachspress [89] and Wachspress and Habetler [90] presents the iteration parameters

$$\rho_n^{(W)} = \bar{b}(\bar{a}/\bar{b})^{(n-1)/(m-1)} \quad m \geqslant 2, \quad n = 1, 2, \cdots, m. \tag{7.216}$$

In practice m is chosen as the smallest positive integer such that

$$(\bar{a}/\bar{b}) \geqslant (\sqrt{2} - 1)^{2(m-1)}. \tag{7.217}$$

In both cases a variable number of parameters may be used. Further discussion and derivations of Eqs. (7.214)–(7.217) can also be found in Birkhoff *et al.* [87]. The number of steps for convergence, that is the number of iterations required to reduce the error to a specified factor ϵ of its original value, for the PR methods, are compared in the following table. Again the comparison is made for the Dirichlet problem of Poisson's equation in a square with mesh dimension h.

Method	Steps	Reference
Fixed number m of PR parameters	$\dfrac{m}{4}\left(\dfrac{2}{h}\right)^{1/m}$	[1, 87]
Variable number of PR parameters	$-\dfrac{\ln(h/2)}{1.55}$	[87]
Fixed number m of W parameters	$\dfrac{m}{8}\left(\dfrac{2}{h}\right)^{1/(m-1)}$	[1, 87]
Variable number of W parameters	$-\dfrac{\ln(h/2)}{3.11}$	[87]

c. Other Procedures and Comments

(1) The PR method is superior to the DR method in certain cases, including the solution of Laplace's equation in two dimensions.

(2) The Wachspress parameters are superior to the Peaceman-Rachford iteration parameters by a factor of two. Numerical experiments confirm this superiority according to Birkhoff *et al.* [87].

(3) The determination of optimum iteration parameters seems to be not far off. Wachspress [91] has in fact devised an algorithm for calculating these when m is a power of two.

(4) For nonrectangular regions the use of an enclosing rectangle allows one to estimate convergence rates. Young and Frank [3] report that solutions of the Dirichlet problem for the Helmholtz equation in nonrectangular regions had never less than one-half the convergence rate for the circumscribing rectangular area.

7.19 EXAMPLES OF NONLINEAR ELLIPTIC EQUATIONS

The numerical solution of nonlinear elliptic equations and the treatment of singularities in such nonlinear problems has not been considered in any generality. The situation here is far less developed than in the parabolic or hyperbolic case. Each problem is treated on an *ad hoc* basis and most of the present techniques are direct applications or extensions of the aforementioned linear methods.

a. Laminar Flow of Non-Newtonian Fluids

The problem of determining the laminar steady flow of a non-Newtonian fluid in a unit square duct has the dimensionless formulation

$$\frac{\partial}{\partial x}\left[w\frac{\partial u}{\partial x}\right] + \frac{\partial}{\partial y}\left[w\frac{\partial u}{\partial y}\right] + \frac{f \cdot \text{Re}}{2} = 0 \qquad (7.218)$$

$$w = \left[\left(\frac{\partial u}{\partial x}\right)^2 + \left(\frac{\partial u}{\partial y}\right)^2\right]^{(n-1)/2} \qquad (7.219)$$

$$\int_0^1 \int_0^1 u(x, y)\, dx\, dy = 1 \qquad (7.220)$$

$$u(x, y) = 0 \quad \text{on the boundary } \Gamma \text{ of the unit square,} \quad (7.221)$$

which is taken as $0 \leq x \leq 1$, $0 \leq y \leq 1$. These equations involve the dimensionless variables u = velocity and w = viscosity given by the "power law" equation (7.219) with the non-Newtonian parameter $0 < n \leq 1$. Here f is a friction factor and Re the Reynolds number. The ultimate objective is to find $f \cdot \text{Re}$ so that Eqs. (7.218)–(7.221) hold. The discussion we give is based upon that of Young and Wheeler [88].

A preliminary analysis reduces the problem's complexity. In fact it suffices to solve this problem with $f \cdot \text{Re} = 2$. For suppose we solve

$$[WU_x]_x + [WU_y]_y + 1 = 0$$
$$W = [(U_x)^2 + (U_y)^2]^{(n-1)/2}, \quad U = 0 \quad \text{on} \quad \Gamma. \quad (7.222)$$

Upon setting $u = cU$, Eq. (7.220) becomes

$$\int_0^1 \int_0^1 u \, dx \, dc = c \int_0^1 \int_0^1 U \, dx \, dy = 1$$

or

$$c = \left[\int_0^1 \int_0^1 U \, dx \, dy \right]^{-1}. \quad (7.223)$$

Since $u = cU$,

$$w = [(u_x)^2 + (u_y)^2]^{(n-1)/2} = c^{n-1}[(U_x)^2 + (U_y)^2]^{(n-1)/2}$$
$$= c^{n-1} W \quad (7.224)$$

so that

$$[wu_x]_x + [wu_y]_y = c^n \{[WU_x]_x + [WU_y]_y\}$$
$$= c^n(-1) = -\frac{f \cdot \text{Re}}{2}.$$

From this relation it follows that after the calculation of U in Eq. (7.222) we may calculate $f \cdot \text{Re}$ by

$$\frac{f \cdot \text{Re}}{2} = c^n = \left[\int_0^1 \int_0^1 U \, dx \, dy \right]^{-n}. \quad (7.225)$$

The problem's symmetry means that it suffices to solve the problem in one of the four quarters of the square, say $\frac{1}{2} \leq x \leq 1$, $\frac{1}{2} \leq y \leq 1$,

7.19 EXAMPLES OF NONLINEAR ELLIPTIC EQUATIONS

as shown in Fig. 7-14 together with the new boundary conditions on this (shaded) domain of integration $R.*$

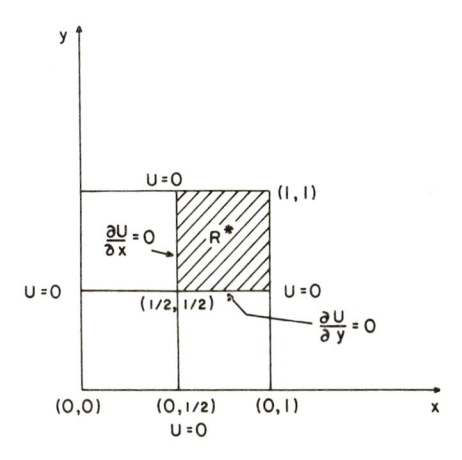

FIG. 7-14. The integration domain for the laminar flow of a non-Newtonian fluid.

To discuss the finite difference analog of the reduced problem (7.222) we proceed in two stages. The first of these is by analog with the *linear* operator $-\partial/\partial x[A(x, y) \partial U/\partial x]$ using the point configuration of Fig. 7-15. One may use Eq. (7.180) or the expression[†]

$$-h^2[AU_x]_x \simeq \left(\frac{A_0 + A_1}{2}\right)(U_0 - U_1) + \left(\frac{A_0 + A_3}{2}\right)(U_0 - U_3) \quad (7.226)$$

FIG. 7-15. Point configuration of the operator used in the non-Newtonian flow problem of Section 7.19.

[†] Young and Wheeler [88] prefer this form. See also Forsythe and Wasow [1] who discuss these finite difference operators.

where the subscripts indicate evaluation at the points of Fig. 7-15. Equation (7.226) can be rewritten in the more convenient form

$$\tfrac{1}{2}(2A_0 + A_1 + A_3)U_0 - \tfrac{1}{2}(A_0 + A_1)U_1 - \tfrac{1}{2}(A_0 + A_3)U_3 \quad (7.227)$$

with the result

$$\tfrac{1}{2}(2A_0 + A_2 + A_4)U_0 - \tfrac{1}{2}(A_0 + A_2)U_2 - \tfrac{1}{2}(A_0 + A_4)U_4 \quad (7.228)$$

for the companion operator $-h^2[AU_y]_y$.

For the nonlinear problem (7.222) we replace the A's in Eqs. (7.227) and (7.228) by the corresponding W's. Thus at *interior* points of R^* the difference equation for Eq. (7.222) becomes

$$\tfrac{1}{2}(2W_0 + W_1 + W_3)U_0 - \tfrac{1}{2}(W_0 + W_1)U_1 - \tfrac{1}{2}(W_0 + W_3)U_3$$
$$+ \tfrac{1}{2}(2W_0 + W_2 + W_4)U_0 - \tfrac{1}{2}(W_0 + W_2)U_2 - \tfrac{1}{2}(W_0 + W_4)U_4 = h^2. \quad (7.229)$$

Special considerations must be given to the boundaries at $x = \tfrac{1}{2}$ and $y = \tfrac{1}{2}$. On each of these boundaries false boundaries must be introduced whereupon the satisfaction of $\partial U/\partial x = 0$ requires that $U_1 = U_3$.[†] Since W is defined in terms of U_x and U_y and since the system is symmetric it follows that $W_1 = W_3$ when the point configuration is centered on the boundary $x = \tfrac{1}{2}$. Thus for a point on the line $x = \tfrac{1}{2}$, Eq. (7.229) becomes

$$(W_0 + W_1)U_0 - (W_0 + W_1)U_1 + \tfrac{1}{2}(2W_0 + W_2 + W_4)U_0 - \tfrac{1}{2}(W_0 + W_2)U_2$$
$$- \tfrac{1}{2}(W_0 + W_4)U_4 = h^2. \quad (7.230)$$

By a similar argument we can show that for a point on the line $y = \tfrac{1}{2}$, Eq. (7.229) becomes

$$\tfrac{1}{2}(2W_0 + W_1 + W_3)U_0 - \tfrac{1}{2}(W_0 + W_1)U_1 - \tfrac{1}{2}(W_0 + W_3)U_3$$
$$+ (W_0 + W_2)U_0 - (W_0 + W_2)U_2 = h^2 \quad (7.231)$$

by virtue of the necessary relations $U_2 = U_4$, $W_2 = W_4$. At the point $(\tfrac{1}{2}, \tfrac{1}{2})$ we have Eq. (7.229) reducing to

$$(W_0 + W_1)U_0 - (W_0 + W_1)U_1 + (W_0 + W_2)U_0 - (W_0 + W_2)U_2 = h^2. \quad (7.232)$$

[†] This is also true by the symmetry condition thereby suggesting a more complicated approximation for $\partial U/\partial x$, $\partial U/\partial y$ than that usually used.

7.19 EXAMPLES OF NONLINEAR ELLIPTIC EQUATIONS

As we have done many times before we set

$$U_{i,j} = U(x_i, y_j), \quad x_i = \tfrac{1}{2} + ih, \quad y_j = \tfrac{1}{2} + jh$$

and consider the region $\bar{R}$ which includes R^* and extends two intervals beyond the boundary of R^* as shown in Fig. 7-16. The values at the

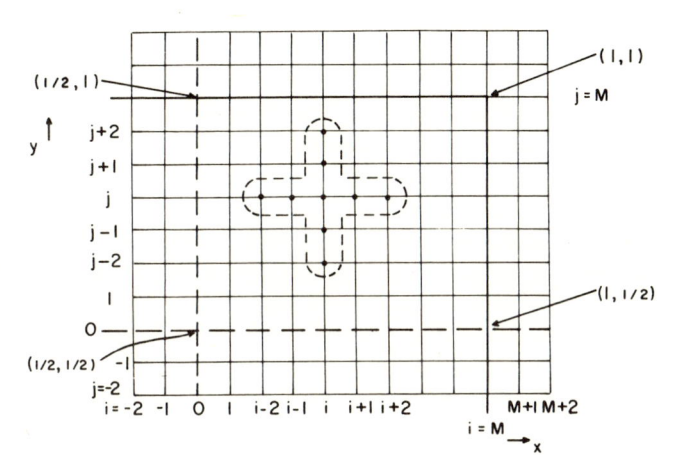

FIG. 7-16. The rectangular mesh for the non-Newtonian flow problem. Double false boundaries are introduced to improve accuracy.

false boundaries of $\bar{R}$ (that is outside R^* and its boundary) are determined as follows:

(a) By symmetry with respect to the lines $x = \tfrac{1}{2}$ and $y = \tfrac{1}{2}$

$$\begin{aligned} U_{-1,j} &= U_{1,j}; & U_{-2,j} &= U_{2,j}, & j &= 0, 1, \cdots, M \\ U_{i,-1} &= U_{i,1}; & U_{i,-2} &= U_{i,2}, & i &= 0, 1, \cdots, M. \end{aligned} \quad (7.233)$$

(b) The values outside the boundaries $x = 1$ and $y = 1$ are determined from the forward extrapolation formulas (see Hildebrand [61])

$$\begin{aligned} U_{M+1,j} &= 5U_{M,j} - 10U_{M-1,j} + 10U_{M-2,j} - 5U_{M-3,j} + U_{M-4,j} \\ U_{M+2,j} &= 15U_{M,j} - 40U_{M-1,j} + 45U_{M-2,j} - 24U_{M-3,j} + 5U_{M-4,j} \\ U_{i,M+1} &= 5U_{i,M} - 10U_{i,M-1} + 10U_{i,M-2} - 5U_{i,M-3} + U_{i,M-4} \\ U_{i,M+2} &= 15U_{i,M} - 40U_{i,M-1} + 45U_{i,M-2} - 24U_{i,M-3} + 5U_{i,M-4}. \end{aligned} \quad (7.234)$$

The reason for extending the intervals outside R^* as shown in Fig. 7-16 is to get a more accurate calculation of $W_{i,j}$ which is calculated from the 9 point star of Fig. 7-16 by means of

$$W_{i,j} = \left\{ \frac{[8(U_{i+1,j} - U_{i-1,j}) - (U_{i+2,j} - U_{i-2,j})]^2}{144h^2} \right.$$

$$\left. + \frac{[8(U_{i,j+1} - U_{i,j-1}) - (U_{i,j+2} - U_{i,j-2})]^2}{144h^2} \right\}^{(n-1)/2}, \quad (7.235)$$

for all points in the interior and on the boundary of R^*.

Before discussing the actual iterative scheme we must note that at the points $(\frac{1}{2}, \frac{1}{2})$ and $(1, 1)$ we have $(\partial U/\partial x)^2 + (\partial U/\partial y)^2 = 0$. This follows from Eqs. (7.235) and (7.233), i.e., from symmetry considerations. Thus for $n < 1$, it follows from Eq. (7.222) that $W(\frac{1}{2}, \frac{1}{2})$ and $W(1, 1)$ become arbitrarily large. To avoid this complication a large value $\bar{W}$ was chosen. If $W_{i,j}$ as computed from Eq. (7.235) was less than $\bar{W}$ then it is accepted—otherwise $\bar{W}$ is used for $W_{i,j}$. This bases the computation upon a truncated function W.

The iterative procedure for solving Eqs. (7.229)–(7.232) and (7.235) is as follows:

(a) one first chooses an initial approximation (for all points of R^*)

$$U^{(0)} = \sin \pi x \sin \pi y \quad (7.236)$$

suggested by the analysis of Schechter [92], whose form is a crude approximation having the advantage that it satisfies the boundary conditions;

(b) the values of $U^{(0)}$ are extended to $\bar{R}$ by means of Eqs. (7.233) and (7.234);

(c) using Eq. (7.235) one computes $W^{(0)}$ for all points of $\bar{R}$;

(d) Holding W fixed compute $U^{(1)}$ via Eqs. (7.229)–(7.232). In general given $U^{(k)}$ at all points R^* we extend to $\bar{R}$ by means of Eqs. (7.233) and (7.234). Then compute $W^{(k)}$ from Eq. (7.235). With W fixed, as $W^{(k)}$, compute $U^{(k+1)}$, etc. The values of $U^{(k+1)}$ are obtained by performing a fixed number of Peaceman-Rachford iterations with iteration parameters determined by Wachspress [see Eq. (7.216)]. The process of going from $U^{(k)}$, $W^{(k)}$ to $U^{(k+1)}$ $W^{(k+1)}$ is called an *outer* iteration. The determination of $U^{(k+1)}$ from $U^{(k)}$ involves several iterations of the Peaceman-Rachford method called *inner* iterations. The bounds $\bar{a}$ and $\bar{b}$ of the iteration parameters were calculated exactly and by an approximate method of Varga [57].

Computations were carried out with $h^{-1} = 20$, 40, and 80 and for $n = 1.0$, 0.7 and 0.5 with various required accuracies. The use of the crude estimates of Varga with one variant of the method lead to divergence of the process for the highly non-Newtonian case $n = 0.5$. In all other cases convergence was good.

After convergence one computes $f \cdot \mathrm{Re}$ by means of Eq. (7.225). This computation is done with some numerical integration procedure like Simpson's rule as in Wheeler [93].

b. Mildly Nonlinear Elliptic Equations (Douglas)

Douglas [94] considers the applicability of alternating direction methods to the Dirichlet problem for the "mildly" nonlinear elliptic equation

$$u_{xx} + u_{yy} = Q(x, y, u) \tag{7.237}$$

in a rectangle R, under the assumption that $0 < m \leqslant \partial Q/\partial u \leqslant M < \infty$ and $u = g(x, y)$ on the boundary of R.

The procedure used by Douglas involves a two level iteration similar to the inner-outer iteration discussed in the previous example. The outer iteration is a modified Picard iteration and the inner iteration an alternating direction (Peaceman-Rachford) method. Specifically the five point computational molecule given by Eq. (7.211) is used for $u_{xx} + u_{yy}$ which we label (after Douglas) $\Delta_x^2 + \Delta_y^2$. Thus the finite difference analog of Eq. (7.237) is

$$(\Delta_x^2 + \Delta_y^2)u_{i,j} = Q(x_i, y_j, u_{i,j}) \quad \text{in R,}$$
$$u_{i,j} = g_{ij} \quad \text{on the boundary of } R, \tag{7.238}†$$

which are nonlinear algebraic equations.

The solution of the Eqs. (7.238) is done by the Picard type *outer iteration*

$$(\Delta_x^2 + \Delta_y^2)u_{i,j}^{(n+1)} - Au_{i,j}^{(n+1)} = Q(x_i, y_j, u_{i,j}^{(n)}) - Au_{i,j}^{(n)} \tag{7.239}$$

and on the boundary of R, $u_{i,j}^{(n+1)} = g_{i,j}$. The solution of this *linear*

---

† Bers [95] proved uniform convergence of Eq. (7.238) to u under very general conditions.

system is accomplished by an *inner iteration* which is carried out by means of an alternating direction method. The optimum value of A, in Eq. (7.239), is shown to be

$$A = \tfrac{1}{2}(M + m). \tag{7.240}$$

The choice of such a procedure as Eq. (7.239) improves the operation of the process. The number of sweeps of the alternating direction process required for each outer iteration is $O(-\ln h)$ which leads to an estimate of the total number of calculations required to obtain a uniformly good approximation as

$$O[h^{-2}(\ln h)^2].$$

A very important result of this analysis is that the number of outer iterations is independent of h!

Direct generalization of the Douglas approach to

$$\nabla \cdot [a(x, y) \nabla u] = Q(x, y, u) \tag{7.241}$$

is difficult using ADI for the inner iterations. The cited paper of Young and Wheeler [88] treats a different but related case. Douglas discusses the use of SOR for the inner iterations, thereby extending his scheme (7.239) to Eq. (7.241), but at the expense of additional labor. Douglas' method is easily generalized to three dimensions.

c. The Newton Second-Order Process for $Lu = f(x, t, u)$

With L a linear operator, the general discussion of the Newton versus Picard approximation for $Lu = f(x, t, u)$, was given in Section 7.9. Here we sketch the results of Bellman, Juncosa, and Kalaba [24] when the second-order Newton process is applied to the elliptic equation[†]

$$u_{xx} + u_{yy} = e^u \tag{7.242}$$

in the region $0 \leqslant x \leqslant \tfrac{1}{2}$, $0 \leqslant y \leqslant \tfrac{1}{4}$. Two sets of boundary conditions are considered, $u \equiv 0$ in the first case and $u \equiv 10$ in the second.

The cited paper compares the Picard linearization of Eq. (7.242) with the Newton linearization. These are respectively *outer iterations* of the form

$$u^{(k+1)}_{xx} + u^{(k+1)}_{yy} = \exp(u^{(k)}) \tag{7.243}$$

[†] This equation is of physical interest in diffusion-reaction problems, vortex problems and electric space charge considerations. These are discussed in Chapter 4, Section 4.16.

7.19 EXAMPLES OF NONLINEAR ELLIPTIC EQUATIONS

and [see Eq. (7.85)]

$$u_{xx}^{(k+1)} + u_{yy}^{(k+1)} = \exp(u^{(k)}) + [u^{(k+1)} - u^{(k)}]\exp(u^{(k)})$$

where the latter is more conveniently written as

$$u_{xx}^{(k+1)} + u_{yy}^{(k+1)} - [\exp(u^{(k)})]u^{(k+1)} = [1 - u^{(k)}]\exp(u^{(k)}). \qquad (7.244)$$

In each case the standard five point molecule is employed for the Laplacian, i.e.,

$$h^{-2}[u_{i+1,j} + u_{i,j+1} + u_{i-1,j} + u_{i,j-1} - 4u_{i,j}]$$

with $\Delta x = \Delta y = h$. As has been previously discussed both equations have finite difference systems whose solutions require the solution of linear algebraic equations whose matrix form is[†]

$$(I - L_k - U_k)u^{(k+1)} = b^{(k)} \qquad (7.245)$$

where I = identity matrix and L_k, U_k are respectively lower triangular and upper triangular matrices.

The inner iteration, i.e., the solution of Eq. (7.245), was carried out by means of the SOR method. The combined process is of the form

$$(I - \omega L_k)u_{r+1}^{(k+1)} = [\omega U_k - (\omega - 1)I]u_r^{(k+1)} + b^{(k)} \quad r = 0, 1, 2, \cdots \qquad (7.246)$$

where k is the outer iteration index, r is the inner iteration index and ω is the relaxation factor. The iteration is carried out in the order imposed. It is clear that not much effort should be expended in obtaining a very accurate solution to Eq. (7.245) if $u^{(k+1)}$ is not a very good approximation to the solution of Eq. (7.242). Similarly, there is little point to iterating on the index k (outer iteration) if the approximations given by Eq. (7.246) are too rough as a consequence of terminating the inner iteration on r too soon. Thus there exists the open question as to when one should iterate on k or on r at each step. [The results of Douglas [94] and Young and Wheeler [88] are important here. Although their problems are different they find that only *one cycle* of the Peaceman-Rachford (ADI) iterations (i.e., only m inner iterations), were required for each outer iteration. Additional inner iterations did not hasten the convergence.]

[†] An ordering has been introduced and the equations have been divided by the main diagonal coefficient.

The authors' goal was not to answer the question raised in the preceding paragraph so they used a simultaneous iteration based on a *blend* formula

$$(I - \omega L_k)u^{(k+1)} = [\omega U_k - (\omega - 1)I]u^{(k)} + \omega b^{(k)}. \qquad (7.247)$$

An "optimum" value of ω was chosen by using the linear theory (Section. 717) which yields 1.732. Hence the range $\omega = 1.70$ to 1.74 was used initially for the boundary condition $u \equiv 0$. In this case no advantage was obtained from the use of the quadratically convergent Newton procedure over the results of the first order Picard method. In fact, the same value of $\omega = 1.74$ gave the fastest convergence in both cases.

In the case of the boundary condition $u \equiv 10$ a number of runs for different values of ω were made to experimentally determine an optimal value of ω. This approach was suggested as appropriate in Section 7.17. The results of this experiment are given in the accompanying tabulation (note that the same accuracy is required in both cases).

Value of ω	Number of iterations	Value of ω	Number of iterations
Picard method		Newton method	
1.00	134	1.00	148
1.10	106	1.40	66
1.20	78	1.50	50
1.30	76	1.55	41
1.35	71	1.60	42
1.40	66	1.65	46
1.50	53	1.70	51
1.55	63	1.74	58

With boundary conditions at 10, rapid changes near the boundary are suggested. Here the Newton method was better requiring 41 iterations as compared to 53 for the Picard approach. The optimal value of ω is slightly smaller for the Picard method.

As a by-product of this analysis we observe that the Gauss-Seidel method ($\omega = 1$) compares rather poorly with SOR, requiring from two and one-half to four times as many iterations. In the case of the boundary condition $u \equiv 0$ (not shown), where $\omega = 1.74$ is optimal, the GS scheme takes up to seven times as many iterations as SOR.

As in the parabolic case, the advantage accruing to the Newton linearization becomes significant when steep gradients occur. We also note that the Newton linearization is easily generalized to the case where L is quasi-linear.

d. The Equation $\nabla \cdot [F \nabla u] = 0$

In Chapter 2 the equation

$$\nabla \cdot [F \nabla u] = 0 \tag{7.248}$$

is discussed with $F = F(u)$. For certain classes of boundary conditions the solution of this equation is most easily attempted by applying the Kirchhoff transformation thereby linearizing the equation.

The case of interest to us here is that when

$$F = F[|\nabla u|] = F[(u_x^2 + u_y^2)^{1/2}]. \tag{7.249}$$

Problems leading to such equations occur in heat conduction where the thermal conductivity depends upon $|\nabla u|$ as discussed by Slattery [96] and Serrin [97]. Vertical heat transfer from a horizontal surface by turbulent free convection (see Priestley [98]) and turbulent flow of a liquid with a free surface over a plane (see Philip [99]) have a steady state mathematical model of the form Eqs. (7.248)–(7.249).

The partial differential equations in magnetostatics as they apply in highly saturated rotating machinery have the form $\nabla \cdot [\mu \nabla V] = 0$ where $V =$ scalar potential, $\mathbf{H} = -\nabla V$. The magnetic permeability $\mu = \mu(H)$[†] establishes the relation between B and H. Determination of μ has proceeded by means of fitting experimental data with mathematical expressions of the form

$$B = \mu(H)H, \qquad H = [V_x^2 + V_y^2]^{1/2}. \tag{7.250}$$

Fischer and Moser [100] have investigated the fitting of the magnetization curve and tabulate fifteen different fitting functions some of which are

$$(a + bH)^{-1}, \qquad aH^{-1} \tanh bH, \qquad H^{-1} \exp[H/(a + bH)]. \tag{7.251}$$

Several numerical studies in rectangular and curvilinear geometries have been undertaken at the University of Delaware by Erdelyi and his

---
[†] We write $\mathbf{H}$ for the vector magnetic field and H for its magnitude.

students [101a,b,c] to obtain solutions of Eq. (7.248) with $\mu = (a + bH)^{-1}$ The calculation uses the standard five point molecule and the SOR algorithm. The choice of the relaxation parameter ω is based upon that for Laplace's equation [Eq. (7.199) and (7.200)]. Large gradients are anticipated in this problem. Consequently the evidence of the Bellman *et al.* calculation (example c of this section) suggests that numerical experimentation may be helpful in problems of this type if the optimal value of ω is desired.

The calculation proceeded by selecting initial guesses $\mu^{(0)}$ and $V^{(0)}$ over the domain. At the kth step $V^{(k)}$ and $\mu^{(k)}$ are known. $V^{(k+1)}$ is calculated using SOR with $\mu = \mu^{(k)}$. Then $\mu^{(k+1)}$ is calculated using $V = V^{(k+1)}$ with SOR. The domain of integration is the region of Fig. 7-17 containing an air gap between the two iron "fields." In addition

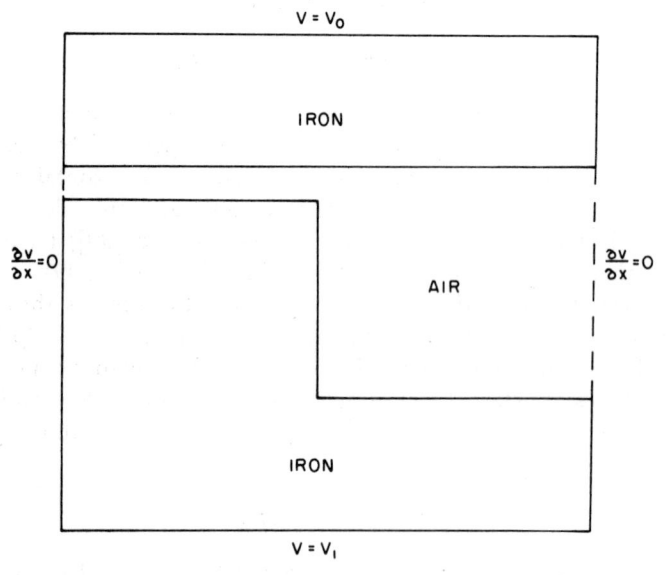

FIG. 7-17. A typical electromagnetic field problem with corner singularities.

the domain contains corners. The interior corners have singularities at which special treatment becomes necessary. These were treated by the technique of mesh refinement, a procedure which will be discussed in the next section.

This problem has many features in common with the Young-Wheeler example (example a of this section) and the Bellman *et al.* example

7.19 EXAMPLES OF NONLINEAR ELLIPTIC EQUATIONS

(example c). It would be interesting to see what, if any, computational savings accrued by the application of Newton linearization and ADI. In addition the more accurate expression, Eq. (7.235) for H, may be useful.

e. Numerical Solution of the Navier-Stokes Equations

For steady isothermal conditions the flow of an incompressible Newtonian fluid is governed by the four equations

$$uu_x + vu_y + wu_z = -\rho^{-1}p_x + \nu\nabla^2 u$$
$$uv_x + vv_y + wv_z = -\rho^{-1}p_y + \nu\nabla^2 u$$
$$uw_x + vw_y + ww_z = -\rho^{-1}p_y + \nu\nabla^2 u \quad (7.252)$$
$$u_x + v_y + w_z = 0.$$

Whitaker and Wendel [102] consider ways of obtaining the solution to a reduced from of these equations for Reynolds numbers up to about 10. The problem considered is that for flow past an infinite set of parallel flat plates placed perpendicular to and between two infinite parallel planes. When the distance between the plates is greater than two channel depths the problem is reducible to two dimensions by assuming that the two major velocity components (u and v) are parabolic in the channel depth direction (see Fig. 7-18).

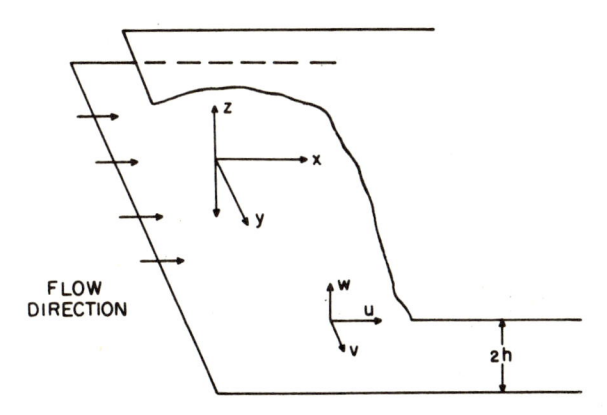

FIG. 7-18. Schematic for the steady incompressible flow of a Newtonian fluid.

Assuming that w is *small* compared to u and v and integrating in the z-direction we get

$$\int_0^h (uu_x + vu_y)\, dz = -\rho^{-1} \int_0^h p_x\, dz + \nu \int_0^h \nabla^2 u\, dz \qquad (7.253a)$$

$$\int_0^h (uv_x + vv_y)\, dz = -\rho^{-1} \int_0^h p_y\, dz + \nu \int_0^h \nabla^2 v\, dz \qquad (7.253b)$$

$$\int_0^h u_x\, dz + \int_0^h v_y\, dz = 0. \qquad (7.253c)$$

The assumed parabolic form in the z-direction

$$\begin{aligned} u &= u_m(1 - z^2/h^2)U(x, y) \\ v &= u_m(1 - z^2/h^2)V(x, y) \end{aligned} \qquad (7.254)$$

allows us to rewrite Eqs. (7.253) in the form

$$UU_X + VU_Y = -P_X + 5/3\, \text{Re}\, [U_{XX} + U_{YY} - 3U] \qquad (7.255a)$$

$$UV_X + VV_Y = -P_X + 5/3\, \text{Re}\, [V_{XX} + V_{YY} - 3V] \qquad (7.255b)$$

$$U_X + V_Y = 0. \qquad (7.255c)$$

The notation in Eq. (7.255) is as follows: U and V are dimensionless velocity components defined by Eq. (7.254);

$$P = 5\bar{p}/6\rho\bar{u}_0^2, \qquad \bar{p} = \int_0^h p\, dz, \qquad \bar{u}_0 = \int_0^h u_m(1 - z^2/h^2)\, dz,$$

u_m is the x-component of velocity as $x \to \infty$; $X = x/h$, $Y = y/h$ and $\text{Re} = $ Reynolds number $= 2\bar{u}_0 h/\nu$.

The finite difference equations for Eq. (7.255) were developed from the standard five point molecule, with $U_{j,n} = U(j\, \Delta Y, n\, \Delta X)$. These are

$$A_n U_{j,n-1} + B_n U_{j,n} + C_n U_{j,n+1} = D_n \qquad (7.256a)$$

$$P_{j+1,n} = P_{j,n} + \Delta Y \left[\frac{5}{3\, \text{Re}}(V_{XX} + V_{YY} - 3V) - UV_X - VV_Y\right]_{j+\frac{1}{2},n} \qquad (7.256b)\dagger$$

$$V_{j+1,n} = V_{j,n} - \Delta Y [U_x]_{j+\frac{1}{2},n} \qquad (7.256c)$$

---

† Rather than write out the finite difference form we only indicate it with the subscript.

where

$$A_n = C_n = \frac{5}{3 \text{ Re}}(\varDelta X)^2, \qquad B_n = -\frac{10}{3 \text{ Re}}[(\varDelta X)^{-2} + (\varDelta Y)^{-2}] - [U_x]_{j,n}$$

$$D_n = (P_X)_{j,n} + V_{j,n}(U_Y)_{j,n} + [5/3 \text{ Re }(\varDelta Y)^2][U_{j+1,n} + U_{j-1,n}].$$

The iterative calculation took the following course: (a) A $U^{(0)}$ field was assumed; (b) Eq. (7.256c) was used to calculate the $V^{(0)}$ field; (c) the $U^{(0)}$ and $V^{(0)}$ fields were used to calculate the $P^{(0)}$ field via Eq. (7.256b); and (d) the assumed values of $U^{(0)}$ together with the calculated values of $V^{(0)}$ and $P^{(0)}$ are used to obtain the coefficients A_n, B_n, C_n, and D_n so that $U^{(1)}$ can be calculated by applying the Thomas algorithm (Section 7.8) to Eq. (7.256a). Direct use of the newly calculated values lead to instability so an *underrelaxation* procedure was adopted. $U^{(i+1)}$ was calculated by means of

$$U^{(i+1)} = \omega \bar{U}^{(i+1)} + (1 - \omega) U^{(i)} \qquad (7.257)$$

with $0 < \omega < 1$. $\bar{U}^{(i+1)}$ is the calculated value of U making use of Eq. (7.256a) but the accepted value of U, $U^{(i+1)}$, is based on an underrelaxation. The authors found that stability was assured when Re = 0.01 for $\omega < 0.4$ and when Re = 10.0, ω had to be less than 0.2 for stability. Computer size prevented the authors from utilizing more sophisticated methods. The results of physical experiments agree with the numerical results.

f. Numerical Weather Prediction

The physical problem in *numerical weather prediction* is the prediction of atmospheric motions of large size say greater than 500 miles in diameter, moving with horizontal velocities of the order of 30 miles per hour. Air passing through these systems may have significantly higher velocities.

The atmosphere may be regarded as a compressible radiative fluid with external heating. Given the state of this fluid at time t_1, we may, in principle, compute its later state by means of thermodynamics and the equations of fluid dynamics. At least two problems are pertinent here: (a) the state of the fluid after a short time, whose determination depends largely on the initial state; and (b) the long term character of the motion, whose determination depends little on the initial state but largely upon external parameters and other features such as heat source

source and sink location. Numerical weather prediction is concerned with problem (a) with a time scale of approximately one week or less. The weather aspect is an interpretative art based upon the numerical solution for the velocity, pressure and temperature fields.

Many of the numerical calculations to date use pressure as a vertical coordinate, rather than height, and assume the fluid frictionless so the equations become

$$\frac{\partial \mathbf{V}}{\partial t} + (\mathbf{V} \cdot \nabla)\mathbf{V} + \omega \frac{\partial \mathbf{V}}{\partial p} + f\mathbf{k} \times \mathbf{V} + g\,\nabla h = 0 \quad \text{(motion)} \quad (7.258)$$

$$\frac{\partial h}{\partial p} + \frac{RT}{gp} = 0 \quad \text{(hydrostatic)} \quad (7.259)$$

$$C_p\,dT - \frac{RT\,dp}{p} - dQ = 0 \quad \text{(thermodynamics)} \quad (7.260)$$

$$\nabla \cdot \mathbf{V} + \frac{\partial \omega}{\partial p} = 0 \quad \text{(continuity)}. \quad (7.261)$$

where p = pressure, t is time, $\mathbf{V} = (u, v)$, $\omega = dp/dt$ is the vertical velocity (corresponding in these coordinates to w in Cartesian coordinates), ∇ is the gradient operator in the horizontal direction (acting on u and v), $\mathbf{k}$ is the unit vertical vector, $f = 2\alpha \sin \phi$ = Coriolis parameter (α = earth's angular velocity, ϕ = the latitude), g is the gravitational constant, h = height of a pressure surface, T = temperature, R = gas constant, C_p = specific heat at constant pressure and dQ the heat input. (See Bushby and Whitelam [103].)

If the horizontal wind $\mathbf{V}$ is assumed to be nondivergent, i.e.,

$$\mathbf{V} = \mathbf{k} \times \nabla \psi \quad (7.262)$$

we find that the horizontal divergence of Eq. (7.258) becomes

$$2[\psi_{xx}\psi_{yy} - \psi_{xy}^2] + f\,\nabla^2\psi + \nabla f \cdot \nabla \psi = g\,\nabla^2 h. \quad (7.263)$$

This equation is of Monge-Ampere type which has been previously studied (in different form) in Chapters 2 and 3. Equation (7.263) is known in the meteorological literature as the *Balance equation* (see Fox [2], Chapt. 36, and Charney and Phillips [104]) because it gives the nondivergent wind exactly balancing the pressure forces. To numerically integrate this equation we assume h is known from observations.

7.19 EXAMPLES OF NONLINEAR ELLIPTIC EQUATIONS

The condition [see (7.156)] that the Balance equation (7.263) be elliptic is that

$$g \nabla^2 h - \nabla f \cdot \nabla \psi + f^2/2 > 0 \qquad (7.264)$$

which cannot be evaluated since it contains ψ. An excellent approximation, according to E. E. Knighting (see Fox [2], Chapt. 36), is obtained by noting that $f \nabla \psi \sim g \nabla h$ so the *approximate* condition for ellipticity is

$$g \nabla^2 h - g/f \nabla f \cdot \nabla h + f^2/2 > 0. \qquad (7.265)$$

This form allows the test to be made since f and h are known. In what follows the ellipticity of Eq. (7.263) is assumed.

Various numerical methods for solving the Balance equation have been proposed. The computational molecule used is the nine point one shown in Fig. 7-19. The occurrence of ψ_{xy} makes such a molecule advisable.

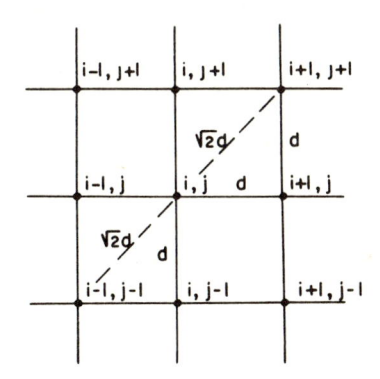

FIG. 7-19. Alternate computational mesh for the cross partial derivative u_{xy}.

The simplest *outer iteration* would probably have the form

$$\nabla^2 \psi^{(n+1)} = \frac{1}{f} \{g \nabla^2 h - \nabla f \cdot \nabla \psi^{(n)} - 2[\psi_{xx}^{(n)} \psi_{yy}^{(n)} - (\psi_{xy}^{(n)})^2]\} \qquad (7.266)$$

but the convergence is questionable in some cases according to Knighting (*op. cit.*) and Bolin [105] although no present convergence proof has been developed. The inner iteration can be any of those previously discussed. The difficulty here appears to be in the nature of the nonlinearity—all previous cases involved nonlinearities in the first derivative or function itself. Here the "Monge-Ampere" term has the *same order* as the *Laplacian*!

The possible divergence of the simple outer iteration has spurred other investigations of the Balance equation. One of these is due to Arnason [106]. The Arnason procedure uses an alternative linearization yielding the outer iteration

$$\psi_{xx}^{(n+1)}\psi_{yy}^{(n)} + \psi_{yy}^{(n+1)}\psi_{xx}^{(n)} - 2\psi_{xy}^{(n+1)}\psi_{xy}^{(n)} + f\,\nabla^2\psi^{(n+1)} + \nabla f \cdot \nabla\psi^{(n+1)} = g\,\nabla^2 h. \quad (7.267)$$

Arnason finds oscillatory solutions which can be accelerated by various methods, e.g., SOR and ADI. The matrices of the finite difference approximation obtained from Eq. (7.267) are functions of the iteration number n. The test of a solution ψ, is carried out by treating Eq. (7.263) as a Poisson equation for h and comparing the computed values and the original values of h.

Special problems occur in the inner iteration of the Balance equation which lead to a *systematic error* in the calculation of ψ. Bolin [105] noted that this systematic error was due to the use of two different grid lengths in computing derivatives. To see this we refer to Fig. 7-19 and note that

$$\psi_{xx} \approx \frac{\psi_{i+1,j} - 2\psi_{i,j} + \psi_{i-1,j}}{d^2}$$

with a similar expression for ψ_{yy} in the other direction where both use the same grid length d. On the other hand ψ_{xy} is approximated using a grid length of $2d$, i.e.,

$$\psi_{xy} \approx \frac{1}{(2d)^2}[\psi_{i+1,j+1} - \psi_{i-1,j+1} - \psi_{i+1,j-1} + \psi_{i-1,j-1}].$$

Theoretically *both* of these approximations have a truncation error which is $O(d^2)$ but in the sparse patchy meteorological fields ψ_{xy} is underestimated compared with ψ_{xx} and ψ_{yy}. The end result is an underestimate of the values of ψ.

The systematic error of the previous paragraph can be eliminated by computing ψ_{xx}, ψ_{yy}, and ψ_{xy} of the "Monge-Ampere" term $\psi_{xx}\psi_{yy} - \psi_{xy}^2$ with an interval length of $\sqrt{2}\,d$ (see Fig. 7-19, we use the same notation as illustrated there) so that

$$\psi_{xx}\psi_{yy} - \psi_{xy}^2 \approx \frac{1}{2d^2}\{(\psi_{i+1,j+1} - 2\psi_{i,j} + \psi_{i-1,j-1})(\psi_{i-1,j+1} - 2\psi_{i,j}$$
$$+ \psi_{i+1,j-1}) - (\psi_{i+1,j} + \psi_{i-1,j} - \psi_{i,j+1} - \psi_{i,j-1})^2\}. \quad (7.268)$$

7.19 EXAMPLES OF NONLINEAR ELLIPTIC EQUATIONS

It would seem appropriate to compute $\nabla^2 \psi$ in the same way thus obtaining

$$\psi_{xx} + \psi_{yy} \approx \frac{1}{2d^2}[\psi_{i+1,j+1} + \psi_{i-1,j+1} + \psi_{i+1,j-1} + \psi_{i-1,j-1} - 4\psi_{i,j}]. \quad (7.269)$$

This representation has the disadvantage that $\psi_{i,j}$ is only related to $\psi_{i+1,j}$, $\psi_{i-1,j}$, $\psi_{i,j+1}$, and $\psi_{i,j-1}$ through the (usually) weak link ψ_{xy} and not throught the Laplacian. If Eq. (7.269) is not used for $\nabla^2 \psi$ but an alternate field, with $\sqrt{2}d$ as the mesh size, is used then the result is a checkerboard pattern of two fields for ψ. The usual compromise is to use Eq. (7.268) with the standard five point molecule for $\nabla^2 \psi$, i.e.,

$$\nabla^2 \psi \approx [\psi_{i+1,j} + \psi_{i-1,j} + \psi_{i,j+1} + \psi_{i,j-1} - 4\psi_{i,j}]/d^2.$$

g. An Extension of Young-Frankel Linear Successive Overrelaxation

When we apply finite difference approximations directly to a nonlinear elliptic equation, nonlinear algebraic equations result. Several methods have been proposed to solve these systems—perhaps the earliest and best known is an extended Liebmann method which is a direct generalization of the (linear) Gauss-Seidel procedure (see Bers [95] and Schechter [107]). Schechter discusses a number of iteration methods for nonlinear problems. While the extended Liebmann method has a sound theoretical basis the following method of *nonlinear overrelaxation* has only numerical experiments and semi-theoretical arguments to justify it as of this writing.

In all the previous methods the solution of the large nonlinear algebraic system was obtained by an outer iteration (Newton's method say) which linearizes, followed by some iterative technique (say SOR). This process is repeated thus constructing a cascade of outer (Newton) iterations alternated with a large sequence of inner linear iterations.

The complexity of the strategy discussed above suggests that alternatives would be highly desirable. One direct and simple method which is particularily well adapted for solving algebraic systems associated with nonlinear elliptic equations is due to Lieberstein [108, 109]—a method called nonlinear overrelaxation (NLOR).

Consider a system of k algebraic equations each having continuous first derivatives

$$f_p(x_1, x_2, ..., x_k) = 0 \quad p = 1, 2, ..., k. \quad (7.270)$$

For convenience we set $f_{pq} = \partial f_p/\partial x_q$. The basic idea in this NLOR is a different use of the Newton process from that carried out by Bellman, Juncosa, and Kalaba [24]. Here we introduce a relaxation factor ω and take

$$x_1^{n+1} = x_1^n - \omega \frac{f_1(x_1^n, x_2^n, \ldots, x_k^n)}{f_{11}(x_1^n, \ldots, x_k^n)}$$

$$x_2^{n+1} = x_2^n - \omega \frac{f_2(x_1^{n+1}, x_2^n, \ldots, x_k^n)}{f_{22}(x_1^{n+1}, x_2^n, \ldots, x_k^n)} \quad (7.271)$$

$$x_3^{n+1} = x_3^n - \omega \frac{f_3(x_1^{n+1}, x_2^{n+1}, x_3^n, \ldots, x_k^n)}{f_{33}(x_1^{n+1}, x_2^{n+1}, x_3^n, \ldots, x_k^n)}.$$

$$\vdots$$

This method has a feature of the Gauss-Seidel method in that it uses corrected results immediately upon becoming available. In addition if the f_p are linear functions of the x_j this method reduces to SOR.

The convergence criteria for NLOR can be shown to be the same as those for SOR (Section 7.17) with the coefficient matrix A replaced by the Jacobian of the Eqs. (7.270) $J(f_{pq}^n)$. This is accomplished by use of the Taylor Series so that the vector form for $x^{n+1} - x^n$ is

$$x^{n+1} - x^n \approx -\omega[D_n^{-1}L_n\epsilon^{n+1} + (I + D_n^{-1}U_n)\epsilon^n] \quad (7.272)$$

where ϵ^n stands for the error vector, $\epsilon^n = x^n - x$ and L_n, D_n, U_n are lower triangular, diagonal and upper triangular matrices formed from

$$J(f_{pq}^n) = L_n + D_n + U_n. \quad (7.273)$$

From Eq. (7.272) it is an easy matter to show that

$$\epsilon^{n+1} \approx L_{\omega_n}\epsilon^n = -\left[\frac{1}{\omega}D_n + L_n\right]^{-1}\left[\left(1 - \frac{1}{\omega}\right)D_n + U_n\right]\epsilon^n$$

which is exactly the same form as the error matrix in SOR (see Young and Frank [3]). For convergence, the Jacobian $J(f_{pq}^n)$, at each stage of the iteration, must have properties required for A in SOR, e.g., property (A) (see Section 7.17). To make the process most efficient an ω_{opt} should be calculated minimizing the spectral norm of L_{ω_n} at *each* iteration n. This may be expensive and not rewarding since one does

7.19 EXAMPLES OF NONLINEAR ELLIPTIC EQUATIONS

better to overestimate ω_{opt} than underestimate. Further the author suggests that for small systems one usually runs with $\omega = 1$ and for large systems with a constant ω slightly less than 2.

We now consider application of this method to closed region, boundary value problems for the elliptic equation

$$F(x, y, u, u_x, u_y, u_{xx}, u_{yy}, u_{xy}) = 0 \qquad (7.274)$$

where F has at least one continuous derivative with respect to *each* of its eight arguments. At an interior point we use a nine point molecule, as in Fig. 7-19, and write for *each* interior point of the region

$$F\left[x_i, y_j, u_{ij}, \frac{u_{i+1,j} - u_{i-1,j}}{2d}, \frac{u_{i,j+1} - u_{i,j-1}}{2d}, \frac{u_{i+1,j} + u_{i-1,j} - 2u_{i,j}}{d^2},\right.$$
$$\left.\frac{u_{i+1,j+1} + u_{i-1,j-1} - u_{i+1,j-1} - u_{i-1,j-1}}{2d^2}, \frac{u_{i,j+1} + u_{i,j-1} - 2u_{i,j}}{d^2}\right] = 0. \qquad (7.275)$$

Equation (7.275) supplies us with the equation

$$f[x_i, y_j, u_{i,j}, u_{i+1,j}, u_{i-1,j}, u_{i,j+1}, u_{i,j-1},$$
$$u_{i-1,j+1}, u_{i-1,j-1}, u_{i+1,j+1}, u_{i+1,j-1}] = 0 \qquad (7.276)$$

which must be satisfied at every point (i, j) of a grid of square mesh d. This supplies the Eqs. (7.270).

To compute with NLOR we read in values of the initial guess u_{ij}^0 and then scan the mesh replacing the value of u_{ij} at each point by

$$u_{ij}^0 - \omega \frac{f}{\partial f/\partial u_{ij}}$$

where f and $\partial f/\partial u_{ij}$ are evaluated as in Eq. (7.271) by using corrected results immediately upon becoming available.

If a *consistent* ordering is chosen, e.g., scanning from left to right and down is one example (see Young [71]) and u_{xy} is not present then the Jacobian $J(f_{ij}^n)$ will have property (A). Otherwise it does not and the molecule of Fig. 7-19 cannot be used and still have property (A). In view of the recent results that SOR is valid on some matrices without property (A) we conjecture that it is also true that NLOR is valid for a large class of systems not having property (A).

Several numerical experiments (Lieberstein [109] and Greenspan and Yohe [110]) have been reported in which excellent results were obtained

410　　　　　　　　7. NUMERICAL METHODS

with NLOR. Greenspan and Yohe compare NLOR to an extended Liebmann method (using a generalized Newton-Raphson procedure) for the three dimensional problem

$$u_{xx} + u_{yy} + u_{zz} = e^u$$

in the spherical sector G bounded by $x^2 + y^2 + z^2 = 1$; $x = 0$; $y = 0$; $z = 0$ and with $u = x + 2y + z^2$ on the boundary of G. With $h = 0.1$ the NLOR took only one-half the computation time of the extended Liebmann method.

A problem of interest in gas dynamics (see Wise and Ablow [111]) is the Dirichlet problem for the nonlinear elliptic equation

$$\nabla^2 u = u^2. \tag{7.277}$$

Iteration methods for this problem have previously been discussed in Chapter 6, Section 6.4. The Newton method presented there and

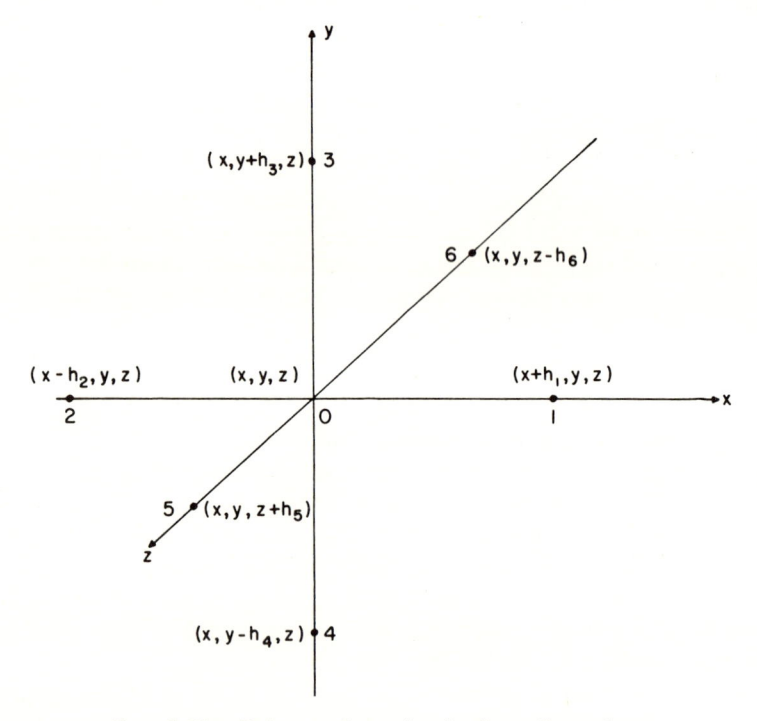

FIG. 7-20. Point configuration in three dimensions.

discussed, more generally, by Bellman *et al.* [24] is applied by Greenspan to the three dimensional problem of Eq. (7.277). The Newton linear approximation of Eq. (7.277) gives the outer iteration

$$\nabla^2 u^{(k)} - 2u^{(k-1)}u^{(k)} = -[u^{(k-1)}]^2. \tag{7.278}$$

Greenspan [112] applies the *seven point* three dimensional molecule, shown in Fig. 7-20, to develop the finite difference analogue of equation (7.278) at the point O. The result is

$$2\left[-u_0^{(k-1)} - \frac{1}{h_1 h_2} - \frac{1}{h_3 h_4} - \frac{1}{h_5 h_6}\right]u_0^{(k)} + \frac{2}{h_1(h_1 + h_2)}u_1^{(k)}$$
$$+ \frac{2}{h_2(h_1 + h_2)}u_2^{(k)} + \frac{2}{h_3(h_3 + h_4)}u_3^{(k)} + \frac{2}{h_4(h_3 + h_4)}u_4^{(k)}$$
$$\frac{2}{h_5(h_5 + h_6)}u_5^{(k)} + \frac{2}{h_6(h_5 + h_6)}u_6^{(k)} = -[u_0^{(k-1)}]^2. \tag{7.279}$$

Greenspan reports four successful calculations with these methods although little theoretical justification is in existence.

7.20 SINGULARITIES

Some elliptic problems are such that singularities occur at points p_i inside the integration domain or on its boundary. Interior singularities occur when one or more coefficients of the partial differential equation become singular there. In such problems the solution will ordinarily also have a singularity at p_i and the finite difference scheme will not be applicable without modification. Physical problems where such singularities occur are source-sink and concentrated point load problems.

The *accepted technique* in such interior singularity problems is to subtract out the singularity, where possible, thereby generating a new problem with different boundary conditions but with a "well behaved" solution. In practice this procedure works well with some linear problems. The more difficult nonlinear cases require individual treatment. The generation of "local" solutions can sometimes be useful if the character of the singularity is known. Weinstein [113] discusses problems associated with the singular equation

$$u_{xx} + \frac{k}{y}u_y + u_{yy} = 0 \tag{7.280}$$

and its generalization. A bibliography of sixty-eight papers constitutes an excellent guide to the literature of some of the linear equations—for example the Tricomi equation, PED equation etc.

The second large class of singularities are those which occur on the boundary of the domain of integration. These arise as a result of discontinuities in boundary conditions or at a sudden change in boundary direction (corners) such as the "re-entrant corner" problem of fluid mechanics.

The treatment of both classes of problems is similar. We shall examine several methods for treating singularities as they apply to linear problems in the hope that they will suggest approaches (or actually be useful) in the more difficult nonlinear problems.

a. Subtracting Out the Singularity

Suppose u has a discontinuity at a point p. The goal is to calculate a solution U of the same partial differential equation such that $u - U$ is well behaved at the point p. To illustrate we suppose the solution of Laplace's equation $\nabla^2 u = 0$ is desired in the region $y > 0$ subject to the boundary conditions $u = f(x)$ for $x > 0$ and $u = g(x)$ for $x < 0$.[†] Let the discontinuities at $x = 0$ be defined as

$$f^{(n)}(0) - g^{(n)}(0) = a_n, \quad n = 0, 1, 2, 3, 4 \quad (7.281)$$

where the a_n are finite. Milne [8] constructs the *local* solution w of $\nabla^2 w = 0$ as

$$w(x, y) = \left\{ \sum_{j=0}^{4} a_j p_j(x, y) \right\} \frac{1}{\pi} \tan^{-1}(y/x) + \left\{ \sum_{j=1}^{4} a_j q_j(x, y) \right\} \frac{1}{2\pi} \ln(x^2 + y^2) \quad (7.282)$$

where $p_0 = 1$, $(p_j + iq_j) = 1/j! \, (x + iy)^j$, $j = 1, 2, 3, 4$, obtained by selecting real and imaginary parts. w satisfies Laplace's equation everywhere except at the origin since it is the imaginary part of the *analytic* function

$$\frac{1}{\pi} \sum_{j=0}^{4} (a_j z^j \ln z)/j!$$

[†] By a translation of the axis the point $x = 0$ can be moved to any other boundary point. There is no loss in generality in considering only $x = 0$.

For $y > 0$ we now take the limit as $y \to 0$ for the two cases $x > 0$ and $x < 0$. We easily show that

$$w(x, 0) = \begin{cases} 0, & x > 0 \\ a_0 + a_1 x + a_2 x^2/2 + a_3 x^3/6 + a_4 x^4/24, & x < 0 \end{cases} \quad (7.283)$$

so that w has the proper jumps at the origin as given by Eq. (7.281). The required solution is then obtained by solving $\nabla^2 v = 0$ where $v = u - w$ with boundary conditions

$$v(x, 0) = \begin{cases} f(x), & x > 0 \\ g(x) - a_0 - a_1 x - a_2 x^2/2 - a_3 x^3/6 - a_4 x^4/24, & x < 0 \end{cases}$$

which are *continuous* in v and its derivatives up to *fourth* order at the origin.

For the Poisson equation $\nabla^2 u = h(x, y)$ we need only use standard methods or their generalization (see Cobble and Ames [114]) and combine any particular integral of this equation with the Laplace equation solution.

We observe from this example that the success of the method of subtracting out the singularity depends strongly upon our ability to construct suitable solutions to the homogeneous differential equation. This may be very difficult in the nonlinear case. Sometimes local solutions may be found by asymptotic or series methods analogous to that given in Section 7.7 for the boundary layer equation.

b. Mesh Refinement

A very common but brutish method of dealing with discontinuities is effectively to ignore them and hopefully diminish their effect by using a refined mesh (smaller interval size) in the region near to and surrounding the singularity. A rather elegant refinement procedure is reported by Trutt *et al.* [101a] for the nonlinear elliptic equation (7.248). The two interior corners (Fig. 7-17) require the use of mesh refinement. The mesh refinement procedure appears to have the effect of minimizing the "area of infection" created by the singularity.

Milne [8] examines the area of infection problem for the case when the boundary turns through a right angle, with the conditions $u = 0$ on the x-axis and $u = 1$ on the y-axis. For Laplace's equation he shows how the effect of the singularity decreases with distance and what interval size is necessary near the singularity to achieve a required precision. Unfortunately these results apply only to the special case considered.

The procedure of subtracting out the singularity appears to be preferable to the refined net method, if the analysis can be performed.

c. The Method of Motz and Woods

The effect of a sudden change of direction of the boundary of a region gives a common form of disturbance. If the angle exceeds π the corner is called are-entrant corner (Fig. 7-21). Motz [115] considers the Laplace

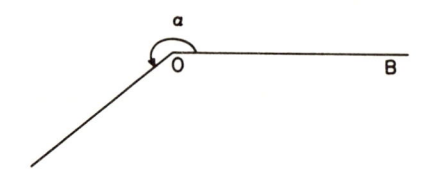

FIG. 7-21. A typical corner.

equation when $\alpha = 2\pi$, although the procedure applies to more general cases. Woods [116] uses similar methods for Poisson's equation $\nabla^2 u = f$ but his calculation requires iteration and is difficult to program for an electronic computer. We discuss herein the approach taken by Motz for Laplace's equation.

Examination of the effect of such a corner as that of Fig. 7-21 is facilitated by writing Laplace's equation in circular cylindrical coordinates (r, θ) as

$$u_{rr} + \frac{1}{r} u_r + \frac{1}{r^2} u_{\theta\theta} = 0. \tag{7.284}$$

By separation the elemental solutions are found to be

$$r^k \cos k\theta, \qquad r^k \sin k\theta. \tag{7.285}$$

When u is required to vanish on the boundary curve B we find the solution to be

$$u(r, \theta) = \sum_{n=1}^{\infty} a_n r^{n\pi/\alpha} \sin \frac{n\pi\theta}{\alpha} \tag{7.286}$$

7.20 SINGULARITIES

If $\partial u/\partial n = 0$ on B the solution is

$$u(r, \theta) = \sum_{n=0}^{\infty} b_n r^{n\pi/\alpha} \cos \frac{n\pi\theta}{\alpha}. \tag{7.287}$$

For $\alpha > \pi$ it is clear in both cases that for $n = 1$ the exponent of r is less than one so that $\partial u/\partial r$ has a singularity at the origin. If $\alpha \leqslant \pi$ this is not the case and in this sense the reentrant corner is more treacherous.

The specific problem of Motz is that of finding the solution to Laplace's equation $\nabla^2 u = 0$ in the finite rectangular region of Fig. 7-22 with $\alpha = 2\pi$. The boundary conditions, shown in Fig. 7-22, are

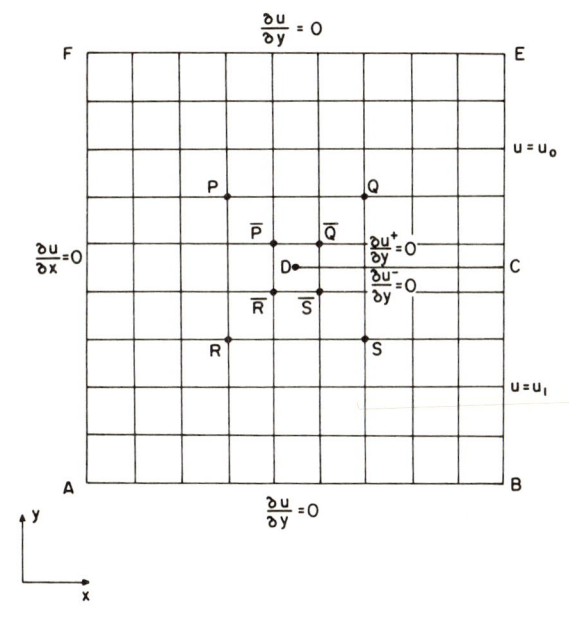

FIG. 7-22. The rectangular mesh illustrating the Motz-Woods procedure for a re-entrant corner.

$\partial u/\partial y = 0$ on AB and FE, $\partial u/\partial x = 0$ on AF, $u = u_1$ on BC, $u = u_0$ on CE and $u_y^+ = u_y^- = 0$ (the $\pm$ are u labels above and below the line DC). At most mesh points the standard five point star is used. However, in the neighborhood of the singularity D, the first few terms of the appropriate series [here (7.287)] are used—thus near D, using four coefficients, we have

$$u = b_0 + b_1 r^{\frac{1}{2}} \cos \tfrac{1}{2}\theta + b_2 r \cos \theta + b_3 r^{\frac{3}{2}} \cos \tfrac{3}{2}\theta. \tag{7.288}$$

Improved accuracy is obtained, not by iteration, but by use of more terms of the series.

Determination of the four coefficients in (7.288) is carried out, at the kth iteration step in the over-all solution, by using the u values at four "remote" points say P, Q, R, and S. Thus four linear equations of the form

$$u^k(P) = b_0 + b_1 r_p^{\frac{1}{2}} \cos \tfrac{1}{2}\theta_p + b_2 r_p \cos \theta_p + b_3 r_p^{\frac{3}{2}} \cos \tfrac{3}{2}\theta_p$$

result. These are solved for the b_i thereby furnishing a relation (7.288) for finding the values at the mesh points $\bar{P}, \bar{Q}, \bar{R}$, and $\bar{S}$. These special equations are used only at $\bar{P}, \bar{Q}, \bar{R}$, and $\bar{S}$.

Generalization of this method to $\pi < \alpha < 2\pi$ is easily carried out.

The Motz idea is simple, can produce accuracy in solutions which are unattainable by finite difference methods, and when high precision and an economic interval size is required is easily improved by selecting more terms in the series.

Some specific investigations have been reported in papers by Lieberstein [117] and Greenspan and Warten [118].

C. HYPERBOLIC EQUATIONS

7.21 METHOD OF CHARACTERISTICS

The concept of characteristics was introduced and discussed in some detail in Chapter 3 for quasi-linear hyperbolic equations. The basic rationale underlying the use of characteristics is that by an appropriate choice of coordinates the original system of hyperbolic first-order partial differential equations can be replaced by a system expressed in characteristic coordinates. These are the "natural" coordinates of the system. In terms of these coordinates differentiation is much simplified. Further we observed that the reduction becomes particularily simple when applied to systems involving only *two* independent variables.

A knowledge of the characteristics is important to the development and understanding of finite difference methods for hyperbolic systems. Indeed the so called "method of characteristics" is in reality the natural numerical procedure for hyperbolic systems. We shall therefor discuss the method in some detail for the case of two simultaneous quasi-linear equations.

7.21 METHOD OF CHARACTERISTICS

Let the ten coefficient functions $a_1, a_2, \ldots, f_1, f_2$ be functions of $x, y, u,$ and v and consider the simultaneous quasi-linear equations

$$a_1 u_x + b_1 u_y + c_1 v_x + d_1 v_y = f_1$$
$$a_2 u_x + b_2 u_y + c_2 v_x + d_2 v_y = f_2. \quad (7.289)$$

We now suppose that the solution for u and v is known from the initial state out to some curve Γ[†]. At any boundary point P, of this curve Γ, continuously differentiable values of u and v and directional derivatives of u and v in directions *below* the curve are known. (See Fig. 7-23).

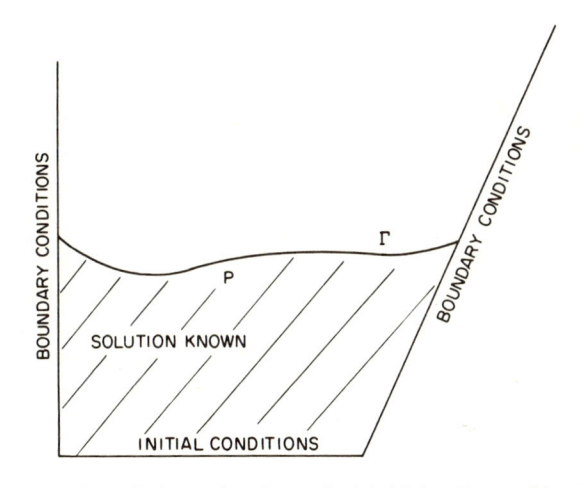

FIG. 7-23. Schematic of a typical initial value problem.

We now seek the answer to the question—is the behavior of the solution just above P uniquely determined by the information below and on the curve? Stated alternatively, is this data sufficient to determine the directional derivatives at P in directions that lie above the curve C? By way of reducing this question suppose θ (an angle with the horizontal) specifies a direction along which σ measures distance. If u_x and u_y are known at P then the directional derivative

$$u_\sigma \bigg|_\theta = u_x \cos \theta + u_y \sin \theta = u_x \frac{dx}{d\sigma} + u_y \frac{dy}{d\sigma} \quad (7.290)$$

[†] For the moment we restrict our discussion to a finite interval over which discontinuities do not occur. Later we shall discuss the degeneration of smooth solutions into discontinuities.

is also known. So we restate the question in the simpler form—under what conditions are the derivatives u_x, u_y, v_x, and v_y uniquely determined at P by values of u and v on Γ? At P we have the four relations (7.289) and

$$\begin{aligned} du &= u_\sigma \, d\sigma = u_x \, dx + u_y \, dy \dagger \\ dv &= v_\sigma \, d\sigma = v_x \, dx + v_y \, dy \end{aligned} \qquad (7.291)$$

whose matrix form is

$$\begin{bmatrix} a_1 & b_1 & c_1 & d_1 \\ a_2 & b_2 & c_2 & d_2 \\ dx & dy & 0 & 0 \\ 0 & 0 & dx & dy \end{bmatrix} \begin{bmatrix} u_x \\ u_y \\ v_x \\ v_y \end{bmatrix} = \begin{bmatrix} f_1 \\ f_2 \\ du \\ dv \end{bmatrix}. \qquad (7.292)$$

With u and v known at P the coefficient functions a_1, a_2, ..., f_1, f_2 are known. With the direction of Γ known, dx and dy are known, and if u and v are known along Γ, du and dv are also known. Thus the four equations (7.292) for the four partial derivatives have known coefficients. A unique solution for u_x, u_y, v_x, and v_y exists if the determinant of the 4×4 matrix in Eq. (7.292) is *not zero*. If the determinant is not zero then the directional derivatives have the same value above and below Γ.

The exceptional case when the determinant is zero implies a multiplicity of solutions are possible. Thus the system of equations (7.292) does not determine the partial derivatives uniquely. Consequently, discontinuities in the partial derivatives may occur as we cross Γ. Exploitation of this observation leads to better understanding of wavelike phenomena as discussed by Jeffrey and Taniuti [119] in their recent treatise on nonlinear wave propagation. In addition a numerical method is generated from this procedure.

Upon equating to zero the determinant of the matrix in Eq. (7.292) we find the *characteristic equation*

$$(a_1 c_2 - a_2 c_1)(dy)^2 - (a_1 d_2 - a_2 d_1 + b_1 c_2 - b_2 c_1) \, dx \, dy + (b_1 d_2 - b_2 d_1)(dx)^2 = 0. \qquad (7.293)$$

which is a quadratic equation in dy/dx. If the curve Γ (Fig. 7-23) at P has a slope such that Eq. (7.293) is satisfied then the derivatives u_x, u_y, v_x, and v_y are not uniquely determined by the values of u and

---

† σ is now measuring distance along L.

7.21 METHOD OF CHARACTERISTICS

v on Γ. The directions specified by Eq. (7.293) are called *characteristic directions* which may be real and distinct, real and identical or not real according as the discriminant

$$(a_1 d_2 - a_2 d_1 + b_1 c_2 - b_2 c_1)^2 - 4(a_1 c_2 - a_2 c_1)(b_1 d_2 - b_2 d_1) \qquad (7.294)$$

is positive, zero, or negative. This is also the criterion for classifying Eqs. (7.289) as hyperbolic, parabolic, or elliptic. These are hyperbolic if Eq. (7.294) is positive, i.e., has two *real* characteristic directions; parabolic if Eq. (7.294) is zero, and elliptic if there are no real characteristic directions.

Such a classification as that given above was stated for the linear second-order system described by Eq. (7.155). For the quasi-linear second-order equation

$$a u_{xx} + b u_{xy} + c u_{yy} = f \qquad (7.295)$$

where a, b, c are functions of x, y, u, u_x, and u_y, a similar process to that above may be applied without the preliminary reduction of Eq. (7.295) to Eq. (7.289). The details are given in numerous references (see for example Jeffrey and Tanuiti [119], p. 9). The equation analogous to Eq. (7.292) is

$$\begin{bmatrix} a & b & c \\ dx & dy & 0 \\ 0 & dx & dy \end{bmatrix} \begin{bmatrix} u_{xx} \\ u_{xy} \\ u_{yy} \end{bmatrix} = \begin{bmatrix} f \\ d(u_x) \\ d(u_y) \end{bmatrix}. \qquad (7.296)$$

Thus the vanishing of

$$a(dy)^2 - b\, dx\, dy + c(dx)^2 \qquad (7.297)$$

is seen to be the condition for the unique determination of the second-order derivatives and hence also for higher derivatives. Accordingly the analog of Eq. (7.293), the *characteristic equation*,

$$a(dy)^2 - b\, dx\, dy + c(dx)^2 = 0 \qquad (7.298)$$

defines two real characteristic directions (the hyperbolic case) if $b^2 - 4ac > 0$. These characteristic curves have the property that along them the higher order derivatives are indeterminate so that a discontinuity (perhaps infinitesimal) can occur across the curve.

For the general second-order equations

$$F(x, y, u, p, q, r, s, t) = 0 \qquad (7.299)$$

with $p = u_x$, $q = u_y$, $r = u_{xx}$, $s = u_{xy}$, and $t = u_{yy}$ we have the three equations

$$\begin{bmatrix} F_r & F_s & F_t \\ dx & dy & 0 \\ 0 & dx & dy \end{bmatrix} \begin{bmatrix} u_{xx} \\ u_{xy} \\ u_{yy} \end{bmatrix} = \begin{bmatrix} G(x, y, u, p, q) \\ d(u_x) \\ d(u_y) \end{bmatrix}$$

From this equation the characteristic equation

$$F_r(dy)^2 - F_s dx\, dy + F_t(dx)^2 = 0 \qquad (7.300)$$

is easily obtained. Equation (7.298) is certainly a special case of Eq. (7.300). Equation (7.299) is hyperbolic if $F_s^2 - 4F_rF_t > 0$, parabolic if $F_s^2 - 4F_rF_t = 0$, and elliptic if $F_s^2 - 4F_rF_t < 0$.

The quasilinear equation (7.295) and general nonlinear equation (7.299) can also be examined by reduction to a first-order equation system. The argument we give shall be for the general second-order nonlinear partial differential equation in one dependent and two independent variables but it is capable of immediate extension to higher order systems. We must emphasize that this "factorization" of Eq. (7.299) into simultaneous first-order equations is *not unique*.[†]

Let us consider Eq. (7.299) with initial data specified on a *noncharacteristic curve* at $x = 0$ (the reason for this will be explained later)

[†] The nonuniqueness is easily demonstrated by means of the example $aw_{xx} + cw_{yy} + w_x = 0$. Both substitutions (i) $u = w_x$, $v = w_y$, and (ii) $u = w_x$, $v = w_x + w_y$ reduce the second-order equation to the required form. In the matrix notation $U_x + AU_y + B = 0$ we find for (i) that

$$U = \begin{bmatrix} u \\ v \\ w \end{bmatrix}, \quad A = \begin{bmatrix} 0 & c/a & 0 \\ -1 & 0 & 0 \\ 0 & 0 & 1 \end{bmatrix}, \quad B = \begin{bmatrix} u/a \\ 0 \\ -u-v \end{bmatrix}.$$

For (ii) we have

$$U = \begin{bmatrix} u \\ v \\ w \end{bmatrix}, \quad A = \begin{bmatrix} -c/a & c/a & 0 \\ -(1+c/a) & c/a & 0 \\ 0 & 0 & 1 \end{bmatrix}, \quad B = \begin{bmatrix} u/a \\ u/a \\ -v \end{bmatrix}.$$

Of these equations, the first two are required for computation while the third is introduced for definition purposes. Some forms may have computational convenience over others. Such a case occurred in a paper by Swope and Ames [120].

7.21 METHOD OF CHARACTERISTICS

$u(0, y) = f(y)$, $u_x(0, y) = g(y)$. Differentiating Eq. (7.299) with respect to x gives

$$F_x + F_u p + F_p r + F_q s + F_r r_x + F_s s_x + F_t t_x = 0.$$

However, we note that $p_y = q_x = s$, $s_x = r_y$, $t_x = s_y$ so that Eq. (7.299) can be written as the first order system

$$F_x + F_u p + F_p r + F_q s + F_r r_x + F_s r_y + F_t s_y = 0$$

$$u_x = p, \quad p_x = r, \quad q_x = p_y, \quad s_x = r_y, \quad t_x = s_y. \qquad (7.301)$$

The initial conditions are $u(0, y) = f(y)$, $p(0, y) = g(y)$, $q(0, y) = f'(y)$, $s(0, y) = g'(y)$, $t(0, y) = f''(y)$ and $r(0, y)$ is obtained from the original equation

For definiteness we now concentrate our attention on the first-order quasi-linear system (7.289) which is hyperbolic. Therefore there are two real characteristic directions, at each point, given by $(dy/dx)_\alpha$ and $(dy/dx)_\beta$, say, as the roots of the quadratic (7.293). Two families of characteristic curves cover the domain of integration in a *not necessarily orthogonal* net. If the curves have the slope $(dy/dx)_\alpha$ or $(dy/dx)_\beta$ we call them α or β characteristics respectively.

In our previous discussion the characteristics were obtained as loci of possible discontinuities in the partial derivatives. We can exploit the vanishing of the determinant of the 4×4 matrix in Eq. (7.292) once more to complete this method. Suppose then that we are considering a characteristic direction so that the determinant of Eq. (7.292) is zero. If there are to be any solutions at all for the first derivatives then the right-hand column matrix must be compatible with this vanishing. That is to say when the right-hand column vector replaces *any* of the columns in the 4×4 matrix the resulting determinant must also vanish.[†] Upon replacing the first column (any one may be used) on the left by the right-hand column and setting the determinant equal to zero we find

$$\left[(b_1 c_2 - b_2 c_1)\frac{dy}{dx} - (b_1 d_2 - b_2 d_1)\right] du + \left[(d_1 c_2 - d_2 c_1)\frac{dy}{dx}\right] dv$$
$$+ \left[(c_1 f_2 - c_2 f_1)\frac{dy}{dx} - (d_1 f_2 - d_2 f_1)\right] dy = 0. \qquad (7.302)$$

[†] This result follows from the following elementary theorem. Let A be a square matrix, x and c be column vectors. If in the matrix equation $Ax = c$ the determinant of A vanishes then a necessary condition for finite solutions to exist for x is that when c is substituted for any column of A the resulting determinant vanishes.

Knowledge of the characteristic directions $(dy/dx)_\alpha$, $(dy/dx)_\beta$ allows one to eliminate the derivatives but not the dy except in theory or numerically (since we can calculate the characteristic curves as $y_\alpha = f(x)$, $y_\beta = g(x)$). In general we see that along the α and β characteristics the functions u and v are related by equations of the form

$$A_\alpha \, du + B_\alpha \, dv + C_\alpha \, dy = 0$$
$$A_\beta \, du + B_\beta \, dv + C_\beta \, dy = 0. \qquad (7.303)$$

For the important special case where $f_1 = f_2 \equiv 0$ the use of the characteristic directions in Eq. (7.302) yields two ordinary differential equations for u and v of the form

$$A_\alpha \, du + B_\alpha \, dv = 0, \qquad A_\beta \, du + B_\beta \, dv = 0 \qquad (7.304)$$

along the characteristics.

The *method of characteristics* is expressible succinctly as: first locate the characteristic curves (solve Eq. (7.293)) and second integrate the ordinary differential equations (7.302) along the characteristics. In practice these calculations are carried out numerically since only very special cases can be integrated exactly by this method. One such case is the linear wave equation $u_x - v_y = 0$, $u_y - v_x = 0$. Another such case is detailed for the threadline equation $u_{tt} + \alpha u_{xt} + \beta u_{xx} = 0$ by Swope and Ames [120].

If one wishes to treat the second-order equation (7.295) directly, the characteristic directions are found from Eq. (7.298) and the equations for $p = u_x$ and $q = u_y$, along the characteristics, are found by the argument leading to Eq. (7.302) to be derivable from (say)

$$\begin{vmatrix} a & f & c \\ dx & dp & 0 \\ 0 & dq & dy \end{vmatrix} = 0$$

as

$$a\left(\frac{dy}{dx}\right)_\alpha dp + c \, dq - f \, dy = 0$$
$$a\left(\frac{dy}{dx}\right)_\beta dp + c \, dq - f \, dy = 0. \qquad (7.305)$$

7.22 THE SUPERSONIC NOZZLE

The governing equations of steady two-dimensional irrotational isentropic flow are (see for example Shapiro [121])

$$uu_x + vu_y + \rho^{-1}p_x = 0$$
$$uv_x + vv_y + \rho^{-1}p_y = 0$$
$$(\rho u)_x + (\rho v)_y = 0, \quad v_x - u_y = 0 \quad (7.306)$$
$$p\rho^{-\gamma} = \text{constant}, \quad \frac{dp}{d\rho} = c^2$$

where u and v are velocity components, p is pressure, ρ is density, c is the velocity of sound, and γ is the ratio of specific heats for air ($\gamma = 1.4$).

Upon multiplying the first of Eqs. (7.306) by ρu, the second by ρv, using $dp = c^2 d\rho$ and adding the two resulting equations we find that Eqs. (7.306) is equivalent to the following pair of first-order equations for u and v,

$$(u^2 - c^2)u_x + (uv)u_y + (uv)v_x + (v^2 - c^2)v_y = 0$$
$$-u_y + v_x = 0 \quad (7.307)$$

where

$$5c^2 = 6c^{*2} - (u^2 + v^2)$$

and the quantity c^* is a reference sound velocity chosen as the sound velocity when the flow velocity $(u^2 + v^2)^{\frac{1}{2}} = c$. This problem can be put in dimensionless form by setting

$$u^1 = u/c^*, \quad v^1 = v/c^*, \quad c^1 = c/c^*, \quad x^1 = x/l, \quad y^1 = y/l$$

where l is one half the nozzle width. Inserting these values in Eq. (7.307) and dropping the primes the dimensionless equations are

$$(u^2 - c^2)u_x + (uv)u_y + (uv)v_x + (v^2 - c^2)v_y = 0$$
$$-u_y + v_x = 0 \quad (7.308)$$

with $c^2 = 1.2 - 0.2(u^2 + v^2)$.

The characteristic directions are obtained from the particular form of Eq. (7.292)

$$\begin{bmatrix} (u^2 - c^2) & uv & uv & (v^2 - c^2) \\ 0 & -1 & 1 & 0 \\ dx & dy & 0 & 0 \\ 0 & 0 & dx & dy \end{bmatrix} \begin{bmatrix} u_x \\ u_y \\ v_x \\ v_y \end{bmatrix} = \begin{bmatrix} 0 \\ 0 \\ du \\ dv \end{bmatrix}$$

as

$$\frac{dy}{dx}\bigg|_\alpha = \frac{uv + c[u^2 + v^2 - c^2]^{\frac{1}{2}}}{u^2 - c^2} \qquad (7.309a)$$

$$\frac{dy}{dx}\bigg|_\beta = \frac{uv - c[u^2 + v^2 - c^2]^{\frac{1}{2}}}{u^2 - c^2} \qquad (7.309b)$$

which are *always real* in the supersonic case. The equations for u and v along the characteristics are obtained, for example via

$$\begin{vmatrix} (u^2 - c^2) & uv & uv & 0 \\ 0 & -1 & 1 & 0 \\ dx & dy & 0 & du \\ 0 & 0 & dx & dv \end{vmatrix} = 0$$

as

$$(u^2 - c^2)\,du + [uv \mp c(u^2 + v^2)^{\frac{1}{2}}]\,dv = 0 \qquad (7.310)$$

where the $(-)$ sign applies to the α characteristic and the $(+)$ sign to the β characteristic.

The normal course of affairs would be to integrate Eqs. (7.309) and (7.310) stepwise (numerically) to successively build up the characteristic

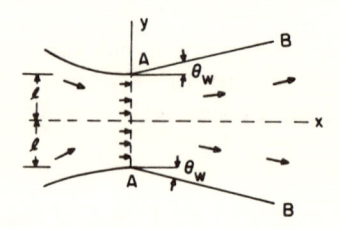

FIG. 7-24a. Supersonic nozzle configuration.

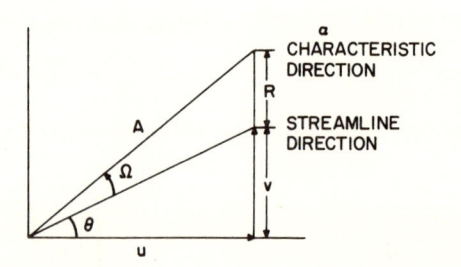

FIG. 7-24b. Relation between the streamline direction and α characteristic direction for a supersonic nozzle.

7.22 THE SUPERSONIC NOZZLE

net and the solutions along the characteristics. In the present case analysis can carry us further. We consider the nozzle shown in Fig. 7-24. At the throat AA ($x = 0$) the flow is assumed to be *parallel sonic flow* so that the following relations hold:

at $x = 0$, $\quad -l < y < l$, $\quad u = c^*$, $\quad v = 0$;

at $y = 0$, $\quad v = 0$; $\quad$ at AB, $\quad v = u \tan \theta_w$.

The streamline direction is defined as

$$\tan \theta = v/u \tag{7.311}$$

and by elementary trigonometry (see Fig. 7-24) we can easily show that both characteristic directions make angles η with the streamline direction θ where

$$\eta = \sin^{-1}[c(u^2 + v^2)^{-\frac{1}{2}}]. \tag{7.312}$$

We now transform the problem into (q, θ) coordinates where

$$q^2 = u^2 + v^2$$
$$u = q \cos \theta, \; v = q \sin \theta \tag{7.313}$$
$$c = q \sin \eta = [(6 - q^2)/5]^{\frac{1}{2}}.$$

Elementary operations on Eqs. (7.309) yield the "polar" form for the *characteristic directions*

$$\left.\frac{dy}{dx}\right|_{\alpha,\beta} = \frac{\sin \theta \cos \theta \pm \sin \eta \cos \eta}{\cos^2 \theta - \sin^2 \eta}$$
$$= \tan(\theta \pm \eta). \tag{7.314}$$

Since $du = \cos \theta \, dq - q \sin \theta \, d\theta$ and $dv = \sin \theta \, dq + q \cos \theta \, d\theta$ we can transform Eqs. (7.310) for u and v *along* the characteristics into

$$q \, d\theta = \pm \cot \eta \, dq. \tag{7.315}$$

Integrating Eq. (7.315) in the α-direction (+ sign) and β-direction (− sign) we find respectively

$$\theta - \omega = \alpha, \quad \theta + \omega = \beta \tag{7.316}$$

where the constants α and β are used as mnemonic symbols denoting the particular characteristics and

$$\omega = \int_1^q \cot \eta \, \frac{dq}{q} = \int_1^q \left[\frac{6(q^2-1)}{6-q^2}\right]^{\frac{1}{2}} \frac{dq}{q}$$
$$= \int_\eta^{\pi/2} \frac{5(1-\sin^2 \eta)}{1+5\sin^2 \eta} \, d\eta \quad (7.317)$$

by virtue of the relations (7.313).

From Eq. (7.316) we can obtain the relations

$$\theta = \frac{\alpha+\beta}{2}, \quad \omega = \frac{\beta-\alpha}{2} \quad (7.318)$$

which will yield the complete solution *after the characteristics have been found*.

Similar analyses can be carried out for the one dimensional expansion of a gas behind a piston, described in Chapter 3 (see Section 7.24), compression of a plastic bar (see Chapter 1), drawing of a plastic strip between rigid walls by Sokolovskii [122], magnetohydrodynamics studies by Gundersen [123], [124], water waves in shallow water [125], and others (see for example Jeffrey and Taniuti [119]).

7.23 PROPERTIES OF HYPERBOLIC SYSTEMS

a. Uniqueness

Our discussion of "factoring" of the general nonlinear second-order equation (7.299) required that the initial data be specified on a curve which was *not* a characteristic curve. The reason for this is that if the initial curve is a characteristic we shall in general have no solution at all and in any case no unique solution. To illustrate the difficulties which arise, and hence motivate the need for some *uniqueness* ideas we consider two elementary examples.

First consider the simple first-order equation

$$2u_x + u_y = 1 \quad (7.319)^\dagger$$

---

† The characteristics for the quasi-linear first-order equation $au_x + bu_y = c$ are found from the subsidiary equations $dx/a = dy/b$ and along the characteristics $dx/a = dy/b = du/c$.

7.23 PROPERTIES OF HYPERBOLIC SYSTEMS

for which the characteristics are $dx/2 = dy$ and along the characteristics $\frac{1}{2}dx = dy = du$. Thus the characteristics are the straight lines $y = x/2 + e$ and along these lines we have $u = \frac{1}{2}x + g = y + g - e$ (Fig. 7-25).

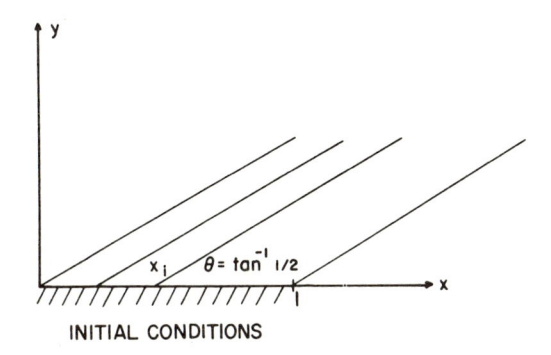

FIG. 7-25. Characteristics for a first-order partial differential equation.

Let us suppose the initial data for u is given on the noncharacteristic line segment $y = 0$, $0 < x < 1$ as $u(x_i, 0) = u_i$. The value of u at other points is obtained by integrating along the characteristics drawn from the points x_i on the initial line. Thus $u(x, y)$ from the solution on the characteristics is

$$u(x, y) = u_i + \tfrac{1}{2}(x - x_i) = u_i + y \tag{7.320}$$

on the line $y = \frac{1}{2}(x - x_i)$ for each x_i, $0 < x_i < 1$. Further we can prove (see for example Bernstein [126]) that the solution *exists* and is *unique* in the region bounded by the terminal characteristics originating at $x = 0$ and $x = 1$ in Fig. 7-25.

On the other hand, if the initial curve is one of the characteristics, say the line $x/2 = y$, passing through the origin, the situation is quite different. The "nonuniqueness condition" associated with the characteristics makes further investigation mandatory. There is a possibility of discontinuous first partial derivatives u_x and u_y on $y = x/2$. From the relations along the characteristics we see that there is a solution only if that initial data is $u = x/2 = y$ on $y = x/2$, Elsewhere the solution is not unique since we can take, for example,

$$u = x/2 + A(y - x/2) \tag{7.321}$$

which is a solution *for any value of A*. The nonuniqueness results from the fact that effectively the "terminal characteristics" are coincidental.

The preceding arguments can be applied to second-order NLPDE and first order systems with qualitatively similar results. In general if the initial curve is a characteristic we may have no solution at all (if the initial data are not properly chosen) and in any case *no unique* solution.

Some guide to existence and uniqueness is advisable if not absolutely necessary before attempting numerical (and analytic) solution. Ignoring the available theorems may lead to no adverse effects in 90% of the problems. It is this possible 10% that suggests we spend some effort on the uniqueness-existence question. The extent to which initial and boundary conditions determine *unique* solutions for the quasi-linear system (7.289) can be obtained in most cases from the following theorems[†] (see Bernstein [126], Garabedian [54]).

(1) Suppose continuously differentiable values of u and v are specified on the *noncharacteristic* curve CD of Fig. 7-26a. We also assume CD is continuously differentiable. A solution to Eq. (7.289), assuming these prescribed values, is uniquely determined in the region CDE bounded by the initial curve CD, the β characteristic CE and the α characteristic DE. The direction of propagation is assumed upward but if it were reversed there would be a corresponding "triangle" of uniqueness below CD.

(2) Suppose CD is a noncharacteristic curve which is continuously differentiable. A unique solution is determined in the region $CDEF$ of Fig. 7-26b where DE and EF are characteristics, provided that u and v are known at C and continuously differentiable values of u or v are given along each of the segments CD and CF. The values at C must be compatible with the characteristics. A unique solution can sometimes be assured even when a discontinuity appears at C.

(3) In the case, sketched in Fig. 7-26,c with CE and CD characteristics, a unique solution is determined in the region $CDFE$ where EF and FD are characteristics when u and v are known at C and continuously differentiable values of u or v are given along CE and CD. The values at C must be compatible with the characteristics.

In the results just stated it is assumed that no boundary interference or other obstruction is present in the considered domain. It is quite possible that unanticipated boundaries such as shock waves, flame fronts, and other discontinuities may appear within the solution domain.

[†] Intensive study of existence-uniqueness problems continues. We shall review some recent results in Chapter 8.

7.23 PROPERTIES OF HYPERBOLIC SYSTEMS

Such discontinuities of properties are propagated by their own special laws and represent boundaries between regions where different equations must be solved. Usually these locations *are not* known in advance and must be determined by a simultaneous computation with the continuous solutions.

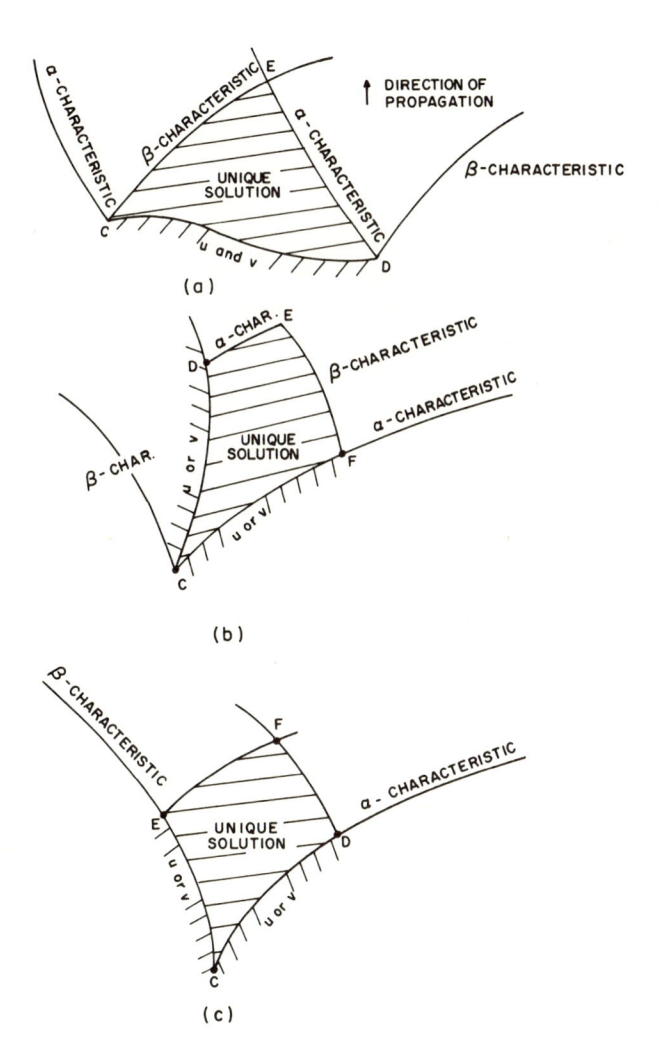

FIG. 7-26. Uniqueness domains for hyperbolic systems.

b. Propagation of Discontinuities

We have seen that the characteristic curves are loci of possible discontinuities in the derivatives of u and v. They are therefore propagation paths for small disturbances. The possibility of propagation of discontinuous initial values into the field can be discussed in the light of the example previously given [Eq. (7.319)]. Suppose in that example that the data on the initial line is prescribed as

$$u = f(x), \quad 0 < x < x_1, \quad u = g(x), \quad x_1 < x < 1 \quad (7.322)$$

such that $f(x_1) \neq g(x_1)$ so that u is double valued at this point. From the solution (7.320), along the characteristics, this double-valued nature will persist all along the specific characteristic $y = \frac{1}{2}(x - x_1)$. The values to the left of this characteristic will be determined by $u = f(x)$ and to the right by $g(x)$.

If the initial data are such that there is a discontinuous slope this will also propagate into the integration field. For example if on $y = 0$

$$u = x^2 \quad \text{for} \quad 0 < x < \tfrac{1}{2}, \quad u = -x + \tfrac{1}{2} \quad \text{for} \quad \tfrac{1}{2} < x < 1 \quad (7.323)$$

the solution of the differential equation (7.319) is

$$u(x, y) = x_i^2 + y \text{ on the line } y = \tfrac{1}{2}(x - x_i) \text{ to the left of } y = \tfrac{1}{2}(x - \tfrac{1}{2})$$

and

$$u(x, y) = \tfrac{1}{2}[x - 3x_i + 1] \text{ on the line } y = \tfrac{1}{2}(x - x_i) \text{ to the right of } y = \tfrac{1}{2}(x - \tfrac{1}{2}).$$

There are discontinuities in u_x and u_y along the characteristic $y = \tfrac{1}{2}(x - \tfrac{1}{2})$.

In more general problems if a discontinuity exists across a certain characteristic at one point, there will be a discontinuity across that characteristic along its entire length. We have also noted that characteristic curves are the natural boundaries for determining which portions of a solution domain are influenced by which boundary conditions. Thus we see that propagated discontinuities and the segmenting by the characteristics restricts the use of finite difference methods. Care must be exercised that discrete models of the continuous system reflects all these facts. The characteristics may not pass through many points of, say, a rectangular grid and any propagated discontinuities would give rise to difficult computational problems.

c. The Constant State and Simple Wave

We return now to a re-examination of the quasi-linear system (7.289) with characteristic directions specified by Eq. (7.293) and the differential equations for u and v along the characteristics given by Eqs. (7.302). The important reducible case (see Chapter 2) has coefficients which depend only on u and v and not upon x and y. In this case the coefficients $a_1, a_2, \ldots$ may be complicated functions of u and v and still have parallel straight lines as characteristics. This occurs in a region of *constant state* in which both u and v remain constant. At first glance this may appear to be trivial but a number of important physical problems have solutions which consist of several regions of constant state interconnected by regions in which u and v change. In a region of constant state all characteristics (of each family) bear the same label.

Now suppose that u and v remain constant along *any one* β characteristic but vary in value from one β characteristic to the next. Generally the slopes will be constant for each β characteristic but will be different between β characteristics so that the β family will be straight nonparallel lines. They often resemble a fan as shown in Fig. 7-27. Since u and v

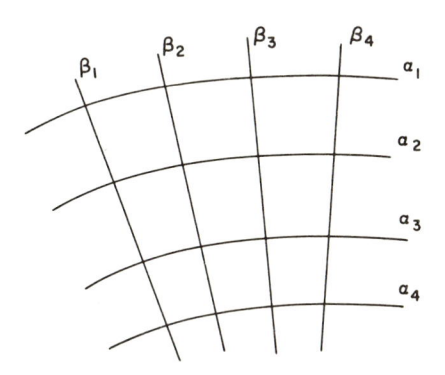

FIG. 7-27. The configuration of a simple wave. The β characteristics are straight lines with different slopes. The α characteristics are curves.

are constant along any specific β characteristic each α characteristic will have the same slope as it crosses this β line. This slope varies from β line to β line so the α family consists of curves. In this case the β characteristics carry a different label but all the α characteristics carry the same label. This results since as the curves marked α_i cross (say) β_2 they take on the same value of u and v. Therefore the integrated relation between u and v along any α_i must be identical with all others. The configuration just described is called a *simple wave*.

Apart from their relative simplicity simple waves are important because of the following theorem:

In a solution containing constant state regions the regions adjacent to constant states are always simple waves. A proof of this theorem is found in Jeffrey and Taniuti [119 pp. 69–71]. This important character of simple waves plays a fundamental role in the construction of solutions of hyperbolic systems.

7.24 ONE-DIMENSIONAL ISENTROPIC FLOW

A typical example where constant states and simple waves play a role is that of the one dimensional isentropic flow of an inviscid gas expanding behind a piston. We have previously discussed the equation in Chapter 3. Suppose for $t' < 0$ the gas is at rest with density ρ_0, the piston is at $x' = a$ and the sound velocity of gas is c_0 in its initial state. A dimensionless formulation can be obtained by introducing the new variables

$$u = u'/c_0, \quad c = c'/c_0, \quad \rho = \rho'/\rho_0, \quad x = x'/a, \quad t = t'c_0/a$$

so that the equations become

$$u_t + uu_x + c^2\rho^{-1}\rho_x = 0, \quad \rho_t + u\rho_x + \rho u_x = 0, \quad c = \rho^{(\gamma-1)/2} \quad (7.324)$$

with u as velocity. At $t = 0$ the piston is withdrawn with (dimensionless) velocity ϵ ($0 < \epsilon < 1$). The problem is therefore characterized by Eqs. (7.324) with the *initial* conditions

$$u(x, 0) = 0, \quad \rho(x, 0) = 1, \quad 0 < x < 1 \quad (7.325a)$$

and boundary conditions

$$u(0, t) = 0, \quad u(1 + \epsilon t, t) = \epsilon. \quad (7.325b)$$

The last condition in Eq. (7.325b) describes the piston motion—its position at time t is given by $1 + \epsilon t$ and it is moving with velocity ϵ.

Upon forming the matrix of Eq. (7.292) we find the equations for the characteristics of Eq. (7.324) to be

$$\left.\frac{dx}{dt}\right|_\alpha = u + c, \quad \left.\frac{dx}{dt}\right|_\beta = u - c \quad (7.326)$$

7.24 ONE-DIMENSIONAL ISENTROPIC FLOW

and the equations along the characteristics as

$$c\rho^{-1}d\rho + du = 0 \quad \text{along } \alpha \text{ characteristics}$$
$$c\rho^{-1}d\rho - du = 0 \quad \text{along } \beta \text{ characteristics.} \quad (7.327)$$

We cannot obtain the characteristic curves in advance here since Eqs. (7.326) depend upon u and c. However, we can integrate Eqs. (7.327) when we recall that $c = \rho^{(\gamma-1)/2}$. Thus we have (with $\gamma = 1.40$)

$$\frac{2}{\gamma - 1}c + u = 5c + u = \alpha$$
$$\frac{2}{\gamma - 1}c - u = 5c - u = \beta \quad (7.328)$$

so that *after the characteristics are found* these relations, when solved for u and c, yield the complete solution as

$$c = \frac{\alpha + \beta}{10}, \quad u = \frac{\alpha - \beta}{2} \quad (7.329)$$

a form which resembles Eqs. (7.318). These equations yield the values of u and c at a point in terms of the labels on the two characteristics passing through the point.

We now examine the configuration of the characteristic curves in the integration domain of this problem as sketched in Fig. 7-28. The initial data are $c = 1$, $\rho = 1$, and $u = 0$ at $t = 0$. Consequently we see from Eqs. (7.328) that characteristic curves originating from the initial line OA all bear the labels $\alpha = 5$ or $\beta = 5$. Further all these characteristic curves are straight lines with slopes $+1$ for the α characteristics and -1 for the β characteristics (see Eqs. (7.326)). The boundary condition $u = 0$ on OB coupled with the second equation of (7.329) implies that the α characteristics reflected from OB carry the same label as the incident β characteristics. Therefore the solution in the triangular shaped region OAB is a region of *constant state* having $\alpha = 5$, $\beta = 5$ and from Eqs. (7.329) $u = 0$, $c = 1$. The boundary curve AB is the line $t = 1 - x$ so the coordinates of B are $(0, 1)$. The physical meaning of this constant state is that of portions of the gas that have not yet been reached by the disturbance wave created by the moving piston. A unique solution in this triangle OAB is guaranteed by Theorems (1) and (2)

434   7. NUMERICAL METHODS

of Section 7.23a. (The reader may wish to examine the remaining discussion in the light of the other theorems of Section 7.23a).

The constant state just discussed cannot continue beyond AB since the boundary condition $u = \epsilon$ on AD disagrees with $u = 0$ of the constant state. Adjacent to the constant state will be a simple wave whose

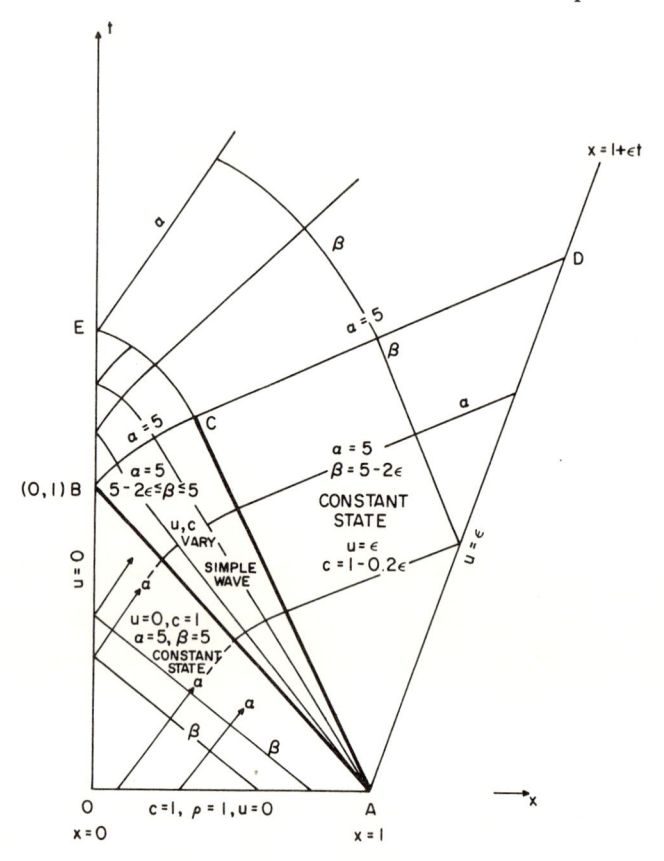

FIG. 7-28. The integration domain illustrating the solution by characteristics of isentropic one-dimensional compressible flow of a gas.

extent will be determined by the condition of the solution on its right. The extent of the simple wave is determined by the requirement that along the last of the β characteristics AC the value of u must be ϵ. The β characteristics are straight lines whose slope and label vary from line to line. The α characteristics are all curved and bear the same label

$\alpha = 5$ (compare Fig. 7-27), inherited from the region of constant state. In this simple wave region u and c as functions of β are

$$u = \frac{5-\beta}{2}, \qquad c = \frac{5+\beta}{10}. \tag{7.330}$$

From Eq. (7.326) the slope of the β characteristic is

$$\left.\frac{dt}{dx}\right|_\beta = \frac{1}{u-c} = \frac{5}{10-3\beta}. \tag{7.331}$$

Now the last β characteristic requires $u = \epsilon$. From Eq. (7.330) we find $\beta = 5 - 2\epsilon$ and hence $c = 1 - 0.2\epsilon$, and $dt/dx\,|_\beta = 5/(6\epsilon - 5)$. The resulting solution is shown in Fig. 7-28.

By analysis analogous to that above we can obtain the constant state $\alpha = 5$, $\beta = 5 - 2\epsilon$ to the right of AC in the triangular region ACD. Here $u = \epsilon$ and $c = 1 - 0.2\epsilon$. Consequently, in the region below the curve BCD the complete solution is obtainable at any point via Eqs. (7.330) as modified for the specific domains.

Further analysis will yield the following results for this example:

(a) the equations of the α characteristics in the simple wave;

(b) a solution in BCE can be carried out by a Picard iteration;

(c) the state directly above ECD is a simple wave with $\beta = 5 - 2\epsilon$ and α variable.

However, the region BCE, where *both* characteristics are curved is more difficult than the simple wave or constant state case and it is here where numerical integration is usually used. We describe the general attack on such problems in the next section.

7.25 METHOD OF CHARACTERISTICS: NUMERICAL COMPUTATION

We shall consider here the basic computational step in advancing the solution from a curve Γ on which we know x, y, u, and v. We suppose our problem is governed by the simultaneous first-order equations (7.289). As previously described we develop the four ordinary differential equations, two describing the characteristics and two relating the dependent variables u and v along the characteristics as

$$\left.\frac{dy}{dx}\right|_\alpha = A(x, y, u, v) \qquad \left.\frac{dy}{dx}\right|_\beta = B(x, y, u, v) \qquad (7.332a)$$

$$\left.\frac{du}{dv}\right|_\alpha = C(x, y, u, v) \qquad \left.\frac{du}{dv}\right|_\beta = D(x, y, u, v). \qquad (7.332b)$$

As in Fig. 7-29 we consider any two adjacent points P and Q on Γ and suppose the α characteristic from P intersects the β characteristic from Q at the point R. Since the coefficients of Eq. (7.332a) depend

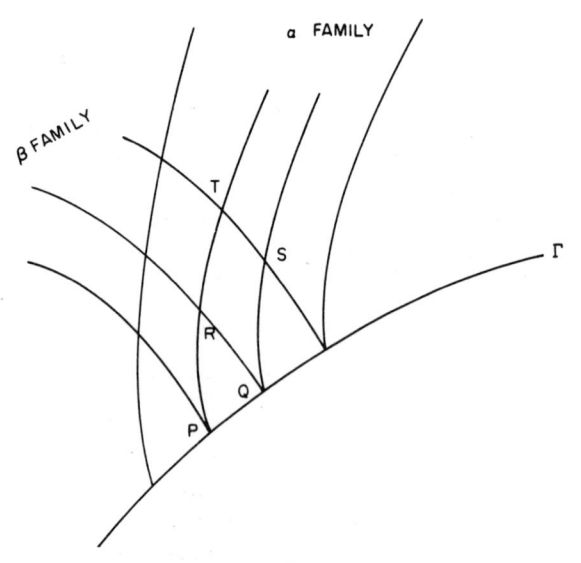

FIG. 7-29. The numerical "method of characteristics."

upon the solution we must determine the (x, y) coordinates of R as well as the values of u and v at this point. In principle any method of numerically integrating ordinary differential equations can be used to integrate the system (7.332). The most common methods are probably the Runge-Kutta method, Adams method and corrected Euler method (see Collatz [127]).

With Euler's method PR can be taken as a straight line but this leads to a discretization error which is $O(h)$. Use of parabolic arcs gives a higher order discretization error ($O(h^2)$). Consequently we use the relations

$$\frac{y_R - y_P}{x_R - x_P} = \tfrac{1}{2}[A(R) + A(P)]^\dagger, \qquad \frac{y_R - y_Q}{x_R - x_Q} = \tfrac{1}{2}[B(R) + B(Q)]$$

$$\frac{u_R - u_P}{v_R - v_P} = \tfrac{1}{2}[C(R) + C(P)], \qquad \frac{u_R - u_Q}{v_R - v_Q} = \tfrac{1}{2}[D(R) + D(P)]$$

(7.333)

as approximations for the four differential equations (7.332). We usually have to solve these by iteration which we shall not discuss. When a boundary is reached the equations are modified to insert the given boundary condition.

By this method we can compute the required values at other grid points of the characteristic net adjacent to the curve Γ, such as at S and from R and S to T, etc.

In the important special reducible case where A and B are independent of x and y we can solve the first pair of equations (7.333) as

$$x_R = \frac{y_Q - y_P + \tfrac{1}{2}[A(P) + A(R)]x_P - \tfrac{1}{2}[B(Q) + B(R)]x_Q}{\tfrac{1}{2}[A(P) + A(R)] - \tfrac{1}{2}[B(Q) + B(R)]}$$

$$y_R = y_P + \tfrac{1}{2}[A(P) + A(R)](x_R - x_P) = y_Q + \tfrac{1}{2}[B(Q) + B(R)](x_R - x_Q).$$

The calculation of y_R by both equations acts as a numerical check.

Adam's method (see Collatz [127]) has been adapted by Thomas [128] to the solution of hyperbolic systems representing one dimensional flows including shocks. Thomas asserts that the complications inherent in this method are more than offset by the possibility of taking larger steps than in the previous approaches. His computations use the canonical form of the hyperbolic system in which α and β are the new independent variables.

7.26 FINITE DIFFERENCE METHODS: GENERAL DISCUSSION

The method of integrating along characteristics is usually the most accurate and most convenient process. For equations of no great complexity, whose solutions are known to be well behaved, we can alter-

---

† We use $A(R)$ to mean evaluation of $A(x, y, u, v)$ at the point R.

natively employ finite difference procedures. For finite difference methods to be useful they must take into account the characteristics of the system. We shall amplify this point subsequently.

One of the main advantages of the use of characteristics and a similar disadvantage in the use of finite differences, is the fact that discontinuities in the prescribed initial values may be propagated along the characteristics. Dealing with such phenomena, other than on the characteristic grid, will be complicated.

With these limitations in mind we now examine the problem of adapting finite difference techniques to hyperbolic systems.

7.27 EXPLICIT METHODS

Under certain circumstances, which depend upon the configuration of the characteristics, it is possible to march out the solution to a hyperbolic system on a fixed rectangular grid (instead of the characteristic grid).

For the simple linear problem

$$u_{xx} = u_{yy}, \quad u(x, 0) = f(x), \quad u_y(x, 0) = g(x) \quad (7.334)$$

and appropriate boundary conditions, with $0 \leqslant x \leqslant 1$, $0 \leqslant y < \infty$ an explicit method is easily constructed. We take the usual rectangular grid with constant intervals Δx and Δy and write $u_{i,j} = u(i\,\Delta x, j\,\Delta y)$. Both second partial derivatives are approximated by Eq. (7.15) whose truncation error is $O[(\Delta x)^2]$. Thus

$$0 = u_{xx} - u_{yy} \approx (\Delta x)^{-2}[u_{i+1,j} - 2u_{i,j} + u_{i-1,j}] - (\Delta y)^{-2}[u_{i,j+1} - 2u_{i,j} + u_{i,j-1}]. \quad (7.335)$$

If we set $m = \Delta x \Delta y$ and solve for $u_{i,j+1}$ there results

$$u_{i,j+1} = m^2 u_{i-1,j} + 2(1 - m^2)u_{i,j} + m^2 u_{i+1,j} - u_{i,j-1}. \quad (7.336)$$

The second initial condition is introduced by means of a false boundary. The computational molecule for Eq. (7.336) is shown in Fig. 7-30. Superimposed on this figure are the characteristics of Eq. (7.334), namely $y = \pm x + \eta$ whose slopes are ± 1, represented by the lines AC and BC. By Theorem (1) of Section 7.23a the solution is uniquely determined in the triangle ABC provided we know the solution up to AB. If the slope of the "finite difference characteristics" is greater than

1 in absolute value (i.e., $m = \Delta y/\Delta x > 1$) Eq. (7.336) would provide a "solution" in a region *not reached* by the continuous solution. Such a result can hardly be correct.

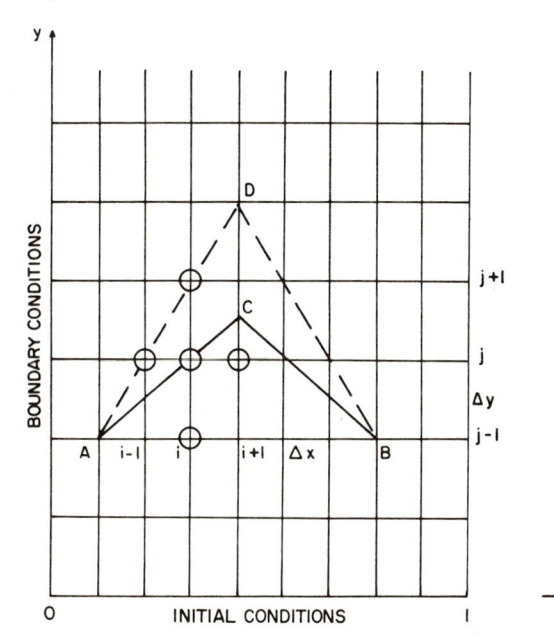

FIG. 7-30. The finite difference characteristics (AD and BD) and the true characteristics of the wave equation.

This argument suggests that the application of Eq. (7.336) will yield a convergent, stable process as Δy and $\Delta x \to 0$, for fixed m, only when $m \leqslant 1$. This has been proved, originally by Courant, Friedrichs, and Lewy [129]. A more accessible proof is found in Collatz [127].

The behavior of the recurrence relation (7.336) for various values of m is revealing:

For $m > 1$, violent instability sets in becoming worse as m increases;
For $m = 1$, the finite difference solution is *identical* with the continuous solution (see Milne [8]);
For $m < 1$, the discrete equation is stable but their accuracy decreases with decreasing m.

From these results we see that *optimum* accuracy is achieved when the "finite difference characteristics" exactly coincide with the true characteristics. By inference we can then anticipate that optimum accuracy will not be attained by applying a fixed rectangular grid to a hyperbolic system whose true characteristics are *curved*. For in such cases our uniqueness theorems require the finite difference characteristics to slope *less steeply* than the continuous characteristics.

Alternate explicit methods are possible. We previously stumbled upon one in Eqs. (7.37) and (7.38).

We should remark that Eq. (7.334) can be treated by introducing the additional variables $w = u_x$, $v = u_t$ whereupon Eq. (7.334) becomes

$$v_t = w_x \qquad w_t = v_x. \tag{7.337}$$

The rather obvious finite difference approximation

$$v_{i,j+1} - v_{i,j} = \frac{\Delta t}{2\,\Delta x}(w_{i+1,j} - w_{i-1,j})$$

$$w_{i,j+1} - w_{i,j} = \frac{\Delta t}{2\,\Delta x}(v_{i+1,j} - v_{i-1,j}) \tag{7.338}$$

is unstable unless $\Delta t/(\Delta x)^2$ is bounded as Δt and $\Delta x \to 0$. This is an impractical requirement so the method is not used (Richtmyer [4]).

7.28 EXPLICIT METHODS IN NONLINEAR SECOND-ORDER SYSTEMS

There is no current theory of a general nature for studying stability in the nonlinear case. In practice some linearization is carried out based on bounds for the function and or its derivatives. The development of global bounds on these functions has been carried out for a few examples as discussed in Chapter 8 but such "theory" is still in its development stage. Determination of stability bounds, for particular examples, can be accomplished by numerical experimentation.

As an example we consider the nonlinear vibrations of a string whose governing equations, due to Carrier, were discussed in Chapter 1. These were

$$[T \sin \theta]_x = \rho A u_{tt}, \qquad [T \cos \theta]_x = \rho\,A v_{tt},$$

$$\theta = \tan^{-1} u_x/(1 + v_x), \qquad T = T_0 + EA\{[(1 + v_x)^2 + u_x^2]^{\frac{1}{2}} - 1\}. \tag{7.339}$$

7.28 EXPLICIT METHODS IN SECOND-ORDER SYSTEMS

For this study we consider the *artificial* example obtained by neglecting the displacement v in the x-direction and eliminate the second equation (7.339). Thus the considered system is

$$[T \sin \theta]_x = \rho A u_{tt}, \qquad \theta = \tan^{-1} u_x$$
$$T = T_0 + EA(1 + u_x^2)^{\frac{1}{2}} - EA. \tag{7.340}$$

By introducing the dimensionless variables

$$U = u/L, \qquad X = x/L, \qquad T = \frac{t}{L[\rho A/T_0]^{\frac{1}{2}}}$$

and setting $B = EA/T_0$ we obtain the dimensionless equation

$$\frac{1 - B + B[1 + U_x^2]^{\frac{1}{2}}}{[1 + U_x^2]^{\frac{3}{2}}} U_{XX} = U_{TT}. \tag{7.341}$$

Equation (7.341) is subject to the auxiliary conditions

$$U(0, T) = U(1, T) = 0, \qquad U_T(X, 0) = 0, \qquad U(X, 0) = 4X(1 - X).$$

Using the same difference approximations as those of Eq. (7.335) and a forward difference for U_X the stability threshold was approximately $(\Delta t/\Delta x) = 0.55$. Violent instability occurred beyond this limit.

The rapid development of highly accurate electronic analog computing equipment and hybrid systems of a digital-analog nature bears mention. These devices are particularly useful for the solution of difference-differential approximations to partial differential equations of propagation problems (hyperbolic and parabolic systems). In this explicit marching method, called the "method of lines" in Russia, one of the variables (usually the finite range space variable(s)) say x is discretized, while the other variable t is left continuous. When suitable finite difference expressions are substituted for the x derivatives the partial differential equation is converted into a system of coupled ordinary differential equations in the continuous variable t. These are then integrated on the analog.

No theoretical analyses of these semi-discrete methods have been conducted for nonlinear problems. However, recognition of the importance of these methods has prompted some error, stability and convergence studies for *linear* problems by Douglas [130], Friedmann [131] and Landau [132]. Friedmann gives a bibliography of examples. In all these papers error estimates are obtained by comparing the exact solution with the computer solution.

Retention of one of the continuous variables appears to reduce some of the convergence-stability problems of the finite difference methods. Considerable future development and study of these analog methods is both warranted and necessary before the results for equations can be accepted without question.

A number of analog studies of both linear and nonlinear hyperbolic (and parabolic) systems have been carried out at Delaware on the PACE equipment of Electronic Associates. A typical problem is given below.

The dimensionless, damped wave equation

$$u_{xx} = u_{tt} + \gamma |u_t| u_t \tag{7.342}$$

with initial conditions

$$u_t(x, 0) = 0, \qquad u(x, 0) = x(x-1)$$

and boundary conditions

$$u(0, t) = u(1, t) = 0$$

is semidiscretized and solved for a range of γ from 1 to 15. Approximating u_{xx} at $x = i\,\Delta x$ with

$$u_{xx}|_i \approx (\Delta x)^{-2}[u_{i+1}(t) - 2u_i(t) + u_{i-1}(t)],$$

having a truncation error of $O[(\Delta x)^2]$, the difference-differential approximation to Eq. (7.342) becomes

$$\frac{d^2 u_i(t)}{dt^2} + \gamma \left|\frac{du_i}{dt}\right| \frac{du_i}{dt} = (\Delta x)^{-2}[u_{i+1} - 2u_i + u_{i-1}] \tag{7.343}$$

where $u_i(t) = u(i\,\Delta x, t)$. To be specific we set $(\Delta x) = 0.2$ thereby approximating the continuous system with a discrete system having four (interior) degrees of freedom. The resulting system, incorporating the boundary conditions, is

$$u_0 = 0$$
$$u_1'' + \gamma |u_1'| u_1' = -50u_1 + 25u_2$$
$$u_2'' + \gamma |u_2'| u_2' = 25u_1 - 50u_2 + 25u_3$$
$$u_3'' + \gamma |u_3'| u_3' = +25u_2 - 50u_3 + 25u_4 \tag{7.344}$$
$$u_4'' + \gamma |u_4'| u_4' = +25u_3 - 50u_4$$
$$u_5 = 0$$

with the initial conditions

$$u_i(0) = i\,\Delta x\,(i\,\Delta x - 1) = 0.2i(0.2i - 1), \quad \frac{du_i(0)}{dt} = 0.$$

The analog solutions resulting from Eqs. (7.344) were compared to a digital solution obtained by the method of characteristics and to the linearly damped wave equation over a range of values of γ ($1 \leqslant \gamma \leqslant 15$). When compared with the linear solution all amplitude and frequency shifts were as expected. No discernable difference existed (to the accuracy of working) between the digital and analog solution.

Skinner [133] considers the large oscillations of an airplane arresting cable by the aforementioned method. The calculated results agreed very well with the experiments.

7.29 IMPLICIT METHODS FOR SECOND-ORDER EQUATIONS

We have previously noted the advantage of implicit finite difference formulas with respect to stability, in parabolic systems. The same general observation holds for hyperbolic systems.

The simplest implicit system for the *wave equation* is obtained by approximating u_{tt} with the usual one column second order difference relation

$$u_{tt}\,|_{i,j} \approx \frac{u_{i,j+1} - 2u_{i,j} + u_{i,j-1}}{(\Delta t)^2}. \tag{7.345}$$

However, if the space derivative is approximated by either of the approximations ($\delta^2 u_n = u_{n+1} - 2u_n + u_{n-1}$)

$$u_{xx}\,|_{i,j} \approx \frac{\delta^2 u_{i,j+1} + \delta^2 u_{i,j-1}}{2(\Delta x)^2} \tag{7.346}$$

or

$$u_{xx}\,|_{i,j} \approx \frac{\delta^2 u_{i,j+1} + 2\delta^2 u_{i,j} + \delta^2 u_{i,j-1}}{4(\Delta x)^2} \tag{7.347}$$

the resulting implicit finite difference approximations are unconditionally stable, i.e., for any positive value of $\Delta t/\Delta x$ the system is stable. The first approximation arises by replacing $u_{i,j}$ by the quantity

$$u_{i,j} \approx \tfrac{1}{2}(u_{i,j+1} + u_{i,j-1})$$

and the second from

$$u_{i,j} \approx \tfrac{1}{4}(u_{i,j+1} + 2u_{i,j} + u_{i,j-1}).$$

The implicit nature for Eq. (7.346) is seen by writing out the expression to be solved in the $(j+1)$ line in dependence on the two preceding lines as

$$-\frac{m}{2}u_{i+1,j+1} + (1+m)u_{i,j+1} - \frac{m}{2}u_{i-1,j+1}$$
$$= 2u_{i,j} + \frac{m}{2}u_{i+1,j-1} - (1+m)u_{i,j-1} + \frac{m}{2}u_{i-1,j-1}. \quad (7.348)$$

Upon writing Eq. (7.348) for each i, $i = 1, ..., N-1$ and inserting the prescribed boundary conditions which we assume to be discretized (false boundaries, etc.), its tridiagonal form becomes clear. Thus the Thomas algorithm [22] (Section 7.8), for tridiagonal matrices, may be applied. We also note that any of the various methods discussed in the Elliptic Equations Sections could be used to solve this system. In fact in each row the diagonal coefficient is $1+m$ and the other two are $-m/2$ so the matrix is diagonally dominant.

Both of the approximations, Eqs. (7.346) and (7.347) are special cases of a general form involving a relaxation factor λ [compare Eq. (7.72)] where $u_{i,j}$ is replaced by the three (time) level form

$$u_{i,j} \approx \lambda u_{i,j+1} + \lambda u_{i,j-1} + (1 - 2\lambda)u_{i,j}$$

to give

$$u_{xx}|_{i,j} \approx \frac{\lambda \delta^2 u_{i,j+1} + (1-2\lambda)\delta^2 u_{i,j} + \lambda \delta^2 u_{i,j-1}}{(\Delta x)^2}.$$

Richtmyer [4] proves that this approximation generates an implicit method which has *unrestricted stability* if $\lambda \geq \tfrac{1}{4}$.[†]

Von Neumann (cf. O'Brien *et al.* [10]) introduced the difference equation

$$k^{-2}\delta_j^2 u_{i,j} = h^{-2}\delta_i^2 u_{i,j} + \omega k^2[h^{-2}k^{-2}\delta_j^2\delta_i^2 u_{i,j}] \quad (7.349)$$

as an approximation for the wave equation. Here we use $k = \Delta t$, $h = \Delta x$, $\omega \geq 0$ and $\delta_i^2 u_{i,j} = u_{i+1,j} - 2u_{i,j} + u_{i-1,j}$. When $\omega = 0$ this is the classical explicit scheme, Eq. (7.335). Otherwise it is an

---

[†] Note: $\lambda = 0$ gives the explicit method, $\lambda = \tfrac{1}{2}$, Eq. (7.346) and $\lambda = \tfrac{1}{4}$, Eq. (7.347). Proof of unrestricted stability for $\lambda = \tfrac{1}{4}$ relates to a periodic condition $u(x, t) = u(x + L, t)$ in lieu of other boundary conditions.

7.30 "HYBRID" METHODS FOR A NONLINEAR FIRST-ORDER SYSTEM

implicit difference equation solvable at each step by using the tridiagonal algorithm. Von Neumann proved that Eq. (7.349) is unconditionally stable if $4\omega > 1$, and conditionally stable if $4\omega \leqslant 1$, the stability condition in the latter case being $kh^{-1} \leqslant (1 - 4\omega)^{-\frac{1}{2}}$.

Friberg [134] and Lees [135] generalized von Neumann's result to linear hyperbolic equations with variable coefficients of the form

$$w_{tt} = a(x, t)w_{xx} + b(x, t)w_x + c(x, t)w_t + d(x, t)w + e(x, t). \quad (7.350)$$

In this case a term identical to the second term on the right-hand side of Eq. (7.349) is added. The stability requirements are the same.

The results of Friberg and Lees can be extended to cover the von Neumann type difference approximation to certain linear multidimensional systems. However, the linear equations that arise are no longer tridiagonal. Lees [136] develops two modifications of Eq. (7.349) for multidimensional hyperbolic systems by applying the alternating direction procedure to the standard von Neumann scheme. These modified von Neumann type difference equations are shown to be unconditionally stable if $4\omega > 1$.

Extension of the von Neumann method to quasi-linear[†] systems of the form

$$w_{tt} = a(x, t)w_{xx} + f(x, t, w_x, w_t, w)$$

seems promising. The stability criterion $4\omega > 1$ is a good starting point although numerical experimentation may be required.

7.30 "HYBRID" METHODS FOR A NONLINEAR FIRST-ORDER SYSTEM

Previously the solution of the equations of one-dimensional unsteady inviscid flow was obtained by the method of characteristics. This natural method has the disadvantage that if one requires spatial distributions of the dependent variables at a *fixed* time then messy two-dimensional interpolation in the characteristic grid is required. A method due to Hartree [137] avoids this difficulty by defining the mesh points in advance in both space and time and interpolating as the computation proceeds. The interpolation is thus one dimensional.

[†] Some authors call Eq. (7.350) quasi-linear. When this term is used herein a nonlinearity such as $(w)^2$ or ww_x is present.

446   7. NUMERICAL METHODS

Since this method is "hybrid" and is related to a similar but simpler scheme of Courant, Issacson, and Rees [138] we consider it here. Both methods are very good ones but have not been extensively used since they require the reduction of the equations to characteristic form (Eqs. (7.293) and (7.303)).

Suppose that a fixed rectangular grid is imposed on the integration domain with $\Delta x = h$ and $\Delta y = k$ as in Fig. 7-31. If the governing

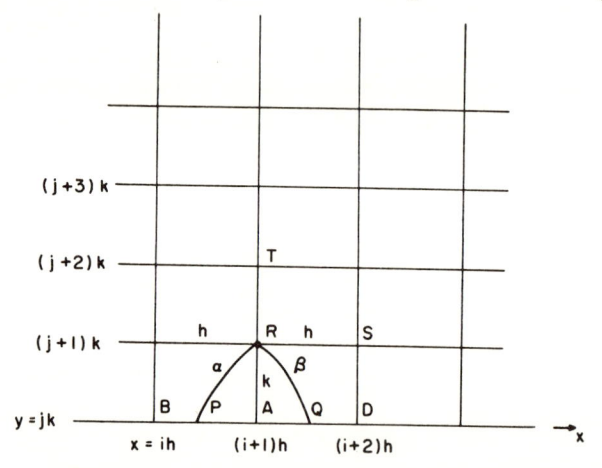

FIG. 7-31. The "hybrid" method of Hartree.

equations are those of Eq. (7.289)[†] then the equations for the characteristics may be written as

$$\frac{dx}{dy} = F_\alpha, \quad \frac{dx}{dy} = F_\beta \qquad (7.351)$$

and the relations along the characteristics are (see Eq. (7.303))

$$dv + G_\alpha du + H_\alpha dy = 0, \quad dv + G_\beta du + H_\beta dy = 0. \qquad (7.352)$$

With reference to Fig. 7-31, suppose the solution is known at the mesh points on the line $y = jk$ (this could be the initial line) and the $R, S, \ldots$, are equally spaced along the next line $y = (j+1)k$. We draw the α and β characteristics, through R back to their intersection with the

---

[†] The Hartree method can also be applied to second-order systems with only minor changes (cf. Fox [2], p. 216).

7.30 "HYBRID" METHODS FOR A NONLINEAR FIRST-ORDER SYSTEM

first line $y = jk$. These two points of intersection are not known at this point. The equations relating the solution values at R, P, and Q are still given by equations similar to Eq. (7.333) provided some obvious changes in format are made. However, we now *know* the values of the coordinates x_R, y_R, y_A, and y_B and wish to calculate u_R, v_R, x_P, and x_Q. The four equations (7.333) suffice for the determination of these unknowns except that interpolation for the values of u and v at P and Q are necessary at each step.

Hartree's computational form is

$$x_R - x_P = \tfrac{1}{2}[F_\alpha(R) + F_\alpha(P)]k, \qquad x_R - x_Q = \tfrac{1}{2}[F_\beta(R) + F_\beta(Q)]k$$
$$v_R - v_P + \tfrac{1}{2}[G_\alpha(R) + G_\alpha(P)](u_R - u_P) + \tfrac{1}{2}[H_\alpha(R) + H_\alpha(P)]k = 0 \qquad (7.353)$$
$$v_R - v_Q + \tfrac{1}{2}[G_\beta(R) + G_\beta(Q)][u_R - u_Q] + \tfrac{1}{2}[H_\beta(R) + H_\beta(Q)]k = 0$$

with a second-order truncation error. This system must be solved by iteration using interpolated values at P and Q.

Courant, Isaacson, and Rees [138] suggested two schemes which on the rectangular grid of Fig. 7-31 are less accurate, first-order systems. The first of these is essentially a simple Euler method:

$$x_R - x_P = F_\alpha(A)k, \qquad x_R - x_Q = F_\beta(A)k \qquad (7.354)$$

$$v_R - v_P + G_\alpha(A)(u_R - u_P) + H_\alpha(A)k = 0$$
$$v_R - v_Q + G_\beta(A)(u_R - u_Q) + H_\beta(A)k = 0. \qquad (7.355)$$

In this system x_P and x_Q are obtained immediately from Eqs. (7.354) (note the evaluation of F_α and F_β at A). After interpolation to get function values of u and v at P and Q (using, say, linear interpolation with the three adjacent mesh points B, A, D) we may calculate u_R and v_R from the simultaneous equations (7.355). Thus *no* iteration is required.

The second method of Courant et al.[†] of comparable accuracy to the first (i.e., is first order) involves rewriting Eq. (7.352) in the form

$$\left[\frac{\partial v}{\partial y} + F_\alpha \frac{\partial v}{\partial x}\right] + G_\alpha\left[\frac{\partial u}{\partial y} + F_\alpha \frac{\partial u}{\partial x}\right] + H_\alpha = 0$$
$$\left[\frac{\partial v}{\partial y} + F_\beta \frac{\partial v}{\partial x}\right] + G_\beta\left[\frac{\partial u}{\partial y} + F_\beta \frac{\partial u}{\partial x}\right] + H_\beta = 0 \qquad (7.356)$$

and replacing the derivatives by differences.

[†] Called the method of "forward and backward space differences" when applied to the transport equation (see Richtmyer [4], p. 142).

Thus we have

$$\left[\frac{v_R - v_A}{k} + F_\alpha(A)\frac{v_A - v_B}{h}\right] + G_\alpha(A)\left[\frac{u_R - u_A}{k} + F_\alpha(A)\frac{u_A - u_B}{h}\right] + H_\alpha(A) = 0 \quad (7.357)$$

and

$$\left[\frac{v_R - v_A}{k} + F_\beta(A)\frac{v_D - v_A}{h}\right] + G_\beta(A)\left[\frac{u_R - u_A}{k} + F_\beta(A)\frac{u_D - u_A}{h}\right] + H_\beta(A) = 0. \quad (7.358)$$

These equations are so constructed that in Eq. (7.357), corresponding to the forward facing (or α) characteristic, the space derivative is replaced by a *backward* difference. In the backward facing (or β) characteristic, the space derivative is replaced by a *forward* difference. Having obtained x_P and x_Q from Eqs. (7.354) we may now immediately obtain, without iteration, v_R and u_R from Eqs. (7.357) and (7.358).

Both of the Courant-Isaacson-Rees *explicit* methods are first order. Stability and convergence have been proven provided $\Delta y/\Delta x$ nowhere exceeds the slope of any characteristic—thus the stability condition would be both

$$|F_\alpha|\frac{\Delta y}{\Delta x} \leqslant 1, \qquad |F_\beta|\frac{\Delta y}{\Delta x} \leqslant 1.$$

7.31 FINITE DIFFERENCE SCHEMES IN ONE-DIMENSIONAL FLOW

The hybrid methods of the previous section have the advantage of attempting to follow the characteristics as closely as possible but are somewhat wasteful of computation. For example with Eqs. (7.357) and (7.358) we compute the characteristic directions at A but really do not exploit this knowledge. Therefore there is some justification for operating with finite differences on the *original* equations of motion without reducing them to characteristic form. Several methods have been introduced for such problems. Fluid flows are often characterized by internal discontinuities, such as shocks, on which special boundary conditions (jumps) are necessitated. Such jump conditions are provided by the Rankine-Hugoniot relations but their application in practice is beset with the difficulty that the surfaces on which the conditions are to be applied are in motion. Further, this motion is not known in advance but is determined by the differential equations and the jump conditions themselves. The resulting numerical work is highly implicit. In this

7.31 FINITE DIFFERENCE SCHEMES IN ONE-DIMENSIONAL FLOW

section we briefly discuss several methods for the *smooth* portion of the flow.

We recall that the equations of one-dimensional inviscid unsteady flow may be expressed in several ways, the two fundamental being the Eulerian and Lagrangian.[†] The Euler form with no heat conduction is

$$\rho_t + u\rho_r + \rho r^{-\alpha}(r^\alpha u)_r = 0, \quad u_t + uu_r + \rho^{-1}p_r = 0$$
$$e_t + ue_r + p\rho^{-1}r^{-\alpha}(r^\alpha u)_r = 0, \quad p = p(e, \rho). \tag{7.359}$$

The notation here uses r for the space coordinate, α is a constant depending on the problem geometry ($\alpha = 0$ for plane flow, $\alpha = 1$ for cylindrical symmetry and $\alpha = 2$ for spherical symmetry), ρ, u, p, and e are respectively density, velocity, pressure, and internal energy.

The Lagrange form of Eq. (7.359) has been developed in Chapter 4. It is repeated here in slightly different notation. With x as the position of a particle at time zero ($x = r(t = 0)$), ρ_0 the initial density and $V = 1/\rho$ the equations are

$$\rho_0 x^\alpha \, dx = \rho r^\alpha \, dr \tag{7.360}$$

$$\left. \begin{array}{l} V_t - \rho_0^{-1} x^{-\alpha}(r^\alpha u)_x = 0 \\ u_t + \rho_0^{-1}(r/x)^\alpha p_x = 0 \end{array} \right\} \tag{7.361}$$

$$\left. \begin{array}{l} e_t + pV_t = 0 \\ r_t = u \\ p = p(e, V) \end{array} \right\} \tag{7.362}$$

One can simplify this system by introducing the alternate Lagrange coordinate y as a mass instead of a length so that

$$dy = \rho_0 x^\alpha \, dx = V^{-1} r^\alpha \, dr. \tag{7.360a}$$

Hence the two Eq. (7.361) take the simpler forms

$$V_t - (r^\alpha u)_y = 0, \quad u_t + r^\alpha p_y = 0. \tag{7.361}$$

[†] The fundamental difference between these approaches was discussed in Chapter 4. Basically the Eulerian equations describe conditions at a point fixed in space and the time variations of conditions at that point. The Eulerian coordinate of the particle becomes one of the dependent variables. On the other hand the Lagrange form describes the motion of particular particles of fluid. The Lagrange coordinate system is fixed in the flow and moves with it. Of course the two forms are equivalent *but* this may not be true of the two truncated finite difference approximations.

In the Lagrange form there are six equations (7.360), (7.361), and (7.362) to solve for the five unknowns V(or ρ), u, e, p, and r. Of course one of these equations is redundant. Most calculations have been carried out eliminating the first equation of (7.361) obtaining V from Eq. (7.360) as $V = \rho_0^{-1}\, dr/dx$. r is obtained by integrating $r_t = u$. With this equation for V we can be sure of automatic conservation of mass. The use of the alternate equation for V usually gives rise to small variations in mass.

Both systems have roughly the same complexity. The major disadvantage of the Eulerian system arises when interfaces occur separating fluids of different density. The Lagrange system does not have the "spatial" coordinate mesh fixed in advance and may *require* refinement of the mesh as the computation goes on. This possibility of "regridding" arises since the Lagrange form is constructed so that the mass between two successive mesh points is (approximately) conserved.

The most commonly used techniques for both systems have been designed after the procedure (7.336) for the wave equation. That method can best be examined if one introduces the variables $v = u_t$, $w = cu_x$ into the wave equation $u_{tt} = c^2 u_{xx}$ obtaining $v_t = cw_x$, $w_t = cv_x$. The finite difference scheme, stable if and only if $c\,\Delta t \leqslant \Delta x$ [cf. 129],

$$v_{i,j+1} - v_{i,j} = c\left(\frac{\Delta t}{\Delta x}\right)(w_{i+\frac{1}{2},j} - w_{i-\frac{1}{2},j})$$

$$w_{i-\frac{1}{2},j+1} - w_{i-\frac{1}{2},j} = c\left(\frac{\Delta t}{\Delta x}\right)(v_{i,j+1} - v_{i-1,j+1}) \qquad (7.363)$$

is equivalent to Eq. (7.336) if we identify

$$v_{i,j} \quad \text{with} \quad (\Delta t)^{-1}(u_{i,j} - u_{i,j-1})$$

and

$$w_{i-\frac{1}{2},j} \quad \text{with} \quad c(\Delta x)^{-1}(u_{i,j} - u_{i-1,j}).$$

For the Eulerian equations, direct transfer of these schemes may lead to an unstable procedure. We must exercise care in the manner of discretizing the terms involving $u(\ )_r$. Richtmyer [4] discusses the approximation

$$u_{i,j+1}(2\Delta r)^{-1}(\rho_{i+1,j} - \rho_{i-1,j})$$

for $u\,\partial \rho/\partial r$ and shows that this leads to an *always unstable* situation where the von Neumann stability condition $\Delta t/\Delta r \leqslant 1$ cannot be satisfied (unless $u \equiv 0$.). He also gives the *Lelevier* method of remedying this situation.

7.31 FINITE DIFFERENCE SCHEMES IN ONE-DIMENSIONAL FLOW

Lelevier corrected the difficulty discussed above by using a forward difference for terms in $u(\ )_r$ when $u < 0$ and a backward difference when $u > 0$. It is generally assumed (proof still lacking) that the stability condition for Lelevier's system is

$$(|u| + c)\frac{\Delta t}{\Delta r} \leqslant 1 \tag{7.364}$$

which is borne out by calculations. In Eq. (7.364) the quantity c is

$$\frac{(p/\rho^2) - (\partial F/\partial \rho)}{\partial F/\partial p} = c^2$$

where the equation of state $p = p(e, \rho)$ is assumed solved for e as

$$e = F(p, \rho).$$

The Lelevier finite difference equations for slab symmetry (Eq. (7.359) with $\alpha = 1$) are

$$\rho_{i+\frac{1}{2},j+1} - \rho_{i+\frac{1}{2},j} + \left(\frac{\Delta t}{\Delta r}\right) u_{i,j+1}(\rho_{i+\frac{1}{2},j} - \rho_{i-\frac{1}{2},j})$$

$$= -\rho_{i+\frac{1}{2},j}\left(\frac{\Delta t}{\Delta r}\right)(u_{i+1,j+1} - u_{i,j+1}) \quad \text{if} \quad u_{i,j+1} \geqslant 0$$

$$\rho_{i,j}[u_{i,j+1} - u_{i,j} + \left(\frac{\Delta t}{\Delta r}\right) u_{i,j}(\underline{u_{i,j} - u_{i-1,j}})]$$

$$= -\frac{\Delta t}{\Delta r}(p_{i+\frac{1}{2},j} - p_{i-\frac{1}{2},j}) \quad \text{if} \quad u_{i,j} \geqslant 0 \tag{7.365}$$

$$\rho_{i,j}\left[e_{i+\frac{1}{2},j+1} - e_{i+\frac{1}{2},j} + \left(\frac{\Delta t}{\Delta r}\right) u_{i,j+1}(\underline{e_{i+\frac{1}{2},j} - e_{i-\frac{1}{2},j}})\right]$$

$$= -p_{i+\frac{1}{2},j}\left(\frac{\Delta t}{\Delta r}\right)(u_{i+1,j+1} - u_{i,j+1}) \quad \text{if} \quad u_{i,j+1} \geqslant 0.$$

If the velocity ($u_{i,j}$ or $u_{i,j+1}$) is negative, a forward space difference is used in the underlined places on the left-hand side of Eqs. (7.365). Coupled with Eq. (7.365) is the energy relation

$$e_{i+\frac{1}{2},j+1} = F(p_{i+\frac{1}{2},j+1}, \rho_{i+\frac{1}{2},j+1}). \tag{7.366}$$

As it is the system is explicit. More accuracy can be obtained by replacing p in the third equation with $\frac{1}{2}(p_{i+\frac{1}{2},j} + p_{i+\frac{1}{2},j+1})$ but the system then usually requires iteration. Note that in the Lelevier system we define

ρ, p, and e at the *midpoints of horizontal lines* while u is defined at the mesh points.

For the Lagrangian system, composed of Eqs. (7.360) the second equation of (7.361), and (7.362), the variables are V, r, u, p and e. Using the midpoint idea we define p, V, and e at the midpoint of horizontal lines, r at the mesh points and u at the midpoints of vertical lines. The general system appears to have the stability criterion

$$\frac{\rho}{\rho_0}\left[\frac{r(x,t)}{x}\right]^\alpha \frac{c\,\Delta t}{\Delta x} \leq 1.$$

The finite difference equations are

$$V_{i+\frac{1}{2},j+1} = \frac{(r_{i+1,j+1})^{\alpha+1} - (r_{i,j+1})^{\alpha+1}}{\rho_0[x_{i+1}^{\alpha+1} - x_i^{\alpha+1}]}$$

$$\frac{u_{i,j+\frac{1}{2}} - u_{i,j-\frac{1}{2}}}{\Delta t} = -\frac{1}{\rho_0}\left(\frac{r_{i,j}}{x_i}\right)^\alpha \frac{p_{i+\frac{1}{2},j} - p_{i-\frac{1}{2},j}}{\Delta x}$$

$$\frac{e_{i+\frac{1}{2},j+1} - e_{i+\frac{1}{2},j}}{\Delta t} = -p_{i+\frac{1}{2},j+\frac{1}{2}} \frac{V_{i+\frac{1}{2},j+1} - V_{i+\frac{1}{2},j}}{\Delta t} \qquad (7.367)$$

$$\frac{r_{i,j+1} - r_{i,j}}{\Delta t} = u_{i,j+\frac{1}{2}}$$

$$p_{i+\frac{1}{2},j+1} = p(e_{i+\frac{1}{2},j+1}, V_{i+\frac{1}{2},j+1})$$

and the introduction of p at the mesh cell center, i.e.,

$$p_{i+\frac{1}{2},j+\frac{1}{2}} = \tfrac{1}{2}[p_{i+\frac{1}{2},j+1} + p_{i+\frac{1}{2},j}],$$

gives a second-order truncation error.

Fuller heuristic discussion of stability is given in Richtmyer [4] and From [139]. Harlow [140] examines various linear and nonlinear equations with stability as one of the major considerations. (Noh and Protter [141] develop a finite difference scheme for the Eulerian hydrodynamic equations based on their soft solutions (see Chapter 4).) Implicit procedures do not have as much appeal here as in other situations since the physical system usually changes markedly in a time $\Delta t = \Delta r/c$. There is therefore little motivation for using large time steps.

However, there are situations where changes occur slowly so that large time steps are justified. To take full advantage of this situation, one needs to adopt implicit schemes which have the feature that the stability conditions are independent of the Δt and Δx chosen. Such a problem is considered by Rouse [142] in the calculation of coupled hydrodynamic flow and radiation diffusion. This method will be briefly discussed in the fourth part of this chapter on mixed systems.

7.32 CONSERVATION EQUATIONS

The formulation of the equations of motion of fluid mechanics is based upon the conservation of mass, momentum, and energy. It therefore seems reasonable to try to preserve these conservation properties in the finite difference approximation. Lax [143] appears to have first discussed the advantages accruing if finite difference approximations are based on "conservation" equations. Lax [144] gave a brief survey of the subject of nonlinear hyperbolic systems of conservation laws.

A system of conservation laws is a system of equations of the form

$$w_t + f(w)_r = 0 \tag{7.368}$$

where w is a vector function of x and t with n components and f is a nonlinear vector function, with n components, of the vector w. This definition can be easily extended to the case of more than one space variable.

The three dimensional Eulerian equations can be written in the form [121]

$$\rho_t = -\nabla \cdot (\rho \mathbf{u})$$
$$(\rho \mathbf{u})_t = -\mathbf{u} \nabla \cdot (\rho \mathbf{u}) - \rho(\mathbf{u} \cdot \nabla)\mathbf{u} - \nabla p \tag{7.369}$$
$$(\rho E)_t = -\nabla \cdot (\rho E \mathbf{u}) - \nabla \cdot (p \mathbf{u})$$

with $E = e + \tfrac{1}{2} |u|^2 =$ total energy per unit mass. For one-dimensional flow these can only be put in conservation form (7.368) if $\alpha = 0$ (slab symmetry) and in that case they become

$$\rho_t = -(\rho u)_r, \quad (\rho u)_t = -(\rho u^2 + p)_r, \quad (\rho E)_t = -(\rho E u + p u)_r. \tag{7.370}$$

One can also develop conservative[†] Lagrange equations which for slab symmetry ($\alpha = 0$) are

$$V_t = u_y, \quad u_t = -p_y, \quad E_t = -(pu)_y \tag{7.371}$$

[†] Authors in fluid mechanics sometimes use the term "divergence" equations in place of the term "conservation" equations used herein.

with y defined by Eq. (7.360a) as a mass coordinate. Of course, in each case, we must add an appropriate equation of state to complete the system.

Various finite difference approximations to conservation equations have been given by Lax [144, 145], Longley [146], Trulio and Trigger [147], Fromm [139], Douglas [148] and Lax and Wendorff [149]. Various discussions and examples of successful calculations are given in these papers. These studies confirm that approximations to the conservative equations give better representations to the correct physical solutions than nonconservative schemes.

The scheme of Lax [143] is to approximate

$$w_t + f_r = 0$$

by the finite difference form

$$(\Delta t)^{-1}[w_{i,j+1} - \tfrac{1}{2}(w_{i+1,j} + w_{i-1,j})] + (2\Delta r)^{-1}(f_{i+1,j} - f_{i-1,j}) = 0, \quad (7.372)$$

where the mean of neighboring values replaces the pivotal value $w_{i,j}$. These procedures are generally stable if

$$\frac{c \, \Delta t}{\Delta r} \leqslant 1$$

where c is the local sound speed. Equation (7.372) is a staggered scheme enabling central space differences and forward time differences to be used without getting instability.

The Lax technique has been applied to the equations in cylindrical geometry, which cannot be put in conservative form, by Payne [150]. The momentum equation has the additional term rp_r. To preserve the staggered array we might try to replace this by

$$r_i(p_{i+1,j} - p_{i-1,j})/2\Delta r$$

which leads to an *unstable* scheme. Roberts [151] attempted to use the Lax scheme on a problem with spherical symmetry. He, too, found that the staggered scheme had to be abandoned.

7.33 INTERFACES

By interfaces we shall mean boundaries separating media whose physical, chemical, and/or thermodynamic properties differ. For example interfaces separate two different fluids, elastic and plastic subdomains,

7.33 INTERFACES

a liquid and solid, or two adjacent portions of the same fluid whose thermodynamic states are different. Moving but unknown boundaries (interfaces) have previously been discussed in parabolic systems [cf. Sections 4.9, 7.13 (example 3)].

If the position of the interface is known, as in some problems of solid mechanics then one can usually adjust the grid so that mesh points lie on these boundaries. In such cases the basic finite difference equations can be applied, as before, if care is used to change the physical parameters or use the appropriate equation of state on each side of the interface. Such an approach may lead to inaccuracies and some special treatment at interfaces is often necessary. Inaccuracies may result from large changes in some system property across the interface (e.g., density in going from a gas to liquid, conductivity from solid to gas, permeability from gas to solid, etc). In such cases we may wish to use quite different spacing Δr on the two sides of the interface. Such a problem is seen in Fig. 7-17 where the air gap is small.

We shall list below a few treatments of interface problems in fluid mechanics since these are usually the more difficult to treat.

(1) At interfaces, ρ and e, but not u and p, are discontinuous. The positions of discontinuities usually will not coincide with points of the Eulerian mesh, especially if more than one contact surface is present. One can attempt to keep track of all discontinuities and construct special schemes to handle them in terms of their closest mesh points. The problem complexity is usually increased by such methods. Longley [152] devised such a procedure using the Euler conservative form of the one dimensional equations. His program prevents diffusion of the interface.

(2) In Lagrange coordinates the interfacial space coordinate remains unchanged in time. The mesh may therefore be defined so the interface is always a mesh point or halfway between mesh points or for that matter lies anywhere we wish. Richtmyer [4, p. 202] notes that inaccuracies may arise if one uses Eq. (7.367) (or a first-order form) for the Lagrange system because while $\partial p/\partial r$ varies continuously[†] across the interface

---

[†] This results from Richtmyer's definition of the Lagrangian coordinate as

$$x = \frac{1}{\rho_0} \int \rho(\eta)\, d\eta$$

the derivative of the pressure gradient *may not*. He suggests using the more accurate expression

$$\left.\frac{\partial p}{\partial x}\right|_{x=x_I} = \frac{3(p_{I+\frac{1}{2},j} - p_{I-\frac{1}{2},j}) - \frac{1}{3}(p_{I+\frac{3}{2},j} - p_{I-\frac{3}{2},j})}{\Delta_L x + \Delta_R x} \quad (7.373)$$

where the notation $i = I$ represents the interface spatial index and $\Delta_L x$ and $\Delta_R x$ are the increments of x used on the left-and right-hand sides of the interface. Equation (7.373) is to be used to $i = I$ in place of

$$(p_{i+\frac{1}{2},j} - p_{i-\frac{1}{2},j})/\Delta x$$

in the second equation of (7.367) (the equation of motion).

(3) Trulio and Trigger [147] choose the interface to be at the center of the mesh and define two values of ρ and e, at each time step, for each side of the interface. Values of u and r must be calculated at the interface points and of course $p_L = p_R$.

7.34 SHOCKS

The interface or contact discontinuity was characterized by discontinuities in ρ and e, but p and u are continuous. The motion of an ideal fluid is characterized by curves in the (x, t) plane across which all quantities ρ, e, p, and u are discontinuous but possess one sided limits on both sides. At such discontinuities, called shocks, the differential equations which no longer have meaning, must be replaced by certain jump conditions. These conditions serve as internal boundary conditions thereby providing for unique solutions. Background theory is amply discussed in general by Courant and Friedrichs [153], propagation in ducts by Chester [154] and in magneto-gasdynamics by Hide and Roberts [155]. Extensive bibliographies are included in each reference.

In reality the shock wave is not a discontinuity at all but a narrow zone, a few mean free paths in thickness through which the variables change continuously, even though very steeply. Across a shock wave the Rankine-Hugoniot conditions hold—they are discrete expressions of mass, momentum, and energy conservation. These conditions are respectively [153]

$$\begin{aligned}
\text{mass:} \quad & \rho_1(U - u_1) = \rho_2(U - u_2) = m \\
\text{momentum:} \quad & m(u_1 - u_2) = p_1 - p_2 \\
\text{energy:} \quad & m(e_1 + \tfrac{1}{2}u_1^2 - e_2 - \tfrac{1}{2}u_2^2) = p_1 u_1 - p_2 u_2
\end{aligned} \quad (7.374)$$

where m is the mass crossing unit area of the shock front in unit time and subscripts 1 and 2 refer to conditions of the unshocked fluid (ahead of the shock) and behind the shock respectively. U is the velocity of the shock front.

In the shock layer the heat conduction and viscosity effects are important and there is an entropy S increase across the shock ($S_2 > S_1$). Consequently the Eulerian equation for plane motion become

$$\rho_t + (\rho u)_r = 0$$
$$(\rho u)_t + (\rho u^2 + p - \bar{\mu} u_r)_r = 0$$
$$(\rho E)_t + [\rho u E + u(p - \bar{\mu} u_r) - kT_r]_r = 0 \qquad (7.375)$$
$$\rho T(S_t + uS_r) - \bar{\mu}(u_r)^2 - (kT_r)_r = 0$$

where $E = e + \frac{1}{2}u^2$ and $\bar{\mu} = \frac{4}{3}\mu$ and k are coefficients of viscosity and conductivity respectively.

Methods of treating shocks by finite difference approximations fall basically into two areas—those which add artificial dissipative terms and those which use characteristics.

a. Artificial Dissipative Methods

1. ARTIFICIAL VISCOSITY

The modification of the one-dimensional equations to include a viscosity term q was first proposed by von Neumann and Richtmyer [156]. They solve the Lagrange equations in slab geometry ($\alpha = 0$), see Eqs. (7.360)–(7.362),

$$V_t = u_y, \qquad u_t = -(p + q)_y, \qquad e_t = -(p + q)V_y$$

where $dy = \rho_0\, dx = \rho\, dr$ (y is the mass coordinate of Eq. (7.360a)). The original form of q,

$$q = -\frac{b(\Delta y)^2}{V} \frac{\partial u}{\partial y} \left|\frac{\partial u}{\partial y}\right|$$

has been found to unnecessarily smear out rarefaction waves and has evolved into the limited form

$$q = \begin{cases} \dfrac{b(\Delta y)^2}{V}\left(\dfrac{\partial u}{\partial y}\right)^2 & \dfrac{\partial u}{\partial y} < 0 \\ 0, & \dfrac{\partial u}{\partial y} \geqslant 0. \end{cases} \qquad (7.376)$$

This "viscosity" dissipative term has the effect of changing the shock from a discontinuity to a narrow region across which the fluid variables *change rapidly but continuously*. Narrow here means only a few Δy's in thickness.[†] Across this narrow region the Rankine-Hugoniot relations hold. The dissipative term q is added over the whole integration field so that the position of the shock does not need to be known.

In practice the shock spreads out over approximately $[8b/(\gamma + 1)]^{\frac{1}{2}}$ mesh points and Richtmyer [4] reports that the choice of b can be critical. The addition of the q viscosity term necessitates the extra stability criterion in the shock region

$$\left[a\frac{\Delta t}{\Delta r}\right]^2 + \frac{4b \mid \Delta V \mid}{V} \leqslant 1 \qquad (7.377)$$

or

$$\left[a\frac{\Delta t}{\Delta r}\right]\left[a\frac{\Delta t}{\Delta r} + \frac{2b}{\gamma^{1/2}}\right] \leqslant 1 \qquad \text{(strong shock)}$$

where a is the local sound speed and ΔV is the change in V in a time cycle.

Brode [157, 158] has utilized the von Neumann-Richtmyer method successfully in cylindrical and spherical symmetry. His calculations concerned the determination of blast waves and explosions. More recently Schulz [159] presents arguments that in *higher dimensions* a tensor artificial viscosity is a more suitable quantity to use than the scalar extension of Eq. (7.370). Schulz reports successful calculations by this method and asserts that the shock stability criterion of the difference equations resulting from using the tensor viscosity is less severe than that deduced from a generalized scalar viscosity.

2. THE METHOD OF LAX

Lax [143] found that when the equations are expressed in conservation form and the difference scheme (7.372) applied the solution of problems with shocks may be calculated without the addition of the dissipative term just discussed. The reason for this is that the finite difference approximation (7.372) has the effect of introducing a "diffusion" term. We may see this by rewriting Eq. (7.372) as

$$(\Delta t)^{-1}[w_{i,j+1} - w_{i,j}] + (2\Delta r)^{-1}(f_{i+1,j} - f_{i-1,j})$$
$$= (2\Delta t)^{-1}[w_{i,1,j} - 2w_{i,j} + w_{i-1,j}] \qquad (7.378)$$

[†] This is much larger than the thickness of the real shock.

which is an approximation to the diffusion equation in w

$$w_t + f_r = D w_{rr}, \qquad D = (\Delta r)^2/2\Delta t.$$

If the difference scheme of Lax (7.372) is applied to develop finite difference approximation for the Eulerian system (7.370), in slab symmetry, we find that the process in effect adds not only an artificial viscosity *but an artificial heat conduction and mass diffusion as well.* In fact the resulting finite difference scheme can be thought of as an approximation to the "new" system

$$\rho_t = -(\rho u)_r + D\rho_{rr}$$
$$(\rho u)_t = -(\rho u^2 + p)_r + D(\rho u)_{rr}$$
$$(\rho E)_t = -(\rho E u + pu)_r + D(\rho E)_{rr}$$

which is equivalent to

$$\rho_t + (\rho u + m)_r = 0$$
$$(\rho u)_t + [p + q + u(\rho u + m)]_r = 0 \qquad (7.379)$$
$$(\rho E)_t + [u(p + q) + (\rho u + m)E + h]_r = 0.$$

where

$$m = -D\rho_r \qquad \text{is a mass diffusion term;}$$
$$q = -D\rho u_r \qquad \text{is a viscosity term;} \qquad (7.380)$$
and
$$h = -D\rho e_r \qquad \text{is a heat conduction term.}$$

Equations (7.379) should be related to the true viscous-heat conduction equations (7.375) where the true dissipative terms are

$$m = 0, \qquad q = -\mu u_r, \qquad h = -kT_r!.$$

The Lax method has the usual stability condition $a\,\Delta t/\Delta r \leqslant 1$ and does not require the additional condition (7.377). Since $D = (\Delta r)^2/2\Delta t$, too small a time step diffuses the shock over may time intervals. A smooth change of pressure is characteristic of this calculation as compared with an oscillation behind the shock in von Neumann's method.

Other successful calculations by the Lax method are reported by

Payne [150], Zovko and Macek [160] (transition from a deflagration to a detonation wave where oscillation behind the shock could not be tolerated) and Filler and Ludloff [161] who solved the proper flow-conduction equations (7.375) by both explicit and implicit means.

3. LANDSHOFF'S METHOD (see Stein [162])

Landshoff [see 162] proposes an artificial viscosity term Q as

$$Q = q_{NR} + \omega q_1 + (1 - \omega) q_2$$

where q_{NR} is the von Neumann viscosity (7.376) and

$$q_1 = -\frac{1}{2} \frac{a \, \Delta x}{V} u_x$$

$$q_2 = -\frac{1}{2} \left(\frac{a}{V}\right)^2 \frac{\Delta t}{\rho_0} u_x$$

with $0 \leqslant \omega \leqslant 1$. This method is applied to the Lagrange form in slab geometry.

Comparison of the Lax, von Neumann, and Landshoff procedures are given in Fox [2, p.355]. Only Lax's method appears to eliminate the oscillation behind the shock although the relaxation of Landshoff reduces the oscillation amplitude.

4. OTHER METHODS

Various other artificial dissipative methods are proposed by Longley [152] and Fromm [139]. A heuristic discussion of the error in these schemes is given in Fox [2], Lax and Wendorff [149] and Richtmyer [4].

b. Characteristics

A numerical solution of the shock problem is obtained, *without artificial additions*, by means of the Hartree method (Section 7.30). Modifications of the method are carried out by Stein [162] in spherical blast wave calculations and Keller et al. [163] for a shallow water bore problem.

7.35 ADDITIONAL METHODS

The rapid progress of research in numerical analysis makes it impossible to discuss all methods of finite difference approximation. The primary journals publishing new ideas in this area are *Communications of Pure and Applied Mathematics, Mathematics of Computation, SIAM Journal, Numerische Mathematik,* and *ACM Journal.*

Some additional promising new ideas are listed herein. Keller and Thomee [164] have developed an *unconditionally stable explicit* finite difference method for quasi-linear hyperbolic systems in two dimensions. The mixed problem in the quasi-linear case, in terms of the vector **u**, is

$$\mathbf{u}_t - D(x, t, \mathbf{u})\mathbf{u}_x = F(x, t, \mathbf{u})$$

$$\mathbf{u}(x, 0) = f_0(x)$$

$$\mathbf{u}^-(0, t) = f^-(t, u^+(0, t))$$

$$\mathbf{u}^+(0, t) = f^+(t, u^-(0, t))$$

(7.381)

where the dichotomy u^+ and u^- is associated with a splitting of the matrix D. The method has first-order truncation error.

Thomee [165] extends the previous method to a second-order scheme which is unconditionally stable but requires iterative solution. His basic nonlinear problem is again the quasi-linear system (7.381). Numerical examples are calculated in both papers. Thomee [166] considers the same problem (7.381) and develops a predictor-corrector scheme. One numerical example of an unrealistic problem is given.

The theory of characteristics in more than two independent variables is known (see e.g. Courant and Hilbert [167]). Computation based upon this theory has been developed by Thornhill [168] and Coburn and Dolph [169]. Both of these methods are generalizations of the two dimensional methods discussed in Section 7.25. An example using the Coburn-Dolph method has been detailed by Bruhn and Haack [170] who were concerned with the starting flow in a nozzle. A third procedure, more analogous to the Hartree method, has been given by Butler [171]. Butler applies his ideas to the case of unsteady flow.

D. MIXED SYSTEMS

7.36 THE ROLE OF MIXED SYSTEMS

Problems which incorporate ideal fluid motion with some other transport process such as heat conduction have mathematical models which are coupled equations of mixed parabolic-hyperbolic type. Such a system occurs in hydrodynamic flow coupled with radiation diffusion (see Rouse [142]) and in (MGD) magneto-gas dynamics (see Jeffrey and Taniuti [119]).

One of the consequences of mixed parabolic-hyperbolic equations is that there are *two time constants*. In the case of practical MGD problems the time constant for the hyperbolic equations (i.e., the hydrodynamic equations of the previous sections) is considerably smaller than that for the parabolic equations. This occurs since hydromagnetic shocks and other phenomenon are relatively quick as compared with diffusion (the energy equation is parabolic) which is relatively slow. We cannot, of course, neglect the diffusion because it counteracts the confinement of the plasma. What we must be careful about is *not* to let the diffusion be swamped by errors of the finite difference process.

7.37 HYDRODYNAMIC FLOW AND RADIATION DIFFUSION

When analyzing such phenomena as exploding wires, initial phases of blasts, etc., the mathematical model must represent coupled hydrodynamic flow and radiation diffusion. Numerical calculations of such systems, prior to the paper of Rouse [142], were usually carried out by an explicit method for the hydrodynamics and an implicit method for the diffusion. Only one stability criterion was therefore required. To take full advantage of the attractive feature of implicit methods, i.e., independence of stability conditions on time and space steps Rouse uses implicit finite difference methods for both.

Consider the Lagrange form of the hydrodynamic equations in slab geometry (see Eqs. (7.361))

$$V_t = V u_r \tag{7.382}$$

$$u_t = -V p_r \tag{7.383}$$

$$e_t = -p V_t + Q_t \tag{7.384}$$

7.37 HYDRODYNAMIC FLOW AND RADIATION DIFFUSION

where the quantity Q represents a specific energy input. Let T be temperature and the equations of state be

$$p = p(T, V), \quad e = e(T, V), \quad V = \rho^{-1}. \tag{7.385}$$

Using these equations of state we may write Eq. (7.383) as

$$\frac{\partial u}{\partial t} = - V\left(\frac{\partial p}{\partial T}\right)_V \frac{\partial T}{\partial r} - V\left(\frac{\partial p}{\partial R}\right)_T \frac{\partial V}{\partial r}. \tag{7.386}$$

In addition we may write Eq. (7.384) as

$$\left(\frac{\partial e}{\partial T}\right)_V \frac{\partial T}{\partial t} + \left(\frac{\partial e}{\partial V}\right)_T \frac{\partial V}{\partial t} = -p\frac{\partial V}{\partial t} + \frac{\partial Q}{\partial t}$$

which can be transformed, by substituting Eq. (7.382), into

$$\left(\frac{\partial e}{\partial T}\right)_V \frac{\partial T}{\partial t} = -\left[p + \left(\frac{\partial e}{\partial V}\right)_T\right] V \frac{\partial u}{\partial r} + \frac{\partial Q}{\partial t}. \tag{7.387}$$

With radiation diffusion *and* thermal conduction included we have

$$\frac{\partial Q}{\partial t} = V \frac{\partial}{\partial r}\left[\lambda \frac{\partial}{\partial r}\left(\frac{a}{3} T^4\right) + k \frac{\partial T}{\partial r}\right] \tag{7.388}$$

with λ = mean free path, $a = \frac{4}{3}\sigma$, σ = Stefan-Boltzmann constant. Setting this result into Eq. (7.387) yields

$$\left(\frac{\partial e}{\partial T}\right)_V \frac{\partial T}{\partial t} = -\left[p + \left(\frac{\partial e}{\partial V}\right)_T\right] V \frac{\partial u}{\partial r} + V \frac{\partial}{\partial r}\left[\left(\frac{4a\lambda}{3} T^3 + k\right) \frac{\partial T}{\partial r}\right]. \tag{7.389}$$

With the given equations of state (7.385), Eqs. (7.386) and (7.389) are prepared for numerical solution except for the term containing V_r (Eq. (7.386)). For the moment we ignore this and implicitly difference Eq. (7.386) *with* T = *constant* as

$$u_{i,j+1} - u_{i,j} = -\frac{\Delta t}{\rho \Delta r}\left(\frac{\partial p}{\partial V}\right)_{i,j+\frac{1}{2}} [\tfrac{1}{2}(V_{i+\frac{1}{2},j} - V_{i-\frac{1}{2},j})$$

$$+ \tfrac{1}{2}(V_{i+\frac{1}{2},j+1} - V_{i-\frac{1}{2},j-1})] \tag{7.390}$$

where $\rho \, \Delta r =$ mass in zones (assumed constant) and the notation

$$V_{i+\frac{1}{2},j} = \tfrac{1}{2}[V_{i,j} + V_{i+1,j}]$$
$$V_{i,j+\frac{1}{2}} = \tfrac{1}{2}[V_{i,j} + V_{i,j+1}].$$

As in all implicit techniques let us assume a knowledge of the solution at the level $t = j \, \Delta t$. On the right-hand side of Eq. (7.390), $V_{(\ ),j+1}$ and $(p_V)_{(\ ),j+1}$ are unknown. The latter is obtained by iteration and the former by using a difference approximation for Eq. (7.382) at $i + \tfrac{1}{2}$ and $i - \tfrac{1}{2}$

$$V_{i+\frac{1}{2},j+1} - V_{i+\frac{1}{2},j} = \frac{\Delta t}{\rho \, \Delta r}(u_{i+1,j+\frac{1}{2}} - u_{i,j+\frac{1}{2}})$$

$$V_{i-\frac{1}{2},j+1} - V_{i-\frac{1}{2},j} = \frac{\Delta t}{\rho \, \Delta r}(u_{i,j+\frac{1}{2}} - u_{i-1,j+\frac{1}{2}}).$$

When the second expression is subtracted from the first we see that

$$V_{i+\frac{1}{2},j+1} - V_{i-\frac{1}{2},j+1} = V_{i+\frac{1}{2},j} - V_{i-\frac{1}{2},j} + \frac{\Delta t}{\rho \, \Delta r}[u_{i+1,j+\frac{1}{2}} - 2u_{i,j+\frac{1}{2}} + u_{i-1,j+\frac{1}{2}}].$$

When this expression is set into Eq. (7.390) we find the relation

$$u_{i,j+1} - u_{i,j} = -\frac{\Delta t}{\rho \, \Delta r}\left(\frac{\partial p}{\partial V}\right)_{i,j+\frac{1}{2}}[V_{i+\frac{1}{2},j} - V_{i-\frac{1}{2},j} \\ + \frac{\Delta t}{2\rho \, \Delta r}(u_{i+1,j+\frac{1}{2}} - 2u_{i,j+\frac{1}{2}} + u_{i-1,j+\frac{1}{2}})] \quad (7.391)$$

which is independent of V in the time level $j + 1$.

The calculation of Eq. (7.389) in implicit form is carried out as in our discussion of parabolic systems. The calculations are completed from the information $u_{i,j+1}$, $T_{i,j+1}$ via Eqs. (7.367) with $\alpha = 0$, $\rho_0 \, dx = \rho \, dr$.

7.38 NONLINEAR VIBRATIONS OF A MOVING THREADLINE

Recent studies initiated by the author have concerned the large oscillations of a moving threadline. Some of the results of this study

7.38 NONLINEAR VIBRATIONS OF A MOVING THREADLINE

are reported in the theses of Zaiser [172] and Hansell [173]. In the first of these a mathematical model is developed for the system. These nonlinear partial differential equations are equations of motion

$$(\alpha^2/4)V^2 u_{xx} + \alpha V u_{xt} + u_{tt} = \frac{T u_{xx}}{M(1 + u_x^2)} \tag{7.392}$$

$$(\alpha^2/4)V V_x + (\alpha/2)V_t = \frac{T_x(1 + u_x^2) - T u_x u_{xx}}{M(1 + u_x^2)^2}; \tag{7.393}$$

a continuity equation

$$M_t = -\frac{\alpha}{2}\left[V M_x + M\left(\frac{V u_x u_{xx}}{1 + u_x^2} + V_x\right)\right]; \tag{7.394}$$

and a tension-mass equation ("equation of state")

$$M(T + N) = BN \tag{7.395}$$

where N and B are constants. The equations are dimensionless relations among the dimensionless variables u = vertical displacement, V = velocity along the thread, M = mass per unit length and T = tension.

This system of equations is probably hyperbolic but the presence of the first-order equations for V and M creates some unusual difficulties. The range of interest for α lies in the transition region around $\alpha = 2$ which separates a "subsonic" from a "supersonic" domain. Various degrees of complexity for the system (7.392)–(7.395) are considered.

a. The Linear Case

Here the *threadline equation* becomes

$$\left(\frac{\alpha^2}{4} - 1\right) u_{xx} + \alpha u_{xt} + u_{tt} = 0. \tag{7.396}$$

Solution of Eq. (7.396) with auxiliary conditions

$$u(0, t) = u(1, t) = 0, \quad u(x, 0) = A \sin \pi x, \quad u_t(x, 0) = 0$$

is easily carried out by an implicit method with the Thomas tridiagonal algorithm. No difficulties are encountered.

b. Nonlinear Case 1

We now assume T, M, and V constant so that the basic equations becomes

$$\frac{\alpha^2}{4} u_{xx} + \alpha u_{xt} + u_{tt} = \frac{u_{xx}}{1 + u_x^2}. \tag{7.397}$$

With the same boundary conditions as before there is no difficulty in computing via an implicit finite difference approximation and the Thomas algorithm.

c. Nonlinear Case 2

For this case we assume T and M constant but let V be the form

$$V = (1 + u_x^2)^{-\frac{1}{2}} \tag{7.398}$$

so that the governing equation is

$$\frac{\alpha^2}{4} V^2 u_{xx} + \alpha V u_{xt} + u_{tt} = \frac{u_{xx}}{1 + u_x^2} \tag{7.399}$$

and the implicit method applies with no difficulty.

d. Nonlinear Case 3

Here we let T and M be constant but u and V are to be governed by the equations (from. 7.392) and (7.393))

$$\frac{\alpha^2}{4} V^2 u_{xx} + \alpha V u_{xt} + u_{tt} = \frac{u_{xx}}{1 + u_x^2}$$

$$\frac{\alpha^2}{4} V V_x + \frac{\alpha}{2} V_t = -\frac{u_x u_{xx}}{(1 + u_x^2)^2}. \tag{7.400}$$

In this case the *implicit scheme* adopted in the same manner as the previous cases displayed the generation of a discontinuity after the passage of time $t = 2.5$. Up to that point the solution was smooth and gave favorable agreement with previous solutions (for $\alpha = 0.8$). Enlargement of the system to include variable T and M also displayed similar discontinuities. Experimental evidence [173] predicts such "shocks" near $\alpha = 2$ but not at $\alpha = 0.8$. Present efforts are aimed at introducing damping effects to examine the system's action under more realistic circumstances.

Zabusky [174] shows that the development of a discontinuity is possible in a nonlinear vibrating (fixed) string and calculates when it should occur. He uses the analogy of the hydrodynamic shock. His explanation, while for a different nonlinearity, may be the explanation for the difficulty here. Lax [175] predicts the breakdown of oscillation by means of a general theory.

REFERENCES

1. Forsythe, G. E., and Wasow, W. R., "Finite Difference Methods for Partial Differential Equations." Wiley, New York, 1960.
2. Fox, L. (ed.), "Numerical Solution of Ordinary and Partial Differential Equations." Macmillan (Pergamon), New York, 1962.
3. Young, D. M., and Frank, T. G., A survey of computer methods for solving elliptic and parabolic partial differential equations, *Intern. Computation Center Bull.* **2**, No. 1 (1963).
4. Richtmyer, R. D., "Difference Methods for Initial-Value Problems." Wiley (Interscience), New York, 1957
5. Todd, J. (ed.), "Survey of Numerical Analysis." McGraw-Hill, New York, 1962.
6. Bickley, W. G., *Quart. J. Mech. Appl. Math.* **1**, 35 (1948).
7. Wasow, W. R., *Z. Angew. Math. Phys.* **6**, 81 (1955).
8. Milne, W. E., "Numerical Solution of Differential Equations," p. 122. Wiley, New York, 1953.
9. Richardson, L. F., *Trans. Roy. Soc. (London)* **A210**, 307 (1910).
10. O'Brien, G. G., Hyman, M. A., and Kaplan, S., *J. Math. Phys.* **29**, 223 (1951).
11. Dufort, E. C., and Frankel, S. P., *Math. Tables Aids Computation* **7**, 135 (1953).
12. John, F., *Commun. Pure Appl. Math.* **5**, 155 (1952).
13. Crandall, S. H., *J. Assoc. Computing Machinery* **2**, 42 (1955).
14. Carslaw, H. S., and Jaeger, J. C., "Conduction of Heat in Solids," 2nd ed. Oxford Univ. Press, London and New York, 1959.
15. Blasius, H., *Z. Math. Phys.* **56**, 4 (1908).
16. Luckert, H., *Schriften Math. Sem. Inst. Angew. Math. Univ. Berlin* **1**, 245 (1934).
17. Goldstein, S., *Proc. Cambridge Phil. Soc.* **26**, 1 (1930).
18. Rosenhead, L., and Simpson, J. H., *Proc. Cambridge Phil. Soc.* **32**, 385 (1936).
19. Crank, J., and Nicolson, P., *Proc. Cambridge Phil. Soc.* **43**, 50 (1947).
20. Crandall, S. H., *Quart. Appl. Math.* **13**, 318 (1955).
21. Juncosa, M. L., and Young, D. M., *Proc. Cambridge Phil. Soc.* **53**, 448 (1957).
22. Thomas, L. H., Elliptic problems in linear difference equations over a

network, Rept. Watson Sci. Computing Lab. Columbia Univ. (New York), 1949.
23. Bruce, G. H., Peaceman, D. W., Rachford, H. H., and Rice, J. D., *Trans. AIME (Petrol. Div.)* **198**, 79 (1953).
24. Bellman, R., Juncosa, M., and Kalaba, R., Some numerical experiments using Newton's method for nonlinear parabolic and elliptic boundary-value problems, *Rand Corp. (Santa Monica, Calif.) Rept.* No. P-2200 (1961).
25. Douglas, J., Jr., and Jones, B. F., Jr., *J. Soc. Ind. Appl. Math.* **11**, 195 (1963).
26. Douglas, J., Jr., *Numer. Math.* **4**, p. 41 (1962).
27. Douglas, J., Jr., A survey of numerical methods for parabolic differential equations, *Advan. Computers* **2**, 1 (1961).
28. Flugge-Lotz, I., and Eichelbrenner, E. A., La Couche limite laminare dans l'écoulement compressible le long d'une surface courbe, *Off. Natl. Etudes Rech. Aeron. Rept.* No. 1/694-A (1948); see also Eichelbrenner, E. A., Méthodes de calcul de la Couche limit laminaire bidimensionelle en régime compressible, *ONERA Rept.* No. 83 (1956).
29. Flugge-Lotz, I., The computation of the laminar compressible boundary layer, ARDC Contract AF 18 (600)-586, *Dept. Mech. Eng. Stanford Univ. (Stanford, Calif.) Proj.* No. R-352-30-7 (1954).
30. Crocco, L., "Lo Strato Limite Laminare nei-Gas," Monogr. Sci. Aeron. No. 3. Assoc. Cult. Aeron., Rome, 1946.
31. Kennard, E. H., "Kinetic Theory of Gases." McGraw-Hill, New York, 1938.
32. Flugge-Lotz, I., and Baxter, D. C., The solution of compressible laminar boundary layer problems by a finite method (Part I), *Dept. Mech. Eng. Stanford Univ. (Stanford, Calif.) Tech. Rept.* No. 103 (1956); see also Part II, *Tech. Rept.* No. 110 (1957).
33. Baxter, D. C., The solution of compressible laminar boundary layer problems by a finite difference method, Ph.D. Dissertation, Stanford University (1957).
34. Kramer, R. F., and Lieberstein, H. M., *J. Aerospace Sci.* **26**, 508 (1959).
35. Flugge-Lotz, I., and Yu, E. Y., Development of a finite difference method for computing a compressible laminar boundary layer with interaction, *Div. Eng. Mech. Stanford Univ. (Stanford, Calif.) Tech. Rept.* No. 127 (1960).
36. Wu, J. C., The solution of laminar boundary layer equations by the finite difference method, *Douglas Aircraft Co. (Santa Monica, Calif.) Rept.* No. SM-37484 (1960).
37. Pallone, A., *J. Aerospace Sci.* **28**, 449 (1961).
38. Blottner, F. G., Computation of the compressible laminar boundary-layer flow including displacement thickness interaction using finite difference methods, Ph.D. Dissertation, Stanford University (1962).
39. de Santo, D. F., and Keller, H. B., *J. Soc. Ind. Appl. Math.* **10**, 569 (1962).
40. Hicks, B. L., Kelso, J. W., and Davis, J., Mathematical theory of the ignition process considered as a thermal reaction, *Ballistic Res. Lab. Aberdeen Proving Ground (Maryland) Rept.* No. 756 (1957).
41. Eddy, R. P., *U.S. Naval Ordnance Lab. (White Oak, Maryland) Rept.* No. NAVORD-2725 (1952).
42. Crank, J., *Quart. J. Mech. Appl. Math.* **10**, 220 (1957).

REFERENCES

43. Butler, R., Lloyd, E. C., Michel, J. G. L., and Sully, E. D., *Ann. Assoc. Intern. Calcul Analogique* (1962).
44. Albasiny, E. L., *IEE Convention Digital Computers, 1956.*
45. Ehrlich, L. W., *J. Assoc. Computing Machinery* **5**, 161 (1958).
46. Douglas, J., Jr., and Gallie, J. M., Jr., *Duke Math. J.* **22**, 557 (1955).
47. Ting, T. C. T., *Quart. Appl. Math.* **21**, 133 (1963).
48. Fromm, J. E., *Los Alamos Sci. Lab. (New Mex.) Rept.* No. LA-2910 (1963).
49. Fromm, J. E., and Harlow, F. H., *Phys. Fluids* **6**, 975 (1963).
50. Philip, J. R., *Australian J. Phys.* **10**, 29 (1957).
51. Philip, J. R., *Soil Sci.* **83**, 345 (1957).
52. Philip, J. R., *Soil Sci.* **83**, 435 (1957).
53. Rubin, J., and Steinhardt, R., *Soil Sci. Soc. Am. Proc.* **27**, 246 (1963).
54. Garabedian, P. R., "Partial Differential Equations." Wiley, New York, 1964.
55. MacNeal, R. H., *Quart. Appl. Math.* **11**, 295 (1953).
56. Varga, R. S., Numerical solution of the two-group diffusion equation in x–y geometry, *Westinghouse Elec. Corp. (Bettis Plant, Pittsburgh, Pa.) Rept.* No. WAPD-159 (1956).
57. Varga, R. S., "Matrix Iterative Numerical Analysis." Wiley, New York, 1962.
58. Mikeladze, Sh., *Izv. Akad. Nauk SSSR, Ser. Matem.* **5**, 57 (1941).
59. Gerschgorin, S., *Z. Angew. Math. Mech.* **10**, 373 (1930).
60. Collatz, L., *Z. Angew. Math. Mech.* **13**, 55 (1933).
61. Hildebrand, F. B., "Introduction to Numerical Analysis." McGraw-Hill, New York, 1956.
62. Shaw, F. S., "An Introduction to Relaxation Methods." Dover, New York, 1950.
63. Allen, D. N., de G., "Relaxation Methods." McGraw-Hill, New York, 1954.
64. Batschelet, E., *Z. Angew. Math. Phys.* **3**, 165 (1952).
65. Viswanathan, R. V., *Math. Tables Aids Computation* **11**, 67 (1957).
66. Householder, A. S. "Principles of Numerical Analysis." McGraw-Hill, New York, 1953.
67. Bodewig, E., "Matrix Calculus." Wiley (Interscience), New York, 1956.
68. Collatz, L., *Math. Z.* **53**, 149 (1950).
69. Reich, E., *Ann. Math. Statist.* **20**, 448 (1949).
70. Frankel, S. P., *Math. Tables Aids Computation* **4**, 65 (1950).
71. Young, D. M., *Trans. AMS* **76**, 92 (1954).
72. Young, D. M., The numerical solution of elliptic and parabolic partial differential equations, *in* "Survey of Numerical Analysis" (J. Todd, ed.), p. 380. McGraw-Hill, New York, 1962.
73. Friedman, B., The iterative solution of elliptic difference equations, *New York Univ. Inst. Math. Sci. Rept.* No. NYO-7698 (1957).
74. Keller, H. B., *Quart. Appl. Math.* **16**, 209 (1958).
75. Arms, R. J., Gates, L. D., and Zondek, B., *J. Soc. Ind. Appl. Math.* **4**, 220 (1956).
76. Kahan, W., Gauss-Seidel methods of solving large systems of linear equations, Ph.D. Thesis, University of Toronto (1958).

77. Garabedian, P. R., *Math. Tables Aids Computation* **10**, 183 (1956).
78. Young, D. M., *J. Assoc. Computing Machinery*, **2**, 137 (1955).
79. Golub, G. H., and Varga, R. S., *Numer. Math.* **3**, 147 (1961).
80. Varga, R. S., *J. Soc. Ind. Appl. Math.* **5**, 39 (1957).
81. Sheldon, J. W., *Math. Tables Aids Computation* **9**, 101 (1955).
82. Sheldon, J. W., *J. Assoc. Computing Machinery* **6**, 494 (1959).
83. Cuthill, E. H., and Varga, R. S., *J. Assoc. Computing Machinery* **6**, 236 (1959).
84. Varga, R. S., Factorization and normalized iterative methods, *in* "Boundary Problems in Differential Equations" (R. E. Langer, ed.), p. 121. Univ. of Wisconsin Press, Madison, Wisconsin, 1960.
85. Peaceman, D. W., and Rachford, H. H., Jr., *J. Soc. Ind. Appl. Math.* **3**, 28 (1955).
86. Douglas, J., Jr., and Rachford, H. H., Jr., *Trans. AMS* **82**, 421 (1956).
87. Birkhoff, G., Varga, R. S., and Young, D. M., Alternating direction implicit methods, *Advan. Computers* **3**, 189 (1962).
88. Young, D. M., and Wheeler, M. F., Alternating direction methods for solving partial difference equations, *in* "Nonlinear Problems of Engineering" (W. F. Ames, ed.), p. 220. Academic Press, New York, 1964.
89. Wachspress, E. L., CURE: A generalized two-space-dimension multigroup coding for the IBM-704, *Gen. Elec. Co. (Schenectady, N. Y.) Rept.* No. KAPL 1724 (1957).
90. Wachspress, E. L., and Habetler, G. J., *J. Soc. Ind. Appl. Math.* **8**, 403 (1960).
91. Wachspress, E. L., *J. Soc. Ind. Appl. Math.* **10**, 339 (1962).
92. Schechter, R. S., *AIChE Journal* **7**, 445 (1961).
93. Wheeler, J. A., Jr., Laminar flow of a non-Newtonian fluid in a rectangular duct, M.A. Thesis, University of Texas (1963).
94. Douglas, J., Jr., *Numer. Math.* **3**, 92 (1961).
95. Bers, L., *J. Res. Natl. Bur. Std.* **51**, 229 (1953).
96. Slattery, J. C., *Appl. Sci. Res.* **A12**, 51 (1963).
97. Serrin, J., *in* "Handbuch der Physik" (S. Flügge, ed.), Vol. 8, Part I, p. 255. Springer, Berlin, 1959.
98. Priestley, C. H. B., *Australian J. Phys.* **7**, 176 (1954).
99. Philip, J. R., *Australian J. Phys.* **9**, 570 (1956).
100. Fischer, J., and Moser, H., *Arch. Elektrotech.* **42**, 286 (1956).
101a. Trutt, F. C., Erdelyi, E. A., and Jackson, R. F., *IEEE Trans. Aerospace* **1**, 430 (1963).
101b. Landgraff, R. W., and Erdelyi, E. A., Influence of airgap curvature on the flux distribution in saturated homopolar alternators, Rept. Dept. Elec. Eng. Univ. of Delaware (Neward, Del.), 1963.
101c. Ahamed, S. V., and Erdelyi, E. A., Nonlinear vector potential equations for highly saturated heteropolar electrical machines, Dept. Elec. Eng. Univ. of Delaware (Newark, Del.), 1963.
102. Whitaker, S., and Wendel, M. M., *Appl. Sci. Res.* **12**, 91 (1963).
103. Bushby, F. H., and Whitelam, C. J., *Quart. J. Roy. Meteorol. Soc.* **87** (1961).
104. Charney, J. G., and Phillips, N. A., *J. Meteorol.* **10**, 71 (1953).

105. Bolin, B., *Tellus* **8**, (1956).
106. Arnason, G., A convergent method for solving the balance equation, Internal Rept. Joint Numer. Weather Prediction Unit, U.S. Weather Bur. (Washington, D.C.), 1957.
107. Schechter, S., *Trans. AMS* **106**, 179 (1962).
108. Lieberstein, H. M., Overrelaxation for non-linear elliptic partial differential equations, *Univ. of Wisconsin Math. Res. Center (Madison, Wis.) Tech. Sum. Rept.* No. MRC-TSR-80 (1959).
109. Lieberstein, H. M., A numerical test case for the nonlinear overrelaxation algorithm, *Univ. of Wisconsin Math. Res. Center (Madison, Wis.) Tech. Rept.* No. MRC-TR-122 (1960).
110. Greenspan, D., and Yohe, M., On the approximate solution of $\nabla^2 u = F(u)$, *Univ. of Wisconson Math. Res. Center (Madison, Wis.), Tech. Rept.* No. MRC-TR-384 (1963); see also *Commun. Assoc. Computing Machinery* **6**, 564 (1963).
111. Wise, H., and Ablow, C. M., *J. Chem. Phys.* **35**, 10 (1961).
112. Greenspan, D., Recent computational results in the numerical solution of elliptic boundary value problems, *Univ. of Wisconsin Math. Res. Center (Madison, Wis.) Tech. Rept.* No. MRC-TR-408 (1963).
113. Weinstein, A., Singular partial differential equations and their applications, in "Fluid Dynamics and Applied Mathematics" (J. B. Diaz and S. I. Pai, eds.), p. 29. Gordon & Breach, New York, 1962.
114. Cobble, M. H., and Ames, W. F., *J. Appl. Mech.* **30**, 415 (1963).
115. Motz, H., *Quart. Appl. Math.* **4**, 371 (1946).
116. Woods, L. C., *Quart. J. Mech.* **6**, 163 (1953).
117. Lieberstein, H. M., Singularity occurrence and stably posed problems for elliptic equations, *Univ. of Wisconsin Math. Res. Center (Madison, Wis.) Rept.* No. MRC-TR-81 (1959).
118. Greenspan, D., and Warten, R., On the approximate solution of Dirichlet-type problems with singularities on the boundary, *Univ. of Wisconsin Math. Res. Center (Madison, Wis.) Tech. Rept.* No. MRC-TR-254 (1961).
119. Jeffrey, A., and Taniuti, T., "Nonlinear Wave Propagation with applications to Physics and Magneto-hydrodynamics." Academic Press, New York, 1964.
120. Swope, R. D., and Ames, W. F., *J. Franklin Inst.* **275**, 36 (1963).
121. Shapiro, A. H., "The Dynamics and Thermodynamics of Compressible Fluid Flow," Vol. 1, Chapter 9. Ronald Press, New York, 1954.
122. Sokolovskii, V. V., A problem of plasticity theory, in "Problems of Continuum Mechanics (In Honor of the 70th Birthday of Muskhelishvili)," p. 513. Soc. Ind. Appl. Math., Philadelphia, Pennsylvania, 1961.
123. Gundersen, R. M., The non-isentropic perturbation of a centered magneto-hydrodynamic simple wave, *Univ. of Wisconsin Math. Res. Center (Madison, Wis.) Tech. Rept.* No. MRC-TR-280 (1961); see also Am. Inst. Aeron. Astronautics **1**, 1191 (1963).
124. Gundersen, R. M., Non-uniform magnetohydrodynamic shock propagation, with special reference to cylindrical and spherical shock waves, *Univ. of Wisconsin Math. Res. Center (Madison, Wis.) Tech. Rept.* No. MRC-TR-310 (1962).
125. Stoker, J. J., "Water Waves." Wiley (Interscience), New York, 1957.

126. Bernstein, D. L., "Existence Theorems in Partial Differential Equations." Princeton Univ. Press, Princeton, New Jersey, 1950.
127. Collatz, L., "The Numerical Treatment of Differential Equations," 3rd ed. Springer, Berlin, 1960.
128. Thomas, L. H., *Commun. Pure Appl. Math.* **7**, 195 (1954).
129. Courant, R., Friedrichs, K., and Lewy, H., *Math. Ann.* **100**, 32 (1928).
130. Douglas, J., Jr., *Quart. Appl. Math.* **14**, 333 (1956).
131. Friedmann, N. E., *J. Math. Phys.* **35**, 299 (1956).
132. Landau, H. G., *J. Soc. Ind. Appl. Math.* **11**, 564 (1963).
133. Skinner, A., Large oscillations of a heavy cable, M.M.E. Thesis, University of Delaware (1961).
134. Friberg, J., *Nord. Tidsskr. Inform. Behandl.* **1**, 69 (1961).
135. Lees, M., *Pacific J. Math.* **10**, 213 (1960).
136. Lees, M., *J. Soc. Ind. Appl. Math.* **10**, 610 (1962).
137. Hartree, D. R., "Numerical Analysis," 2nd ed. Oxford Univ. Press, London and New York, 1958.
138. Courant, R., Isaacson, E., and Rees, M., *Commun. Pure Appl. Math.* **5**, 243 (1952).
139. Fromm, J. E., Lagrangian difference approximations for fluid dynamics, *Los Alamos Sci. Lab. (New Mex.) Rept.* No. LA-2535 (1961).
140. Harlow, F., Stability of difference equations—selected topics, *Los Alamos Sci. Lab. (New Mex.) Rept.* No. LA-2452 (1960).
141. Noh, W. F., and Protter, M. H., *J. Math. Mech.* **12**, No. 2, 149 (1963).
142. Rouse, C. A., *J. Soc. Ind. Appl. Math.* **9**, 127 (1961).
143. Lax, P. D., *Commun. Pure Appl. Math.* **7**, 159 (1954).
144. Lax, P. D., Nonlinear hyperbolic systems of conservation laws, *in* "Nonlinear Problems" (R. E. Langer, ed.), p. 3. Univ. of Wisconsin Press, Madison, Wisconsin, 1963.
145. Lax, P. D., *Commun. Pure Appl. Math.* **10**, 537 (1957).
146. Longley, H. J., Methods of differencing in Eulerian hydrodynamics, *Los Alamos Sci. Lab. (New Mex.) Rept.* No. LA-2379 (1960).
147. Trulio, J. G., and Trigger, K. R., Numerical solution of the one dimensional Lagrangian hydrodynamic equations, *Univ. of California (Berkeley) Lawrence Radiation Lab. Rept.* No. UCRL-6267 (1961).
148. Douglas, A., *Commun. Pure Appl. Math.* **14**, 267 (1961).
149. Lax, P. D., and Wendorff, B., *Commun. Pure Appl. Math.* **13**, 217 (1960).
150. Payne, R. B., *J. Fluid Mech.* **2**, 185 (1957).
151. Roberts, L., *J. Math. Phys.* **36**, 329 (1958).
152. Longley, H. J., Methods of differencing in Eulerian hydrodynamics, *Los Alamos Sci. Lab. (New Mex.) Rept.* No. LAMS-2379 (1960).
153. Courant, R., and Friedrichs, K. O., "Supersonic Flow and Shock Waves." Wiley (Interscience), New York, 1948.
154. Chester, W., The propagation of shock waves along ducts of varying cross section, *Advan. Appl. Mech.* **6**, 120 (1960).
155. Hide, R., and Roberts, P. H., Some elementary problems in magnetohydrodynamics, *Advan. Appl. Mech.* **7**, 216 (1962).
156. von Neumann, J., and Richtmyer, R. D., *J. Appl. Phys.* **21**, 232 (1950).
157. Brode, H. L., *J. Appl. Phys.* **26**, 766 (1955).

REFERENCES

158. Brode, H. L., Point source explosion in air, *Rand Corp.* (*Santa Monica, Calif.*) *Res. Memo.* No. RM-1824-AEC (1956).
159. Schulz, W. D., *J. Math. Phys.* **5**, 133 (1964).
160. Zovko, C. T., and Macek, A., 3rd *Symp. Detonation*, *1959* Vol. II, p. 606. Office of Naval Res., Washington, D.C. (1960).
161. Filler, L., and Ludloff, H. F., *Math. Computation* **15**, 261 (1961).
162. Stein, L. R., A numerical solution of a spherical blast wave utilizing a completely tabular equation of state, *Los Alamos Sci. Lab.* (*New Mex.*) *Rept* No LA-2277)1959)
163. Keller, H. B., Levine, D. A., and Whitham, G. B., *J. Fluid Mech.* **7**, 302 (1960).
164. Keller, H. B., and Thomee, V., *Commun. Pure Appl. Math.* **16**, 184 (1963).
165. Thomee, V., *J. Soc. Ind. Appl. Math.* **10**, 229 (1962).
166. Thomee, V., *J. Soc. Ind. Appl. Math.* **11**, 964 (1963).
167. Courant, R., and Hilbert, D., "Methods of Mathematical Physics," Vol. 2. Wiley (Interscience), New York, 1962.
168. Thornhill, C. K., The numerical method of characteristics for hyperbolic problems in three independent variables, *A.R.C. Rept.* No. ARC-2615 (1948). (See Fox, L. [2].)
169. Coburn, N., and Dolph, C. L., *Proc. 1st Appl. Math. Symp.*, Providence, Rhode Island, 1947 p. 55.Am. Math. Soc., Providence ,Rhode Island, 1949.
170. Bruhn, G., and Haack, W., *Z. Angew. Math. Phys.* **9**, 173 (1958).
171. Butler, D. S., *Proc. Roy. Soc.* **A255**, 232 (1960).
172. Zaiser, J. A., Nonlinear vibrations of a moving threadline, Ph.D. Dissertation, University of Delaware (1964).
173. Hansell, G. A., Moving threadline oscillation experiments, M.M.E. Thesis, University of Delaware (1964).
174. Zabusky, N. J., *J. Math. Phys.* **3**, 1028 (1962).
175. Lax, P. D., *J, Math. Phys.* **5**, 611, (1964).

CHAPTER **8**

Some Theoretical Considerations

8.0 INTRODUCTION

In the preceding chapters mathematical questions of existence, uniqueness, convergence, stability, and error estimates received uneven treatment. If this were a book intended primarily for the mathematical community, then these topics (if they are available, which is a doubtful conjecture) would form a major portion of this volume. The first two questions of existence and uniqueness were considered in some detail late in the book in the chapter on numerical methods. Chief among the reasons for this spotty treatment is the nonexistence of many general results for the equations the engineer encounters. There are areas where existence-uniqueness questions have been successfully answered—witness the importance to quasi-linear hyperbolic systems of the existence—uniqueness theorems exemplified in Fig. 7-26. As we shall see, progress is being made in the difficult questions associated with the Navier-Stokes equations and the related boundary layer equations.

The partial differential equations encountered in practical problems are often very complicated and the practitioner is often unskilled in the difficult subject of existence and uniqueness proofs. The urgencies of the engineering problem usually induces the engineer to assume (either knowingly or unknowingly) the existence and uniqueness of the "solution" to this problem.

Numerical method development for partial differential equations has also been based to a considerable degree on intuition and empiricism. Existence and uniqueness is often ignored as are convergence proofs and error estimates. We should not hastily condemn these attempts for if the practitioner waited for existence and uniqueness proofs for his nonlinear boundary value problem and convergence proofs and error estimates for any new numerical methods, much of engineering and most of the computers in use would come to a halt! Further, he cannot be sure if he waited that mathematicians *would* or *could* attempt his problem.

The answer to these contradictory desires probably lies in collaboration between the two groups. Both must leave their inflexible positions. The engineer must attach more importance to fundamental understanding of the methods he uses and the mathematician must try to understand the importance of the empirical. Beginnings in this rapproachment have been made but the urgency is acute given the present state (or lack of it) of nonlinear theories.

In this chapter we cannot discuss the methods for investigating the difficult problems of the above mathematical problems. Indeed, some of these questions are themselves the subjects of entire books (see, e.g., Bernstein [1] and Ladyzhenskaya [2]). What we shall do is summarize some known results for some physically important systems of equations and associated boundary conditions.

8.1 WELL-POSED PROBLEMS

Exact physical laws are idealization of reality. As our knowledge progresses we see that a given physical situation can be idealized in mathematical form in a number of different ways. Consequently, it is important to characterize those ideal formulations which are reasonable. Hadamard [3] (see for example Courant and Hilbert [4]) discusses this problem asserting that a physical problem is *well posed* if its solution *exists*, is *unique* and depends continuously on the auxiliary data. He asserts that the discontinuous dependence on the initial data precludes any physical meaning because physical data are by their nature only approximate. We should observe here that these criteria are for general orientation only and may not be literally true. The improperly posed problems are receiving increased attention.

These criteria are physically reasonable in most cases. The existence and uniqueness criteria are an affirmation of the principle of determinism. Without determinism experiments could not be repeated with the expectation of consistent data. The continuous dependence is an expression of the stability of the solution, i.e., a small change in any of the problem auxiliary data should produce only a correspondingly small change in the solution.[†]

Much progress in partial differential equations has been achieved by series developments. For the differential equation of second order

$$u_{xy} = G(x, y, u, u_x, u_y, u_{xx}, u_{yy}) \tag{8.1}$$

subject to the auxiliary data

$$u(x, 0) = f(x), \qquad u_y(x, 0) = g(x) \tag{8.2}$$

we may try to obtain a solution in the form

$$u(x, y) = \sum_{i=0}^{\infty} a_i(x) y^i \tag{8.3}$$

and determine the coefficients $a_i(x)$ from the auxiliary conditions (8.2) and Eq. (8.1). It is not a difficult task to show that this procedure works if $G, f(x)$, and $g(x)$ are analytic[‡] functions of their arguments. Further, the series (8.3) converges for small enough y. The next theorem is an existence theorem for rather general systems (including (8.1)). For a proof see Garabedian [6].

Cauchy-Kowalewski Theorem. *Let the component functions $a_{ij}(\xi, \eta, u)$ of the matrix A and the elements $h_j(\eta)$ of the column vector h be analytic at a given point. Then we can find a neighborhood of that point in which there exists a unique solution vector, u, with analytic components, u_k, of the differential system:*

$$u_\xi = A(u) u_\eta, \qquad u(0, \eta) = h(\eta). \tag{8.4}$$

This theorem seems to be quite general even though it is restricted to problems involving only analytic functions. This requirement of

---

[†] There are limitations in stability investigations. See for example LaSalle and Lefschetz [5].

[‡] We use the term *analytic* to mean that the function has all derivatives at a point and in some neighborhood, i.e., all derivatives exist.

8.1 WELL-POSED PROBLEMS

analyticity may be regarded as not too serious. For we know by the Weierstrass approximation theorem that any continuous function can be approximated arbitrarily closely by analytic functions, in fact by polynomials. Such reasoning would be valid *if we could be sure that close approximation of the boundary values always implied close approximation of the solution for $y > 0$.* That this is not the case follows from the simple considerations below.

Consider Laplace's equation $u_{xx} + u_{yy} = 0$ and ask for a solution such that

$$u(x, 0) = p^{-\alpha} \sin px, \quad u_y(x, 0) = 0$$

where $\alpha > 0$ is fixed. A solution exists by the above theorem (we must "factor" the equation and put the problem in the form (8.4)) and in fact is

$$u(x, y) = p^{-\alpha} \sin px \cosh py. \qquad (8.5)$$

With $\alpha > 0$ fixed, let $p \to \infty$, whereupon the initial data converge uniformly to

$$u(x, 0) = 0, \quad u_y(x, 0) = 0$$

thus determining the solution $u(x, y) = 0$. But for $y \neq 0$ the functions (8.5) *do not converge* to zero but become very large, as $p \to \infty$, in an arbitrary neighborhood of the x-axis. Thus approximating the initial data arbitrarily closely does not guarantee a corresponding approximation for the solution.

While no engineering problem is known to lead to a problem of this type the possibility exists—*thus the warning*.

Not well-posed problems (in the Hadamard sense) have received some attention since such problems can represent realizable physical phenomena. For example, problems in which a portion of the boundary is inaccessible for the measurement of the boundary data, fall into this class. Laurentiev [7] and Payne [8] have examined such problems. For Laplace's equation Laurentiev demonstrates that an additional condition to the Cauchy data[†] is required, namely that the harmonic function be uniformly bounded by some constant, in order for the solution to be stable.

[†] This term is used for data of the form (8.2) which are initial data in propagation problems.

8.2 EXISTENCE AND UNIQUENESS IN VISCOUS INCOMPRESSIBLE FLOW

The basic problem of the unique solvability "in the large" of the boundary value problem for the general three dimensional time dependent Navier-Stokes equations (with no assumptions other than a certain regularity of the initial state and external forces) has not been resolved. A second question concerns the quality of the description of real flows by the solutions of the Navier-Stokes equations and the equation of continuity (1.35) and Eq. (1.36)

$$\frac{\partial u_i}{\partial t} + u_j \frac{\partial u_i}{\partial x_j} = -\rho^{-1} \frac{\partial p}{\partial x_i} + \nu \nabla^2 u_i + F_i \qquad i = 1, 2, 3 \qquad (8.6)$$

$$\frac{\partial u_i}{\partial x_i} = 0.$$

The difficulties in obtaining answers to these questions have been compounded by the lack of many useful exact solutions to Eq. (8.6). Most of the known solutions involve simplifications which disregard the inherent nonlinearities.

Experiments and approximate calculations have pointed out various discrepancies between the idealized mathematical model of a viscous fluid and the actual phenomena. Two of these, which raise questions about the *validity* of the idealization and the uniqueness are as follows.

(1) For any Reynolds number Re, the only possible solutions of Eqs. (8.6) in an infinitely long circular cylindrical pipe, which are symmetric with respect to its axis (say the z-axis, with velocity component w) are

$$w = a(r_0^2 - r^2), \qquad u_r = 0, \qquad u_\theta = 0 \qquad (8.7)$$

where r_0 is the pipe radius and a is a parameter. However, the so-called Poiseuille flows corresponding to these results are *only* observed for values of Re which do not exceed a critical value. The flows become turbulent when that critical value is reached.

(2) The second discrepancy appears in Couette flow.[†] The symmetry of the problem suggests that the flow ought to be symmetric. In fact

---

[†] Couette flow is a steady flow between symmetric rotating concentric cylinders and any plane perpendicular to the cylinder axis.

solutions of Eq. (8.6) possessing this symmetry exist for all Re. But actual symmetric flow is observed only for small values of Re; for large values of Re the physical flow is still laminar *but not* symmetric. (This leads us to question whether the widely held belief "symmetric causes produce symmetric effects" is valid.)

In both cases, it is not known if the Navier-Stokes equations have solutions for large Re which correspond to the observed flows; but if they could be found this would lead to a *violation* of the uniqueness for steady solutions to Eq. (8.6).

An example by Ladyzhenskaya [2] pertains to the uniqueness question. She shows that in an unbounded plane domain the problem of flow with sources can have infinitely many solutions. Consequently, additional conditions must be imposed to single out a *unique* solution. Consider the steady Navier-Stokes and continuity equation in plane polar coordinates r and θ,

$$u_r \frac{\partial u_r}{\partial r} + \frac{u_\theta}{r} \frac{\partial u_r}{\partial \theta} - \frac{u_\theta^2}{r} = -\frac{1}{\rho} \frac{\partial p}{\partial r} + \nu \left[\frac{\partial^2 u_r}{\partial r^2} + \frac{1}{r} \frac{\partial u_r}{\partial r} - \frac{u_r}{r^2} \right.$$

$$\left. + \frac{1}{r^2} \frac{\partial^2 u_r}{\partial \theta^2} - \frac{2}{r^2} \frac{\partial u_\theta}{\partial \theta} \right]$$

$$u_r \frac{\partial u_\theta}{\partial r} + \frac{u_\theta}{r} \frac{\partial u_\theta}{\partial \theta} + \frac{u_r u_\theta}{r} = -\frac{1}{\rho r} \frac{\partial p}{\partial \theta} + \nu \left[\frac{\partial^2 u_\theta}{\partial r^2} + \frac{1}{r} \frac{\partial u_\theta}{\partial r} - \frac{u_\theta}{r^2} \right.$$

$$\left. + \frac{1}{r^2} \frac{\partial^2 u_\theta}{\partial \theta^2} + \frac{2}{r^2} \frac{\partial u_r}{\partial \theta} \right]$$

$$\frac{\partial u_r}{\partial r} + \frac{u_r}{r} + \frac{1}{r} \frac{\partial u_\theta}{\partial \theta} = 0.$$

(8.8)

The functions

$$u_r = \frac{a}{r}, \qquad u_\theta = a_1 \left[\frac{1}{r} - r^{(a/\nu)+1} \right]$$

$$p = -\frac{a^2 + a_1^2}{2r^2} - \frac{2a_1^2 \nu r^{a/\nu}}{a} + \frac{a_1^2}{(2a/\nu) + 2} r^{(2a/\nu)+2}$$

(8.9)

with a and a_1 arbitrary constants, satisfy Eqs. (8.8). For fixed $a < -2\nu$, these functions give infinitely many solutions in $r \geq 1$ which go to zero as $r \to \infty$ and satisfy the boundary conditions

$$u_r = a, \qquad u_\theta = 0 \qquad \text{at} \qquad r = 1.$$

In spite of these discrepancies (and others) it may be that satisfactory explanations lie in the system of Navier-Stokes (Eq. 8.6). The reasoning behind this statement is based primarily on the *nonlinearity* of Eq. (8.6). It is known that for *nonlinear equations* a well-behaved solution to the unsteady problem may not exist on the entire time interval $t \geqslant 0$. Outside of a finite interval the solution may become *arbitrarily large* or split up by forming branches. Even if a solution exists for all $t \geqslant 0$ it may not approach the steady state solution.

The second paradox has been considered by Goldshtik [9] who examined the problem of interaction between an infinite vortex filament and a plane. He proves there is a unique solution with the same symmetry as the problem if $\text{Re} < R_1$, but if Re exceeds a certain numer $R_2 > R_1$ *there are no such solutions*.

Comparison of boundary value problems for the Navier-Stokes equations with those of other problems suggests the following conjecture: because of the nonlinearity of the Navier-Stokes equations the steady problem has a unique solution for $Re < R_1$, several solutions (split up) for $R_1 < \text{Re} < R_2$, and no solutions for $Re > R_2$. This is only a conjecture but some results, in addition to Goldshtik, are known in this direction and are summarized, in general terms, below.

Research described here is given in detail, with a lengthy bibliography, by Ladyzhenskaya [2]. We shall discuss only the results pertinent to the *nonlinear* system (8.6).

(1) A steady boundary-value problem for Eq. (8.6) has solutions for any Reynolds number Re where both the boundaries of the object past which the flow occurs and the external force field can be quite irregular. However, these solutions are stable for only small values of Re (Chapter 5 of Ladyzhenskaya [2]).

(2) A nonsteady boundary-value problem for Eq. (8.6) has a unique solution for all time $\geqslant 0$ if the data of the problem are independent of one of the space coordinates or if the problem has axial symmetry.

(3) A nonsteady boundary-value problem for Eq. (8.6) in three dimensions, has a unique solution if the external forces can be derived from a potential and if the number Re is small at $t = 0$.

(4) In the general case, where the conditions in (2) and (3) are not satisfied, for all time $t \geqslant 0$ there exists at least one *weak solution* (in the sense of Bers—see Chapter 4) $\mathbf{v}(x, t)$ which has derivatives appearing in Eq. (8.6) $\mathbf{v}_t$, $\mathbf{v}_{x_i x_j}$ which are summable with respect to space and time,

with exponent $\frac{5}{4}$.† The uniqueness of the solution on the whole time axis has not been established.

For nonsteady problems the following results on stability are known.

(5) If as $t \to \infty$ the external forces F_i die out and the boundary conditions correspond to a steady state then the motion also dies out, regardless of the motion at $t = 0$.

(6) If as $t \to \infty$ the external forces $F_i \to F_{0i}(x)$ (steady values), for which the corresponding boundary value problem has a solution $\mathbf{u}_0(x)$ (x is the vector (x_1, x_2, x_3)) when Re is small, then the solutions $\mathbf{u}(x, t)$ of the nonsteady problem, with initial data $\mathbf{u}(x, 0)$, approach $\mathbf{u}_0(x)$ as $t \to \infty$. If Re is large, then in general, the solutions $\mathbf{u}(x, t)$ do not approach any definite limit as $t \to \infty$.

The results of the extensive studies reported by Ladyzhenskaya [2], and continuing in the work of R. Finn, J. Serrin, E. Hopf, J. Lions, K. Golovkin, and others, support the belief that it is reasonable to use the Navier-Stokes equation to describe the motions of a viscous fluid for Renolds numbers not exceeding certain limits. These equations are probably invalid for large Re. In seeking explanations for observed phenomena in real fluids one should not rest one's entire case on the Navier-Stokes equations. Enlargement of the system to include variable dissipative effects, heat transfer, etc., is necessary. But the existence-uniqueness-stability theory for these enlarged systems is in even poorer shape than the Navier-Stokes theory.

In addition to the Ladyzhenskaya reference, Serrin [10] gives an excellent readable summary of the existence-uniqueness theory for the initial value problem of the Navier-Stokes equations.

The role of "providing confidence" in his solutions, for the engineering analyst, is the major one for these theories. In addition to this the methods of proof may suggest approximate and numerical methods of solution. Among these the iteration methods, such as that of Picard, are common methods of proof which were later adapted to the development of approximate solutions. Yet a third use accrues to the engineer, that of solution and stability bounds, which are a direct consequence of

---

† The methods of linear function spaces, specifically Hilbert Space, are most often used in this subject. By the term summable with exponent 5/4 we mean that the quantity

$$\iint\limits_{0 \leqslant t \leqslant T} \left| \sum_k v_k \frac{\partial v_i}{\partial x_k} \right|^{5/4} dx\, dt \quad \text{is finite.}$$

the development of basic inequalities. Inequalities have long been a powerful tool in mathematics but only recently have begun to play a role in engineering studies. A good elementary introduction is by Kazarinoff [11], while advanced treatment is available in Hardy, Littlewood, and Polya's classic "Inequalities" [12]. Ladyzhenskaya [2] summarizes some of the inequalities useful in her treatise.

A sufficient condition for the stability of a viscous incompressible flow in a bounded domain D has been given by Serrin [13]. The condition depends upon the maximum velocity, the viscosity and the geometric quantity

$$\lambda = \min\left\{\iiint_D \sum_{i=1}^{3} \left(\frac{\partial u_i}{\partial x_j}\right)^2 dV \Big/ \iiint_D \sum_{i=1}^{3} u_i^2 \, dV\right\} \tag{8.10}$$

where the minimum is with respect to all vector fields u_i defined in D, zero on the boundary and satisfying the continuity equation. The quantity λ is a *nonincreasing domain functional* so that the value of λ for a domain D gives a stability bound for any domain contained in D. When D is a cube, a lower bound is obtained by Velte [14]. Payne and Weinberger [15] consider flow in a sphere D of radius R. They announce [15] the *exact* value of λ as the smallest positive root of

$$\tan \sqrt{\lambda} R = \sqrt{\lambda} R$$

i.e.,
$$\lambda \approx 20 R^{-2} \tag{8.11}$$

with minimizing flow

$$u_1 = x_2 r^{-2}[\sqrt{\lambda} r \cos(\sqrt{\lambda} r) - \sin(\sqrt{\lambda} r)]$$
$$u_2 = -x_1 r^{-2}[\sqrt{\lambda} r \cos(\sqrt{\lambda} r) - \sin(\sqrt{\lambda} r)]$$
$$u_3 = 0$$

where $r^2 = x_1^2 + x_2^2 + x_3^2$. The beginning of turbulence is expected to resemble this flow.

8.3 EXISTENCE AND UNIQUENESS IN BOUNDARY LAYER THEORY

The mathematical questions of existence and stability of solutions for *steady laminar two-dimensional boundary layer flow* are still open. Some results have been obtained concerning uniqueness, separation, etc.,

8.3 BOUNDARY LAYER THEORY

by Nickel [16] using the Nagumo [17], Westphal [18] theory of parabolic differential inequalities. This theory appears to be applicable to the Prandtl equations only after such transformations as the von Mises and Crocco change of variables.

A uniqueness theorem is proved by Nickel [16] for the boundary layer equations under rather mild conditions. It seems certain that this result can be extended. In the same paper Nickel gives bounds for the number of maxima, minima, and turning points of the velocity profiles which depend only on the pre-assigned boundary values. These results are only fragmentary but do allow us to make some assertions about the possible shape of the velocity profiles prior to integration.

The question of boundary layer separation has led to the following result: separation cannot occur for arbitrary initial velocity profiles if the pressure gradient dp/dx has the bound

$$\frac{dp}{dx} \leqslant Ax^{-1.1808} \tag{8.12}$$

where x is the wall coordinate and $A \geqslant 0$ depends on the initial velocity profile. If the initial velocity profile is *monotone*, separation takes place only at the wall.

Witting [19] has considered the question of stability under steady perturbations. He finds that in the region of pressure rise steady perturbations may grow locally but are bounded for all x. For a finite x interval there is always stability up to the separation point.

In nonsteady flow uniqueness of boundary layer solutions has been proved by Nickel [20]. Typical results for *two component* fluids including a generalized von Mises transformation is given by Nickel [21]. The mathematical attack has some interesting features which we give here.

Rachmatulin [22] gives the incompressible two-dimensional steady boundary layer equations for a *two component* fluid in rectangular geometry as

$$u_1 u_{1x} + v_1 u_{1y} = -p_x/\rho_1 + \nu_1 u_{1yy} + k_1(u_2 - u_1)$$

$$u_2 u_{2x} + v_2 u_{2y} = -p_x/\rho_2 + \nu_2 u_{2yy} + k_2(u_1 - u_2)$$

$$u_{1x} + v_{1y} = 0, \qquad u_{2x} + v_{2y} = 0. \tag{8.13}$$

The variables x and y denote distance along and normal to the wall; u_i and v_i are velocity components inside the boundary layer in the x- and y-directions; p_x is the known pressure gradient; ρ_i, ν_i, and k_i

are positive density, viscosity and friction respectively. For a solid wall we would have initial and boundary conditions

$$u_1(x_0, y) = \bar{u}_1(y), \qquad u_2(x_0, y) = \bar{u}_2(y)$$
$$u_1(x, 0) = u_2(x, 0) = v_1(x, 0) = v_2(x, 0) = 0 \qquad (8.14)$$
$$\lim_{y \to \infty} u_1(x, y) = U_1(x), \qquad \lim_{y \to \infty} u_2(x, y) = U_2(x).$$

The velocity components of the outer flow, designated by U_1 and U_2, are calculated from the equations

$$U_1 U_1' = -p_x/\rho_1 + k_1(U_2 - U_1)$$
$$U_2 U_2' = -p_x/\rho_2 + k_2(U_1 - U_2). \qquad (8.15)$$

Nickel subjects Eqs. (8.13) to a generalized von Mises transformation in which the variables x and y are replaced by ξ, η, and θ defined by

$$\xi = x$$
$$\eta = \frac{\psi_1(x, y)}{1 + \psi_1(x, y)}, \qquad \psi_1 = \int_0^y u_1(x, t)\, dt$$
$$\theta = \frac{\psi_2(x, y)}{1 + \psi_2(x, y)}, \qquad \psi_2 = \int_0^y u_2(x, t)\, dt$$
(8.16)†

where ψ_1, ψ_2 are stream functions. We further set

$$u(\xi, \eta) = u_1(x, y), \qquad w(\xi, \theta) = u_2(x, y)$$
$$U(\xi) = U_1(x), \qquad W(\xi) = U_2(x). \qquad (8.17)$$

The transformation (8.16) maps the region $x_0 \leqslant x \leqslant x_1$, $0 \leqslant y < \infty$ onto the *two finite domains*

$$x_0 \leqslant \xi \leqslant x_1, \qquad 0 \leqslant \eta < 1$$
$$x_0 \leqslant \xi \leqslant x_1, \qquad 0 \leqslant \theta < 1$$

in a one-to-one fashion if $u_1 > 0$, $u_2 > 0$.

The new variables η, θ are related. They must agree if $u_1 = u_2$. The relation between them is unknown in the beginning and is to determined during the calculation by means of the equation

$$\int_0^\eta \frac{1 - 2t}{(1 - t)^2 u(\xi, t)}\, dt = \int_0^\theta \frac{1 - 2t}{(1 - t)^2 w(\xi, t)}\, dt. \qquad (8.18)$$

---

† ψ_1 and ψ_2 are *stream functions*. For example $\psi_{1y} = u_1$, $\psi_{1x} = -v_1$ and therefore continuity is satisfied.

8.3 BOUNDARY LAYER THEORY

The formula (8.18) follows from the definitions (8.16) which yield the information that

$$\psi_1\left(\frac{1-\eta}{\eta}\right) = \psi_2\left(\frac{1-\theta}{\theta}\right).$$

Upon setting Eqs. (8.16) and (8.17) into Eq. (8.13), we have after using Eq. (8.15), the new system

$$u_\xi - \nu_1(1-\eta)^3[(1-\eta)(uu_{\eta\eta} + u_\eta^2) - 2uu_\eta]$$
$$- \frac{1}{u}[UU' + k_1(U - W + w - u)] = 0$$
$$w_\xi - \nu_2(1-\theta)^3[(1-\theta)(ww_{\theta\theta} + w_\theta^2) - 2ww_\theta] \qquad (8.19)$$
$$- \frac{1}{w}[WW' + k_2(W - U + u - w)] = 0.$$

This system must be solved for u and w in the two regions $x_0 < \xi \leqslant x_1$, $0 < \eta < 1$, $0 < \theta < 1$ where η and θ are related by Eq. (8.18). The initial and boundary conditions are

$$u(x_0, \eta) = \bar{u}(\eta), \qquad w(x_0, \theta) = \bar{w}(\theta)$$
$$u(\xi, 0) = w(\xi, 0) = 0$$
$$u(\xi, 1) = U(\xi), \qquad w(\xi, 1) = W(\xi).$$

Care must be exercised because u_η, $u_{\eta\eta}$, w_θ, and $w_{\theta\theta}$ are singular for $\eta = \theta = 0$. This follows, for example, from the transformation since

$$u_\eta = (1 + \psi_1)^2 \frac{u_{1y}}{u_1}$$

and

$$u_{\eta\eta} = (1 + \psi_1)^4 \frac{u_1 u_{1yy} - u_{1y}^2}{u_1^3} + 2(1 + \psi_1)^3 \frac{u_{1y}}{u_1}.$$

At this writing, Eqs. (8.19) appear to be more useful theoretically. However, they have the advantage of a finite domain—the singularity at the origin has been successfully handled previously in the one component case (see Chapter 7, Section 7.7). For computation, the relation (8.18) between θ and η is a stumbling block.

Nickel [21] uses the transformed set (8.19) to obtain a rather general uniqueness theorem. His separation theorem is of interest to us here. It is: if the initial profiles $\bar{u}_1(x)$, $\bar{u}_2(x)$ are not separation profiles then for $p_x \leqslant 0$ separation cannot occur. That is, for $p_x \leqslant 0$ we have $u_i(x, y) > 0$ in $0 < y < \infty$ and $u_{1y}(x, 0) > 0$, $u_{2y}(x, 0) > 0$ for all $x > x_0$ if these inequalities hold only for $x = x_0$.

8.4 EXISTENCE AND UNIQUENESS IN QUASI-LINEAR PARABOLIC EQUATIONS

We consider now the quasi-linear parabolic equation

$$u_t = a(x, t, u)u_{xx} + b(x, t, u)u_x + c(x, t, u) \tag{8.20}$$

subject to the auxiliary conditions

$$\alpha_1(t)u + \beta_1(t)u_x = \gamma_1(t) \quad \text{at } x = 0$$
$$\alpha_2(t)u + \beta_2(t)u_x = \gamma_2(t) \quad \text{at } x = x_0 \tag{8.21}$$
$$u(0, x) = f(x).$$

Such a system has only a single family of characteristics, the lines $t =$ constant (see Chapter 7, Section 7.21). In the linear case existence and uniqueness of solution has been established for this data (Eq. 8.21) below any characteristic, i.e., for $0 \leqslant t \leqslant T$.

Hermes [23] has reported results on the system

$$u_t = A(u)u_{xx}, \quad 0 \leqslant t \leqslant T, \quad -\infty < x < \infty, \quad u(0, x) = f(x). \tag{8.22}$$

He assumes

(i) $|f(x)| \leqslant M_2$;

(ii) $|f'(x)| \leqslant M_2$;

(iii) $|f'(x) - f'(\tilde{x})| \leqslant B_2 |x - \tilde{x}|$ (Lipschitz condition)

and

(iv) $0 < \delta \leqslant A(u) \leqslant M_1$;

(v) $A'(u)$ is continuous on $-M_2 \leqslant u \leqslant M_2$, $|A'(u)| \leqslant N_1$;

(vi) $|A'(u) - A'(v)| \leqslant N_2 |u - v|$.

Under these finiteness conditions he shows that the system has a unique solution which can be extended for all time. The result generalizes to parabolic equations of the form $u_t = A(t, x, u)u_{xx}$.

While the methods of Hermes are not directly applicable it seems clear that the problem characterized by Eqs. (8.20) and (8.21) is well-posed in the sense of Hadamard.

8.5 UNIQUENESS QUESTIONS FOR QUASI-LINEAR ELLIPTIC EQUATIONS

We consider quasi-linear elliptic equations of the form

$$\sum_{i,j=1}^{n} a_{ij}(x, u, \nabla u) \frac{\partial^2 u}{\partial x_i \, \partial x_j} = 0, \qquad x = (x_1, \ldots, x_n). \tag{8.23}$$

It is well known (see Garabedian [6]) that in two dimensions ($n = 2$) the Dirichlet problem has solutions for quasi-linear, uniformly elliptic equations of the form (8.23). The uniqueness question is still open except in the linear case. Can Eq. (8.23) have two solutions with the *same* boundary values?[†]

Meyers [24] gives examples in *all dimensions* of equations with non-unique solutions. Consider the nonlinear equation

$$\sum_{i,j=1}^{n} \left(\delta_{ij} + g(r, u) \frac{x_i}{r} \frac{x_j}{r} \right) \frac{\partial^2 u}{\partial x_i \, \partial x_j} = 0 \tag{8.24}$$

with $r = \sum_{i=1}^{n} x_i^2$. Equation (8.24) is uniformly elliptic if and only if $-1 + \epsilon \leqslant g(r, u) \leqslant 1/\epsilon$ for some $\epsilon > 0$. The notation $\delta_{ij} = 0$ if $i \neq j$, and 1 if $i = j$.

If $u = u(r)$ one can easily show that Eq. (8.24) is equivalent to

$$u'' + \left[\frac{n-1}{r(1+g)} \right] u' = 0, \qquad u' = \frac{du}{dr}. \tag{8.25}$$

Can we find an ordinary differential equation of the form

$$u'' + u'f(r, u) = 0 \quad \text{on} \quad 1 \leqslant r \leqslant 2 \tag{8.26}$$

with $\delta \leqslant f \leqslant 1/\delta$ for some $\delta > 0$, and such that two distinct solutions of Eq. (8.26), $v = v(r)$, $w = w(r)$ agree at the end points? We can then set $g = -1 + (n-1)/rf$ to return to Eq. (8.25).

First, we find a *polynomial* $v = v(r)$ which for $1 \leqslant r \leqslant 2$ satisfies

$$v' > 0, \qquad v'' < 0, \qquad v(1) = 0, \qquad v(2) = 1. \tag{8.27}$$

[†] As engineers, we often "pooh-pooh" such possibilities as, pathological (nonsense) but the example here raises questions.

488   8. SOME THEORETICAL CONSIDERATIONS

Second, we find a polynomial $z = z(r)$ such that on $1 \leqslant r \leqslant 2$ it satisfies

$$
\begin{aligned}
&z > 0 \quad \text{in } 1 < r < 2, \quad z = 0 \quad \text{at the end points} \\
&z'(1) = v'(1), \quad z''(1) = v''(1) \\
&z'(2) = -v'(2), \quad z''(2) = -v''(2).
\end{aligned}
\tag{8.28}
$$

Now let α be a positive constant such that

$$w = v + \alpha z. \tag{8.29}$$

For sufficiently small α we have in $1 \leqslant r \leqslant 2$

$$
\begin{aligned}
&w > v \quad \text{for } 1 < r < 2, \quad w' > 0, \quad w'' < 0 \\
&w = v, \quad w' \neq v', \quad \frac{v''}{v'} = \frac{w''}{w'} \quad \text{at the end points.}
\end{aligned}
\tag{8.30}
$$

Let R be the *closed* domain shown in Fig. 8-1.

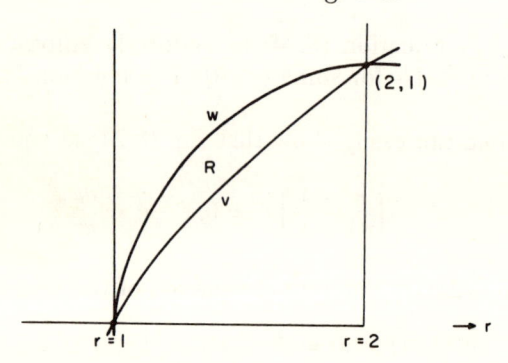

FIG. 8-1. Construction of a nonunique solution of a nonlinear elliptic equation.

On R define

$$f(r, u) = (u - v) \frac{\left[\dfrac{v''}{v'} - \dfrac{w''}{w'}\right]}{w - v} - \frac{v''}{v'} \tag{8.31}$$

which is analytic in R since $w' \neq v'$ at the end points. On the v curve, $f(r, v) = -v''/v'$, while on the w curve, $f(r, w) = -w''/w'$. Therefore

v and w are analytic solutions of Eq. (8.26) where $f(r, u)$ is Eq. (8.31). If we write f as

$$f(r, u) = \frac{u - v}{w - v}\left(-\frac{w''}{w'}\right) + \left(1 - \frac{u - v}{w - v}\right)\left(-\frac{v''}{v'}\right)$$

we see that the uniformity condition $\delta \leqslant f \leqslant 1/\delta$ holds.
Thus Eqs. (8.24) are uniformly elliptic and analytic, the coefficients depend upon u and the solutions are analytic but *not unique*.

References

1. Bernstein, D. L., "Existence Theorems in Partial Differential Equations." Princeton Univ. Press, Princeton, New Jersey, 1950.
2. Ladyzhenskaya, O. A., "The Mathematical Theory of Viscous Incompressible Flow." Gordon & Breach, New York, 1963.
3. Hadamard, J., "Lectures on Cauchy's Problem in Linear Partial Differential Equations." Yale Univ. Press, New Haven, Connecticut, 1923.
4. Courant, R., and Hilbert, D., "Methods of Mathematical Physics," Vol. 2. Wiley (Interscience), New York, 1962.
5. LaSalle, J. P., and Lefschetz, S., "Stability by Liapunov's Direct Method with Applications." Academic Press, New York, 1961.
6. Garabedian, P. R., "Partial Differential Equations." Wiley, New York, 1964.
7. Laurentiev, M. M., *Izv. Akad. Nauk SSSR, Ser. Mat.* **20**, 19 (1956).
8. Payne, L. E., *Arch. Rational Mech. Anal.* **5**, 34 (1960).
9. Goldshtik, M. A., *Appl. Math. Mech.* (English Transl. of *Prikl. Math. Mekh.*) **24**, 913 (1960).
10. Serrin, J. B., The initial value problem for the Navier-Stokes equations, *in* "Nonlinear Problems" (R. E. Langer, ed.), p. 69. Univ. of Wisconsin Press, Madison, Wisconsin, 1963.
11. Kazarinoff, N. D., "Analytic Inequalities." Holt, New York, 1961.
12. Hardy, G. H., Littlewood, J. E., and Polya, G., "Inequalities," 2nd ed. Cambridge Univ. Press, London and New York, 1952.
13. Serrin, J. B., *Arch. Rational Mech. Anal.* **3**, 1 (1960).
14. Velte, W., *Arch. Rational Mech. Anal.* **9**, 9(1962).
15. Payne, L. E., and Weinberger, H. F., An exact stability bound for Navier-Stokes flow in a sphere, *in* "Nonlinear Problems" (R. E. Langer, ed.), p. 311. Univ. of Wisconsin Press, Madison, Wisconsin, 1963.
16. Nickel, K., *Arch. Rational Mech. Anal.* **2**, 1 (1958).
17. Nagumo, M., and Simoda, S., Note sur l'inéqualité différentiel concernant les équations du type parabolique, *Proc. Japan Acad.* **28**, 536 (1951).
18. Westphal, H., *Math. Z.* **51**, 690 (1949).
19. Witting, H., Über die Instabilitäten der Prandtlschen Grenzschichtgleichungen, *in* "50 Jahre Grenzschichtforschung" (H. Görtler and W. Tollmien, eds.), p. 334. Vieweg, Braunschweig, 1955.
20. Nickel, K., *Math. Z.* **74**, 209 (1960).

21. Nickel, K., Parabolic equations with applications to boundary layer theory, *in* "Partial Differential Equations and Continuum Mechanics" (R. E. Langer, ed.), p. 319. Univ. of Wisconsin Press, Madison, Wisconsin, 1961.
22. Rachmatulin, H. A., *Symp. Boundary Layer Theory, Freiberg, 1957* p. 335. Springer, Berlin, 1958.
23. Hermes, H., On the initial value problem for the quasilinear parabolic equation, *in* "Nonlinear Problems" (R. E. Langer, ed.), p. 289. Univ. of Wisconsin Press, Madison, Wisconsin, 1963.
24. Meyers, N. G., *Arch. Rational Mech. Anal.* **14**, 177 (1963).

APPENDIX

Elements of Group Theory

A.1 BASIC DEFINITIONS

A group G is a set of elements $a, b, c, \ldots$[†] together with a well-defined binary operation (mapping, transformation) T. The operation is often written $T(a, b) = ab$, and when this notation is used, it is called multiplication. The following properties must hold:

(i) *Closure*: $T(a, b)$ is defined for all members of G and $T(a, b)$ is a member of G.

(ii) *Associative Law*: $aT(b, c) = T(a, b)c$

(iii) *Identity*: There is a unique element e in G, called the identity (unit) of G, such that $T(e, a) = T(a, e) = a$ for every element of G.

(iv) *Unique Inverse*: Corresponding to every element of G there is a unique element a^{-1}, called the inverse of a, such that $T(a, a^{-1}) = T(a^{-1}, a) = e$.

A *subgroup* H of a group G is a subset of G which is itself a group relative to the operation T. A *proper subgroup* is a subgroup other than $\{e\}$ or G. If $T(a, b) = T(b, a)$ the group is *commutative* (Abelian).

[†] These elements may be integers, functions, matrices, etc.

A transformation between groups G and G' is called an *isomorphism I*, denoted by $I(a) = a'$, if this correspondence is one-to-one and the group operation is preserved. That is if $I(a) = a'$ and $I(b) = b'$ then $I[T(a, b)] = T'(a', b')$.

Example. Consider the set of *positive* real numbers under the operation $T(a, b) = ab$ (ordinary multiplication). It is easy to verify that this is a group with identity 1, and the inverse of a is $a^{-1} = 1/a$. The transformation $I(x) = \log x$ is an isomorphism. It is well known that as x increases through $0 < x < +\infty$, $\log x$ continuously varies in $-\infty < \log x < \infty$. That is the logarithmic mapping is one-to-one[†] between the group of positive real numbers (with multiplication as the operation) and the group of all real numbers (with addition as the group operation). Moreover $I(xy) = \log xy = \log x + \log y$ and the group operations are preserved.

If the mapping I is not one-to-one, but preserves the group operation it is called a homomorphism. If I is an isomorphism and $I(A) = B$ then A and B are said to be *isomorphic*. An *isomorphism* of G with itself is called an automorphism. If a is an element of G the transformation h_a taking G into itself, defined by $h_a(x) = a^{-1}xa$ is an automorphism of G called an *inner automorphism*. All other automorphisms are labelled *outer*.

A subgroup A of a group G is said to be *invariant* (normal) in G, if $x^{-1}Ax = A$ for *every* x in G.

A.2 GROUPS OF TRANSFORMATIONS

A transformation ϕ of a (nonempty) set R into a set S ($\phi: R \to S$) is a rule which assigns to each element p in R a *unique* image element ϕp in S. The set R is called the *domain* of ϕ, and S is its *codomain*. The set ϕR of all images of R under ϕ, called the *range* of ϕ, may comprise only *part* of the codomain S. If the range and codomain are equal so that every q in S is the image $q = \phi p$ of a least one p in R we say ϕ transforms R *onto* T. One-to-one transformations are defined in the preceding footnote.

Two transformation $\phi : R \to S$ and $\phi' : R \to S$ are equal if they have the same effect upon every point of R; i.e., $\phi = \phi'$ means that $\phi p = \phi' p$ for every p in R.

[†] A transformation ϕ of a set S into a set T is said to be *one to one* from S to T if *each* element of T is the image of at most one element of S under ϕ.

A.2 GROUPS OF TRANSFORMATIONS

The "product" operation $\phi\psi$ of two transformations is defined as that of *composition*, i.e., is the result of performing them in succession, first ψ, then ϕ. Of course we must assume that the domain of ϕ is the codomain of ψ. In other words if

$$\psi : R \to S, \quad \phi : S \to T$$

then $\phi\psi : R \to T$ given by $(\phi\psi)p = \phi(\psi(p))$ for every p in R.

The set G of all *one-to-one* transformations of any space S *onto* itself is a group of transformations.

Any abstract group G' is isomorphic with a group of transformations.

Author Index

Numbers in parentheses are reference numbers and indicate that an author's work is referred to although his name is not cited in the text. Numbers in italic show the page on which the complete reference is listed.

Abbot, D. E., 134, 144, *192*
Ablow, C. M., 285, *313*, 410, *471*
Acrivos, A., 227, 236, 237, *269*, 280, *313*
Adkins, J. E., *19*
Ahamed, S. V., 400(101c), *470*
Albasiny, E. L., 361, *469*
Allen, D. N., 369, *469*
Ames, W. F., 143, 150, *194*, 208, 210, 232, 237, *268*, *269*, 413, 420, 422, *471*
Arms, R. J., 378, 385, *469*
Arnason, G., 406, *470*
Ayers, Jr., A., 66
Aziz, A. K., 279, *313*

Banach, S., 257, *270*
Bankoff, S. G., 158, *193*
Barua, S. N., 275, *312*
Bateman, H., 180, *194*
Batschelet, E., 369, *469*
Baxter, D. C., 320(32, 33), 352, *468*
Bellman, R. E., 196, *268*, 287, *313*, 343, 344, 396, 408, 411, *468*
Benney, D. J., 198, *268*
Bergman, S., 173, 178, *193*, 286, *313*
Bernstein, D. L., 427, 428, *472*, 475, *489*
Bers, L., 109, *122*, 173, *193*, 395, 407, *470*
Bickley, W. G., 249, *269*, 316, *467*
Bird, R. B., 7(8), *17*
Birkhoff, G., *19*, 135, 144, 156(7), *192*, 387, 388, 389, *470*
Blasius, H., 296, *314*, 335, *467*
Blottner, F. G., 354, 355, *468*
Bodewig, E., 374, *469*
Bodner, S. R., 261(87), *270*
Bohm, D., 204, *268*

Bolin, B., 405, 406, *471*
Boltzmann, L., 33, *69*
Boyer, R. H., 150, 168, 169, *192*
Brenner, H., 227, *269*
Brian, P. L. T., 241, *269*
Bridgman, P. W., 156, *192*
Brode, H. L., 458, *472*, *473*
Bromberg, E., 227, *269*
Brown, O. E., 66
Bruce, G. H., 341, *468*
Bruhn, G., 461, *473*
Bückner, H. F., 219, *313*
Budiansky, B., 258(85), 261(87), *270*
Burgers, J. M., 24, *69*
Bushby, F. H., 404, *470*
Butler, D. S., 461, *473*
Butler, R., 360, *469*

Carrier, G. F., 10, *18*, 211, 214(21), *268*
Carslaw, H. S., 158, *192*, 278, 293, *313*, 334, *467*
Chambré, P. L., 180, *194*, 280, *313*
Chandrasekhar, S., 10, *18*, *19*, 200, *268*, 278, 306, *313*, *314*
Chapman, S., *19*
Charney, J. G., 404, *470*
Cherry, T. M., 43(13), *70*, 172, 173, 179, *193*
Chester, W., 456, *472*
Chien, W. Z., 204, *268*, 304, *314*
Christianovich, S. A., 178, *193*
Coan, J. M., 296, *314*
Cobble, M. H., 413, *471*
Coburn, N., 109, *122*, 461, *473*
Cole, J. D., 24, 25, *69*, 220, 221, 226, 227(24), *268*

Collatz, L., 243, 257, *269*, *270*, 367, 376, 436, 437, 439, *469*, *472*
Collings, W. Z., 249, *269*
Cooley, I. D., 204, 207(12), *268*
Cope, W. F., 275, *312*
Copson, E. T., 85, *122*
Courant, R., 9, 13(19), *18*, 50(20), *70*, 75, 78(2, 3), 85, 109, 113, *121*, 172, 173, 175, *193*, 276, 286(9), *313*, 439, 446, 447, 450(129), 456, 461, *472*, *473*, 475, *489*
Cowling, T. G., *19*
Cox, R. G., 227, *269*
Crandall, S. H., 243, 249, 256, *269*, 332, 340, *467*
Crank, J., 14, *18*, 35, *69*, 191, *194*, 339, 355, 360, *467*, *468*
Crocco, L., 320(30), 350(30), *468*
Curle, N., 141, 185(15), 188, *192*, 277, *313*
Curtiss, C. F., 7(8), *17*
Cuthill, E. H., 386, *470*

Darboux, G., 84(6), 88, *121*
Datta, S. K., 162, 166, *193*
Davis, J., 358, *468*
de G., 369(63), *469*
de la Cuesta, H., 232, *269*
de la Vallée-Poussin, Ch. J. 265, *270*
de Santo, D. F., 355, *468*
Doetsch, G., 293, *313*
Dolph, C. L., 109, *122*, 198, *268*, 305, *314*, 461, *473*
Dorodnitsyn, A. A., 187, *194*
Douglas, A., 454, *472*
Douglas, J., Jr., 15, *18*, 346, 348(27), 361, 386, 395, 397, 441, *468*, *472*
Dufort, E. C., 328, *467*
Dunford, N., 257, *270*

Eddy, R. P., 359, 365, *468*
Ehrenfest-Afanassjewa, T., 157, *192*
Ehrlich, L. W., 361, *469*
Eichelbrenner, E. A., 349, 350, *468*
Emde, F., 88, *122*
Epstein, B., 178, *193*
Erdelyi, A., 88(15), *122*, 226(32, 33), 237, *268*, *269*,
Erdelyi, E. A., 400(101a, b, c), *314*(101a), *470*

Eringen, C. E., *18*, 162, *193*
Eyring, H., 14, *18*
Fainzil'ber, A. M., 184, 185, *194*
Falkner, V. M., 45, *70*
Falkovich, S. V., 109, *122*
Feeny, H. F., 237, *269*
Ferron, J., 13, *18*, 241, *269*
Feshbach, H., 207, *268*
Filler, L., 460, *473*
Fischer, J., 399, *470*
Fisher, D. D., 177, *193*
Flugge-Lotz, I., 320(29, 37, 35), 349, 352, 353, *468*
Foppl, A. L., 304, *314*
Forsyth, A. R., 50(19), 54, 65, *66*, *70*
Forsythe, G. E., 23, *69*, 315, 320, 328(1), 330, 359, 366, 373(1), 375(1), 379(1), 383, 330, 359, 366, 373(1), 375(1), 379(1), 383, 385, 388(1), 391, *467*
Fox, 279, *313*, 315, 328(2), 346, 361, 375(2), 379(2), 404, 405, 446, 460, *467*
Frank, T. G., 315, 327, 328, 340, 382, 386(3), 389, 408, *467*
Frankel, S. P., 328, 378, *467*
Frankl, F. I., 109, *122*, 173, *193*
Frazer, R. A., 249, *269*
Frederiksen, E., 258, *270*
Friberg, J., 445, *472*
Friedman, B., 257, *270*, 378, 385, 386(73), *469*
Friedmann, N. E., 441, *472*
Friedrichs, K. O., 9, *18*, 78(2, 4), 109, 113, *121*, *122*, 172, 173, 175, *193*, 227, *268*, *269*, 439, 450(129), 456, *472*
Fromm, J. E., 364, 452, 454, 460, *469*, *472*
Fujita, H., *19*, 249, 254, *270*

Galerkin, B. G., 243, 249, *269*
Galimou, K. Z., 258, *270*
Gallie, J. M., Jr., 361, *469*
Garabedian, P. R., 365, 379, 385, 428, *469*, *470*, 476, 487, *489*
Gates, L. D., 378, 385(75), *469*
Geis, T., 136, *192*
Gelfand, I. M., 78(5), *121*
Gerschgorin, S., 367, *469*
Goertler, H., 121(39), *122*, 277, 296, 299, 300(50), 301, *313*, *314*

AUTHOR INDEX

Goldshtik, M. A., 480, *489*
Goldstein, S., 6, 8, *17*, 45, *70*, 221, 232, 233, *268*, 277, 281(27), 296, *313*, 338, *467*
Golub, G. H., 383, *470*
Gottschalk, W. M., 237, *269*
Green, A. E., *19*
Greenspan, D., 409, 411, 416, *471*
Gross, E. P., 204, *268*
Gundersen, R. M., 109, *122*, 426, *471*

Haack, W., 461, *473*
Habetler, G. J., 388, *470*
Hadamard, J., 475, *489*
Halmos, P. R., 257, *270*
Hamill, T. D., 158, *193*
Hansell, G. A., 465, 466(173), *473*
Hansen, A. G., *19*, 134, *194*, 136, 143, *192*
Hardy, G. H., 482, *489*
Harlow, F., *472*
Harlow, F. H., 364, 452, *469*
Hartree, D. R., 275, *312*, 445, *472*
Hassan, H. A., 296, *313*
Hasseltine, E. H., 241, *269*
Hermes, H., 486, *490*
Hetenyi, M., 249, *269*
Hicks, B. L., 358, *468*
Hide, R., 456, *472*
Hiemenz, K., 296, *314*
Hilbert, D., 13(19), *18*, 50(20), *70*, 75, *121*, 276, 286(9), *313*, 461, *473*, 475, *489*
Hildebrand, F. B., 168, 170, *193*, 276, *313*, 369, 393, *469*
Hill, R., 37, *69*
Hirschfelder, J. O., 7, *17*
Hoffman, O., 12, *18*, 109, *122*
Hopf, E., 24, *69*
Horvay, G. J., 158, *193*
Householder, A. S., 374, *469*
Howarth, L., 296, 299, *314*
Humphreys, J. S., 261(87), *270*
Hurley, J. F., 241, *269*
Hyman, M. A., 327, 339(10), 363, 444(10), *467*

Isaacson, E., 446, 447, *472*
Ismail, I. A., 196, *268*
Jackson, R. F., 400(101a), 413(101a), *470*

Jaeger, J. C., 158, *192*, 278, 293, *313*, 334, *467*
Jahnke, E., 88, *122*
Jain, M. K., 162, *193*, 256, 258, *270*
Jeffrey, A., 418, 419, 426, 432, 462, *471*
Jeffreys, H., 237, 245(50), *269*
Jensen, E., 7(7), *17*
John, F., 329, 330, *467*
Jones, B. F. Jr., 15, *18*, 346, *468*
Jones, J. R., 162, *193*
Jones, W. P., 249, *269*
Juncosa, M. L., 340, 343, 344(24), 396, 408, 411(24), *467, 468*

Kahan W., 379, *469*
Kalaba, R., 287, 289, 291, *313*, 343, 344(24), 396, 408, 411(24), *468*
Kalikhman, B. L., 275, *312*
Kalman, G., 26(7), 31(7), *69*
Kamke, E., 56, *70*
Kantorovich, L. V., 243, 247, 265(58), *269*, 286, *313*
Kaplan, S., 327, 339(10), 363, 444(10), *467*
Kaplun, S., 220, 227, *268*
Katz, Ş. M., 14, *18*
Kazarinoff, N. D., 482, *489*
Keller, J. B., 112, *122*, 290, *313*
Keller, H. B., 355, 378, 385, 460, 461, *468, 469, 473*
Kelso, J. W., 358, *468*
Kennard, E. H., 320(31), *468*
Klamkin, M. S., 281, 282, *313*
Kline, S. J., 134, 144, *192*
Knuth, E. L., 158, *192*
Kramer, R. F., 320(34), 353, *468*
Kruskal, M., *19*
Krylov, V. I., 243, 265, *269*
Kuba, E. T., 14(22), *18*
Kuo, Y. H., 217, 219, *268*

Ladyzhenskaya, O. A., *19*, 475, 479, 480, 481, 482, *489*
Laganelli, T. L., 208, *268*
Lagerstrom, P. A., 24, *69*, 220, 221, 226, 227, *268*
Laitone, E. V., 173, 178, 179, *193*
Landau, H. G., 158, *193*, 441, *472*
Landgraff, R. W., 400(101b), *470*

AUTHOR INDEX

LaSalle, J. P., 476, *489*
Laurentiev, M. M., 477, *489*
Lax, P. D., 53, *70*, 78(3), *121*, 127, *192*, 453, 454, 458, 460, *472*, 467, *473*
Lees, M., 445, *472*
Lefschetz, S., 476, *489*
Lessmann, F., 170, *193*
Levine, D. A., 460, *473*
Levy, H., 170, *193*
Levy, S., 296, 301(44), 302, 304(45), *314*
Lewy, H., 439, 450(129), *472*
Lidov, M. L., 112, *122*
Lieberstein, H. M., 320(34), 353, 407, 409, 416, *468, 471*
Lighthill, M. J., 43(14), *70*, 172, 173, 178, *193*, 217, 219, *268*
Lin, C. C., 173, *193*, 278, 306, *313, 314*
Littlewood, J. E., 482, *489*
Lloyd, E. C., 360(43), *469*
Longley, H. J., 454, 455, 460, *472*
Luckert, H., 337, *467*
Ludford, G. S. S., 84(7, 8), 87, 88, 94(7), 103, 112, *122*
Ludloff, H. F., 460, *473*
Lutz, B. C., 237, *269*

McLachlan, N. W., 259, *270*
MacNeal, R. H., 366, *469*
Macek, A., 460, *473*
Magnus, W., 85, *127*
Mangler, W., 45, *70*
Mann, W. R., 278, *313*
Manohar, R., 136, 143, *192*
Martin, C. J., 227, *269*
Martin, M. H., 66, 84(7), 88, 94, 103, 112, *122*
Meksyn, D., 141, 150(16), 184, 188, *192*
Meyers, N. G., 487, *490*
Michel, J. G. L., 360(43), *469*
Mikeladze, Sh., 367, *469*
Millikan, C. B., 212(19), 222, 223, *268*
Milne, W. E., 324, 346, 412, 413, 439, *467*
Milne-Thomson, L. M., 170, *193*
Minorsky, N., 196, *267*
Moon, P., 207(13, 14), *268*
Morgan, A. J. A., 136, 185(13), *192*
Morgan, G. W., 237, *269*
Morris, M., 66

Moser, H., 399, *470*
Morse, P. M., 207, *268*
Motz, H., 414, *471*
Munk, W. H., 214, *268*
Mushtari, Kh. M., 258, *270*
Muskat, M., 5, *17*

Nagumo, M., 483, *489*
Nash, W. A., 204, 209, *268*
Nehari, Z., 191, *194*
Nickel, K., 483, 485, *489, 490*
Nicolson, P., 14, *18*, 339, 355, *467*
Noh, W. F., 54, *70*, 84, *127*, 127, *192*, 452, *472*
Nowinski, J. L., 12, *18*, 196, 258, *268, 270*

Oberhetinger, F., 85(12), *122*
O'Brien, G. G., 327, 339, 363, 444, *467*
Oplinger, D. W., 109, *122*
Oseen, C. W., 220, *268*

Pai, S. I., 6, 8, *17, 18*, 109, *122*, 123, 125, 141, 185(14), 188, *192*, 223, 227, *268*, 273, 275, 299, *312*
Pallone, A., 320(37), 354, *468*
Pattle, R. E., 150, *192*
Payne, L. E., 477, 482, *489*
Payne, R. B., 454, 460, *472*
Payton, R. G., 261(87), *270*
Peaceman, D. W., 346, 386, 387, *468, 470*
Pearson, J. R. A., 13, *18*, 220, 221, 227, 241, *268, 269*
Perry, Cih., 285, *313*
Philip, J. R., 190, 191, *194*, 364, 365(52), 399, *469, 470*
Phillips, N. A., 404, *470*
Pipkin, A. C., 237, *269*
Pogorelov, A. V., 66
Pohlhausen, K., 273, *312*
Poincaré, H., 196, *267*
Polskii, N. F., 144, *192*
Polya, G., 482, *489*
Prager, S., *18*
Prandtl, L., 277, *313*
Priestley, C. H. B., 399, *470*
Protter, M. H., 54, *70*, 84, 127, *192*, 452, *472*
Protusevich, Ya. A., 258, *270*
Proudman, I., 220, 221, 227, *268*

AUTHOR INDEX

Rachford, H. H., Jr., 341, 386, 387, *470*, *468*
Rachmatulin, H. A., 483, *490*
Radowski, P. P., 261, *270*
Rees, M., 446, 447, *472*
Reese, C. E., 14, *18*
Reich, E., 378, *469*
Reissner, E., 227, *269*
Rice, J. D., 341, *468*
Richardson, L. F., 327, *467*
Richardson, O. W., 181(63), *194*
Richtmyer, R. D., 315, 323, 326, 328, 341, 440, 444, 447, 450, 452, 455, 457, 460, *467*, *472*
Riemann, B., 84(8), *122*
Riley, N., 154, 155, *192*
Ringleb, 173, 176, *193*
Ritz, W., 243, *269*
Roberts, L., 454, *472*
Roberts, P. H., 456, *472*
Rosenhead, L., 273, 283, 296, 299, *312*, 338, *467*
Rosseland, S., 7, *17*
Roth, R. S., 258, *270*
Rouse, C. A., 453, 462, *472*
Ruark, A. E., 157, *192*
Rubin, J., 365, *469*
Ruoff, A. L., 158, *192*

Sachs, G., 12, *18*, 109, *122*
Sandri, G., *18*
Schechter, R. S., 394, *470*,
Schechter, S., 407, *471*
Schetz, J. A., 258, *270*
Schiffer, M., 286, *313*
Schlichting, H., 6(5), *17*, 299, *314*
Schuh, H., 136, *192*, *313*
Schulz, W. D., 458, *473*
Schwartz, J. T., 257(70), *270*
Sedov, L. I., 112, *122*
Segel, L. A., 309, *314*
Serrin, J. B., 399, *470*, 481, 482, *489*
Severnyi, A. B., 7, *17*
Shapiro, A. H., 423, 453(121), *471*
Sheldon, J. W., 383, 384, 386(81, 82), *470*
Show, F. S., 369, *469*
Shvets, I. T., 144, *192*
Siekmann, J., 283, 284, *313*
Simoda, S., 483(17), *489*

Simpson, J. H., 338, *467*
Skan, S. W., 45, *70*, 249, *269*
Skinner, A., 443, *472*
Slattery, J. C., 399, *470*
Smith, P., 91, *122*
Sneddon, I. N., 50(18), 66, *70*
Snyder, L. J., 258, *270*
Sokolouskii, V. V., 426, *471*
Spencer, D. E., 207(13, 14), *268*
Spitzer, Jr., J. R., 7, *18*
Spriggs, T. W., 258, *270*
Srivastava, A. C., 162, 166, *193*
Stein, L. R., 460, *473*
Steinhardt, R., 365, *469*
Stewart, W. E., 258, *270*
Stoker, J. J., 196, *267*, 227, *269*, 426, *471*
Stuart, J. T., 305, 308, *314*
Sully, E. D., 360(43), *469*
Swope, R. D., 420, 422, *471*
Sylvester, R. J., 258, *270*

Tamada, K., 112, 117, 120, *122*
Tandberg-Hanssen, E., 7(7), *17*
Tani, I., 299, *314*
Taniuti, T., 418, 419, 426, 432, 462, *471*
Taylor, G. I., 121(38), *122*
Taylor, T. D., 227(41), 236, *269*
Tetervin, N., 275, *312*
Thomas, L. H., 437, 444, *467*, *472*
Thomee, V., 341, 463, *473*
Thornhill, C. K., 461, *473*
Tifford, A. N., 299, *314*
Timoshenko, S. P., 302, *314*
Ting, T. C., 361, *469*
Tipei, N., 296, *314*
Todd, J., 315, 328(5), 373(5), *467*
Tomotika, S., 112, 117, 120, *122*
Tricomi, F. G., 43, *70*, 279, *313*
Trigger, K. R., 454, 456, *472*
Trilling, L., 24, *69*
Trulio, J. G., 454, 456, *472*
Trutt, F. C., 400(101a), 413, *470*
Tsien, H. S., 173, *193*
van Dusen, M. S., 21(1), *69*
Van Dyke, M., *19*, 227(42, 43), *269*
Varga, R. S., 366, 379, 383, 386, 387, 388(87), 389(87), 395, *469*, *470*

Vette, W., 482, *489*
Viswanathan, R. V., 369, *469*
von Kármán, T., 12, 27, *18*, 40(11), *70*, 173, *193*, 212, 222, 223, *268*, 271, 273, *312*
von Mises, R., 6, *17*, 109, *122*, 173, *193*
von Neumann, J., 457, *472*

Wachspress, E. L., 388, 389, *470*
Wagner, C., 280, *313*
Wakelin, J. H., 14(22), *18*
Walker, G. W., 181, *194*
Warner, W. H., 237, *269*
Warten, R., 416, *471*
Wasow, W. R., 23, *69*, 226, *268*, 315, 318, 320, 328(1), 330, 359, 366, 373(1), 375(1), 379(1), 383, 385, 388(1), 391, *467*
Watson, J., 305, 309, *314*
Way, S., 304, *314*
Weinberger, H. F., 482, *489*
Weinstein, A., 411, *471*
Weir, D., 109, *122*
Wendel, M. M., 401, *470*
Wendorff, B., 454, 460, *472*
Westphal, H., 483, *489*
Weyl, H., 383, *313*
Wheeler, J. A., Tr., 395, *470*
Wheeler, M. F., 387, 390, 391, 396, 397, *470*
Whitaker, S., 401, *470*

Whitehead, A. N., 222, *268*
Whitham, G. B., 460, *473*
Whitelam, C. J., 404, *470*
Wise, H., 410, *471*
Witting, H., 483, *489*
Woinowsky-Krieger, S., 302, *314*
Wolf, F., 278, *313*
Woods, L. C., 414, *471*
Wu, C. S., 162, *193*, 320(36), 353, 354, *468*

Yakura, J. K., 227, *269*
Yamada, H., 249, *270*
Yeh, K. V., 304, *314*
Yohe, M., 409, *471*
Young, D. M., 315, 327, 328, 340, 378, 380, 382, 386(3), 387, 388(87), 389(87), 390, 391, 396, 397, 408, 409, *467*, *469*, *479*
Yu, E. V., 320(35), 353, *468*
Yusuff, S., 304, *314*

Zabusky, N. J., 11, *18*, 87, 103, *122*, 467, *473*
Zaiser, J. A., 465, *473*
Zarantonello, E. H., *19*
Zerna, W., *19*
Zondek, B., 378, 385(75), *469*
Zovko, C. T., 460, *473*
Zwick, S. A., 133, *192*

Subject Index

A

Acceleration of convergence, by successive overrelaxation, 378
Acoustic impedance, 114
Ad-hoc methods, 47–49, 112, 117, 120
  in magneto-gas dynamics, 123–126
Adjoint equation, 85
Alfven's wave, local speed of, 125
Algebraic equations,
  direct methods of solution, 373
  iterative methods of solution, 373
  sparse, 373
Alternating direction implicit methods, 386-388
  comparison of, 388, 389
  Douglas-Rachford, 386
  Peaceman-Rachford, 386-388
Analog solution of partial differential equations, 441-443
Anisentropic flow,
  auxiliary function for, 90
  equations, 90
  general solutions, 100-103
Arnason's method for balance equation, 406
Arrhenius relation, 14, 180
Asymptotic,
  approximations, 237-243, 334
  methods, 195
  solution, 338
Auxiliary function, 26
  stress function, 26
  stream function, 26
  two, 94

B

Backward difference, 347, 359
Balance equation,
  approximate condition of ellipticity, 405
  condition for ellipticity, 405
  numerical methods, 405-407

Bernoulli's equation, 26, 138
  differential form, 118
Bessel functions, completeness of, 258
Biharmonic equation, 230
Blasius,
  function, 188
  problem, 149
  solution, 335
Blasius series,
  for steady boundary layer flow, 296–299
  limitations, 299
Block iterative methods, 374, 384–389
Block tridiagonal matrices, 385
Boltzmann-Landau-Vlasov equation, 198
Boltzmann transformation, 33, 34, 133
Boundary, unknown, 16
Boundary conditions,
  consolidation of, 34
  essential, 245
  natural, 245
Boundary conditions, nonlinear, 15, 16, 278–280
  heat conduction, 278
  laminar flow with reaction, 280
  radiation, 278
  superfluidity, 278
Boundary layer, integration across, 272
Boundary layer assumption, 212
Boundary layer equations,
  conversion to integral equation, 283, 284
  Crocco transformation of, 351, 352
  finite difference methods, 349-355
  flow over flat plate, 6
  Howarth-Dorodnitsyn transformation of, 354
  physical plane, 350
  properties of solutions, 59
  radial flow, 154
  similarity solutions, 143, 154
  stability criterion, 353
Boundary layer flow, similarity variables, 44–47, 137–141
Boundary layer profiles, development, 277
Boundary layer separation, 483

Boundary layer stretching, 237, 238
Boundary layer theory,
　as singular perturbation problem, 215–219
　higher approximations, 227
　three dimensional stream functions, 28
Boundary method, 262
Boundary singularity, 331, 412
Boundary value problems, conversion to initial value problems, 105, 281–283
Buckling of shells, solution by Galerkin's method, 258–261
Burgers' equation, 24, 347
　modified, 210

C

Canonical equations, 75, 77
Canonical form,
　elliptic equations, 365
　quasi-linear system, 73
Cantilever beam, deformation of, 361–363
Cavity ratios, for compressible flow past flat plates, 177–179
Change of state, 360
Chaplygin's equation, 43
Characteristic,
　coordinates, 416
　directions, 419, 425
　equation, 418, 419
　net, 76
　plane, 102
Characteristic length, in similarity solutions, 167
Characteristics, 75, 96
　first order equations, 426, 427
　isentropic flow, 82
　method of, 315, 416–422
　numerical computation, 435–437
Charpit's,
　equations, 56–58
　method, 54–56
Chebyshev polynomials, 382
Chemical reaction,
　diffusion controlled, 284–287
　heat generated in, 180
Christianovich transformation, 178
Circulatory flow, 175

Collocation, 246, 256, 257
　extremal point, 257
Collatz condition, for essential boundary conditions, 245
Compatibility conditions, 54, 329
Compatible equations, 54
Complete set,
　of functions, 244
　relatively, 264
Complete solution, 50
Completeness, importance in weighted residual methods, 264
Composition, operation of, 493
Compressible fluid equations, general solution, 51–53
Compressible gases, similarity solutions of, 186–189
Computational molecule, 316–318
　irregular, 366–370
　nine-point, 394
　seven-point, 411
Concave function, 289
Conduction,
　and chemical reaction, 14
　and radioactive decay, 15
　explicit finite difference method, 325
Conduction equation,
　linear, 2
　nonlinear, 3, 4
Conservation equations, 53, 453, 454
　finite difference approximations, 454
　general solutions, 54
Conservation laws,
　integral form, 148, 149
　and similarity, 150
Consistency conditions, for iteration methods, 328, 329, 375
Consistent ordering, 409
Constant quantity condition, 151, 169
Constant state, 431
Constitutive equations, boundary layer flow, 162
Convergence,
　finite difference scheme, 319
　iteration process, 170
　Ritz method, 263, 264
Convergence theorems,
　equations of elliptic type, 265–267

SUBJECT INDEX

weighted residual methods, 262–267
Conversion, boundary value problem to initial value problem, 105
Conversion of differential equation to integral form, 168
Convex function, 287, 288
Convolution kernel, 279
Coordinate stretching, 213, 354
Corrector formulae, 346, 347
Crank-Nicolson method, 339, 346, 356, 358
  convergence, 340
  predictor-corrector modification, 347
  stability, 357
Crocco equations, 352, 353
Crocco transformation, 350, 351
Cyclic Chebyshev method, 383

D

D'Arcy's law, 5
Derived quantities, 156
Developable surfaces, 39
Diagonal dominant matrices, 373
Difference equation, first order, 170
Diffusion,
  nonlinear equation, 4, 150, 347
  short time dependence, 255
  similarity solution, 33, 143, 144, 190, 191
Diffusion coefficient, concentration dependent, 189
Diffusion with reaction,
  asymptotic solutions, 240–243
  stretched variables, 242
Diffusion with second-order reaction, 13, 361
  Crank-Nicolson method, 361
Diffusion problem, solution by method of moments, 249–256
Diffusion through a fluid, 184
Dimensional analysis, principles of, 156–158
Direct methods, calculus of variations, 196
Directional derivative, approximations of, 368
Dirichlet,
  boundary conditions, 318
  problem, 22

Discontinuity,
  expansion into line or curve, 332–334
  generation in calculation, 466, 467
  violent, 331
  weak, 331, 332
Discretization error, 318
  for explicit methods, 322
Dispersion relation, 203
Displacement thickness, 273
Douglas-Rachford method, 386–388
Dorodnitsyn transformation, 187
Dufort-Frankel explicit method,
  for parabolic equations, 327–329
  stability criterion, 328
Dynamic similitude of fluid flows, 157

E

Earth collapse equation, 12
Elastic string, nonlinear vibrations of, 11, 103–108, 464–467
Elasticity, perturbation solution in, 204–207
Elimination of two variables, 141
Elliptic equations,
  finite difference methods, 365–416
  mildly nonlinear, 395, 396
  nonlinear examples, 389–411
Energy equation, theorems on, 232
Entropy function, 95
  polytropic gas, 96
Equation of state, for anisentropic flow, 124
Equation splitting, 59
Equations of fluid mechanics, 8–10
Error analysis, 319
Error distribution principles, 261
Euler equations, 8, 113
  conservative form, 453
  one dimensional flow, 449
Exact methods of solution, 71–194
Expansion of gas behind piston, 432–435
Expansion, uniformly valid, 226
Existence theorem of Cauchy-Kowalewski, 476
Existence and uniqueness, 478–489
  boundary layer theory, 482–485
  elliptic systems, 487–489
  quasilinear parabolic equations, 486
  viscous flow, 478–482

Explicit methods,
  generalized, 329
  hyperbolic equations, 438–442
  parabolic equations, 320–330
  stability bounds, 326
  unconditional stability, 461
Extrapolation formulae, forward, 393

F

Factorization, of partial differential equations, 420
False boundaries, 321, 323, 338, 341, 438
Finite difference approximations, construction of, 366
Finite difference equations of Lelevier, 451
Finite difference methods,
  comparison with characteristics, 430, 437, 438
  explicit, 320
  implicit, 320
  hyperbolic, 438–442
  parabolic, 320-330
First integrals, 60
First order equations,
  complete solutions, 54
  general solutions, 50, 56
  separable form, 58
  special types, 57-58
Floating boundary, 362
Flow past sphere, higher approximations for, 231–236
  singular perturbation solution for, 227–236
Fluid mechanics, irrotational equations of motion, 15
Forced convection, energy equation, 237
Forward and backward difference method, 447, 448
Fourier series, double, 302
Free boundary, flow problems with, 15
Functional method, stationary, 244
Functional transformation, 23, 25
Fundamental units, 156

G

Galerkin's method, 244, 247, 249
  advantages, 249
  applications, 258–261
Galilei-Newton group, 157
Gas, isentropic flow, 26
Gas combustion in rockets, 359
Gauss-Seidel method, 377
  convergence, 378
  ordering in, 377
Gaussian elimination, 342
General solutions, 49–69
  applications, 180
  implicit character, 65
  first order equations, 50–58
  quasi–linear equations, 50–51
  second order equations, 58-69
  table, 66-69
Goertler's series, 299
Goertler's transformation, 299
Gradient method, 383
Green's function, 85
  for Laplace's equation, 291
Group, 491
Group theory, elements of, 491–493

H

Hartree's method, 445–447
Heat in chemical reaction, 355–358
  equations, 355
  numerical method, 356, 357
Heat and mass transfer from a sphere, singular perturbations in, 227, 236
Heat and mass transfer, similar solutions, 183-186
Heat conduction,
  melting ice, 360
  shells and rockets, 358, 359
Higher dimensions, parabolic equations in, 364
Hodograph, pseudo-logarithmic, 178
Hodograph plane,
  advantages of, 172
  for nonlinear oscillation, 104
Hodograph transformation, 35–37, 171, 172
  applications, 173–180
Hopf transformation, 24, 210
Howarth-Dorodnitsyn transformation, 354
Hybrid methods, for first order equations, 445–448

SUBJECT INDEX

Hydrodynamics and radiation, solution by Rouse's method, 462-464
Hyperbolic equations, numerical methods, 416–461
 quasilinear, 72–78, 417
Hypergeometric function, 88
Hypersonic flow, singular perturbation, 227

I

Ideal gas, hodograph equations of, 171, 172
Implicit methods,
 comparison with explicit methods, 338, 339
 parabolic equations, 338–343
 hyperbolic equations, 443–445
Implicit solution, occurrence in nonlinear problems, 349
Inertia forces in plastic region, 362
Initial residual, 244, 248
Inner expansion, 222, 228
Inner iteration, 394, 399
 for balance equation, 406
Inner-outer expansion, 219
Inner solution, 212, 218
Inspectional analysis, 157
Instability, "positive test" for, 327, 337
 qualitative description, 320
Integral equations,
 and boundary layer theory, 280
 methods, 167–171
 Volterra type, 279
Integral methods,
 advantages of, 284
 in fluid mechanics, 271–276
 of Polhausen, 273, 274
Integral representation of partial differential equation, 363
Integration domain, Riemann invariant coordinates, 106
Interaction between errors, 344
Interfaces, 454–456
Interior method, 262
Intermediate integrals, 99
Intermediate limits, theory of, 225–226
Invariance, under a transformation, 136
Invariant,
 absolute, 136
 conformally, 136
 constant conformally, 136
Invariant subgroup, 492
Invariants of a group, 136
Inverse method, 182
Irreducible matric, 371
Irregular domain, treatment, 366–370
Isentropic flow,
 characteristics for, 97
 equations, 37, 91
 Riemann invariants, 97
 solution by characteristics, 432–435
Isentropic gas, ad-hoc solutions for, 125, 126
Isobars, 102
Isopycnics, 102
Isothermals, 102
Isovels, 102
Iteration, 169, 195, 374, 375
 inner, 394–399
 integration, 167, 171, 276
 Newton, 396–398
 outer, 394–399
 Picard, 395, 397, 398
 single steps, 377, 378
 total steps, 375-377
Iteration parameters, 387, 388
 Peaceman-Rachford, 387, 388
 Wachspress, 388

J

Jacobi method, 375–377
 convergence, 376
 order of computation, 377

K

Karman-Polhausen method, 275
 limitations of, 276
Karman-Tsien gas, 99
Kinetic temperature, 8, 201
Kirchhoff transformation, 21, 399

L

Lagrange coordinates, 127
 choice of, 113, 114
 definition, 455

SUBJECT INDEX

generalized, 115
mass, 449
utility, 126, 127
Lagrange equations,
  conservative form, 453
  fluid motion, 113
  gas dynamics, 128
  one dimensional flow, 449
Lagrange form, of first order equation, 50
Lagrange subsidiary equations, 50, 62
  generalized form, 56
Laplace's equation, 16, 21, 414
Laplacian, pseudo, 21
Laval nozzle, 121
Lax's staggered method for conservation equations, 454
Least squares, method of, 247
Legendre polynomial, 230
Legendre transformation, 37-40, 174
  application, 39
  for potential function, 37
  for stream function, 40
  generalized, 39
Leibniz rule, for differentiation of integrals, 114
Lelevier method of stabilization, 450, 451
Liebmann's extended method, 407
Linear iterative method, 379
Linear transformation, 72
Liouville's equation, 66
Local solution, 412
  in re-entrant corner problems, 414-416
Longitudinal oscillations, of a string, 29, 103-108

M

Mach number, 178
Matching,
  conditions of Lagerstrom-Kaplun, 225
  of inner and outer solutions, 225, 233
Maximum operation, 287-295
  conservation equations, 293-295
  elliptic equations, 289-291
  parabolic equations, 292, 293
Maximum norm, 291
Maximum principle, 286
Maxwell velocity distribution, 201

Melting problem, 160
Mesh points, irregular, 366-370
Mesh refinement, 400, 413
Mesh size, 319
  optimum for explicit method, 324
Method of lines, 441-443
Minimum surface area, equation, 13
Mixed flow of Ringleb, 176, 177
Mixed methods, 262
Mixed parabolic-hyperbolic systems, 462
  two time constants for, 462
Mixed systems, finite difference methods, 462-467
Mixed transformations, 35
Molenbroek-Chaplygin transformation, 40-43
Moment curvature relation, 362
Moments,
  calculation, 199, 200
  method of, 246
Moments of the plasma equation, 201
Momentum thickness, 273
Monge-Ampere equation, 95, 404
  intermediate integrals, 98
Monge-Ampere nonlinearity, 405
Monge's equations, 60, 61
Monge's method, for general solutions, 60-65
Motz and Woods, method, 414-416
Moving boundary, immobilization, 159-162
Moving boundary problems, similarity solutions, 158-162
Moving threadline, vibrations, 464-467

N

Natural coordinates, 416
  Abbott-Kline form, 147
Natural iteration, 286
Navier-Stokes equations, 10
  numerical solution, 401-403
Nebular theory, distribution of interstellar matter, 181
Neumann problem, 22
Newton iteration, 285, 292, 295, 344, 408
  compared with Picard iteration, 398
  for elliptic equations, 396-398

SUBJECT INDEX

Newton linearization, advantages, 398, 399
Newton's laws, invariance under Galilei-Newton group, 158
Nonlinear overrelaxation, 407–410
Nonlinear vibration of a string, 103-108
Nonlinearity,
  definition, 2
  removal, 334
  weak, 111
Non-Newtonian fluids, 162
  computation of friction factor, 389–395
Normal derivative, boundary conditions involving, 369
Normal form, for quasi-linear systems, 75
Norms, 257
Nusselt number, 238, 239

O

Ocean basin, average motion of fluid, 211–215
One dimensional flow, finite difference methods, 448–453
One to one mapping, 172
Operator,
  linear, 2
  nonlinear, 3
Optimum overrelaxation factor, experimental determination, 398
Order,
  convergence, 266, 267
  error, 267
Orthogonality methods, 261
Oseen,
  approximation, 220, 234
  region of flow, 229
  solution, 221
Outer expansion, 222, 229
Outer iteration, 394–399, 405
  possible divergence, 406
Outer solution, 212, 218, 230
Outer variables, 228
Overestimation of relaxation factor, 380
Overrelaxation, 344
  optimum factor, 380
Overshoot of solution, 351

P

Parabolic equations,
  consistency condition, 330
  examples, 355–365
  explicit methods, 320–330
  finite difference methods, 320–365
  quasilinear, 321, 330
  stability condition, 330
Parabolic velocity profile, 402
Parameter free equations, 148
Parameters, undetermined, 245
Partial derivatives, finite differences for, 316–318
Partial differential equations, classification, 365, 419
Partition of matrices, in point iterative methods, 375
Peaceman-Rachford method, 386–388
Percolation, 364
Perfect gas, isentropic flow, 78
Periodic extension of domain, 105
Permutation matrix, 376, 379
Perturbation, 196-236
  about nonlinear exact solutions, 208-210
  methods, 196, 207, 208
  regular, 196
  singular, 196, 211, 214
  surface, 206
  volume, 206
Pfaffian equation, 55
Picard approximation, 343, 344
Picard iteration, 196, 286, 395, 435
Prandtl number, 238
Plasma oscillations, 10, 31
  equations, 200
  perturbation solution, 198–204
Plastic flow,
  equations, 37
  ideal, 12
Plateau's problem, 13
  general solution, 65
Plates, large deflection, 12, 204-207
Point iterative methods, 374-384
  comparison, 383, 384
Poiseuille flow, 305
  subcritical, 306
  supercritical, 306

Poisson-Euler-Darboux equation, 83–90, 101, 104
Polar coordinates, subsonic flow, 40, 41
Polhausen method, 354
Polynomial solution in elasticity, 206
Polytropic gas, 90
Porous media, flow in, 5
Positive definite matrix, 378
Potential,
 three dimensional equation, 28
 two dimensional equation, 27
 velocity, 26
Power law viscosity, 390
Predictor-corrector methods, 345–348, 363
Predictor formulae, 346, 347
Principal function of Gortler, 300
Propagation of discontinuities, 430
Property (A) of matrices, 379

Q

Quadratic convergence, 344
Quasi-linear elliptic equations, iterative solutions, 284–287
Quasi-linear hyperbolic equations, 29
 examples, 78–84, 109
 general theory of, 72–78

R

Radial flow, 175
Rankine-Hugoniot conditions, 456
Reactors, circulating fuel, 16
Reducible equations, 35, 36, 77
 characteristics, 77, 78
Reducible matrix, 376
Re-entrant corner singularity, 412, 414–416
Regular perturbations, vibration theory, 197, 198
Relaxation factor, 339, 378, 444
 nonlinear overrelaxation, 408
Removal of singularity, 332–334
Resolvent, 89–90
Reynolds number, 238
Richardson's explicit method, instability of, 327
Richardson's point iterative method, 383

Riemann function, 85, 107
 adjoint equation, 88
Riemann invariants, 28, 30, 100
 isentropic flow, 81, 82
 nonlinear vibration, 103
Riemann-Volterra method, 85–87, 105–108
Ritz method, 249
Rocket, combustion 23, 25
Rotating machinery, highly saturated
 computation, 400
 equations for, 399, 400
Round-off error, 318

S

Schroedinger equation, 232
Second order equations, general solution, 58–65
Second order method, quasi-linear equations, 343–345
Self-adjoint equation, 248, 373
Semi-iterative methods, 381–382
Separation of variables, 48, 91, 109, 115, 150
 similarity solutions and, 135, 144–155, 162–164
Series expansion, 296–312, 476
 elasticity, 301–305
 stability of fluids, 305–312
Shear stress at the wall, 273
Shear stress and temperature field, similarity solution, 188
Shock calculation, 457–460
 artificial viscosity, 457, 458
 characteristics, 460
 Landshoff's method, 460
 Lax's method, 458, 460
 tensor artificial viscosity, 458
Shock wave, formation, 126
Shocks, 448, 456, 457
Short time diffusion, method of moments solution, 251–254
Similarity,
 and conservation laws, 150
 general discussion, 166, 167
 mathematical meaning, 134
 physical meaning, 133

SUBJECT INDEX

Similarity solutions, 133-167
  group theory, 136-144
  separation of variables, 144-154
  three dimensions, 162-166
Similarity transformation, 33, 43-47
Similarity variables, 136
Simple wave, 431
  isentropic flow, 98
Simultaneous displacements, method of, 375-377
Single line block iterative methods, 385, 386
  comparison of, 386
Singular integral equation, 279
Singularity,
  area of infection, 413
  at the origin, 336
  difficulties with, 278
  elliptic problems, 411-416
  example in boundary layer flow, 334-338
  in the boundary, 219, 331
  parabolic equations, 330-338
  subtraction of, 411-413
  treatment by asymptotic series, 353
Singularity of perturbation, 224
Small parameter methods, 196
Soft solutions, 294
  gas dynamics, 127-133
  two dimensions, 129
  three dimensions, 130
Soil moisture content, 364
Solar prominences, persistence of, 7
Solidifying steel, 360
Sonic singularity, 179
Sound wave propagation, 78
Source solutions, 152, 153, 334
Space charge of electricity, 181
Sparse algebraic systems, iterative methods, 374
Spectral radius of matrix, 379
Speed of sound,
  effective, 125
  local, 124
Spinning cone, three dimensional boundary layer, 162
Spiral flow, 175, 176
Split of solutions, 480
Stability condition,

explicit schemes, 329
Lelevier, 451
nonlinear vibration, 441
Stability of explicit methods, 319, 320
  hyperbolic equations, 439
  parabolic equations, 323
Stability of velocity profiles, 306
Stability of viscous flow, sufficient condition, 482
Staggered finite difference scheme, conservation equations, 454
Stationary functional method, 245, 261
Stationary iterative method, 376
Steep gradients, influence on convergence, 345
Stefan problem, 158
Stokes,
  paradox, 221
  problem, 219
  region of flow past sphere, 228
  solution, 231
  stream function, 219
  theory, 220
Stream function, hodograph relations for, 172
Streamlines, 171
Stress function, Airy, 12, 301
Stress-strain equation, 11
Stretched coordinate system, 224
Stretching of thermal boundary layer, 238, 239
Stretching transformation, 228, 236
Subdomain, method of, 246
Subsonic flow, 40, 173
Subtracting out the singularity, 411-413
Successive approximations,
  convergence theorem, 279
  method, 279, 363
Successive displacements method, 377
Successive overrelaxation, 378-381
  advantages, 379-381
  convergence, 380
Superposition, 3
Supersonic flow, 173
Supersonic nozzle, 423-426
Symmetric causes and symmetric effects, 479
Symmetric solutions, search for, 135

Symmetry of finite difference approximations, 356
Symmetric successive overrelaxation, 383

T

Tangent plane coordinates, 38
Taylor flow, transition to Meyer flow, 121
Thermal boundary layer equation, 238
Thermal ignition problem, 358
Theta function, 293
Thomas algorithm, tridiagonal matrices, 341, 342
Threadline equations,
  nonlinear, 465
  linear, 465
Trajectory, 94, 95, 102
Transformation groups, 134, 156–158
Transformations, classes of, 20
Transmission line, equations, 168
Transonic flow, 117–121
Transverse oscillations of string, 109–112
Traveling wave,
  series, 305–312
  solutions, 47, 117, 203, 348, 349
Trial solution,
  construction from complete set, 262, 265
  Galerkin's method, 260
  polynomial form, 251
  weighted residual methods, 244
Tridiagonal algorithm, 357
Tridiagonal matrix, in implicit methods, 341
Truncation error, 316–318
  in boundary conditions, 318
  in finite difference approximations, 323
Turbulence, Burgers' model for, 24
Two component fluids, boundary layer equations, 483

U

Unconditional stability, of implicit methods, 443, 444
Underrelaxation method, and Navier-Stokes equations, 403
Undetermined functions, 244, 247
Undetermined parameters, 244
Unequal-interval finite difference schemes, in moving boundary problems, 360
Uniform approximation, 221
Uniqueness and the Navier–Stokes equations, 479
Uniqueness for hyperbolic systems, 426–430
Uniqueness question, method of moments, 253, 254
Unit free relation, 156
Universal function, in Blasius series, 298
  in Gortler's series, 301
Unstable explicit finite difference methods, 326, 327

V

Velocity of sound, 27
Velocity, temperature and concentration fields, similarity integrals, 186
Velocity profile, polynomial approximation of, 274, 275
Vibrating string, large amplitude oscillations, 29
Vibration, perturbation solution of, 197, 198
Vibrations of a nonlinear string, 440–442
Vibrations of circular plates, singular perturbations in, 227
Viscosity,
  concentration dependence of, 189
  temperature dependence of, 187
  Sutherland law, 350
Viscous flow, singular perturbation in, 215–240
Vlasov-Boltzmann equation, 10
von Karman equations, 12, 204, 301
von Mises-Hencky condition, 12
von Mises transformation, 6, 7, 222, 336
  and plasma oscillations, 31
  generalized, 484
von Neumann-Eddy method, 359, 365
von Neumann method for hyperbolic equations, 444, 445
Vortex motion, equation, 180

Vorticity and concentration fields, similar integral, 184
Vorticity and temperature fields, similarity integrals, 185

W

Wake, 335–338
Wave equation, 25
  solution on analog computer, 442–443
Weak solutions, 127

Weather prediction,
  equations for, 404
  numerical computation, 403–407
Weierstrass approximation theorem, 265, 477
Weighted residual methods, 195, 243–268
  comparison of, 261–262
Well posed problems, 134, 475
  similarity solution of, 145–148
Weyl-Siekmann method, 283, 284
Whitehead's method, 222